KU-285-028

Fermentation Microbiology and Biotechnology

Second Edition

LIVERPOOL
JOHN MOORES UNIVERSITY
AVRIL ROBARTS LRC
TITHEBARN STREET
LIVERPOOL L2 2ER
TEL. 0151 231 4022

Fermentation Microbiology and Biotechnology

Second Edition

E.M.T. El-Mansi, *Editor-in-Chief*
C.F.A. Bryce, *Senior Editor*
A.L. Demain, *Associate Editor*
A.R. Allman, *Associate Editor*

Taylor & Francis
Taylor & Francis Group
Boca Raton   London   New York

CRC is an imprint of the Taylor & Francis Group,
an informa business

CRC Press
Taylor & Francis Group
6000 Broken Sound Parkway NW, Suite 300
Boca Raton, FL 33487-2742

© 2007 by Taylor & Francis Group, LLC
CRC Press is an imprint of Taylor & Francis Group, an Informa business

No claim to original U.S. Government works
Printed in the United States of America on acid-free paper
10 9 8 7 6 5 4 3 2

International Standard Book Number-10: 0-8493-5334-3 (Hardcover)
International Standard Book Number-13: 978-0-8493-5334-5 (Hardcover)

This book contains information obtained from authentic and highly regarded sources. Reprinted material is quoted
with permission, and sources are indicated. A wide variety of references are listed. Reasonable efforts have been made to
publish reliable data and information, but the author and the publisher cannot assume responsibility for the validity of
all materials or for the consequences of their use.

No part of this book may be reprinted, reproduced, transmitted, or utilized in any form by any electronic, mechanical, or
other means, now known or hereafter invented, including photocopying, microfilming, and recording, or in any information storage or retrieval system, without written permission from the publishers.

For permission to photocopy or use material electronically from this work, please access www.copyright.com (http://
www.copyright.com/) or contact the Copyright Clearance Center, Inc. (CCC) 222 Rosewood Drive, Danvers, MA 01923,
978-750-8400. CCC is a not-for-profit organization that provides licenses and registration for a variety of users. For organizations that have been granted a photocopy license by the CCC, a separate system of payment has been arranged.

Trademark Notice: Product or corporate names may be trademarks or registered trademarks, and are used only for
identification and explanation without intent to infringe.

Library of Congress Cataloging-in-Publication Data

Fermentation microbiology and biotechnology / E.M.T. El-Mansi, editor-in-chief ; C.F.A. Bryce, senior
    editor ; A.L. Demain, associate editor, A.R. Allman, associate editor. -- 2nd ed.
        p. ; cm.
    Includes bibliographical references.
    ISBN 0-8493-5334-3
    1. Microbial biotechnology. 2. Fermentation. I. El-Mansi, Mansi.
    [DNLM: 1. Industrial Microbiology. 2. Fermentation--physiology. 3. Microbiological Techniques.
QW 75 F3596 2006]

TP248.27.M53F467 2006
660'.28449--dc22   2006009173

Visit the Taylor & Francis Web site at
http://www.taylorandfrancis.com

and the CRC Press Web site at
http://www.crcpress.com

I wish to dedicate my contribution, with affection, to my sons Adam and Sammy, in the hope that it provides a stepping stone in their progress in higher education. Forgive me, boys, for having to spend so much time away from you.

Dr. Mansi El-Mansi

Preface

"Never underestimate the power of the microbe".

Jackson W. Foster

The versatility and diverse array of microbial biosynthetic pathways are currently being exploited by the fermentation industry for the production of important primary metabolites including amino acids, nucleotides, vitamins, solvents, and organic acids, as well as secondary metabolites such as antibiotics, hypocholesterolemic agents, enzyme inhibitors, immunosuppressants, and antitumor compounds. Current advances in proteomics, metabolomics, functional genomics, bioinformatics, and cell immobilization as well as sensor technology are currently being exploited in drug development programs with the view of increasing titres and yields of microbial processes. Microorganisms, both free-living as well as immobilized, are also extremely useful in carrying out biotransformation processes to the extent that they have become essential in the production of single-isomer intermediates. The best is yet to come as we enter a new era in which the conversion of renewable resources to energy, e.g., the conversion of lignocellulosic waste to ethanol, is a major goal.

In recognition of current advances in the field, the second edition builds on the fine pedigree of the first and extends the spectrum of the book to reflect the multidisciplinary and buoyant nature of this subject. The following new chapters, written by eminent scientists in the field, have been introduced:

- Products of primary metabolism
- Metabolic and co-factor engineering and their role in the fermentation and pharmaceutical industries
- The conversion of renewable resources to fine chemicals
- Cell immobilization and its role in the fermentation and pharmaceutical industries

More exciting advances are still to come as the complete sequencing of industrially important microbial genomes is unraveled. Functional genomics and proteomics are already major tools in the search for new drugs and the development of overproducing strains.

In future editions, we will endeavor to keep readers, academics, as well as industrialists, abreast with current advances and future prospects within this exciting and ever-changing field.

The Editorial Team
June 2006

Acknowledgments

The work presented in this book collectively draws on the work of great many researchers including our distinguished authors, and highly regarded academics and industrial experts drawn from different disciplines within the fermentation and the pharmaceutical industries. It also owes much to the influence of many peers and colleagues worldwide, who are too numerous to mention.

I wish to thank my editorial team for being very helpful, supportive and responsive at all times. In particular, I wish to thank Arnold (Arny) Demain not only for helpful advice but also for his many sound contributions to science, which lit the path for many microbial biochemists—the Arny's army.

I am greatly indebted to a good number of colleagues and friends, in particular R. Leake, I. Hunter, M. Arshad, M. Atif and G. Lang for their kind help and support.

It is also fitting to thank the team at the Taylor & Francis Group for their skills in transforming our manuscripts into a high-quality book, which I hope meets with your expectations as a reader.

The credit for producing this book is only partly ours; it is the blame that rests totally with us. Have a good read.

Dr. Mansi El-Mansi
Editor-in-Chief

Editors

Dr. Mansi El-Mansi was educated at the University of Assiut, city of El-Minya, Egypt (BSc, First Hon and MSc Microbiology). He was intrigued and fascinated by the power of micro-organisms and soon realized that understanding the physiology of micro-organisms demanded clear understanding of their biochemistry. He made the conscious decision of undertaking his PhD in the field of microbial biochemistry and was fortunate enough to carry it out at the UCW, Aberystwyth, UK, under the supervision of David J. Hopper, whose meticulous approach to experimental design was a towering influence. During the course of his PhD studies, he became familiar with the work of the late Stanley Dagley, the father of microbial biochemistry as we know it, and this in turn galvanized his resolve to further his understanding of micro-organisms at the molecular level.

Immediately after the completion of his PhD, Dr. El-Mansi joined Harry Holms at Glasgow University, Scotland, and such a happy and stimulating association continued for the best part of a decade, during which their group in Glasgow was the first to clone and that the structural gene encoding ICDH kinase/phosphatase is a member of the glyoxylate bypass operon. During the course of their fascinating discoveries at Glasgow University, he became acquainted with flux control analysis and its immense potential in the fermentation and pharmaceutical industries. In the fall of 1990, he joined Napier University, Edinburgh, as a lecturer in applied microbiology and biotechnology and was privileged to establish a fruitful collaboration with Henrik Kacser, the founder of MCA (Metabolic Control Analysis), at Edinburgh University.

His current research, consultancy, and scholarly activities are focused on flux control analysis and metabolic engineering of micro-organisms with the view of exploiting global and specific regulatory control mechanisms for the diversion of carbon flux to product formation. His research activities, which span over twenty years, yield an extensive list of publications of which four are single-author publications in peer-reviewed journals, a record of which he is very proud.

Professor Charlie Bryce is head of the School of Life Sciences at Napier University, Edinburgh. At the present time he is vice president of the European Federation of Biotechnology (EFB) in addition to being chairman of the EFB Task Group on Education and Mobility and secretary general of the European Association for Higher Education in Biotechnology (the body that promotes and oversees the operation of the  Eurodoctorate in Biotechnology scheme) and visiting professor at Zhengzhou University in China.

In the last 30 years he has published in excess of 90 refereed research publications, about 80 nonrefereed publications, and designed and produced about 30 teaching packages in a wide variety of media formats. His research interests cover such topics as protein structural biochemistry, enzymology, environmental pollution, biotechnology, bioremediation, site-directed mutagenesis, educational technology, and computer applications.

For the last 10 years he has worked extensively with a number of colleagues in India, Bangladesh, China, Hong Kong, Turkey, the United States, and a number of countries in Europe on a variety of issues relating to environmental pollution, curriculum development, teaching innovations, and manpower and training strategies in biotechnology. He has been invited on numerous occasions to deliver keynote addresses and lectures at national and international conferences on a variety of biotechnology-related topics.

Professor Arnold Demain, research fellow in microbial biochemistry at the Charles A. Dana Research Institute for Scientists Emeriti of Drew University in Madison, New Jersey, is an icon synonymous with excellence in the fields of industrial microbiology and biotechnology. Born in Brooklyn, New York City, in 1927, he was educated in the New York public school system and received his BS and MS in bacteriology from Michigan State University in 1949  and 1950, respectively. He obtained his PhD on pectic enzymes in 1954 from the University of California, having divided his time between the Berkeley and Davis campuses. In 1956, he joined Merck Research Laboratories at Rahway, New Jersey, where he worked on fermentation microbiology, β-lactam antibiotics, flavor nucleotides, and microbial nutrition. In 1965, he founded the Fermentation Microbiology Department at Merck and directed research and development on processes for monosodium glutamate, vitamin B$_{12}$, streptomycin, riboflavin, cephamycin, fosfomycin, and interferon inducers. In 1969, he joined MIT, where he set up the Fermentation Microbiology Laboratory. Since then, he has published extensively on enzyme fermentations, mutational biosynthesis, bioconversions, and metabolic regulation of primary and secondary metabolism. His success is evident in a long list of publications (nearly 500), 11 books of which he is co-editor or co-author, and 21 U.S. patents. His ability to "hybridize" basic studies and industrial applications was recognized by his election to the presidency of the Society for Industrial Microbiology in 1990, membership in the National Academy of Sciences in 1994, the Mexican Academy of Sciences in 1997, and in the Hungarian Academy of Science in 2002. In recognition of his outstanding contribution to our current understanding in fermentation microbiology and biotechnology, he has been awarded honorary doctorates from the University of Leon (Spain), Ghent University (Belgium), Technion University (Israel), Michigan State University (USA), and Muenster University (Germany).

Dr. Anthony (Tony) R. Allman is a graduate of the University of Liverpool, completing both his first degree (BSc) and later his PhD with the Department of Medical Microbiology. A member of both the Institute of Biology and the Society for General Microbiology for more than 25 years, he is well known for passion for making biotechnology  accessible through his various activities such as devising practical workshops plus writing and lecturing about fermentation.

His first encounter with fermentors took place during the course of his employment with Glaxo to develop bacterial vaccines. Subsequently, armed with six years of experience of novel fermentation techniques, he acted as a specialist in this area and as the UK agent of a major European fermentor-manufacturer. In the late 1980s, he joined Infors UK as product manager (later technical director) and in 2002 became fermentation product manager for the Swiss parent company. His work involves providing technical support and training both in-house and worldwide.

Contributors

A.R. Allman
Infors UK Ltd.
Rigate, England

Aristos A. Aristidou
Nature Works LLC
Minnetonka, Minnesota

George N. Bennett
Department of Biochemistry and Cell Biology
Rice University
Houston, Texas

C.F.A. Bryce
School of Life Sciences
Napier University
Edinburgh, Scotland

Marco F. Cardosi
Lifescan Scotland Ltd.
Inverness, Scotland

Boutaieb Dahhou
Laboratoire d'Analyse et d'Architecture des Systèmes du CNRS
France

A.L. Demain
Research Institute for Scientists Emeriti
Drew University
Madison, New Jersey

E.M.T. El-Mansi
School of Life Sciences, Applied Microbiology and
Biotechnology Group
Napier University
Edinburgh, Scotland

Brian S. Hartley
Cambridge, England

Iain S. Hunter
• Department of Pharmaceutical Sciences
University of Strathclyde
Glasgow, Scotland

Chris E. French
Institute of Structural and Molecular Biology
University of Edinburgh
Edinburgh, Scotland

Craig J. L.Gershater
Cambridge, England

David M. Mousdale
Beocarta Ltd.
Glasgow, Scotland

Y. Nakkabi
Laboratoire de Biotechnologie-Bioprocédés,
France

Jens Nielsen
Centre for Process Biotechnology
Technical University of Denmark
Copenhagen, Denmark

Gilles Roux
Laboratoire d'Analyse et d'Architecture des Systèmes du CNRS
France

K.-Y. San
Department of Bioengineering
Rice University
Houston, Texas

Sergio Sanchez
Departamento de Biología Molecular y Biotecnología
Instituto de Investigaciones Biomédicas
Universidad Nacional
Autónoma de México
México

Gregory Stephanopoulos
Department of Chemical Engineering
Massachusetts Institute of Technology
Cambridge, Massachusetts

F. Bruce Ward
Institute of Structural and Molecular Biology
University of Edinburgh
Edinburgh, Scotland

Ronnie Willaert
Department of Ultrastructure
Flanders Interuniversity Institute for Biotechnology
Vrije Universiteit Brussel
Brussels, Belgium

Contents

Chapter 13   Control of Fermentations: An Industrial Perspective

Craig J.L. Gershater

1

Fermentation Microbiology and Biotechnology: An Historical Perspective

E.M.T. El-Mansi, C.F.A. Bryce, Brian S. Hartley and Arnold L. Demain

1.1 Fermentation: An ancient tradition

Fermentation has been known and practiced by mankind since prehistoric times, long before the underlying scientific principles were understood. That such a useful technology should arise by accident will come as no surprise to those people who live in tropical and subtropical regions, where, as Marjory Stephenson put it, "every sandstorm is followed by a spate of fermentation in the cooking pot" (Stephenson, 1949). For example, the productions of bread, beer, vinegar, yogurt, cheese, and wine were well-established technologies in ancient Egypt (Figures 1.1 and 1.2). It is an interesting fact that archaeological studies have revealed that bread and beers, in that order, were the two most abundant components in the diet of ancient Egyptians. Everyone, from the Pharaoh to the peasant, drank beer for social as well as ritual reasons. Archaeological evidence has also revealed that ancient Egyptians were fully aware not only of the need to malt the barley or the emmer wheat but also of the need for starter cultures, which at the time may have contained lactic acid bacteria in addition to yeast.

1.2 The rise of fermentation microbiology

With the advent of the science of microbiology, and in particular fermentation microbiology, we can now shed light on these ancient and traditional activities. Consider, for example, the age-old technology of wine making, which relies upon crushing the grapes (Figure 1.2) and letting nature take its course, i.e., fermentation. Many micro-organisms can grow on the grape

Figure 1.1
Bread making as depicted on the wall of an ancient Egyptian tomb dated c. 1400 BC. (Reprinted with the kind permission of the Fitzwilliam Museum, Cambridge, England.)

Figure 1.2
Grape treading and wine making as depicted on the walls of Nakhte's tomb, Thebes, c. 1400 BC. (Reprinted with the kind permission of AKG, London, England/Erich Lessing.)

sugars more readily and efficiently than yeasts, but few can withstand the osmotic pressure arising from the high sugar concentrations. Also, as sugar is fermented, the alcohol concentration rises to a level at which only the osmotolerant, alcohol-tolerant cells can survive. Hence inhabitants of ancient civilizations did not need to be skilled microbiologists in order to enjoy the fruits of this popular branch of fermentation microbiology.

In fact, the scientific understanding of fermentation microbiology and, in turn, biotechnology only began in the 1850s, after Louis Pasteur had succeeded in isolating two different forms of amyl alcohol, of which one was optically active (L, laevorotatory) while the other was not. Rather unexpectedly, the optically inactive form resisted all Pasteur's attempts to resolve it into its two main isomers, i.e., the *laevorotatory* (L) and the *dextrorotatory* (D) forms. It was this observation that led Pasteur into the study of fermentation, in the hope of unraveling the underlying reasons behind his observation, which was contrary to stereochemistry and crystallography understandings at the time.

In 1857, Pasteur published the results of his studies and concluded that fermentation is associated with the life and structural integrity of the yeast cells rather than with their death and decay. He reiterated the view that the yeast cell is a living organism and that the fermentation process is essential for the reproduction and survival of the cell. In his paper, the words "cell" and "ferment" are used interchangeably, i.e., the yeast cell is the ferment. The publication of this classic paper marks the birth of fermentation microbiology, and in turn biotechnology, as a new scientific discipline. Guided by his knowledge and armed with his experimental observations, Pasteur was able to confidently challenge and reject Liebig's perception that fermentation occurs as a result of contact with decaying matter. He also ignored the well-documented view that fermentation occurs as a result of "contact catalysis" though it is possible that this was not suspect in his view. The term "contact catalysis" probably implied that fermentation is brought about by a chain of enzyme-catalyzed reactions, and we might also credit Pasteur with the concept of an enzyme as "in yeast."

Although Pasteur's interpretations were essentially physiological rather than biochemical, they were pragmatically correct. During the course of his further studies, Pasteur was also able to establish not only that alcohol was produced by yeast through fermentation but also that souring was a consequence of contamination with bacteria that were capable of converting alcohol to acetic acid. Souring could be avoided by heat treatment at a certain temperature for a given length of time. This eliminated the bacteria without adversely affecting the organoleptic qualities of the beer or the wine, a process we now know as pasteurization.

A second stage in the development of fermentation microbiology and biotechnology, more collectively known as fermentation biotechnology, began in 1877, when Moritz Traube proposed the theory that fermentation and other chemical reactions are catalyzed by protein-like substances and that, in his view, these substances remain unchanged at the end of the reactions. Furthermore, he described fermentation as a sequence of events in which oxygen is transferred from one part of the sugar molecule to another, culminating in the formation of a highly oxidized product, i.e., CO^2, and a highly reduced product, i.e., alcohol. Considering the limited knowledge of biochemistry in general and enzymology in particular at the time, Traube's remarkable vision was to prove some fifty years ahead of its time.

In 1897, two years after Pasteur died, Eduard Buchner was successful in preparing a cell-free extract that fermented sugar. This discovery was received with a great deal of interest, not only because it was the first evidence of fermentation without a living organism but also because it was in sharp contrast to the theory proposed by Pasteur. In the early 1900s, the views of Pasteur were modified and extended to stress the idea that fermentation is a function of a living, but not necessarily multiplying, cell and that it is not a single step but rather a chain of different reactions, each of which is probably catalyzed by a different enzyme.

1.3   Developments in metabolic and biochemical engineering

The outbreak of the First World War provided an impetus and a challenge to produce certain chemicals that, for one reason or another, could not be manufactured by conventional means. For example, there was a need for glycerol, an essential component in the manufacture of ammunition, because no vegetable oils could be imported due to the naval blockade. German biochemists and engineers were able to adapt yeast fermentation, turning sugars into glycerol rather than alcohol. Although this process enabled the Germans to produce in excess of 100 tones of glycerol per month, it was abandoned as soon as the war was over because glycerol could be made very cheaply as a by-product of the soap industry. There was also, of course, a dramatic drop in the level of manufacture of explosives and, in turn, the need for glycerol.

The diversion of carbon flow from alcohol production to glycerol formation was achieved by adding sodium bisulphite, which reacts with *acetaldehyde* to give an adduct that cannot be converted to alcohol (Figure 1.3). Consequently, NADH accumulates intracellularly, thus perturbing the steady-state redox balance (NAD^+/NADH ratio) of the cell. The drop in the intracellular level of NAD^+ is accompanied by a sharp drop in the flux through *glyceraldehyde-3-phosphate dehydrogenase* which, in turn, allows the accumulation of the two isomeric forms of triose phosphate, i.e., glyceraldehyde-3-phosphate and dihydroxyacetone-3-phosphate. Accumulations of the latter together with high intracellular levels of NADH trigger the expression of glycerol-3-phosphate dehydrogenase, which, in turn, leads to the diversion of carbon flux from ethanol production to glycerol formation, thus restoring the redox

Figure 1.3

Diversion of carbon flux from alcohol production to glycerol formation in the yeast *Saccharomyces cerevisiae*. Note that the functional role of bisulphite is to arrest acetaldehyde molecules, thus preventing the regeneration of NAD⁺ as a consequence of making alcohol dehydrogenase redundant. To redress the redox balance, i.e., the NAD⁺/NADH ratio, *Saccharomyces cerevisiae* diverts carbon flow (dashed route) toward the reduction of dihydroxyacetone-3-phosphate to glycerol-3-phosphate, thus regenerating the much-needed NAD⁺. The glycerol-3-phosphate thus generated is then dephosphorylated to glycerol.

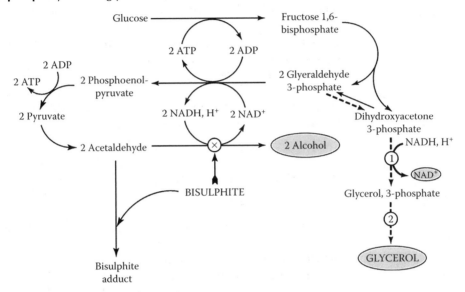

Key enzymes are as follows:
1, Glycerol 3-phosphate dehydrogenase and
2, Glycerol phosphates. X, indicates no flux through ethanol dehydrogenase.

balance within the cells by regenerating NAD⁺ (Figure 1.3). Although this explanation is with the hindsight of modern biochemistry, the process can be viewed as an early example of metabolic engineering.

Following the First World War, research into yeast fermentation was largely influenced by the work of Neuberg and his proposed scheme (biochemical pathway) for the conversion of sugars to alcohol (alcohol fermentation). Although Neuberg's scheme was far from perfect and proved erroneous in many ways, it provided the impetus and the framework for many scientists at the Delft Institute, who vigorously pursued research into oxidation/reduction mechanisms and the kinetics of product formation in a wide range of enzyme-catalyzed reactions. Such studies were to prove important in the development of modern biochemistry as well as fermentation biotechnology.

While glycerol fermentation was abandoned immediately after the First World War, the acetone-butanol fermentation

Box 1.1 **"Fill and Spill"**

This pattern of fermentation is essentially a 'batch fermentation' or a 'fed-batch fermentation' process in which the organism is allowed to grow and once product formation has reached maximum level, the fermentation pot is harvested, leaving some 10% of the total volume as an inoculum for the next batch. This process is repeated until the level of contamination becomes unacceptably high.

process, catalyzed by *Clostridium acetobutylicum* flourished. Production lines were modified to accommodate the new approach of "Fill and Spill," which permitted substantial savings in fuels without adversely affecting the output of solvent production during the course of the Second World War. However, as soon as the production of organic solvents as a by-product of the petrochemical industry became economically viable, the acetone-butanol fermentation process was discontinued.

1.4   Discovery of antibiotics and genetic engineering

The discovery of penicillin in the late 1920s and its antibacterial properties in the early 1940s represent a landmark in the development of modern fermentation biotechnology. This discovery, to a country at war, was both sensational and invaluable. However, *Penicillium notatum*, the producing organism, was found to be susceptible to contamination by other organisms, and so aseptic conditions were called for. Such a need led to the introduction of so-called stirred tank bioreactors, which minimize contamination with unwanted organisms. The demand for penicillin prompted a worldwide screen for alternative penicillin-producing strains, leading to the isolation of *Penicillium chrysogenum*, which produced more penicillin than the original isolate *P. notatum*. *Penicillium chrysogenum* was then subjected to a very intensive program of random mutagenesis and screening. Mutants that showed high levels of penicillin production were selected and subjected to further rounds of mutagenesis, and so on. This approach was successful, as indicated by the massive increase in production from less than $1g\ l^{-1}$ to slightly over $20g\ l^{-1}$ of culture.

Once the antibacterial spectrum of penicillin was determined and found to be far from universal, pharmaceutical companies began the search for other substances with antibacterial activity. This screening programs led to the discovery of many antibacterial agents produced by various members of the actinomycetes. Although the search for new antibiotics is never over, intensive research programs involving the use of genetic and metabolic engineering were initiated with the aim of increasing the productivity and potency of current antibiotics. For example, the use of genetic and metabolic engineering has increased the yield of penicillin many-fold.

1.5 The rise and fall of single cell protein

The latter part of the 1960s saw the rise and fall of single cell protein (SCP) production from petroleum or natural gas. A large market for SCP was forecast as the population in the third world, the so-called underdeveloped countries, continued to increase despite a considerable shortfall in food supply. However, the development of SCP died in its infancy, largely due to the sharp rise in the price of oil, which made it economically non-viable. Furthermore, improvements in the quality and the yields of traditional crops did not help the cause of SCP production.

1.6 Fermentation biotechnology and the production of amino acids

The next stage in the development of fermentation biotechnology was dominated by the success in the use of regulatory control mechanisms for the production of amino acids. The first breakthrough was the discovery of glutamic acid overproduction by *Corynebacterium glutamicum* in the late 1950s and early 1960s, when a number of Japanese researchers discovered that regulatory mutants, isolated by virtue of their ability to resist amino acid analogues, were capable of overproducing amino acids. The exploitation of such a discovery, however, was hampered by the induction of degradative enzymes once the extracellular concentration of the amino acid increased beyond a certain level, e.g., accumulation of tryptophan induced the production of tryptophanase, thus initiating the breakdown of the amino acid. This problem was resolved by the use of penicillin, which, with tryptophan as the sole source of carbon in the medium, eliminated the growing cells, i.e., those capable of metabolizing tryptophan, but not those that were quiescent. Following the addition of penicillin, the mutants that had survived the treatment (3×10^{-4}) were further tested. Enzymic analysis revealed that one mutant was totally devoid of tryptophanase activity. This approach was soon extended to cover the production of other amino acids, particularly those not found in sufficient quantities in plant proteins. The successful use of regulatory mutants stimulated interest in the use of auxotrophic mutants for the production of other chemicals. The rationale is that auxotrophic mutants will negate feedback inhibition mechanisms and in turn allow the accumulation of desired end product. For example, an arginine-auxotroph was successfully used in the production of ornithine, while a homoserine-auxotroph was used for the production of lysine.

1.7   Impact of functional genomics, proteomics, metabolomics and bio-informatics on the scope and future prospects of fermentation microbiology and biotechnology

Modern biotechnology, a consequence of innovations in molecular cloning and overexpression in the early 1970s, has started in earnest in the early 1980s after the manufacture of insulin with the first wave of products hitting the market in the early 1990s. During this early period, the biotechnology companies focused their efforts on specific genes/proteins (natural proteins) that were of well-known therapeutic value and typically produced in very small quantities in normal tissues. Later, monoclonal antibodies became the main products of the biopharmaceutical industry.

In the mid-1980s, Thomas Roderick coined the term "genomics" to describe the discipline of mapping, sequencing, and analyzing genomic DNA with the view to answering biological, medical, or industrial questions (Jones, 2000). Recent advances in genomics, stimulated by the human genome project and computer software technology, led many biotechnologists to venture from the traditional *in vivo* and *in vitro* research into the *in silico* approach, thus establishing a new science in the shape of bio-informatics. In this approach, microbiologists employ computers to store, retrieve, analyze and compare a given sequence of DNA or protein with those stored in data banks from other organisms. Microbial biotechnologists were quick to realize that the key to successful commercialization of a given sequence relied on the development of innovative methodology (bio-informatics) that facilitates the transformation of a given sequence into a diagnostic tool and/or a therapeutic drug, thus bridging the gap between academic research and commercialization.

The new innovations in functional genomics, proteomics, metabolomics, and bio-informatics will certainly play a major role in transforming our world in an unparalleled way, despite political and ethical controversies.

In this book, we have addressed the multidisciplinary nature of this subject and highlighted its many fascinating aspects in the hope that we provide stepping stone in its progress. As we enter a new era in which the use of renewable resources for the production of desirable end products is recognized as an urgent need, fermentation microbiology and biotechnology have a central role to play.

References

Stephenson, M. (1949): *Bacterial Metabolism*. Longmans, Green and Co. Ltd, London.

Jones, PBC (2000). The commercialisation of Bioinformatics: *EJB Electronic Journal of Biotechnology*, vol. 3, no. 2, Issue of August 15, ISSN: 0717-3458.

2

Microbiology of Industrial Fermentation

E.M.T. El-Mansi and F. Bruce Ward

2.1 Introduction

Micro-organisms play a central role in the production of a wide range of primary and secondary metabolites, industrial chemicals, enzymes, and antibiotics. The diversity of fermentation processes may be attributed to many factors including the high surface-to-volume ratio and ability to utilize a wide spectrum of carbon and nitrogen sources. The high surface-to-volume ratio supports a very high rate of metabolic turnover, e.g., the yeast *Saccharomyces cerevisiae* has been reported to be able to synthesize protein several orders of magnitudes higher than plants. On the other hand, the ability of micro-organisms to adapt to different metabolic environments makes them capable of utilizing inexpensive renewable resources such as wastes and by-products of the farming and petrochemical industries as the primary carbon source (see Chapter 9). Industrially important micro-organisms include bacteria, yeasts, molds, and actinomycetes.

While the metabolic route through which glucose is converted to pyruvate, glycolysis, is universally conserved among all organisms, micro-organisms differ from eukaryotes in their ability to process pyruvate through a diverse array of routes (Figure 2.1), giving rise to a multitude of different end products. Such diversity has been fully exploited by fermentation technologists for the production of fine chemicals, organic solvents, and dairy products.

Prokaryotic organisms differ from yeast and fungi as well as other eukaryotes, in a number of ways including cell structure and growth cycle. For example, while DNA is compartmentalized within the nucleus in eukaryotes, it is intricately and neatly folded within the cytoplasm in prokaryotes. Furthermore, the newly synthesized DNA molecules in prokaryotes need no special assembly to form a chromosome as the DNA is already attached to the bacterial membrane, thus ensuring its successful segregations into the two daughter cells. Another example that is pertinent to

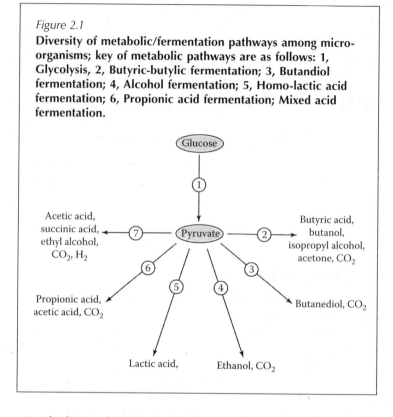

Figure 2.1

Diversity of metabolic/fermentation pathways among micro-organisms; key of metabolic pathways are as follows: 1, Glycolysis, 2, Butyric-butylic fermentation; 3, Butandiol fermentation; 4, Alcohol fermentation; 5, Homo-lactic acid fermentation; 6, Propionic acid fermentation; Mixed acid fermentation.

microbial growth is that prokaryotic organisms, unlike eukaryotes where the electron transport chain and, in turn, oxidative phosphorylation is located within the mitochondria, associate oxidative phosphorylation with cytoplasmic membranes.

2.2   Chemical synthesis of bacterial protoplasm/biomass

While central metabolism (glycolysis, the pentose phosphate pathway and the Krebs cycle) is concerned with the breakdown of growth substrates and its conversion to the twelve biosynthetic precursors, ATP, and reducing powers, intermediary metabolism focuses on the conversion of those biosynthetic precursors to monomers and their subsequent polymerisation and assembly into polymers.

In addition to carbon, nitrogen, phosphate, potassium, sulphur, irons, and magnesium, the chemical input required for growth also include trace elements. The necessity for such a multitude of inputs is paramount as it is required for enzymic activities. The contribution of each additive to biomass and product formation can be assessed quantitatively.

2.2.1   Central and intermediary metabolism

In the following sections we will be addressing the function of enzymes of central and intermediary metabolism (Figure 2.2) with respect to their role in bacterial growth and as such it would be helpful if we were to appreciate the functional role of each.

Figure 2.2

An overview of the metabolic pathways of central and intermediary metabolism employed by *E. coli* for the conversion of glucose and other substrates to biosynthetic precursors; indicated by heavy-dotted arrows, ATP, and reducing powers. Entry of substrates other than glucose into central metabolism is indicated by dashed arrows. Key enzymes are as follows: 1, Glucose-PEP phosphotransferase system; 2, the Pentose phosphate pathway; 3, Pyruvate kinase; 4, PEP carboxylase; 5, PEP carboxytransphosphorylase; 6, Pyruvate carboxylase; 7, Isocitrate lyase; 8, Malate synthase; 9, Malic enzyme; 10, PEP carboxykinase.

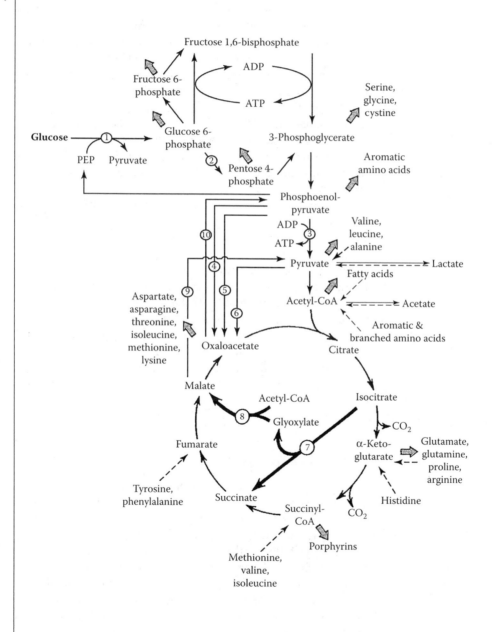

During growth on glucose minimal medium under aerobic conditions, *Escherichia coli* catabolizes glucose through glycolysis and the Krebs cycle (Figure 2.2) to bring about its transformation to biosynthetic precursors; twelve in total, over half of which are phosphorylated, ATP and reducing powers in the shape of NADH, NADPH and $FADH_2$ as well as esterified CoA derivatives, e.g., succinyl CoA and Malonyl CoA. In this process a whole host of different reactions, the fueling reactions, (Figure 2.3) are coordinated in a precise manner to ensure successful adaptation and survival of the organism. Once the biosynthetic precursors, ATP, and reducing powers are generated, the organism employs anabolic enzymes to convert those biosynthetic precursors into monomers (building blocks) in the shape of various sugars, amino acids, and nucleotides (Figure 2.3). The monomers are then polymerized into polymers (macromolecules), which, in turn, are assembled into different structures/organelles and thence to biomass (Figure 2.3). The central metabolic pathways (glycolysis, the pentose phosphate pathway, and the Krebs cycle) fulfill both catabolic (*Cata*, a Greek word for breakdown) and anabolic (*Ana*, a Greek word for building up) functions and as such may be referred to as amphibolic pathways.

Although the central metabolic pathways are highly conserved in all organisms, micro-organisms display a great deal of diversity even within a single species such as *E. coli* in response to growth conditions. As can be seen from Figure 2.2, microorganisms are capable of utilizing a wide range of substrates and afford a different entry point for each into central metabolism. As the points of entry into central metabolism vary according to the chemical structure of the substrate, the make-up of the enzymic machinery necessary for its metabolism changes accordingly (Guest and Russell, 1992). For example, during growth on acetate or fatty acids, *E. coli* displays a unique set of anaplerotic reactions otherwise know as the glyoxylate bypass (Cozzone, 1998), which are not expressed during growth on glucose or glycerol.

2.2.2 *Anaplerotic pathways*

During growth on glucose or other acetogenic substrates, i.e., those that support flux to acetate excretion, metabolites of the Krebs cycle, namely α-ketoglutarate, succinyl CoA, and oxaloacetate (Figure 2.2) are constantly being withdrawn for biosynthesis. It follows; therefore, that such intermediates must be replenished; otherwise, the cycle will grind to a halt. The reactions that fulfill such a function are known as anaplerotic, a Greek word for replenishing, pathways. The make up of the anaplerotic enzymes differs from one phenotype to another, i.e., substrate-dependent. For example, during growth on glucose,

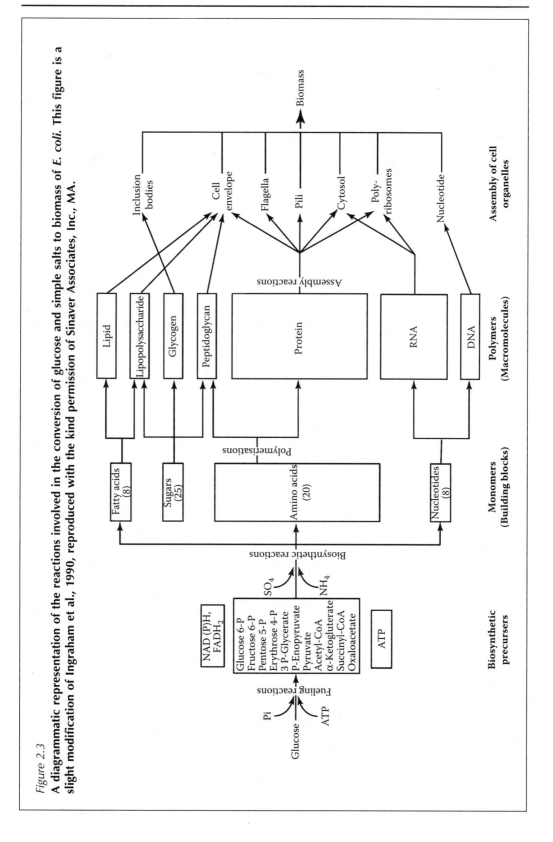

Figure 2.3

A diagrammatic representation of the reactions involved in the conversion of glucose and simple salts to biomass of *E. coli*. This figure is a slight modification of Ingraham et al., 1990, reproduced with the kind permission of Sinaver Associates, Inc., MA.

phosphoenolpyruvate carboxylase, pyruvate carboxylase, and phosphoenolpyruvate carboxytransphosphorylase (Figure 2.2) represent the full complement of anaplerotic enzymes that may be used in full or in part depending on the organism under investigation. During growth on acetate, however, *E. coli* employs the glyoxylate bypass operon enzymes, namely isocitrate lyase and malate synthase, as an anaplerotic sequence. Interestingly, however, while the enzymes of the glyoxylate bypass in *E. coli* form an operon and include the bifunctional regulatory enzyme isocitrate dehydrogenase kinase/phosphatase, these enzymes are not organized in the same way in *Corynebacterium glutamicum*.

2.2.3 *Polymerization and assembly*

Following transcription through the activity of RNA polymerase, mRNA is translated into protein by the ribosomes. In this process ribosomes attach to mRNA together with cofactors, enzymes, and complementary tRNA, thus forming a polysome and, in turn, initiating the synthesis of polypeptides. Polysomes are one of the most abundant organelles in growing cells; each polysome contains approximately twenty subunits of ribosomal RNA (Ingraham et al., 1983).

2.2.4 *Biomass formations*

The rate of product formation in a given industrial process, a significant parameter, is directly related to the rate of biomass formation, which, in turn, is influenced directly or indirectly by a whole host of different environmental factors, e.g., oxygen supply, pH, temperature, accumulation of inhibitory intermediates, etc. It is, therefore, important that we are able to describe growth and production in quantitative terms. The study of growth kinetics and growth dynamics involves the formulation and use of differential equations (for more details, see Chapter 3). While the mathematical derivation of these equations is beyond the scope of this chapter, it is important that we understand how such equations can be used to further our understanding of microbial growth in general and its impact on product formation (yield) in particular.

While growth kinetics focuses on the measurement of growth rates during the course of fermentation, growth dynamics relates the changes in population (biomass) to changes in growth rate and other parameters, e.g., pH, temperature. To unravel such intricate inter-relationships, the description of growth kinetics and growth dynamics relies on the use of differential equations.

2.3 The growth cycle

Now let us consider the basic equation used to describe microbial growth:

$$\frac{dx}{dt} = \frac{ax}{b} \tag{2.1}$$

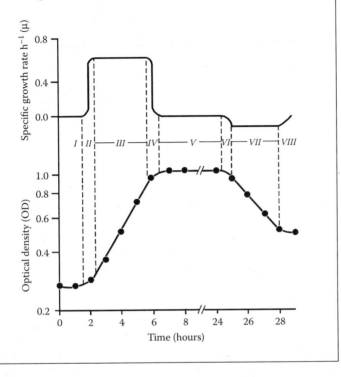

Figure 2.4

Typical pattern of growth cycle during growth of micro-organisms in batch cultures; the vertical dotted lines and the Roman numerals indicate the changes in specific growth rate (µ) throughout the cycle.

The term autocatalytic growth is generally used to indicate that the rate of increase in biomass formation in a given fermentation is proportional to the original number of cells present at the beginning of the process, thus reflecting the positive nature of growth.

Equation (2.1) implies that the rate of biomass (x) formation changes as a function of time (t) and that the rate of change is directly proportional to the concentration of a particular factor (a) such as growth substrate or temperature but is inversely proportional to the concentration of another factor (b) such as inhibitors. In the equation above, a and b are independent of time t, and the proportionality factor in Equation (2.1) can in effect be ignored. In the early stages of any fermentation process, the increase in biomass is unrestricted and, as such, the pattern of growth follows an autocatalytic first-order reaction (autocatalytic growth) up to a point where either side of Equation (2.1) becomes negative, resulting in autocatalytic death.

During batch fermentation, a typical pattern of growth curve, otherwise known as the growth cycle, is observed (Figure 2.4). Clearly, a number of different phases of the growth cycle can be differentiated. These are

   i. Lag phase
   ii. Acceleration phase
   iii. Exponential (logarithmic) phase

The term growth cycle is used to describe the overall pattern displayed by micro-organisms during growth in batch cultures. It is noteworthy that such a cycle is by no means a fundamental property of the bacterial cell, but rather a consequence of the progressive decrease in food supply or accumulation of inhibitory intermediates in a closed system to which no further additions or removals are made.

iv. Deceleration phase
v. Stationary phase
vi. Accelerated death phase
vii. Exponential death phase
viii. Survival/death phase

The changes in the specific growth rate (μ) as the organism progresses through the growth cycle can also be seen in Figure 2.4.

We shall now describe the metabolic events and their implications in as far as growth, survival, and productivity are concerned. Naturally, the scenario begins with the first phase of the growth cycle.

2.3.1 *The lag phase*

In this phase the organism is simply faced with the challenge of adapting to the new environment. While adaptation to glucose as a sole source of carbon appears to be relatively simple, competition with other carbon sources, although complex, is resolved in favor of glucose through the operation of two different mechanisms, namely catabolite inhibition and catabolite repression.

Adaptation to other carbon sources, however, may require the induction of a particular set of enzymes that are specifically required to catalyze transport and hydrolysis of the substrate, e.g., adaptation to lactose, or to fulfill anaplerotic as well as regulatory functions as is the case in adaptation to acetate as the sole source of carbon and energy. Irrespective of the mechanisms employed for adaptation, the net outcome at the end of the lag phase is a cell that is biochemically vibrant, i.e., capable of transforming chemicals to biomass.

Entering a lag phase during the course of industrial fermentation is not desirable as it is very costly and as such should be avoided. The question of whether a particular organism has entered a lag phase in a given fermentation process can be determined graphically by simply plotting log *n* (biomass) as a function of time, as shown in Figure 2.5.

Note that the transition from the lag phase to the exponential phase involves another phase, namely the acceleration phase, as described earlier (Figure 2.4). This difficulty can be easily overcome by extrapolating the lag phase sideways and the exponential phase downwards as shown, with the point of interception (L) taken as the time at which the lag phase ended. What is also interesting about the graph in Figure 2.5 is that if one continues to extrapolate the exponential phase downwards, then the point at which the ordinate is intersected gives the number of cells that were viable and metabolically active at the point of inoculation.

Figure 2.5
Graphical determination of the lag phase and the number of viable cells at the onset of batch fermentation; as the exponential phase is extrapolated downwards, it intercepts the extrapolated line of the lag phase and the ordinate, respectively (see text for details).

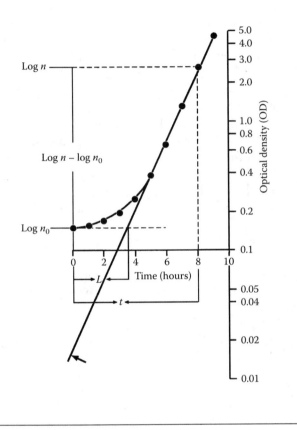

The question of whether a lag (L) has occurred during the course of the fermentation process and for how long can be easily determined from the following equation:

$$\frac{\log n - \log n_0}{t - L} = \frac{\log 2}{T} \tag{2.2}$$

where n is the total number of cells after a given time (t) since the start of fermentation, n_0 is the number of cells at the beginning of fermentation, and MGT is the organism's mean generation time (doubling time). Equation (2.2) describes the exponential growth, taking into consideration a lag phase in the process. In the following section we shall describe the exponential phase in general and the derivation of Equation (2.2) in particular.

Doubling time or mean generation time (MGT) is the time required for a given population (N_0) to double in number ($2N_0$).

2.3.2 *The exponential phase*

In this phase, each cell increases in size, and providing that conditions are favorable, divides into two, which, in turn, grow and divide; and the cycle continues. During this phase, the cells are capable of transforming the primary carbon source into biosynthetic precursors, reducing power and energy, which is generally trapped in the form of ATP, PEP, and proton gradients. The biosynthetic precursors thus generated are then channeled through various biosynthetic pathways for the biosynthesis of various monomers (amino acids, nucleotides, fatty acids, sugars) that, in turn, are polymerized to give the required polymers (proteins, nucleic acids, ribonucleic acids, lipids). Finally, these polymers are assembled in a precise way and the cell divides to give the new biomass characteristic of each organism (Figure 2.3). The time span of each cycle (cell division) is known as generation time or doubling time, but because we generally deal with many millions of cells in bacterial cultures, the term "mean generation time" (MGT) is more widely used to reflect the average generation times of all cells in the culture. Such a rate, providing conditions are favorable, is fairly constant.

If a given number of cells (n_0) is inoculated into a suitable medium and the organism was allowed to grow exponentially, then the number of cells after one generation is $2n_0$; at the end of two generations, the number of cells becomes $4n_0$ (or 2^2n_0). It follows, therefore, that at the end of a certain number of generations (Z), the total number of cells equals 2^Zn_0. If the total number of cells, or its log value, at the end of Z generations is known, then

$$n = n_0 2^Z \tag{2.3}$$

$$\log n = \log n_0 + Z \log 2 \tag{2.4}$$

To determine the number of cell divisions, i.e., number of generations (Z) that have taken place during fermentation; the above equation can be modified to give

$$Z = \frac{\log n - \log n_0}{\log 2} \tag{2.5}$$

If T is the mean generation time required for the cells to double in number and t is the time span over which the population has increased exponentially from n_0 to n, then

$$Z = \frac{t}{T} = \frac{\log n - \log n_0}{\log 2} \tag{2.6}$$

If during the course of a particular fermentation, a lag time has been demonstrated, Equation (2.6) can be modified to take account of this observation. The modified equation is as follows:

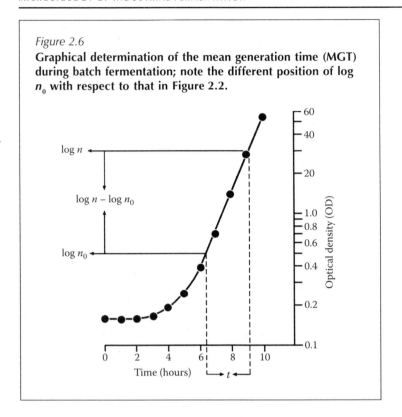

Figure 2.6

Graphical determination of the mean generation time (MGT) during batch fermentation; note the different position of log n_0 with respect to that in Figure 2.2.

$$Z = \frac{t-L}{T} = \frac{\log n - \log n_0}{\log 2} \tag{2.7}$$

The above equations, (2.6) and (2.7), can be rearranged to give the familiar equations governing the determination of the mean generation time (MGT) as follows:

$$\frac{\log n - \log n_0}{t} = \frac{\log 2}{T} \tag{2.8}$$

$$\frac{\log n - \log n_0}{t-L} = \frac{\log 2}{T} \tag{2.9}$$

While Equation (2.8) describes the exponential phase of growth in pure terms, Equation (2.9) takes into consideration the existence of a lag phase in the process.

The exponential pattern (i.e. 2-4-8-16, and so on) demonstrated in Figure 2.6 obeys Equations (2.8) and (2.9) for logarithmic growth. Note the different position of log n_0 to that cited in Figure 2.5.

Although the use of log to the base 2 has the added advantage of being able to determine the number of generations relatively easily, because an increase of one unit in log $2n$ corresponds to one generation, the majority of researchers continue to use log to the base 10. In this case, the slope of the line (Figure 2.6) equals $\mu/2.303$. The relationship between the mean generation time (T) and the specific growth rate (μ) can be described mathematically by the following equation:

$$\ln 2 = \mu T \tag{2.10}$$

or

$$\mu = \ln 2/T \tag{2.11}$$

Because $\ln 2 = 0.693$, either of the above equations can be rearranged to give

$$T = \frac{0.693}{\mu} \tag{2.12}$$

During the course of exponential growth, the culture reaches a steady state. As such, the intracellular concentrations of all enzymes, cofactors, and substrates are considered to be constant. During this phase, one can therefore safely assume that all bacterial cells are identical and that the doubling time is constant with no loss in cell numbers due to cell death. The rate of growth of a given population (N) represents, therefore, the rate of growth of each individual cell in the population multiplied by the total number of cells. Such a rate can be described mathematically by the following differential equation:

$$\frac{dN}{dt} = \mu N \tag{2.13}$$

This equation implies that the rate of new biomass formation is directly proportional to the specific growth rate (μ) of the organism under investigation and the number of cells (N). This pattern of growth may be referred to as autocatalytic: a term described above. If the above equation describes the exponential phase correctly, then a straight line should be obtained when $\ln N$ is plotted as a function of time (t). During the exponential phase, it is generally assumed that all cells are identical and as such the specific growth rate (μ) of individual cells equals that of the whole population. Equation (2.13) can therefore be rearranged to take account of the fact that dN/dt is proportional to the number of cells (N) and that the specific growth rate (μ) is a proportionality factor to give

$$\mu = \frac{1}{N}\frac{dN}{dt} \tag{2.14}$$

The specific growth rate (μ) is usually expressed in terms of units per hour (h^{-1}).

Although Equations (2.9) and (2.10) describe the exponential phase satisfactorily, in some fermentations, as is the case during growth of E. coli on sodium acetate (M El-Mansi, unpublished results), the organism fails to maintain a steady state for any length of time and so μ falls progressively with time until the

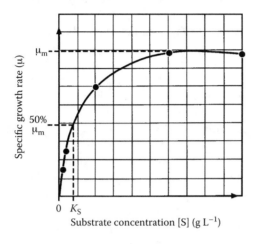

Figure 2.7

Graphical determination of specific growth rate (μ) and saturation constant (*K*ₛ) during batch fermentations; note that this graph was constructed without taking maintenance energy into consideration. With maintenance in mind the curve should slide sideways to the right in direct proportion to the fraction of carbon diverted towards maintenance.

organism reaches the stationary phase. Such a drop in growth rate (μ) can be accounted for by a whole host of different factors including nutrient limitations, accumulation of inhibitors and/or a crowding factor, i.e., as the population increases in size μ decreases.

2.3.2.1 *Metabolic interrelationship between nutrient limitations and specific growth rate (μ)*

To account for the effect of nutrient limitations on growth rate (μ), Monod modified the above equation so that the effect of substrate concentration (limitations) on μ could be assessed quantitatively. Monod's equation is as follows:

$$\mu = \frac{\mu_m S}{K_S + S} \tag{2.15}$$

where S is the concentration of the limiting substrate, μ_m is the maximum specific growth rate, and K_S is the saturation constant, i.e., when $S = K_S$, then $\mu = \mu_m/2$, as illustrated in Figure 2.7.

During growth in a steady state, the following equations may be used to describe growth using biomass (X) rather than number of cells (N) as a measure of growth:

$$\frac{dX}{dt} = \mu X \tag{2.16}$$

$$\frac{dS}{dt} = -\mu \frac{X}{Y} \qquad (2.17)$$

where X is the biomass concentration, Y is the growth yield (e.g., biomass generated per gram substrate utilized). It is noteworthy, however, that during steady-state growth, i.e., in a chemostat or a turbidostat, the terms used to describe the bacterial numbers (N) and biomass (X) in Equations (2.16) and (2.17) are identical. This is not necessarily the case in batch cultures because the size and shape of bacterial cells vary from one stage of growth to another.

While Equation (2.15) addresses the relative change in biomass (the first variable) with respect to time, Equation (2.16) addresses the relative change in substrate concentration (the second variable) as a function of time. Note that following the exhaustion of substrate $dX/dt = dS/dt = 0$. While either Equation (2.16) and (2.17) can surely predict the deceleration and stationary phases of growth, it should be remembered that nutrient limitation is not the only reason for the deceleration of growth and subsequent entry into the stationary phase. In addition to environmental factors, crowding is reported to have an adverse effect on growth rate (μ), i.e., growth rate diminishes as the size of the population increases. Although the equation describing the effect of crowding on growth rate (Verhulst–Pearl logistic equation) predicts a sigmoidal pattern of growth and fits rather well with the growth curve for populations of higher organisms, yeasts, and bacteria, it will not be further discussed in this chapter as it overlooks the effect of other environmental factors on growth.

2.3.3 *Stationary phase and cell death*

As the exponential phase draws to an end, the organism enters the stationary phase and then the death phase. The equations used earlier to describe growth do not adequately describe cell death. However, if we assume that the kinetics of death are similar to that of growth, then the specific rate of cell death (λ) can be described mathematically by the following equation:

$$\frac{dN}{dt} = (\mu - \lambda) N \qquad (2.18)$$

Equation (2.18) clearly indicates that if μ is greater than λ, then the organism will grow at a rate equal to $\mu - \lambda$. If, on the other hand, μ is less than λ, then the population dies at a rate equal to $\lambda - \mu$. Under conditions where μ equals λ, neither growth nor death is observed: a situation thought to prevail throughout the course of the stationary phase. As the energy supply continues to fall, the equilibrium between λ and μ will finally shift in favor of λ, and consequently cell death begins.

Micro-organisms respond differently to nutrient limitations during the stationary phase. For example, while *Bacillus subtilis* and other Gram-positive, spore-forming organisms respond by sporulation, *E. coli* and other Gram-negative, non-spore-forming bacteria cannot respond in the same way and as such other mechanisms must have evolved. Although a fraction of the bacterial population die during this phase of growth, a relatively large number remains viable for a long time despite starvation. The ability of cells to remain viable despite a prolonged period of starvation is advantageous, as most micro-organisms in nature are subject to nutrient limitations in one form or another. Our understanding of the molecular mechanisms employed by micro-organisms for survival during this phase of growth is rather limited. However, contrary to the notion that micro-organisms enter a logarithmic death phase soon after the onset of the stationary phase, recent investigations have revealed that some micro-organisms such as *Saccharomyces cerevisiae* and *E. coli* adapt to starvation, presumably through mutations and induction mechanisms, very effectively.

The ability of some cells to survive prolonged starvation inspired some researchers to pursue the question of whether such cells are biochemically distinguishable from their predecessors, which were actively growing or had just entered the stationary phase. In the case of *Saccharomyces cerevisiae*, analysis of mRNAs, which are specifically expressed following the onset of the stationary phase, revealed the presence of a new family of genes. This family is referred to as SNZ (short for snooze); sequence analysis revealed that they are highly conserved. The functional role of each member of this family, however, remains to be determined. It is interesting that some researchers use the term "viable but nonculturable" (VBNC) to describe cells in the stationary phase. However, as colony-forming ability is our only means of assessing whether a particular cell is alive or not, other microbiologists argue the case for reversibility, i.e., the cell's ability to transform itself from being dormant to being metabolically active. Recent investigations have revealed that resuscitation of stationary phase cells may be aided by the excretion of pheromones as has been demonstrated in *Micrococcus luteus*.

In the case of *E. coli*, an attempt was made to identify the genes that are uniquely turned on in the stationary phase in response to starvation, i.e., not expressed during the exponential phase, with the aid of random transposon mutagenesis. A number of mutants unable to survive in the stationary phase were isolated and subsequently designated as survival negative mutants (Sur⁻). As such, the genes involved were designated as *sur* genes. While some genes, e.g., *surA*, are required for survival during "famine" (starvation), others, e.g., *surB*, enable the organism to exit the stationary phase as conditions change from "famine" to "feast." Recent studies have also revealed that starvation induces the production of a stationary phase-specific transcriptional activator (sigma factor) that is essential for the transcription of *sur* genes by RNA polymerase. The physiological function of the *sur* gene products is to downshift the metabolic demands made on central and intermediary metabolism for maintenance energy.

2.3.4 *Maintenance and survival*

Maintenance can be defined as the minimal rate of energy supply required to maintain the viability of a particular organism without contributing to biosynthesis. The fraction of carbon oxidized in this way is, therefore, expected to end up in the form of carbon dioxide (CO_2). The need for maintenance energy is obvious as the "living state" of any organism, including ourselves, is remote from equilibrium and as such demands energy expenditure. Moreover, in addition to the carbon processed for maintenance, another fraction of the carbon source may be wasted through excretion, e.g., acetate, α-ketoglutarate, succinate, lactate, etc. In this context, excretion of metabolites ought to be seen as an accidental consequence of central metabolism rather than by design to fulfill certain metabolic functions. The fraction of carbon required for maintenance differs from one organism to another and from one substrate to another. For example, maintenance requirements for the lactose phenotype of *E. coli* are greater than those observed for the glucose phenotype as the *lac*-permease is much more difficult to maintain than the uptake system employed for the transport of glucose, i.e., the phosphotransferase system (PTS).

The metabolic interrelationship between maintenance and growth was first described by Pirt (1965). In his theoretical treatment of this aspect, he formulated a maintenance coefficient (m) based on the earlier work of Monod and defined the coefficient as the amount of substrate consumed per unit mass of organism per unit time (e.g., g substrate per g biomass per h). If 's' represents the energy source (substrate), then the rate of substrate consumption for maintenance is governed by the following equation:

$$-(ds/dt)_M = mX \tag{2.19}$$

where m is the maintenance coefficient. The yield can therefore be related to maintenance by the following equation:

$$Y = \frac{\Delta x}{(\Delta s)_G + (\Delta s)_M} \tag{2.20}$$

where Y is the growth yield, Δx is the amount of biomass generated, $(\Delta s)_G$ is the amount of substrate consumed in biosynthesis and $(\Delta s)_M$ is the amount of substrate consumed for maintenance of cell viability.

If maintenance energy $(\Delta s)_M$ was determined and found to be zero, then Equation (2.20) reduces to

$$Y_G = \frac{\Delta x}{(\Delta s)_G} \tag{2.21}$$

Equation (2.21) clearly means that growth yield (Y) is a direct function of the amount of substrate utilized. The growth yield (Y, g dry weight per g substrate) obtained in this case may be referred to as the true growth yield to distinguish it from the Y_G where maintenance $(\Delta s)_M$ is involved. However, if a specific fraction of carbon is diverted towards maintenance energy or survival, then one might conclude that the slower the growth rate (μ), the higher the percentage of carbon that is diverted towards maintenance, the less the growth yield.

On the other hand, however, during unrestricted growth, i.e., no substrate limitation, the interrelationship between substrate concentration and maintenance can be assessed from the following equation (Pirt, 1965):

$$ds/dt = (ds/dt)_M + (ds/dt)_G \tag{2.22}$$

In this case, Equation (2.22) implies that the overall rate of substrate utilization equals the rates of substrate utilization for both maintenance and growth. With growth rate expressed in the usual way, i.e., $dx/dt = \mu x$, and assuming that Y equals Y_G, Equations (2.16) and (2.18) can be rearranged to give:

$$1/Y = m/\mu \; 1/Y_G \tag{2.23}$$

According to Equation (2.23), if m and Y_G are constants, then the plot of $1/Y$ against $1/\mu$ should give a straight line, the slope of which is maintenance (m) and the point at which the ordinate is intercepted is equal to $1/Y_G$, as illustrated in Figure 2.8.

During the course of fermentation, the minimum rate of energy supply may fall below maintenance requirements due to a shortfall in the supply of phosphorylated intermediates, ATP, and/or reducing powers (NADH, NADPH, $FADH_2$, etc.). Although such a drop in energy supply may not be fatal, it is likely to have an adverse effect on the fitness of the organism, i.e., the cells become less capable of taking advantage of favorable changes in the environment and consequently cells resume growth after a lag period. On the other hand, if the cells were able to grow immediately following inoculation or transfer into another medium, i.e., without lag, then the cells must have been left in optimal conditions. However, if the cells were to enter the lag phase prior to growth, then it is fair to suggest that the organism was left in a suboptimal state, i.e., the organism must have suffered a drop in energy supply below its maintenance requirements.

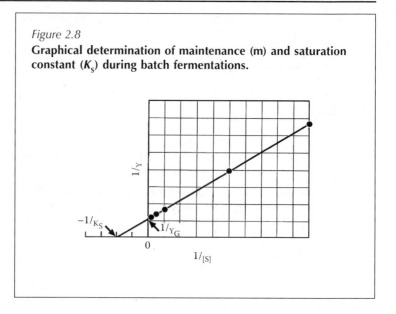

Figure 2.8

Graphical determination of maintenance (m) and saturation constant (K_S) during batch fermentations.

Ribosome particles, which consist of protein and RNA, are the target to which mRNA and amino acyl-tRNA must bind in preparation for translation (the conversion of mRNA to a polypeptide chain).

It follows, therefore, that the physiological state of the cell at the time when energy supply falls below maintenance is very important and that the deeper the shortfall in energy supply, the longer the lag period. A lag period due to the above condition is surely different to that observed when the cells are faced with the challenge of changing phenotype, as is the case when glucose phenotype of *Escherichia coli* is forced to change to lactose phenotype. Individual cells in a given population may respond differently when a given drop of energy supply is exerted. The ability of a given population to respond successfully to "shift-up" or to survive a "shift-down" in nutrients appears to be directly related to the intracellular concentration of ribosomes. For example, in comparison with cells grown in rich medium, restricted growth of *Salmonella typhimurium* was accompanied by a 35-fold drop in the concentration of ribosomes. In the yeast *Saccharomyces cerevisiae*, a drop in growth rate from $0.40 \ h^{-1}$ to $0.10 \ h^{-1}$ was accompanied by a drop in the cellular concentrations of RNA, DNA, and proteins from 12.1%, 0.6%, and 60.1% to 6.3%, 0.4%, and 45%, respectively (Nissen et al., 1997). Furthermore, it was demonstrated that the cells were capable of "scaling up" or "scaling down" the intracellular level of ribosomes in response to the shift-up or the shift-down in nutrients, respectively.

Autogenous regulation: a regulatory control mechanism that is exerted at the level of translation; the conversion of mRNA to protein, rather than transcription.

Recent investigations have revealed that such ability to modulate the cellular content of ribosomes is achieved through a mechanism of autogenous regulation.

To account for such a phenomenon, Nomura et al. (1984) proposed a new hypothesis, translational couplings, which simply implies that binding to and initiation of translation at the first ribosome binding site of the operon is essential in exposing the ribosome binding sites of all other cistrons and that inhibition or prevention of translation at the first binding site means that all other sites within the polycistronic message will remain unexposed and as such will not be translated.

The argument over the question of whether smaller cells are less able to cope with environmental changes, i.e., nutrient limitations and the drop in energy supply below maintenance, because they contain less ribosomes, can now be answered to the satisfaction of everyone as recent research revealed that it is not the number of ribosomes that matter but rather their concentration inside the cell. Furthermore, the need for the cell size to be relatively large during growth under no limitations is not a reflection of high concentration of ribosomes but rather the need to attain a certain mass before replication of the chromosome can be initiated (Donachie, 1968). The essence of Donachie's discovery is that, if initiation of replication is triggered in response to a certain volume of cell mass, then the faster the cell divides, the faster the growth rate, the larger the size of the cell.

In addition to induction and auto-regulatory control mechanisms, successful transition from one phase of growth to another might also involve proteolysis, a mechanism that is particularly significant when the cells have to remove certain proteins at a rate that is significantly higher than their specific growth rate (μ), e.g., key regulatory proteins which fluctuates from one phase of growth to another (Grunenfelder et al., 2001) and those that are specific to a particular phase of growth and might be detrimental to the organism as it switches to another phase (Weichart et al., 2003).

2.4   Diauxic growth

The ability of micro-organisms to display biphasic (diauxic) growth patterns (Figure 2.9) is well documented. For example, E. coli and Aerobacter aerogenes display such a pattern when faced with two different substrates, glucose and lactose in the case of the former (Figure 2.9) and glucose and citrate in the latter. This is generally due to the preferential utilization of glucose to either of the competing substrates. This can be explained on the grounds that glucose has been found to exert catabolite inhibition, i.e., inhibition of other sugars or organic acids uptake systems, and catabolite repression of the enzymes required for the transport as well as hydrolysis of other sugars, as is the case in lactose metabolism.

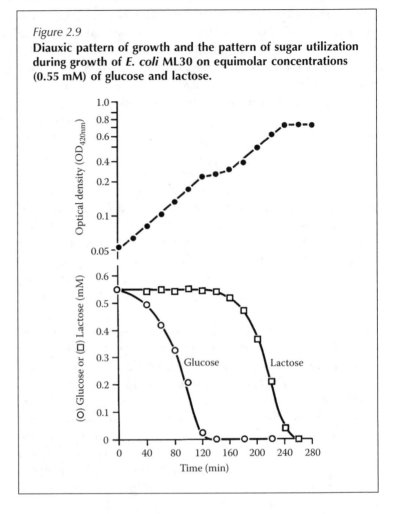

Figure 2.9

Diauxic pattern of growth and the pattern of sugar utilization during growth of *E. coli* ML30 on equimolar concentrations (0.55 mM) of glucose and lactose.

2.5   Growth yield in relation to carbon and energy contents of growth substrates

Unlike growth on sugars and polyhydrated alcohols where a constant yield of 1.1g dry weight biomass per g substrate carbon was observed, growth yield on carboxylic acids gives a much inferior yield. An attempt to resolve this paradox was made (Linton and Stephenson, 1978) by relating the maximum specific growth yield observed on any given carbon source to its carbon and energy content; the latter is defined by the heat combustion (kcal/g substrate carbon). Although the authors recognized the need for using chemostats to achieve steady-state growth and to account for maintenance energy requirements, their analysis of various data in the literature revealed that the growth yield was directly proportional to the heats of combustion up to 11.00 kcal/g substrate carbon and that beyond this level no increase in yield was observed, suggesting the involvement of another mechanism. It is possible, therefore, to argue that during growth on substrates with a heat of combustion value of less than 11.00 kcal/g substrate, the energy generated is insufficient to convert all the available carbon to biomass and as such growth may be described as energy limited. Further analysis revealed that maximum growth yield is inherently set by the ratio between the biologically available energy and the carbon contents of the substrate. Assuming a bacterial carbon content of 48%, the maximum yield of 1.43 g bacterial dry weight per g substrate carbon, representing a maximum carbon (substrate) to carbon (biomass) conversion of 68%, was observed. This treatment enables the prediction of growth yield with a good degree of accuracy and provides a rational basis for growth limitations, e.g., while glycerol and mannitol-limited growth of *Aerobacter aerogenes* may be described as carbon limited, gluconate-limited growth may be viewed as energy limited.

The relative contribution of each substrate to total biomass formation can also be assessed with a good deal of precision providing that growth limitation is a consequence of diminishing carbon supply in a chemostat rather than the accumulation of toxic end products or adverse changes in pH. The relationship between total biomass formation and the concentration of any given substance can therefore be determined from the plot of log *n* (biomass) as a function of substrate concentration and providing that all other components are in excess, the organism will enter the stationary phase upon depletion of the substrate or the substance in question. This experiment should obviously be done over a wide range of substrate concentrations and from the slope obtained, the yield constant or yield coefficient per unit of the substance in question can be determined. If the unit is, say, one mole of a substrate, then the yield coefficient obtained is referred to as the molar growth yield coefficient. This method is used widely as a biological assay for the determination of vitamins, amino acids, purines, and pyridines.

2.6   Fermentation balances

2.6.1 *Carbon balance*

Apart from the fraction of carbon used for biomass formation, the remainder is partitioned between products and by-products, including carbon dioxide. The ratio of recovered carbon to that present at the onset of fermentation is referred to as the carbon balance or carbon recovery index. Such a balance or an index is a measure of efficiency; i.e., the higher the index, the higher the carbon recovery, the more efficient the fermentation process. The carbon balance is generally calculated by working out the number of moles (or millimoles) produced of a given product per 100 moles (or millimoles) of substrate utilized. The values obtained can then be multiplied by the number of carbon atoms in each respective molecule. The resulting values for the products in question can then be totaled and compared with that of the substrate. If the values are equal, i.e., 1: 1 ratio, then a complete recovery of carbon into product formation has been achieved. Although this is theoretically possible, our experience indicates otherwise; as part of this carbon is used for maintenance and assimilation to support growth, however slow. A complete carbon balance can, therefore, be calculated for any given fermentation if the fraction of carbon diverted towards biosynthesis and maintenance is determined. In addition to the carbon balance, some fermentations demand calculation of the redox balance.

2.6.2 *Redox balance*

Fermentations of sugars and other primary carbon sources give rise to a whole host of different intermediates, some of which are phosphorylated (energy-rich) while others are not. While biosynthetic intermediates are utilized for the biosynthesis of monomers, other intermediates are produced in excess and this is balanced by their excretion into the medium. The stoichiometry of product and by-product formations of any given fermentation process can be ascertained by carefully analyzing the culture filtrates at different stages (see Chapter 6). From a physiological standpoint and in order for fermentation to go to completion, the redox balance must be maintained.

2.7 Efficiency of central metabolism

2.7.1 *Futile cycling and efficiency of central metabolism*

The efficiency of carbon conversion to biomass and desirable end products is influenced by many different factors (see Chapters 3, 6, and 7 for more details) of which futile cycling is a major contributor. For example, any transient increase in the intracellular concentration of pyruvate in *Escherichia coli* may trigger a futile cycle in the central metabolic pathways at the junction of phosphoenolpyruvate (PEP). Futile cycling at the junction of PEP (Figure 2.10) involves, in addition to pyruvate dehydrogenase and malic enzyme, PEP carboxykinase and pyruvate kinase (ATP-generating reactions) on one hand and PEP-synthetase and PEP-carboxylase (ATP utilizing reactions) on the other. The operation of this futile cycle may, in addition to wasting ATP, adversely affect the adenylate energy charge within the cell. Transient increase in the intracellular level of pyruvate may also trigger futile cycling in other organisms. For example, *Klebsiella aerogenes* triggers a futile cycle involving two enzymes, the first is the NAD^+-dependent pyruvate reductase, which catalyzes the transformation of pyruvate to D-lactate and the

Figure 2.10
Futile cycle at the level of phosphoenolpyruvate (PEP) in central metabolism and its role in energy dissipation.

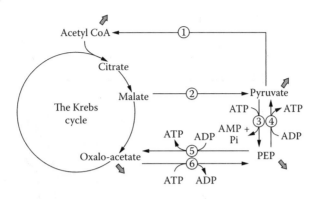

Figure 2.11
The role of pyruvate reductase and D-lactate dehydrogenase in bypassing the first phosphorylation site of the electron transport chain (ETC) in the oxidation of NADH, H⁺.

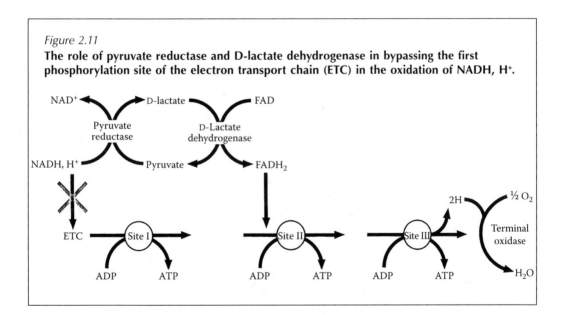

second is the FAD-dependent D-lactate dehydrogenase, which completes the cycle (Figure 2.11). Flux of carbon through these two enzymes provides a futile cycle in which NADH is oxidized at the expense of reducing FAD^+, thus bypassing the first phosphorylation site in the electron transport chain (Figure 2.11). It follows, therefore, that any changes leading to transient increase in the intracellular level of pyruvate may lead to the operation of this futile cycle and, in turn, energy dissipation. The presence

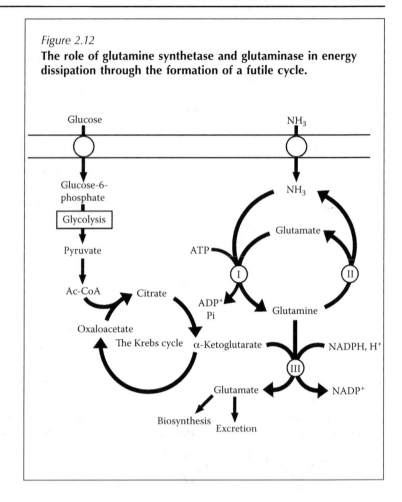

Figure 2.12
The role of glutamine synthetase and glutaminase in energy dissipation through the formation of a futile cycle.

and function of glutamine synthetase and glutaminase in ammonia-limited cultures represents yet another possible futile cycle (Figure 2.12) in which the synthetase generates glutamate while the other affects its hydrolysis with a net loss of one molecule of ATP per turn (Tempest, 1978).

2.7.2 *Metabolite excretion and efficiency of central metabolism*

Flux to metabolite excretions in general and acetate in particular diminishes the efficiency of carbon conversion to desirable end products significantly (El-Mansi, 2004). Acetate, particularly its undissociated form, is a potent uncoupler of oxidative phosphorylation. As such, conditions that promote flux to acetate excretion will, in turn, diminish flux to product formation, as a large fraction of the carbon source has to be diverted towards maintenance requirements (El-Mansi, unpublished observations). While such a role is fully appreciated, it should be remembered that flux to acetate excretion is physiologically significant as it allows faster growth rate and facilitates high cell-density growth (El-Mansi, 2004). Flux to acetate excretion is also important as flux through phosphotransacetylase replenishes central and intermediary metabolism with free-CoA, thus fulfilling an anaplerotic function (El-Mansi, 2005) that is central for the smooth operation of carbon flux through enzymes of central and intermediary metabolism. Interestingly, flux through pyruvate oxidase to acetate excretion, which may appear to be wasteful, has recently been shown to be equated with higher efficacy of growth and energy generation (Guest et al., 2004).

Chemostat: a continuous culture in which growth rate is limited by the rate of nutrients supply.

2.8 Continuous cultivation of micro-organisms

Continuous cultures outperform batch cultures economically by eliminating the inherent downtime that is lost for cleaning, sterilization, and the re-establishing of biomass within the bioreactor. Furthermore, unlike batch cultures where the rate of product formation is at its highest for only a limited period of time, continuous cultures sustain such a period over a much longer period. Furthermore, the interpretation of data obtained from continuous cultures is much simpler than batch culture because, unlike batch culture where variations in the concentrations of primary carbon and nitrogen sources, product formation, pH, and the redox balance within the cells vary during the course of growth, continuous cultures afford a steady-state growth in which input and output are in perfect balance.

Turbidostat: a continuous culture in which the organism grows at its maximum specific growth rate (μ_{max}), thus reaching a steady state of fixed turbidity.

With the exception of waste water treatment, continuous industrial microbial processes are much less common than batch processes because contaminations represent a major problem. The earliest continuous process was employed for the production of vinegar by Acetobacter sp. In the process, a sugar solution was trickled down a bioreactor containing Acetobacter cells attached to wood shavings; the earliest form of cell immobilization. The acetic acid thus generated discourages the growth of contaminants.

2.8.1 Types of continuous cultures

There are three main types of continuous culture, namely chemostat, turbidostat, and auxostat; the latter depending on the set-parameter may also be known as nutristat or pHstat.

In an auxostat, the feeding rate is adjusted to match the rate of cellular metabolism. In that sense, a turbidostat is also an auxostat because the turbidity created as a consequence of bacterial growth is maintained at a set point by adjusting the feeding rate of fresh medium. The most popular auxostat, however, is the pH auxostat (pH-stat); a continuous culture in which pH

Auxostat, Nutristat, pHstat (synonymous names) for a continuous culture in which the dilution rate changes to maintain a certain factor constant.

is maintained at a set point by the addition of NaOH to counter-balance the excreted acetate and other acidic by-products.

2.9 Current advances and innovations in the fermentation and pharmaceutical industry

Although mammalian cell and other eukaryotic cell cultures like the yeast *Pichia* represent an attractive alternative to bacteria for the expression and production of recombinant proteins, especially those requiring post-translational modification such as glycosylation, recombinant protein expression in bacteria remains the most cost-effective and as such, from an industrial perspective, it is the most desirable option particularly since the development of tightly regulated promoters such as the T7 polymerase-specific promoter. In this case, effective control through repression ensures that expression and, in turn, production is turned off until significant biomass is formed, which is significant if the product was toxic in nature. Ingenious and effective approaches have recently been developed in bacterial expression systems including modification of ribosome binding sites to increase yield, expression at low temperature on its own and in combination with the expression of molecular chaperones, which reduce misfolding and precipitation of proteins, to aid correct folding of recombinant proteins.

2.9.1 The "quiescent cell factory": A novel approach

A nongrowing bacterial culture in which the primary nutrients are directly and quantitatively converted to products represents the ideal scenario for biotechnology. However, this is very difficult to achieve and an alternative approach is to render transcription and translation processes genes-specific. Although this approach may seem insurmountable, specific expression of certain proteins in eukaryotic organisms is a well-established phenomenon. Attempts to develop a bacterial system that mimics eukaryotic organisms in that direction led to the discovery of the *E. coli* "Quiescent Cell Protein Expression System," otherwise referred to as the "Q-Cells" (Summer, 2002).

Overexpression of Rcd in the mutant strain bns205, which apart from being defective in *bns* is capable of producing an N-terminal fragment of H-NS, led to a complete cessation of growth within 2-3 hours and the cells entered a quiescent (nongrowing) state (Summer, 2002). Fortuitously, the Q-cells appear to preferentially express the plasmid encoded rather than the chromosomally encoded genes. The reason for this odd but useful feature was unravelled by the same author following DAPI–DNA staining and examination of Q-cells using fluorescence microscopy, which revealed that the bacterial nucleoid was

Nucleoid: a structure within the bacterial cytoplasm representing the chromosome together with associated proteins.

highly condensed; a consequence of Rcd/H-NS complex formation, thus preventing its expression in a manner reminiscent of that observed for heterochromatin formation in eukaryotes, which results in global repression of transcription. The Q-cell system has now been shown to function well at the pilot-scale and was not unduly sensitive to media composition or culture density. The Q-cell system is currently employed for the production of recombinant proteins. Further developments are currently underway for its adoption in other systems (Summer, 2002).

2.10   Microbial fermentations and the production of biopharmaceuticals

In the previous sections, we have outlined the principles underlying microbial fermentation. However, these have been dealt with in a general way without marked reference either to specific micro-organisms or industrial processes. The following case study illustrates how understanding of microbial biochemistry in general and metabolic pathways in particular can be applied to microbial fermentation and the production of useful industrial products. Emphasis is placed on the micro-organisms involved and the development of processes using both genetic engineering and strain development for optimization of hyaluronic production. This study is not comprehensive but opens up a discussion on the development of industrial microbiological processes.

2.10.1   *Hyaluronic acid synthesis: a case study*

2.10.1.1   *Composition and applications of hyaluronic acid*

Hyaluronic acid (HA) is a polysaccharide that has applications in treatment of skin disorders, arthritis, and surgical treatments (Andre, 2004; Talke, 2002) with a market value in the region of 700 million US dollars (Fong Chong et al., 2005). While HA can be derived from animal sources, *Streptococcus* spp offers a much brighter scope for its production (see Fong Chong et al., 2005; Yamada and Kawasaki, 2005, for reviews). HA is the most abundant aminoglycan in the extracellular matrix of connective tissue, and the fact that bacterial HA is structurally similar, though smaller in size, has been exploited by the fermentation and pharmaceutical industries.

HA is a heteropolysaccharide composed of D-glucuronic acid (Gluc) and N-acetyl glucosamine (NAG), which are linked together as repeating units of β1-4 Gluc β1-3 NAG. The rheological properties of HA depend on a combination of molecular size and concentration. The greater the interaction between polysaccharide molecules leads to increase in viscosity and ulti-mately formation of gels. For biomedical applications, HA with an overall molecular weight (M_r) of the order of 3–6 MDa is required, whereas typical reported sizes for bacterial HA are in the order of 1–2 MDa (Fong Chong and Nielsen, 2003; Hong et al., 2004). Improvements in M_r have been achieved by cross-linking, optimization of fermentation processes and ran-dom mutation to produce high M_r yielding strains (Kim, 1996; Stangohl, 2003; Fong Chong et al., 2005).

2.10.1.1.1   Choice of microbial host

For fermentation, the bacterial species used have been *Strep-tococcus equi*: either *S. equi* subsp *equi* and *S. equi* subsp *zooepidemicus*; these are group C Streptococci that rarely cause human disease (based on the antigenic properties of cell mem-branes, species of the genus Streptococci are divided into Lance-field groups A-G). HA is also produced by Group A pathogenic Streptococci such as *S. pyogenes*. Bacterial capsules, which con-tain HA, have several physiological roles: prevention of desic-cation, biofilm formation, and for pathogens both as virulence factors in infection and protection against the host immune response. Capsular HA has been shown to be required for colo-nization by *S. pneumoniae* (Magee and Yother, 2001). group C Streptococci have been used commercially because the capsular HA of streptococci has an identical molecular structure to that found in mammalian tissue.

2.10.1.1.2   Factors affecting HA production

The microbial physiology of Streptococci is important for under-standing the factors influencing (and limiting) HA production. A number of interrelated factors affect HA yield. Streptococci lack cytochromes and, in turn, the capacity for aerobic respi-ration; energy metabolism is via fermentation, which may be homolactic or heterolactic. *S. equi* subsp *zooepidemicus* pro-duces lactate as the major fermentation product under anaer-obic conditions with minor amounts of formate, acetate, and lactate (Fong Chong et al., 2005). Under aerobic conditions HA production is increased, which could be a protective mechanism against oxygen toxicity (Table 2.1). An alternative and more likely mechanism relates to energy considerations. HA produc-tion is relatively energy expensive in that 4 ATP are required per disaccharide repeating unit. A typical HA molecule is c 2 MDa and so consists of c. 5,000 repeating units requiring 20,000 ATP. However, for every molecule of glucose converted to glucuronic acid, the organism generates 2 NADH. Under anaerobic condi-tions these reducing equivalents have to be reoxidized by the terminal steps in fermentation, conversion of pyruvate to lactate and ethanol. However, under aerobic conditions, the regenera-tion of NAD^+ can be achieved through the activity of NADH

Table 2.1 Influence of aeration on HA production in batch cultures of *S. equi* subsp *zooepidemicus*

Parameter	Anaerobic	Aerobic
Specific growth rate (h^{-1})	0.85	1.02
Specific glucose uptake rate (g g^{-1} h^{-1})	4.45	4.96
Acetate yield (mg g^{-1})	49	98
Ethanol yield (mg g^{-1})	68	7
Formate yield (mg g^{-1})	122	0
Lactate yield (mg g^{-1})	651	662
HA yield (mg g^{-1})	65	88
Biomass yield yield (mg g^{-1})	192	206
Molecular weight (MDa)	1.07	2.4

Data taken from Fong Chong and Nielsen, 2003 & 2005; with permission.

oxidases. As a result, under aerobic conditions pyruvate can be diverted to other products such as acetate and CO_2 (Table 2.1).

Oxygen has also been shown to influence capsule (HA) production in pathogenic *Streptococcus pneumoniae,* where two different colony types (O and T) are found. O-type strains give opaque colonies and show a 5–10-fold decrease in capsule production when grown aerobically rather than anaerobically, the opposite effect to that described above (Weiser et al., 2001). T-type strains gave transparent colonies and showed no such variation and HA yields were low. This regulation was explained in terms of the action of CspD, a response regulator. Because of the role of HA in pathogencity of group A Streptococci, there has been increasing interest in the regulation of virulence factors, including HA (Bisno et al., 2003; Hynes et al., 2004). While this aspect may be beyond the scope of this chapter, it is pertinent to the overall objectives of the book because an understanding of the factors regulating HA production can help considerably in formulating a rationale to increasing its productivity through metabolic engineering. The two component system CovRS (CsrRS) regulates about 15% of the genes in the streptococcal genome including those for HA biosynthesis (Federle and Scott, 2002; Dalton and Scott, 2004). The sensor protein, CovS, responds to conditions of low pH, high osmolarity, and high temperature; activation of CovR leads to repression of HA biosynthesis.

Kinetic factors will also influence HA production. The precursor of D-glucuronic acid is glucose-6-P, which is utilized also for alternative pathways for antigenic cell wall polysaccharides and teichoic acid biosynthesis; diversion of glucose-6-P from HA biosynthesis to these pathways would decrease HA yield. A second kinetic factor is if there is a rate-limiting (controlling) step in HA production. In recombinant hosts, the

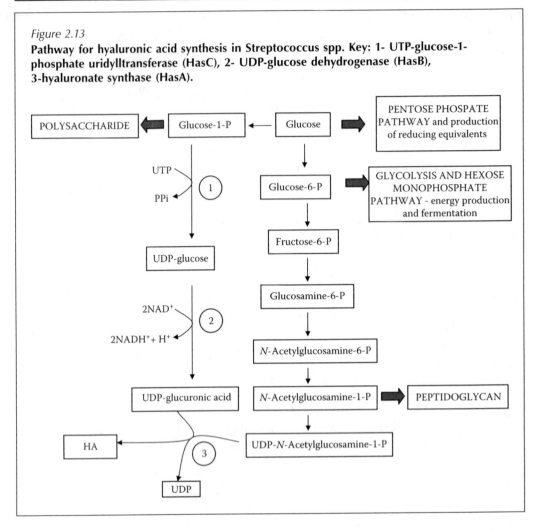

Figure 2.13

Pathway for hyaluronic acid synthesis in Streptococcus spp. Key: 1- UTP-glucose-1-phosphate uridylltransferase (HasC), 2- UDP-glucose dehydrogenase (HasB), 3-hyaluronate synthase (HasA).

step of HA production that involves conversion of UDP-glucose to UDP-glucuronic acid has been shown to be rate limiting (controlling).

2.10.1.1.3   *Genetics of HA production*

The formation of the repeating disaccharide unit of – [D-glucuronic acid-N-acetylglucosamine} - is catalyzed by hyaluronan synthase (HAS), encoded by *hasA* (Figure 2.13, DeAngelis, PL, 2002; Weigel, 2002). This enzyme causes elongation by addition of the repeating unit to the reducing end of the nascent chain (Tlapak-Simmons et al., 2005). It is a well-conserved membrane-bound protein in both prokaryotes and eukaryotes. Cardiolipin and phosphatidylcholine were required for activity for an expressed His-tagged protein in *E. coli*, confirming that HasA requires a specific lipid moieties for activity. The structural gene encoding HasA is a member of the *hasABC* operon

Homologues: compounds with similar structures and evolutionary origin, but not necessarily in function.

where *hasB* and *hasC* encode UDP-glucose dehydrogenase and UTP-glucose-1-phosphate uridylltransferase, respectively. Together HasC and HasB catalyze the conversion of glucose-6-phosphate to UDP-glucuronic acid (Figure 2.13).

In *S. equi* subsp *zooepidemicus* the presence of just hasA and hasB is sufficient for HA synthesis, suggesting a homologue for hasC exists. This simplicity enables HA to be synthesized in alternative hosts. Successful production of HA in both *E. coli* and *B. subtilis* has been demonstrated (Widner et al., 2005), with the latter proving advantageous in being a Gram-positive nonpathogenic species with a well-characterized genetic system. Novozyme have demonstrated HA production in *B. subtilis* using hasA from *S. equisimilis* (Lancefield C, G) and a *B. subtilis* homologue of hasB. HA was secreted and was of high quality, comparable to commercially available sources of HA but of somewhat lower M_r (1 MDa). Despite this simplicity the method of translocation and secretion of HA remain unclear. However, the successful expression in heterologous hosts indicates that these bacteria can provide suitable transporters.

Commercial requirements of HA Early work on optimization of the fermentation parameters such as temperature and stirring rate, showed that the fermentation process was optimal at pH 6.7 and 37 °C with relative low stirring speeds of 90–120 rpm and as explained above that aeration at about 1.5 vvm gave better yields than those obtained by anaerobic growth (Johns et al., 1994; Kim et al., 1996). However, in practice, the yield of HA is limited by the fact that HA solutions are highly viscous and so the energy required for stirring becomes high. Yields of 5–10 g/l HA were observed.

However, both in terms of downstream costs and in terms of product quality, it is important that the properties of HA are controlled. A high M_r product free from contaminants that meets safety standards for medical application is required (Yamada and Kawasaki, 2005). S. equi, a pathogen for horses, produces a number of virulence factors similar to those found in group A Streptococci. These include M-like proteins, IgG binding proteins, and fibronectin-binding proteins (Brisno et al., 2003). Wild-type capsular Streptococci also have HA degrading enzymes such as hyaluronate lyase as a surface antigen (Akhtar and Bhakuni, 2004). This would cause problems in controlling yield and size of HA. Mutants lacking hyaluronate lyase have been isolated to circumvent this particular problem. Hyaluronate lyases play several roles including:

- control of HA size,
- control of HA degradation and
- degradation of mammalian HA during invasion.

The multifactorial nature of virulence factors and protein contamination triggered an interest in using heterologous GRAS (generally regarded as safe) organisms for the production of HA (Yamada and Kawasaki, 2005).

HA is a biodegradable polymer and as such has to be stabilized for skin applications, which may be achieved through cross-linking, a commercially sensitive issue. As well as stabilizing HA, cross-linking also enhances viscosity. A comparative study between Restylane®, a bacterial product, with Hylaform®, an animal product, concluded that Hylaform had better gel properties and less protein than Restylane (Manna et al., 2005). However, there is undoubtedly scope for product improvement in this area and possibilities for improvements both via genetic engineering and downstream modifications.

Guided by the information given above as well as by recent literature, address the following questions:

- What are the desired properties of HA for biomedical application?
- What problems are associated with HA production and how might they be overcome?

Summary

- The diverse array of metabolic networks, together with a very high rate of metabolic turnover, makes micro-organisms ideal candidates for the production of fine chemicals using fermentation biotechnology.
- The conversion of a given carbon source to biomass involves a whole host of different reactions in central and intermediary metabolism. These reactions can be classified according to function as fueling reactions, biosynthetic (anabolic) reactions, anaplerotic reactions, polymerization, and assembly.
- Micro-organisms change in size as the growth rate changes, which is consistent with the view that initiation of replication is dependent on cell mass.
- Micro-organisms change in composition as a function of growth rate, e.g., while the relative concentrations of protein, DNA, and RNA in *Saccharomyces cerevisiae* increase (in ascending order) with growth rate, the intracellular concentration of glycogen, trehalose, and carbohydrates decrease.
- The drop in energy supply below maintenance requirement is not necessarily fatal but may lead to a lag, the severity of which is directly proportional to its magnitude and the relative concentration of ribosomes within the cell.
- The question of whether a particular organism has lagged following inoculation, together with the number of viable cells at the onset of fermentation, can be ascertained graphically.

- The interrelationships among various growth parameters during unrestricted and restricted growth have been described.
- Micro-organisms adopt different tactics to survive starvation. While Gram-positive organisms resort to sporulation, *Saccharomyces cerevisiae* and *E. coli* express specific proteins, the primary function of which is to scale down maintenance requirement on one hand and exiting the stationary phase on the other.

- How might the mechanism of polysaccharide secretion be investigated? (Hint: compare with previously characterized systems)
- In the patent literature, what areas of biotechnology are covered with respect to HA production?

References

Akhtar, S Bhakuni, V. (2004). *Streptococcus pneumoniae* hyaluronate lyase: An overview. Current Science 8, 285–295.

Andre, P (2004). Hyaluronic acid and its use as a "rejuvenation" agent in cosmetic dermatology. Seminars in cutaneous medicine and surgery 23, 218–222.

Bazin, M.J., A.P. Wood, and Paget-Brown, D. (1997). Analysis of microbial growth data, in Rhodes, P.M. and Stanbury, P.F. (eds.) Applied Microbial Physiology, Oxford, New York, Tokyo: IRL Press, pp. 103–211.

Bisno, A.L., M.O. Brito, and C. M. Collins (2003). Molecular basis of group A streptococcal virulence. Lancet Infectious Diseases 3, 191–200.

Cozzone, A. J. Regulation of acetate metabolism by protein phosphorylation in enteric bacteria. Ann. Rev. Microbiol. 1998, 52, 127–164.

Dalton, T. L., J. R. Scott (2004). CovS inactivates CovR and is required for growth under conditions of general stress in *Streptococcus pyogenes*. Journal of Bacteriology 186, 3928–3937.

DeAngelis, P. L. (2002). Microbial glycosaminoglycan glycosyltransferases. Glycobiology 12, 9R–16R.

Donachie, W.D. (1968). Relationship between cell size and time of initiation of DNA replication, Nature 219, 1077–1083.

El-Mansi, E. M. T., W. H. Holms (1989). Control of carbon fux to acetate excretion during growth of *Escherichia coli* in batch and continuous cultures, J. Gen. Microbiol., 135, 2875–2883.

El-Mansi, E.M.T. (2004). Flux to acetate and lactate excretions in industrial fermentations: Physiological and biochemical implications. J. Ind Microbiol. Biotechnol **31**, 295–300.

El-Mansi, E.M.T. (2005). Free-CoA mediated regulation of intermediary and central metabolism: A hypothesis which accounts for the excretion of α-ketoglutarate during aerobic growth on acetate. Research in Microbiology **156**, 874–879.

Federle, M. J., J. R. Scott (2002). Identification of binding sites for the group A streptococcal global regulator CovR. Molecular Microbiology **43**, 1161–1172.

Fong Chong, B., L. M. Blank, R. Mclaughlin, L. K. Nielsen, (2005). Microbial hyaluronic acid production. Appl. Microbiol. Biotechnol. **66**, 341–351.

Fong Chong, B., L. K. Nielsen (2003). Aerobic cultivation of *Streptococcus zooepidemicus* and the role of NADH oxidase. Biochemical Engineering Journal **16**, 153–162.

Grunenfelder, B., G. Rummel, J. Vohradsky, D. Roder, H. Langen, U. Jenal (2001). Proteomic analysis of the bacterial cell cycle. Proc Natl Acad Sci USA **98**, 4681–4686.

Guest, J. R., G. C. Russell (1992). Complexes and Complexities of the citric acid cycle in *Escherichia coli*. Current Topics in Cellular Regulations, **33**, 231–247.

Guest J. R., A. M. Abdel-Hamid, G. A. Auger, L. Cunnigham, R. A. Henderson, R. S. Machado, M. M. Attwood (2004). Physiological effects of replacing the PDH complex of *E. coli* by genetically engineered variants or by pyruvate oxidase. In Thiamine: Catalytic Mechanisms and Role in Normal and Disease States, edited by Frank Gordon and Mulchand S. Patel, pp. 389–407. Marcel Dekker Inc, New York. ISBN 0-8247-4062-9.

Hong, S. S., J. Chen, J. G. Mang, Y. C. Tao, L. Y. Liu (2004). Purification and structure identification of hyaluronic acid. Chinese Chemical Letters **15**, 811–812.

Hynes, W. (2004). Virulence factors of the group A streptococci and genes that regulate their expression. Frontiers in Bioscience **9**, 3399–3433.

Ingraham, J. L., O. Maaloe, F. C. Neidhardt (1983). Growth of the Bacterial Cell. Sinauer associates Inc, Sunderland, MA USA. ISBN 0-87893-352-2.

Johns, M. R., L. T. Goh, A. Oeggerli (1994). Effect of pH, agitation and aeration on hyaluronic-acid production by *Streptococcus zooepidemicus*. Biotechnology Letters **16**, 507–512.

Kim, J. H., S. J. Yoo, D. K. Oh, Y. G. Kweon, D. W. Park, C. H. Lee, G. H. Gil (1996). Selection of a *Streptococcus equi* mutant and optimization of culture conditions for the production of high molecular weight hyaluronic acid. Enzyme and Microbial Technology **19**, 440–445.

Linton, J. D., R. J. Stephenson (1978). A preliminary study of growth yields in relation to the carbon and energy content of various organic growth substrates. FEMS Microbiol Lett. **3**, 95–98.

Magee, A. D., J. Yother (2001). Requirement for capsule in colonization by *Streptococcus pneumoniae*. Infection and Immunity **69**, 3755–3761.

Manna, F., M. Dentini, P. Desideri, O. De Pita, E. Mortilla, B. Maras (1999). Comparative chemical evaluation of two commercially available derivatives of hyaluronic acid (Hylaform((R)) from rooster combs and Restylane((R)) from streptococcus) used for soft tissue augmentation. Journal of the European Academy of Dermatology and Venereology **13**, 183–192.

Neidhardt, F.C., J. L. Ingraham, M. Schaechter (1990). Physiology of the Bacterial Cell: A Molecular Approach, Sunderland, MA: Sinauer Associates.

Nissen, L. N., U. Schulze, J. Nielsen, J. Villadsen (1997). Flux distribution in anaerobic, glucose-limited continuous cultures of *Saccharomyces cerevisiae*, Microbiology, **143**, 203–218.

Nomura, M., R. Gourse, G. Baughman (1984). Regulation of synthesis of ribosomes and ribosomal components, Ann. Rev. Biochem., **53**, 75–89.

Pirt, S. J. (1965). The maintenance energy of bacteria in growing culture, Proc. R. Soc. London, Ser. B, **163**, 224–231.

Saier, M.H. Jr. (1996). Cyclic AMP-independent catabolite repression in bacteria, FEMS Microbiol. Lett., **138**, 97–103.

Stangohl, S. (2003). Methods and means for production of hyaluronic acid. US Patent 6090596.

Summer, D. (2002) A quiet revolution in the bacterial cell factory. Microbiology Today **29**, 76–78.

Talke, M. (2002). Intraarticular hyaluronic acid in osteoarthritis of the carpometacarpal joint. Aktuelle rheumatologie **27**, 101–106.

Tlapak-Simmons, V. L., C. A. Baron, R. Gotschall, D. Haque, W. M. Canfield, P. H. Weigel (2005). Hyaluronan biosynthesis by class I streptococcal hyaluronan synthases occurs at the reducing end. Journal of Biological Chemistry **280**, 13012–13018.

Tempest, D.W. (1978) The biochemical significance of microbial growth yield, Trends Biochem. Sci., **3**, 180–184.

Weichart, D., N. Querfurth, M. Dreger, R. Hengge-Aronis (2003). Global role for ClpP-containing proteases in stationary phase adaptation of *Escherichia coli*. J. Bacteriol **185**, 115–125.

Suggested reading

Weigel, P. H. (2002). Functional characteristics and catalytic mechanisms of the bacterial hyaluronan synthases. IUBMB Life 54, 201–211.

Weiser, J. N., D. Bae, H. Epino, S. B. Gordon, M. Kapoor, L. A. Zenewicz, M. Shchepetov (2001). Changes in availability of oxygen accentuate differences in capsular polysaccharide expression by phenotypic variants and clinical isolates of Streptococcus pneumoniae. Infection and Immunity 69, 5430–5439.

Widner, B., R. Behr, S. Von Dollen, M. Tang, T. Heu, A. Sloma, D. Sternberg, C. DeAngelis, P. H. Weigel, S. Brown (2005). Hyaluronic acid production in Bacillus subtilis. Applied and Environmental Microbiology 71, 3747–3752.

Wolfe A. J. (2005). The acetate switch. Microbiology and Molecular Biology Reviews 69, 12–50.

Yamada, T., T. Kawasaki (2005). Microbial synthesis of hyaluronan and chitin: New approaches. Journal of Bioscience and Bioengineering 99, 521–528.

3

Fermentation Kinetics

Jens Nielsen

3.1   Introduction

Growth of microbial cells is the result of many chemical reactions leading to the synthesis of macromolecules such as DNA, RNA, and proteins (see Chapter 2 for more details). Prior to cell division, the cells increase in size (or extend their hyphae in the case of filamentous micro-organisms) as the macromolecules are assembled *en route* to biomass formation. The growth of microbial cells can be quantified by many different methods including total dry weight (see Box 3.1) and cellular contents of RNA, DNA, or protein. Other methods are based on the measurement of turbidity, which is generally linearly related to the dry weight determination, and today there are a number of commercially available sensors that allow *in situ* measurements of the turbidity (Olsson and Nielsen, 1997).

Growth of microbial cells is often illustrated with a batch-wise growth of a unicellular organism (either a bacterium or yeast). Here the growth occurs in a constant volume of medium

Box 3.1   **Standard operating procedure (SOP) for dry weight determination**

Biomass is most frequently determined by dry weight measurements. This can be done either using an oven or a microwave oven, with the latter being the fastest procedure. An important prerequisite for the measurement is that the sample is dried completely, and it is therefore important to apply a consistent procedure. A suggested protocol is as follows:

1. Dry the filter (pore size 0.45 μm for yeast or fungi, 0.20 μm for bacteria) on a glass dish in the microwave oven on 150 W for 10 min. Place a tissue paper between the glass and

the filter so that the filter does not stick to the glass.
2. Place the filter in a desiccator and allow to cool for 10–15 min. Weigh the filter.
3. Filter the cell suspension through the filter and wash the cells with demineralized water.
4. Place the filter on the glass dish again and dry in the microwave oven for 15 min at 150 W.
5. Put the filter in a desiccator and allow to cool for 10–15 min. Weigh the filter.
6. If more than 30 mg dry weight is present on the filter, the time in the microwave oven may have to be longer.

with one growth-limiting substrate component that is used by the cells. Cell growth is generally quantified by the so-called **specific growth rate** μ (h^{-1}), which for such a culture is given by

$$\mu = \frac{1}{x}\frac{dx}{dt} \tag{3.1}$$

where x is the biomass concentration (or cell number). The specific growth rate is related to the **doubling time** $t_d(h)$ of the biomass through:

$$t_d = \frac{\ln 2}{\mu} \tag{3.2}$$

The doubling time t_d is equal to the generation time for a cell, i.e., the length of a cell cycle for unicellular organisms, which is frequently used by life scientists to quantify the rate of cell growth.

The design and optimization of a given fermentation process require a quantitative description of the process, which, considering the nature of microbial growth, is generally a complex task. Furthermore, often the product is not the cells themselves but a compound synthesized by the cells, and depending on the type of product the kinetics of its formation may vary from one phase of growth to another. Thus, while primary metabolites (see Chapter 4 for more details) are typically formed in conjunction with cellular growth, an inverse relationship between product formation and cell growth is often found in the case of secondary metabolites (see Chapter 5 for more details), and here flux to product formation may be greatest in the stationary phase.

With these differences in mind, it is clear that quantification of product formation kinetics may be a difficult task. However, with the rapid progress in biological sciences, our understanding of cellular function has increased dramatically, and this may form the basis for far more advanced modeling of cellular growth kinetics than seen earlier. Thus, in the literature one may find mathematical models describing events like gene expression, kinetics of individual reactions in central pathways, together with macroscopic models that describe cellular growth and product formation with relatively simple mathematical expressions. These models cannot be compared directly because they serve completely different purposes, and it is therefore important to consider the aim of the modeling exercise in a discussion of mathematical models.

In this chapter the applications of kinetic modeling to fermentation and cellular processes will be discussed.

Figure 3.1

An overview of the intracellular biochemical reactions in micro-organisms; in addition to the formation of biomass constituents, e.g., cellular protein, lipids, RNA, DNA, and carbohydrates, substrates are converted into primary metabolites, e.g., ethanol, acetate, lactate; secondary metabolites, e.g., penicillin; and/or extracellular macromolecules, e.g., enzymes, heterologous proteins, polysaccharides.

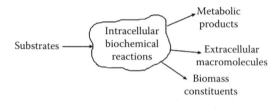

3.2 Framework for kinetic models

The net result of the many biochemical reactions within a single cell is the conversion of substrates to biomass and metabolic end products (see Figure 3.1 and Chapter 2). Clearly the number of reactions involved in the conversion of, say, glucose into biomass and desirable end products is very large, and it is therefore convenient to adopt the structure proposed by Neidhardt *et al.* (1990) for cellular metabolism, which can be summarized as follows:

- **Assembly reactions** carry out chemical modifications of macromolecules, their transport to prespecified locations in the cell, and finally, their assembly to form cellular structures such as cell walls, membranes, the nucleus, etc.
- **Polymerization reactions** represent directed, sequential linkage of activated molecules into long (branched or unbranched) polymeric chains. These reactions form macromolecules from a moderately large number of building blocks.
- **Biosynthetic reactions** produce the building blocks used in the polymerization reactions. They also produce coenzymes and related metabolic factors, including signal molecules. Furthermore, a large number of biosynthetic reactions occur in functional units called biosynthetic pathways, each of which consists of one to twelve sequential reactions leading to the synthesis of one or more building blocks. Pathways are easily recognized and are often controlled *en bloc*. In some cases their reactions are catalyzed by enzymes made from a polycistronic message of mRNA transcribed from a set of contiguous genes forming an operon. All biosynthetic pathways begin with one

Table 3.1   Overall composition of an average cell of *Escherichia coli*

Macromolecule	Percentage of total dry weight	Different kinds of molecules
Protein	55.0	1050
RNA	20.5	
rRNA	16.7	3
tRNA	3.0	60
mRNA	0.8	400
DNA	3.1	1
Lipid	9.1	4
Lippolysaccharide	3.4	1
Peptidoglycan	2.5	1
Glycogen	2.5	1
Metabolite pool	3.9	

Data are taken from Ingraham *et al.* (1983).

of twelve precursor metabolites. Some pathways begin directly with such a precursor metabolite, others indirectly by branching from an intermediate or an end product of a related pathway.

- **Fueling reactions** produce the twelve precursor metabolites needed for biosynthesis. Additionally, they generate Gibbs free energy in the form of ATP, which is used for biosynthesis, polymerization, and assembling reactions. Finally, the fueling reactions produce the reducing power needed for biosynthesis. The fueling reactions include all biochemical pathways referred to as catabolic pathways (degrading and oxidizing substrates).

Thus, the conversion of glucose into cellular protein, for example, proceeds *via* precursor metabolites formed in the fueling reactions, further *via* building blocks (in this case amino acids) formed in the biosynthetic reactions, and finally through polymerization of the building blocks (or amino acids). In the fueling reactions there are many more intermediates than the precursor metabolites, and similarly a large number of intermediates are also involved in the conversion of precursor metabolites into building blocks. The number of cellular metabolites is therefore very large, but still they only account for a small fraction of the total biomass (Table 3.1). The reason for this is the *en bloc* control of the individual reaction rates in the biosynthetic pathways mentioned above. Furthermore, the high affinity of enzymes for the reactants ensures that each metabolite can be maintained at a very low concentration even at a high flux through the pathway (see Box 3.2).

This control of the individual reactions in long pathways is very important for cell function, but it also means that in

Box 3.2 Control of metabolite levels in biochemical pathways

The level of intracellular metabolites is normally very low. This is due to tight regulation of the enzyme levels and to the high affinity most enzymes have towards the reactants. To illustrate this consider two reactions of a pathway — one forming the metabolite X_i and the other consuming this metabolite:

$$....X_{i-1} \xrightarrow{v_i} X_i \xrightarrow{v_{i+1}} X_{i+1}....$$

Assuming that there is no allosteric regulation of the two enzyme-catalyzed reactions, the kinetics can be described with reversible Michaelis–Menten kinetics:

$$v_i = \frac{v_{i,\max}\left(\dfrac{c_{i-1}}{K_{i-1}} - \dfrac{c_i}{K_i}\right)}{1 + \dfrac{c_{i-1}}{K_{i-1}} + \dfrac{c_i}{K_i}}$$

where c_i is the metabolite concentration and $v_{i,\max}$ expresses the enzyme activity. If the rate of the first reaction increases drastically, e.g., due to an increase in the concentration of the metabolite X_{i-1}, the metabolite X_i will accumulate. This will lead to an increase in the second reaction rate and a decrease in the first reaction rate. Consequently, the concentration of metabolite X_i will decrease again. The parameters K_i and K_{i-1} quantify the affinity of the enzyme for the reactant and the product in each reaction, and generally these are in the order of a few µM. Thus, even for low metabolite concentrations (in the order of 10 times K_i) the enzyme will be saturated and the reaction rate will be close to $v_i\max$, but typically the metabolite concentration is of the order of K_i because hereby the enzyme can respond rapidly to changes in the metabolite level, and metabolite accumulations can be avoided.

a quantitative description of cell growth it is not necessary to consider the kinetics of all the individual reactions, and this obviously leads to a significant reduction in the degree of complexity. Consideration of the kinetics of individual enzymes or reactions is therefore necessary only when the aim of the study is to quantify the relative importance of a particular reaction in a pathway.

3.2.1 *Stoichiometry*

The first step in a quantitative description of cellular growth is to specify the stoichiometry for those reactions to be considered for analysis. For this purpose it is important to distinguish between substrates, metabolic products, intracellular metabolites, and biomass constituents (Stephanopoulos *et al.*, 1998):

- A **substrate** is a compound present in the sterile medium, which can be further metabolized or directly incorporated into the cell.
- A **metabolic product** is a compound produced by the cells and excreted to the extracellular medium.
- **Biomass constituents** are pools of macromolecules that make up the biomass, e.g., RNA, DNA, protein, lipids, carbohydrates, but also macromolecular products accumulating inside the cell, e.g., a polysaccharide or a nonsecreted heterologous protein.

- **Intracellular metabolites** are all other compounds within the cell, i.e., glycolytic intermediates, precursor metabolites, and building blocks.

Above there is a distinction between biomass constituents and intracellular metabolites, because the timescales of their turnover in cellular reactions are very different, i.e., intracellular metabolites have a very fast turnover (typically in the range of seconds) compared with that of macromolecules (typically in the range of hours). This means that on the timescale of growth the intracellular metabolite pools can be assumed to be in pseudo steady state.

With the goal of specifying a general stoichiometry for biochemical reactions we consider a system where N substrates are converted to M metabolic products and Q biomass constituents. The conversions are carried out in J reactions in which K intracellular metabolites participate as pathway intermediates. The substrates are termed S_i, the metabolic products are termed P_i, the biomass constituents are termed $X_{macro,i}$, and the intracellular metabolites are termed $X_{met,i}$. With these definitions the general stoichiometry for the jth reaction can be specified as:

$$\sum_{i=1}^{N}\alpha_{ji}S_i + \sum_{i=1}^{M}\beta_{ji}P_i + \sum_{i=1}^{Q}\gamma_{ji}X_{macro,i} + \sum_{i=1}^{K}g_{ji}X_{met,i} = 0; \quad j=1,..,J \quad (3.3)$$

Here α_{ji} is a stoichiometric coefficient for the ith substrate, β_{ji} is a stoichiometric coefficient for the it^{th} metabolic product, γ_{ji} is a stoichiometric coefficient for the ith macromolecular pool, and g_{ji} is a stoichiometric coefficient for the ith intracellular metabolite. All the stoichiometric coefficients are with sign. Thus, all compounds consumed in the jth reaction have negative stoichiometric coefficients whereas all compounds that are produced have positive stoichiometric coefficients. Furthermore, compounds that do not participate in the jth reaction have a stoichiometric coefficient of zero.

If there are many cellular reactions, i.e., J is large, it is convenient to write the stoichiometry for all the J cellular reactions in a compact form using matrix notation:

$$\mathbf{AS} + \mathbf{BP} + \Gamma\mathbf{X}_{macro} + \mathbf{GX}_{met} = 0 \qquad (3.4)$$

where the matrices $\mathbf{A}$, $\mathbf{B}$, Γ and $\mathbf{G}$ are stoichiometric matrices containing stoichiometric coefficients in the J reactions for the substrates, metabolic products, biomass constituents, and pathway intermediates, respectively. In these matrices, rows represent reactions and columns metabolites, i.e., the element in the jth row and the ith column of $\mathbf{A}$ specifies the stoichiometric coefficient for the ith substrate in the jth reaction. Formulation of the stoichiometry in matrix form may seem rather complex, but for large models it is much easier to handle, and especially if the model is to be used for analysis of pathways (e.g., metabolic flux analysis, see Chapter 7) the use of matrices is convenient. However, if the model is simple, i.e., only a few reactions, a few substrates, and a few metabolic products are considered, it is generally more convenient to use the more simple stoichiometric representation in Equation (3.5).

3.2.2 *Reaction rates*

The stoichiometry of the individual reaction is the basis of any quantitative analysis. However, of equal importance is specification of the rate of the individual reactions. Normally the rate of a chemical reaction is given as the forward rate, which if termed v_i, specifies that a compound that has a stoichiometric coefficient β in the ith reaction is formed with the rate βv_i. Normally the stoichiometric coefficient for one of the compounds is arbitrarily set to 1, whereby the forward reaction rate becomes equal to the consumption or production of this compound in this particular reaction. For this reason the forward reaction rate is normally specified with the unit moles (or g) h^{-1}, or if the total amount of biomass is taken as reference (so-called specific rates) with the unit moles (or g) (g DW h^{-1}).

For calculation of the overall production or consumption rate we have to sum the contributions from the different reactions, i.e., the total specific consumption rate of the ith substrate equals the sum of substrate consumptions in all the J reactions:

$$r_{s,i} = -\sum_{j=1}^{J} \alpha_{ji} v_j \tag{3.5}$$

The stoichiometric coefficients for substrates are generally negative, i.e., the specific formation rate of the ith substrate in the jth reaction given by $\alpha_{ji} v_j$ is negative, but the specific substrate uptake rate is normally considered as positive, and a minus sign is therefore introduced in Equation (3.5). For the specific formation rate of the ith metabolic product we have similarly:

$$r_{p,i} = \sum_{j=1}^{J} \beta_{ji} v_j \tag{3.6}$$

Equations (3.5) and (3.6) specify some very important relations between what can be directly measured, namely the specific substrate uptake rates and the specific product formation rates, and the rates of the reactions in the metabolic model. If a compound is consumed or formed in only one reaction, it is quite clear that we can get a direct measurement of this reaction rate – something that is the basis of metabolic flux analysis using metabolite balancing (see Chapter 7 for more details). For the biomass constituents and the intracellular metabolites we can specify similar expressions for the net formation rate in all the J reactions:

$$r_{macro,i} = \sum_{j=1}^{J} \gamma_{ji} v_j \tag{3.7}$$

$$r_{met,i} = \sum_{j=1}^{J} g_{ji} v_j \tag{3.8}$$

These rates are net specific formation rates, because a compound may be formed in one reaction and consumed in another, and the rates specify the net results of consumption and formation in all the J cellular reactions. Thus, if $r_{met,i}$ is positive there is a net formation of the ith intracellular metabolite and if it is negative there is a net consumption of this metabolite. Finally, if $r_{met,i}$ is zero, the rates of formation of the ith metabolite exactly balances its consumption.

If the forward reaction rates for the J cellular reactions are collected in the rate vector $\mathbf{v}$ the summations in Equations (3.5) to (3.8) can be formulated in matrix notation as:

$$\mathbf{r}_s = -\mathbf{A}^T\mathbf{v} \tag{3.9}$$

$$\mathbf{r}_p = \mathbf{B}^T\mathbf{v} \tag{3.10}$$

$$\mathbf{r}_{macro} = \mathbf{\Gamma}^T\mathbf{v} \tag{3.11}$$

$$\mathbf{r}_{met} = \mathbf{G}^T\mathbf{v} \tag{3.12}$$

Here r_s is a rate vector containing the specific uptake rates of the N substrates, r_p a vector containing the specific formation rates of the M metabolic products, r_{macro} a vector containing the net specific formation rate of the Q biomass constituents, and r_{met} is a vector containing the net specific formation rate of the K intracellular metabolites. Notice that what appears in the matrix equations are the transposed stoichiometric matrices, which are formed from the stoichiometric matrices by converting columns into rows and vice versa (see Example 3a). Equation (3.7) gives the net specific formation rate of biomass constituents, and because the intracellular metabolites only represent a small fraction of the total biomass, the specific growth rate μ of the total biomass is given as the sum of formation rates for all the macro-molecular constituents:

$$\mu = \sum_{i=1}^{Q} r_{macro,i} = \mathbf{1}_Q^T \mathbf{r}_{macro} = \mathbf{1}_Q^T \mathbf{\Gamma}^T\mathbf{v} \tag{3.13}$$

where $\mathbf{1}_Q$ is a Q-dimensional row vector with all elements being 1. Equation (3.13) is very fundamental because it links the information supplied by a detailed metabolic model with the macroscopic (and measurable) parameter μ. It clearly specifies that the formation rate of biomass is represented by a sum of formation of many different biomass constituents (or macromolecular pools), a point that will be discussed further in Section 3.4.3.

3.2.3 Yield coefficients and linear rate equations

The overall yield, e.g., how much carbon in the glucose ends up in the metabolite of interest, is a very important design parameter in many fermentation processes. This overall yield is normally represented in the form of **yield coefficients**, which can be considered as relative rates (or fluxes) towards the product of interest with a certain compound as reference, often the carbon source or the biomass. These yield coefficients therefore have the units mass per unit mass of the reference, e.g., moles of penicillin formed per mole of glucose consumed or g protein formed per g biomass formed. An often used yield coefficient in the design and operation of aerobic fermentations is the respiratory quotient RQ, which specifies the moles of carbon dioxide formed per mole of oxygen consumed (see also Example 3a). Several different formulations of the yield coefficients can be found in the literature. Here we will use the formulation of Nielsen et al. (2003), where the yield coefficient is stated with a double subscript Y_{ij} which states that a mass of j is formed or consumed per mass of i formed or consumed. With the ith substrate as the reference compound, the yield coefficients are given by

$$Y_{s_i s_j} = \frac{r_{s,j}}{r_{s,i}} \tag{3.14}$$

$$Y_{s_i p_j} = \frac{r_{p,j}}{r_{s,i}} \tag{3.15}$$

$$Y_{s_i x} = \frac{\mu}{r_{s,i}} \tag{3.16}$$

In the classical description of cellular growth introduced by Monod (1942) (see Section 3.4.2), the yield coefficient Y_{sx} was taken to be constant, and all the cellular reactions were lumped into a single overall growth reaction where substrate is converted to biomass. However, in the late 1950s it was shown (Herbert, 1959) that the yield of biomass with respect to substrate is not constant. In order to describe this he introduced the concept of **endogenous metabolism**, and specified substrate consumption for this process in addition to that for biomass synthesis. Similarly, Luedeking and Piret (1959) found that lactic acid bacteria produce lactic acid at nongrowth conditions, which was consistent with an endogenous metabolism of the cells. Their results indicated a linear correlation between the specific lactic acid production rate and the specific growth rate:

$$r_p = a\mu + b \tag{3.17}$$

In the mid-1960s, Pirt (1965) introduced a similar linear correlation between the specific rate of substrate uptake and the specific growth rate, and suggested the term **maintenance**, which is currently a widely used concept in endogenous metabolism. The linear correlation of Pirt takes the form

$$r_s = Y_{xs}^{true} \mu + m_s \tag{3.18}$$

where Y_{xs}^{true} is referred to as the true yield coefficient and ms as the maintenance coefficient. With the introduction of the linear correlations the yield coefficients can obviously not be constants. Thus for the biomass yield on the substrate:

$$Y_{sx} = \frac{\mu}{Y_{xs}^{true} \mu + m_s} \tag{3.19}$$

which shows that Y_{sx} decreases at low specific growth rates where an increasing fraction of the substrate is used to meet the maintenance requirements of the cell. For large specific growth rates the yield coefficient approaches the reciprocal of Y_{xs}^{true}. A compilation of true yield and maintenance coefficients for various microbial species is given in Table 3.2.

Table 3.2 True yield and maintenance coefficients for different microbial species and growth on glucose or glycerol

Organism	Substrate	Y_{xs}^{true} [g (g DW)$^{-1}$]	m_s [g (g DW h)$^{-1}$]
Aspergillus awamori	Glucose	1.92	0.016
Aspergillus nidulans		1.67	0.020
Candida utilis		2.00	0.031
Escherichia coli		2.27	0.057
Klebsiella aerogenes		2.27	0.063
Penicillium chrysogenum		2.17	0.021
Saccharomyces cerevisiae		1.85	0.015
Aerobacter aerogenes	Glycerol	1.79	0.089
Bacillus megatarium		1.67	—
Klebsiella aerogenes		2.13	0.074

Data are taken from Nielsen and Villadsen (1994).

The empirically derived linear correlations are very useful for correlating growth data, especially in steady-state continuous cultures where linear correlations similar to Equation (3.18) were found for most of the important specific rates. The remarkable robustness and general validity of the linear correlations indicate that they have a fundamental basis, and this basis is likely to be the continuous supply and consumption of ATP, because these two processes are tightly coupled in all cells. Thus the role of the energy producing substrate is to provide ATP to drive biosynthesis and polymerization reactions as well as cell maintenance processes according to the linear relationship:

$$r_{ATP} = Y_{xATP}\mu + m_{ATP} \tag{3.20}$$

which is a formal analogue to the linear correlation of Pirt. Equation (3.20) states that ATP produced balances the consumption for growth and for maintenance, and if the ATP yield on the energy producing substrate is constant, i.e., r_{ATP} is proportional to r_s, it is quite obvious that Equation (3.20) can be used to derive the linear correlation in Equation (3.18), as illustrated in Example 3a. Notice that Y_{xATP} in Equation (3.20) is a true yield coefficient, but it is normally specified without the superscript "true."

The concept of balancing ATP production and consumption can be extended to other cofactors, e.g., NADH and NADPH, and as such it is possible to derive linear rate equations for three different cases (Nielsen and Villadsen, 1994):

- anaerobic growth where ATP is supplied by substrate-level phosphorylation
- aerobic growth without metabolite formation
- aerobic growth with metabolite formation

For aerobic growth with metabolite formation, the specific substrate uptake rate takes the form:

$$r_s = Y_{xs}^{true}\mu + Y_{ps}^{true}r_p + m_s \tag{3.21}$$

This linear rate equation can be interpreted as a metabolic model with three reactions:

- conversion of substrate to biomass with a stoichiometric coefficient for the substrate and a forward reaction rate equal to the specific growth rate
- conversion of substrate to the metabolic product with a stoichiometric coefficient for the substrate and a forward reaction rate equal to the specific product formation rate
- metabolism of substrate to meet the maintenance requirements (normally the substrate is oxidized to carbon dioxide) with the rate m_s.

Consequently, the stoichiometry for these three reactions can be specified as:

$$-Y_{xs}^{true}S + X = 0 \quad ; \quad \mu \tag{3.22}$$

$$-Y_{ps}^{true}S + P = 0 \quad ; \quad r_p \tag{3.23}$$

$$-S = 0 \quad ; \quad m_s \tag{3.24}$$

With this stoichiometry, the linear rate Equation (3.21) can easily be derived using Equation (3.5), i.e., the overall specific substrate consumption rate is the sum of substrate consumption for growth, metabolite formation, and maintenance.

Thus, it is important to distinguish between true yield coefficients (which are rather stoichiometric coefficients) and overall yield coefficients, which can be taken to be stoichiometric coefficients in **one lumped reaction** (often referred to as the black box model; see Section 3.2.4), which represents all the cellular processes, i.e.,

$$-Y_{xs}S + Y_{xp}P + X = 0 \; ; \mu \tag{3.25}$$

Despite the subscript "true," the true yield coefficients are only parameters for a given cellular system as they simply represent overall stoichiometric coefficients in lumped reactions, e.g., reaction (3.22) is the sum of all reactions involved in the conversion of substrate into biomass. If, for example, the environmental conditions change, a different set of metabolic routes may be activated, and this may result in a change in the overall recovery of carbon in each of the three processes mentioned above, i.e., the values of the true yield coefficients change. Even the more fundamental Y_{xATP} cannot be taken to be constant, as illustrated in a detailed analysis of lactic acid bacteria (Benthin *et al.*, 1994).

Example 3a: Metabolic model for aerobic growth of Saccharomyces cerevisiae

To illustrate the derivation of the linear rate equations for an aerobic process with metabolite formation, we consider a simple metabolic model for the yeast *Saccharomyces cerevisiae*. For this purpose we set up a stoichiometric model that summarizes the overall cellular metabolism, and based on assumptions of pseudo steady state for ATP, NADH, and NADPH, linear rate equations can be derived where the specific uptake rates for glucose and oxygen and the specific carbon dioxide formation rate are given as functions of the specific growth rate. Furthermore, by evaluating the parameters in these linear rate equations, which can be done from a comparison with experimental data, information on key energetic parameters, may be extracted.

Table 3.3 Macromolecular composition of *Saccharomyces cerevisiae*

Macromolecule	Content [g (g DW)$^{-1}$]
Protein	0.39
Polysaccharides + trehalose	0.39
DNA + RNA	0.11
Phospholipids	0.05
Triacylglycerols	0.02
Sterols	0.01
Ash	0.03

From an analysis of all the biosynthetic reactions the overall stoichiometry for synthesis of the constituents of a *S. cerevisiae* cell can be specified (Oura, 1983) as follows:

$$CH_{1.62}O_{0.53}N_{0.15} + 0.12\ CO_2 + 0.397\ NADH - 1.12\ CH_2O - $$
$$0.15\ NH_3 - Y_{xATP}\ ATP - 0.212\ NADPH = 0 \qquad (3a1)$$

The stoichiometry (3a1) holds for a cell with the composition specified in Table 3.3, and the substrate is glucose and inorganic salts, i.e., ammonia is the nitrogen source. The stoichiometry is given on a C-mole basis, i.e., glucose is specified as CH_2O, and the elemental composition of the biomass was calculated from the macromolecular composition to be $CH_{1.62}O_{0.53}N_{0.15}$ (see Table 3.3). The ATP and NADPH required for biomass synthesis are supplied by the catabolic pathways, and excess NADH formed in the biosynthetic reactions is, together with NADH formed in the catabolic pathways, re-oxidized by transfer of electrons to oxygen via the electron transport chain. Reactions (3a2) to (3a5) specify the overall stoichiometry for the catabolic pathways. Reaction (3a2) specifies NADPH formation by the PP pathway, where glucose is completely oxidized to CO_2; reaction (3a3) is the overall stoichiometry for the combined EMP pathway and the TCA cycle; reaction (3a4) is the fermentative glucose metabolism where glucose is converted to ethanol (this reaction only runs at high glucose uptake rates); and finally, reaction (3a5) is the overall stoichiometry for the oxidative phosphorylation, where the P/O ratio is the overall (or operational) P/O ratio for the oxidative phosphorylation.

$$CO_2 + 2\ NADPH - CH_2O = 0 \qquad (3a2)$$

$$CO_2 + 2\ NADH + 0.667\ ATP - CH_2O = 0 \qquad (3a3)$$

$$CH_3O_{0.5} + 0.5\ CO_2 + 0.5\ ATP - 1.5\ CH_2O = 0 \qquad (3a4)$$

$$P/O\ ATP - 0.5\ O_2 - NADH = 0 \qquad (3a5)$$

Finally, consumption of ATP for maintenance is included simply as a reaction where ATP is used:

$$-ATP = 0 \tag{3a6}$$

Note that with the stoichiometry given on a C-mole basis, the stoichiometric coefficients extracted from the biochemistry, e.g., formation of 2 moles ATP per mole glucose in the EMP pathway, are divided by six, because glucose contains six C-moles per mole.

Above, the stoichiometry is written as in Equation (3.3), but we can easily convert it to the more compact matrix notation of Equation (3.4):

$$
\begin{pmatrix}
-1.120 & 0 \\
-1 & 0 \\
-1 & 0 \\
-1.5 & 0 \\
0 & -0.5 \\
0 & 0
\end{pmatrix}
\begin{pmatrix} S_{glc} \\ S_{o_2} \end{pmatrix}
+
\begin{pmatrix}
0 & 0.120 \\
0 & 1 \\
0 & 1 \\
1 & 0.5 \\
0 & 0 \\
0 & 0
\end{pmatrix}
\begin{pmatrix} P_{eth} \\ P_{CO_2} \end{pmatrix}
+
$$

$$
\begin{pmatrix} 1 \\ 0 \\ 0 \\ 0 \\ 0 \\ 0 \end{pmatrix} X
+
\begin{pmatrix}
-Y_{xATP} & 0.397 & -0.212 \\
0 & 0 & 2 \\
0.667 & 2 & 0 \\
0.5 & 0 & 0 \\
P/O & -1 & 0 \\
-1 & 0 & 0
\end{pmatrix}
\begin{pmatrix} X_{ATP} \\ X_{NADH} \\ X_{NADPH} \end{pmatrix}
=
\begin{pmatrix} 0 \\ 0 \\ 0 \\ 0 \\ 0 \\ 0 \end{pmatrix}
\tag{3a7}
$$

where X represents the biomass.

We now collect the forward reaction rates for the six reactions in the rate vector $\mathbf{v}$ given by

$$
\mathbf{v} =
\begin{pmatrix}
\mu \\
\upsilon_{PP} \\
\upsilon_{EMP} \\
r_{eth} \\
\upsilon_{OP} \\
m_{ATP}
\end{pmatrix}
$$

In analogy with Equation (3.18), we balance the production and consumption of the three cofactors ATP, NADH, and NADPH. This gives the three equations

$$-Y_{xATP}\mu + 0.667\upsilon_{EMP} + 0.5r_{eth} + P/O\upsilon_{OP} - m_{ATP} = 0 \tag{3a8}$$

$$0.397\mu + 2\upsilon_{EMP} - \upsilon_{OP} = 0 \tag{3a9}$$

$$-0.212\mu + 2\upsilon_{PP} = 0 \tag{3a10}$$

Notice that these balances correspond to zero net specific formation rates for the three cofactors, and the three balances can therefore also be derived using Equation (3.12):

$$
r_{met} = G^T v = \begin{pmatrix} -Y_{xATP} & 0 & 0.667 & 0.5 & P/O & -1 \\ 0.397 & 0 & 2 & 0 & -1 & 0 \\ -0.212 & 2 & 0 & 0 & 0 & 0 \end{pmatrix} \begin{pmatrix} \mu \\ v_{PP} \\ v_{EMP} \\ r_{eth} \\ v_{OP} \\ m_{ATP} \end{pmatrix} = \begin{pmatrix} 0 \\ 0 \\ 0 \end{pmatrix} \tag{3a11}
$$

In addition to the three balances (3a8)–(3a10), we have the relationships between the reaction rates and the specific substrate uptake rates and the specific product formation rate given by Equations (3a5) and (3a6), or using the matrix notation of Equations (3a9) and (3a10):

$$
\begin{pmatrix} r_{glc} \\ r_{O_2} \end{pmatrix} = - \begin{pmatrix} -1.120 & -1 & -1 & -1.5 & 0 & 0 \\ 0 & 0 & 0 & 0 & -0.5 & 0 \end{pmatrix} \begin{pmatrix} \mu \\ v_{PP} \\ v_{EMP} \\ r_{eth} \\ v_{OP} \\ m_{ATP} \end{pmatrix} \tag{3a12}
$$

$$
= \begin{pmatrix} 1.120\mu + v_{PP} + v_{EMP} + 1.5 r_{eth} \\ 0.5 v_{OP} \end{pmatrix}
$$

$$
\begin{pmatrix} r_{eth} \\ r_{CO_2} \end{pmatrix} = \begin{pmatrix} 0 & 0 & 0 & 1 & 0 & 0 \\ 0.120 & 1 & 1 & 0.5 & 0 & 0 \end{pmatrix} \begin{pmatrix} \mu \\ v_{PP} \\ v_{EMP} \\ r_{eth} \\ v_{OP} \\ m_{ATP} \end{pmatrix} \tag{3a13}
$$

$$
= \begin{pmatrix} r_{eth} \\ 0.120\mu + v_{PP} + v_{EMP} + 0.5 r_{eth} \end{pmatrix}
$$

Clearly, the specific ethanol production rate is equal to the rate of reaction (3a5) because the stoichiometric coefficient for ethanol in this reaction is 1, and it is the only reaction where ethanol is involved. Using the combined set of Equations (3a11)–(3a13), the four reaction rates v_{EMP}, v_{PP}, v_{OP}, and m_{ATP} can be eliminated and the linear rate equations (3a14)–(3a16) can be derived:

$$
r_{glc} = (a + 1.226)\mu + (1.5 - b)r_{eth} + c = Y_{xs}^{true}\mu + Y_{ps}^{true}r_{eth} + m_s \tag{3a14}
$$

$$
r_{CO_2} = (a + 0.226)\mu + (0.5 - b)r_{eth} + c = Y_{xc}^{true}\mu + Y_{pc}^{true}r_{eth} + m_c \tag{3a15}
$$

LIVERPOOL
JOHN MOORES UNIVERSITY
AVRIL ROBARTS LRC
TITHEBARN STREET
LIVERPOOL L2 2ER
TEL. 0151 231 4022

$$r_{O_2} = (a + 0.229)\mu - br_{eth} + c = Y_{xo}^{true}\mu + Y_{po}^{true}r_{eth} + m_o \qquad (3a16)$$

The three common parameters a, b, and c are functions of the energetic parameters Y_{xATP}, m_{ATP}, and the P/O ratio according to Equations (3a17)–(3a19):

$$a = \frac{Y_{xATP} - 0.458\text{P/O}}{0.667 + 2\text{P/O}} \qquad (3a17)$$

$$b = \frac{0.5}{0.667 + 2P/O} \qquad (3a18)$$

$$c = \frac{m_{ATP}}{0.667 + 2\text{P/O}} \qquad (3a19)$$

If there is no ethanol formation, which is the case at low specific glucose uptake rates, Equation (3a14) reduces to the linear rate Equation (3.18), but the parameters of the correlation are determined by basic energetic parameters of the cells. It is seen that the parameters in the linear correlations are coupled via the balances for ATP, NADH, and NADPH, and the three true yield coefficients cannot take any value. Furthermore, the maintenance coefficients are the same. This is due to the use of the units C-moles per C-mole biomass per h for the specific rates. If other units are used for the specific rates, the maintenance coefficients will not take the same values but remain proportional. This coupling of the parameters shows that there are only three degrees of freedom in the system, and one actually only has to determine two yield coefficients and one maintenance coefficient – the other parameters can be calculated using the three equations (3a14)–(3a16).

The derived linear rate equations are certainly useful for correlating experimental data, but they also allow evaluation of the key energetic parameters Y_{xATP}, m_{ATP}, and the operational P/O ratio. Thus, if the true yield coefficients and the maintenance coefficients of Equations (3a14)–(3a16) are estimated, the values of a, b, and c can be found and these three parameters relate the three energetic parameters through Equations (3a17)–(3a19). Thus, from one of the ethanol yield coefficients, b can be found and, thereafter, the P/O ratio can be determined. Then m_{ATP} can be found from one of the maintenance coefficients, and finally Y_{xATP} can be found from one of the biomass yield coefficients. In practice, it is, however, difficult to extract sufficiently precise values of the true yield coefficients from experimental data to estimate the energetic parameters – especially since the three parameters a, b, and c are closely correlated (especially b and c). However, if either the P/O ratio or Y_{xATP} is known, Equations (3a14)–(3a16) allow an estimation of the two remaining unknown energetic parameters. Consider the situation where there is no ethanol formation; here the true yield coefficient for biomass is 1.48 C-moles glucose (C-mole biomass)$^{-1}$ and the maintenance coefficient (equal to b) is 0.012 C-moles glucose (C-mole biomass h)$^{-1}$ (both values taken from Table 3.2). Thus, a is equal to 0.254 moles ATP (C-mole^{-1} biomass). If the operational P/O ratio is about 1.5 (which is a reasonable value for *S. cerevisiae*), we find that Y_{xATP} is 1.62 moles ATP (C-mole^{-1} biomass) or about 67 mmoles ATP (g DW^{-1}). Similarly, we find m_{ATP} to be about 2 mmoles (g DW h^{-1}).

In connection with baker's yeast production, it is important to maximize the yield of biomass on glucose:

$$Y_{sx} = \frac{\mu}{(a+1.226)\mu+(1.5-b)r_{eth}+c} \tag{3a20}$$

Clearly, this can best be done if ethanol production is avoided. Thus, the glucose uptake rate is to be controlled below a level where there is respiro-fermentative metabolism. A very good indication of whether there is respiro-fermentative metabolism is the respiratory quotient RQ:

$$RQ = \frac{(a+0.226)\mu+(0.5-b)r_{eth}+c}{(a+0.229)\mu-br_{eth}+c} \tag{3a21}$$

If there is no ethanol production, RQ will be close to 1 (independent of the specific growth rate), whereas if there is ethanol production, RQ will be above 1 and will increase with $r$eth (b is always less than 0.5). From measurements of carbon dioxide and oxygen in the exhaust gas, the RQ can be evaluated. If it is above 1, there is respiro-fermentative metabolism resulting in a low yield of biomass on sugar due to high sugar concentration, and as such the sugar concentration in the feed must be reduced (typically the baker's yeast production is operated as a fed-batch process; see Section 3.4).

3.2.4 *The black box model*

In the black box model of cellular growth, all the cellular reactions are lumped into a single reaction. In this overall reaction the stoichiometric coefficients are identical to the yield coefficients [see also Equation (3.25)], and it can therefore be presented as

$$X+\sum_{i=1}^{M}Y_{xp_i}P_i-\sum_{i=1}^{N}Y_{xs_i}S_i=0 \tag{3.26}$$

Because the stoichiometric coefficient for biomass is 1, the forward reaction rate is given by the specific growth rate of the biomass, which together with the yield coefficients completely specifies the system. As discussed in the previous section, the yield coefficients are not constants, and the black box model can therefore not be applied to correlate, for instance, the specific substrate uptake rate with the specific growth rate. However, it is very useful for validation of experimental data because it can form the basis for setting up elemental balances. Thus, in the black box model there are $(M + N + 1)$ parameters: M yield coefficients for the metabolic products, N yield coefficients for the substrates and the forward reaction rate μ. Because mass is conserved in the overall conversion of substrates to metabolic products and biomass, the $(M + N + 1)$ parameters of the black box model are not completely independent but must satisfy several constraints. Thus, the elements flowing into the system must balance the elements flowing out of the system, e.g., the carbon entering the system via the substrates has to be recovered in the metabolic products and biomass. Each element considered in the black box obviously yields one constraint. Thus a carbon balance gives

$$1+\sum_{i=1}^{M} f_{p,i} Y_{xp_i} - \sum_{i=1}^{N} f_{s,i} Y_{xs_i} = 0 \tag{3.27}$$

where $f_{s,i}$ and $f_{p,i}$ represent the carbon content (C-moles mole^{-1}) in the ith substrate and the ith metabolic product, respectively. In the above equation, the elemental composition of biomass is normalized with respect to carbon, i.e., it is represented by the form CHaObNc (see also Example 3a). The elemental composition of biomass depends on its macromolecular content and, therefore, on the growth conditions and the specific growth rate (e.g., the nitrogen content is much lower under nitrogen limited conditions than under carbon-limited conditions; see Table 3.4). However, except for extreme situations, it is reasonable to use the general composition formula $CH_{1.8}O_{0.5}N_{0.2}$ whenever the biomass composition is not exactly known. Often the elemental composition of substrates and metabolic products is normalized with respect to their carbon content, e.g., glucose is specified as CH_2O (see also Example 3a). Equation (3.27) is then written on a per C-mole basis as

$$1+\sum_{i=1}^{M} Y_{xp_i} - \sum_{i=1}^{N} Y_{xs_i} = 0 \tag{3.28}$$

In Equation (3.28), the yield coefficients have units of C-moles per C-mole biomass. Conversion to this unit from other units is illustrated in Box 3.3. Equation (3.28) is very useful for checking the consistency of experimental data. Thus, if the sum of carbon in the biomass and the metabolic products does not equal the sum of carbon in the substrates, there is an inconsistency in the experimental data.

Table 3.4 Elemental composition of biomass for several micro-organisms

Micro-organism	Elemental composition	Ash content (w/w %)	Growth conditions
Candida utilis	$CH_{1.83}O_{0.46}N_{0.19}$	7.0	Glucose limited, $D = 0.05$ h^{-1}
	$CH_{1.87}O_{0.56}N_{0.20}$	7.0	Glucose limited, $D = 0.45$ h^{-1}
	$CH_{1.83}O_{0.54}N_{0.10}$	7.0	Ammonia limited, $D = 0.05$ h^{-1}
	$CH_{1.87}O_{0.56}N_{0.20}$	7.0	Ammonia limited, $D = 0.45$ h^{-1}
Klebsiella aerogenes	$CH_{1.75}O_{0.43}N_{0.22}$	3.6	Glycerol limited, $D = 0.10$ h^{-1}
	$CH_{1.73}O_{0.43}N_{0.24}$	3.6	Glycerol limited, $D = 0.85$ h^{-1}
	$CH_{1.75}O_{0.47}N_{0.17}$	3.6	Ammonia limited, $D = 0.10$ h^{-1}
	$CH_{1.73}O_{0.43}N_{0.24}$	3.6	Ammonia limited, $D = 0.80$ h^{-1}
Saccharomyces cerevisiae	$CH_{1.82}O_{0.58}N_{0.16}$	7.3	Glucose limited, $D = 0.080$ h^{-1}
	$CH_{1.78}O_{0.60}N_{0.19}$	9.7	Glucose limited, $D = 0.255$ h^{-1}
	$CH_{1.94}O_{0.52}N_{0.25}$	5.5	Unlimited growth
Escherichia coli	$CH_{1.77}O_{0.49}N_{0.24}$	5.5	Unlimited growth
	$CH_{1.83}O_{0.50}N_{0.22}$	5.5	Unlimited growth
	$CH_{1.96}O_{0.55}N_{0.25}$	5.5	Unlimited growth
	$CH_{1.93}O_{0.55}N_{0.25}$	5.5	Unlimited growth
Pseudomonas fluorescens	$CH_{1.83}O_{0.55}N_{0.26}$	5.5	Unlimited growth
Aerobacter aerogenes	$CH_{1.64}O_{0.52}N_{0.16}$	7.9	Unlimited growth
Penicillium chrysogenum	$CH_{1.70}O_{0.58}N_{0.15}$		Glucose limited, $D = 0.038$ h^{-1}
	$CH_{1.68}O_{0.53}N_{0.17}$		Glucose limited, $D = 0.098$ h^{-1}
Aspergillus niger	$CH_{1.72}O_{0.55}N_{0.17}$	7.5	Unlimited growth
Average	$CH_{1.81}O_{0.52}N_{0.21}$	6.0	

Compositions for *P. chrysogenum* are taken from Christensen et al. (1995); other data are taken from Roels (1983).

Box 3.3 **Calculation of yields with respect to C-mole basis**

Yield coefficients are typically described as moles (g DW)$^{-1}$ or g (g DW)$^{-1}$. To convert the yield coefficients to a C-mole basis, information on the elemental composition and the ash content of biomass is needed. To illustrate the conversion, we calculate the yield of 0.5 g DW biomass (g glucose)$^{-1}$ on a C-mole basis. First, we convert the g DW biomass to an ashfree basis, i.e., determine the amount of biomass that is made up of carbon, nitrogen, oxygen, and hydrogen (and, in some cases, also phosphorus and sulphur). With an ash content of 8% we have 0.92 g ashfree biomass (g DW biomass)$^{-1}$,

which gives a yield of 0.46 g ashfree biomass (g glucose)$^{-1}$. This yield can now be directly converted to a C-mole basis using the molecular weights in g C-mole^{-1} for ashfree biomass and glucose. With the standard elemental composition for biomass of $CH_{1.8}O_{0.5}N_{0.2}$ we have a molecular weight of 24.6 g ashfree biomass C-mole^{-1}, and therefore find a yield of 0.46/24.6 = 0.0187 C-moles biomass (g glucose)$^{-1}$. Finally, by multiplication with the molecular weight of glucose on a C-mole basis (30 g C-mole^{-1}), a yield of 0.56 C-moles biomass (C-mole glucose)$^{-1}$ is found.

Similar to Equation (3.27), balances can be written for all other elements participating in the conversion (3.26). Thus, the hydrogen balance will read

$$a_x + \sum_{i=1}^{M} a_{p,i} Y_{xp_i} - \sum_{i=1}^{N} a_{s,i} Y_{xs_i} = 0 \qquad (3.29)$$

where $a_{s,i}$, $a_{p,i}$ and a_x represent the hydrogen content (moles C-mole^{-1} if a C-mole basis is used) in the ith substrate, the ith metabolic product and the biomass, respectively. Similarly, we have for the oxygen and nitrogen balances:

$$b_x + \sum_{i=1}^{M} b_{p,i} Y_{xp_i} - \sum_{i=1}^{N} b_{s,i} Y_{xs_i} = 0 \qquad (3.30)$$

$$c_x + \sum_{i=1}^{M} c_{p,i} Y_{xp_i} - \sum_{i=1}^{N} c_{s,i} Y_{xs_i} = 0 \qquad (3.31)$$

where $b_{s,i}$, $b_{p,i}$ and b_x represent the oxygen content (moles C-mole^{-1}) in the ith substrate, the ith metabolic product and biomass, respectively, and $c_{s,i}$, $c_{p,i}$ and c_x represent the nitrogen content (moles C-mole^{-1}) in the ith substrate, the ith metabolic product and the biomass, respectively. Normally, only these four balances are considered; balances for phosphate and sulphate may also be set up, but generally these elements are of minor importance. The four elemental balances (3.28) to (3.31) can be conveniently written by collecting the elemental composition of biomass, substrates, and metabolic products in the columns of a matrix **E**, where the first column contains the elemental composition of biomass, columns 2 through $M + 1$ contain the elemental composition of the M metabolic products, and columns $M + 2$ to $M + N + 1$ contain the elemental composition of the N substrates. With the introduction of this matrix the four elemental balances can be expressed as

E Y = 0 $\qquad (3.32)$

where **Y** is a vector containing the yield coefficients (the substrate yield coefficients are given with a minus sign). With $N + M + 1$ variables, $N + M$ yield coefficients and the forward rate of reaction (3.26) and four constraints, the degree of freedom is $F = M + N + 1 - 4$. If exactly F variables are measured, it may be possible to calculate the other rates by using the four algebraic equations given by (3.32), but, in this case, there are no redundancies left to check the consistency of the data. For this reason, it is advisable to strive for more measurements than the degrees of freedom of the system.

Example 3b: Elemental balances in a simple black box model

Consider the aerobic cultivation of the yeast *Saccharomyces cerevisiae* on a defined, minimal medium, i.e., glucose is the carbon and energy source and ammonia is the nitrogen source. During aerobic growth, the yeast oxidizes glucose completely to carbon dioxide. However, at very high glycolytic fluxes, a bottleneck in the oxidation of pyruvate leads to ethanol formation. Thus, at high glycolytic fluxes, both ethanol and carbon dioxide should be considered as metabolic products. Finally, water is formed in the cellular pathways. This is also included as a product in the overall reaction. Thus, the black box model for this system is

$$X + Y_{xe} \text{ ethanol} + Y_{xc} CO_2 + Y_{xw} H_2O - Y_{xs} \text{ glucose} - Y_{xo} O_2 - Y_{xN} NH_3 = 0 \tag{3b1}$$

which can be represented with the yield coefficient vector:

$$\mathbf{Y} = (1 \ Y_{xe} \ Y_{xc} \ Y_{xw} \ -Y_{xs} \ -Y_{xo} \ -Y_{xN})T \tag{3b2}$$

We now rewrite the conversion using the elemental composition of the substrates and metabolic products. For biomass we use the elemental composition of $CH_{1.83}O_{0.56}N_{0.17}$, and therefore we have

$$CH_{1.83}O_{0.56}N_{0.17} + Y_{xe} CH_3O_{0.5} + Y_{xc} CO_2 + Y_{xw} H_2O - Y_{xs} CH_2O - Y_{xo} O_2 - Y_{xN} NH_3 = 0 \tag{3b3}$$

Some may find it difficult to identify CH3O0.5 as ethanol, but the advantage of using the C-mole basis becomes apparent immediately when we look at the carbon balance:

$$1 + Y_{xe} + Y_{xc} - Y_{xs} = 0 \tag{3b4}$$

This simple equation is very useful for checking the consistency of experimental data. Thus, using the classical data of von Meyenburg (1969), we find $Y_{xe} = 0.713$, $Y_{xc} = 1.313$, and $Y_{xs} = 3.636$ at a dilution rate of $D = 0.3$ h^{-1} in a glucose-limited continuous culture. Obviously the data are not consistent as the carbon balance is not close. A more elaborate data analysis (Nielsen and Villadsen, 1994) suggests that the missing carbon may be accounted for by ethanol evaporation/stripping due to intensive aeration of the bioreactor.

Similarly, using Equation (3.31) we find that a nitrogen balance gives

$$Y_x N = 0.17 \tag{3b5}$$

If the yield coefficients for ammonia uptake and biomass formation do not conform to Equation (3b5), an inconsistency is identified in one of these two measurements or the nitrogen content of the biomass is different from that specified.

We now write all four elemental balances in terms of the matrix equation (3.32):

$$
E = \begin{pmatrix}
1 & 1 & 1 & 0 & 1 & 0 & 0 \\
1.83 & 3 & 0 & 2 & 2 & 0 & 3 \\
0.56 & 0.5 & 2 & 1 & 1 & 2 & 0 \\
0.17 & 0 & 0 & 0 & 0 & 0 & 1
\end{pmatrix}
\quad
\begin{matrix}
\leftarrow \text{carbon} \\
\leftarrow \text{hydrogen} \\
\leftarrow \text{oxygen} \\
\leftarrow \text{nitrogen}
\end{matrix}
\qquad (3b6)
$$

where the rows indicate, respectively, the content of carbon, hydrogen, oxygen, and nitrogen and the columns give the elemental composition of biomass, ethanol, carbon dioxide, water, glucose, oxygen, and ammonia, respectively. Using Equation (3.32), we find

$$
\begin{pmatrix}
1 & 1 & 1 & 0 & 1 & 0 & 0 \\
1.83 & 3 & 0 & 2 & 2 & 0 & 3 \\
0.56 & 0.5 & 2 & 1 & 1 & 2 & 0 \\
0.17 & 0 & 0 & 0 & 0 & 0 & 1
\end{pmatrix}
\begin{pmatrix}
1 \\
Y_{xe} \\
Y_{xc} \\
Y_{xw} \\
-Y_{xs} \\
-Y_{xo} \\
-Y_{xN}
\end{pmatrix}
=
$$

$$
\begin{pmatrix}
1 + Y_{xe} + Y_{xc} - Y_{xs} \\
1.83 + 3Y_{xe} + 2Y_{xw} - 2Y_{xs} - 3Y_{xN} \\
0.56 + 0.5Y_{xe} + 2Y_{xc} + Y_{xw} - Y_{xs} - 2Y_{xo} \\
0.17 - Y_{xN}
\end{pmatrix}
=
\begin{pmatrix}
0 \\
0 \\
0 \\
0
\end{pmatrix}
\qquad (3b7)
$$

The first and last rows are identical to the balances derived in Equations (3b4) and (3b5) for carbon and nitrogen, respectively. The balances for hydrogen and oxygen introduce two additional constraints. However, because the rate of water formation is impossible to measure, one of these equations must be used to calculate this rate (or yield). This leaves only one additional constraint from these two balances.

3.3   Mass balances for bioreactors

In the previous section we derived equations that relate the rates of the intracellular reaction with the rates of substrate uptake, metabolic product formation, and biomass formation. These rates are the key elements in the dynamic mass balances for the substrates, the metabolic products, and the biomass, which describe the change in time of the concentration of these state variables in a bioreactor. The bioreactor may be any type of device, ranging from a shake flask to a well-instrumented bioreactor. Figure 3.2 is a general representation of a bioreactor.

Figure 3.2

Bioreactor with addition of fresh, sterile medium and removal of spent medium where c_i^f is the concentration of the ith compound in the feed and c_i is the concentration of the ith compound in the spent medium. The bioreactor is assumed to be very well mixed (or ideal), so that the concentration of each compound in the spent medium becomes identical to its concentration in the bioreactor.

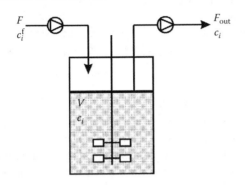

It has a volume V (units: L) and it is fed with a stream of fresh, sterile medium with a flow rate F (units: L h^{-1}). Spent medium is removed with a flow rate of F_{out} (units: L h^{-1}). The medium in the bioreactor is assumed to be completely (or ideally) mixed, i.e., there is no spatial variation in the concentration of the different medium compounds. For small volume bioreactors (< 1 L) (including shake flasks) this can generally be achieved through aeration and some agitation, whereas for laboratory, stirred tank bioreactors (1–10 L) special designs may have to be introduced in order to ensure a homogenous medium (Sonnleitner and Fiechter, 1988; Nielsen and Villadsen, 1993). The bioreactor may be operated in many different modes of which we will only consider the three most common:

- **Batch,** where $F = F_{out} = 0$, i.e., the volume is constant.
- **Continuous,** where $F = F_{out} \neq 0$, i.e., the volume is constant.
- **Fed-batch** (or semibatch), where $F \neq 0$ and $F_{out} = 0$, i.e., the volume increases.

These three different modes of reactor operation are discussed in the following pages, but first we derive general dynamic mass balances for the substrates, metabolic products, biomass constituents, and the biomass.

3.3.1 *Dynamic mass balances*

The basis for derivation of the general dynamic mass balances is the mass balance equation:

Accumulated = Net formation rate + In – Out   (3.33)

where the first term on the RHS is given by Equations (3.5) to (3.8) for substrate, metabolic product, biomass constituents, and intracellular metabolites, respectively. The term "In" represents the flow of the compound into the bioreactor; the term "Out" the flow of the compound out from the bioreactor. In the following we consider substrates, metabolic products, biomass constituents, intracellular metabolites, and the total biomass separately.

We consider the ith substrate, which is added to the bioreactor via the feed and is consumed by the cells present in the bioreactor. The mass balance for this compound is

$$\frac{d(c_{s,i}V)}{dt} = -r_{s,i}xV + Fc_{s,i}^f - F_{out}c_{s,i}$$ (3.34)

where ri is the specific consumption rate of the compound (unit: moles (g DW h^{-1}), $c_{s,i}$ is the concentration in the bioreactor, which is assumed to be the same as the concentration in the outlet (unit: moles L^{-1}), is the concentration in the feed (unit: moles L^{-1}), and x is the biomass concentration in the bioreactor (unit: g DW L^{-1}). The first term in Equation (3.34) is the accumulation term, the second term is the consumption (or reaction term), the third term is accounting for the inlet, and the last term is accounting for the outlet. Rearrangement of this equation gives

$$\frac{dc_{s,i}}{dt} = -r_{s,i}x + \frac{F}{V}c_{s,i}^f - \left(\frac{F_{out}}{V} + \frac{1}{V}\frac{dV}{dt}\right)c_{s,i}$$ (3.35)

For a fed-batch reactor:

$$F = \frac{dV}{dt}$$ (3.36)

and $F_{out} = 0$. So the term within the parentheses becomes equal to the so-called **dilution rate** given by

$$D = \frac{F}{V}$$ (3.37)

For a continuous and a batch reactor, the volume is constant, i.e., $dV/dt = 0$, and $F = F_{out}$, and so for these bioreactor modes also the term within the parentheses becomes equal to the dilution rate. Equation (3.35) therefore reduces to the mass balance (3.38) for any type of operation:

$$\frac{dc_{s,i}}{dt} = -r_{s,i}x + D(c_{s,i}^f - c_{s,i})$$ (3.38)

The first term on the right-hand side of Equation (3.38) is the volumetric rate of substrate consumption, which is given as the product of the specific rate of substrate consumption and the biomass concentration. The second term accounts for the addition and removal of substrate from the bioreactor. The term on the left-hand side of Equation (3.38) is the accumulation term, which accounts for the change in time of the substrate, which in a batch reactor (where $D = 0$) equals the volumetric rate of substrate consumption.

Dynamic mass balances for the metabolic products are derived in analogy with those for the substrates and take the form

$$\frac{dc_{p,i}}{dt} = r_{p,i}x + D(c^f_{p,i} - c_{p,i}) \tag{3.39}$$

where the first term on the right-hand side is the volumetric formation rate of the ith metabolic product. Normally, the metabolic products are not present in the sterile feed to the bioreactor, and $c^f_{p,i}$ is therefore often zero. In these cases the volumetric rate of product formation in a steady-state continuous reactor is equal to the dilution rate multiplied by the concentration of the metabolic product in the bioreactor (equal to that in the outlet).

With sterile feed the mass balance for the total biomass is derived directly:

$$\frac{dx}{dt} = (\mu - D)x \tag{3.40}$$

where μ (unit: h^{-1}) is the specific growth rate of the biomass given by Equation (3.13).

For the biomass constituents, we normally use the biomass as the reference, i.e., their concentrations are given with the biomass as the basis. In this case the mass balance for the ith biomass constituent is derived from (sterile feed is assumed):

$$\frac{d(X_{macro,i}xV)}{dt} = r_{macro,i}xV - F_{out}X_{macro,i}x \tag{3.41}$$

where $X_{macro,i}x$ is the concentration of the ith biomass component in the bioreactor (unit: g L^{-1}) and $r_{macro,i}$ is the specific, net rate of formation of the ith biomass constituent. Rearrangement of Equation (3.41) gives

$$\frac{dX_{macro,i}}{dt} = r_{macro,i} - \left(\frac{F_{out}}{V} + \frac{1}{x}\frac{dx}{dt} + \frac{1}{V}\frac{dV}{dt}\right)X_{macro,i} \tag{3.42}$$

Again we have that for any mode of bioreactor operation:

$$D = \frac{F_{out}}{V} + \frac{1}{V}\frac{dV}{dt} \tag{3.43}$$

which, together with the mass balance (3.40) for the total biomass concentration, gives the mass balance

$$\frac{dX_{\text{macro,i}}}{dt} = r_{\text{macro,i}} - \mu X_{\text{macro,i}} \tag{3.44}$$

where $X_{\text{macro,i}}$ is the concentration of the ith biomass constituent within the biomass. Different units may be applied for the concentrations of the biomass constituents, but they are normally given as g (g DW)$^{-1}$, because then the sum of all the concentrations equals 1, i.e.,

$$\sum_{i=1}^{Q} X_{\text{macro,i}} = 1 \tag{3.45}$$

Furthermore, this unit corresponds with the experimentally determined macromolecular composition of cells, where weight fractions are generally used. In Equation (3.44) it is observed that the mass balance for the biomass constituents is completely independent of the mode of operation of the bioreactor, i.e., the dilution rate does not appear in the mass balance. However, there is indirectly a coupling via the last term, which accounts for dilution of the biomass constituents when the biomass expands due to growth. Thus, if there is no net synthesis of a macromolecular pool, but the biomass still grows, the intracellular level decreases.

For intracellular metabolites it is not convenient to use the same unit for their concentrations as for the biomass constituents. These metabolites are dissolved in the matrix of the cell; therefore, it is more appropriate to use the unit moles per liquid cell volume for the concentrations. The intracellular concentration can then be compared directly with the affinities of enzymes, typically quantified by their Km values, which are normally given with the unit moles L^{-1}. If the concentration is known in one unit, it is, however, easily converted to another unit if the density of the biomass (in the range of 1 g cell per mL cell) and the water content (in the range of 0.67 ml water per ml cell) are known. Even though a different unit is applied, the biomass is still the basis and the mass balance for the intracellular metabolites therefore takes the same form:

$$\frac{dX_{\text{met,i}}}{dt} = r_{\text{met,i}} - \mu X_{\text{met,i}} \tag{3.46}$$

where X_{met},i is the concentration of the ith intracellular metabolite. It is important to distinguish between concentrations of intracellular metabolites given in moles per liquid reactor volume and in moles per liquid cell volume. If concentrations are given in the former unit the mass balance will be completely different.

3.3.2 *The batch reactor*

This is the classical operation of the bioreactor, and many life scientists use it because it can be carried out in a relatively simple experimental setup. Batch experiments have the advantage of being easy to perform and by using shake flasks a large number of parallel experiments can be carried out. The disadvantage is that the experimental data are difficult to interpret because there are dynamic conditions throughout the experiment, i.e., the environmental conditions experienced by the cells vary with time. However, by using well-instrumented bioreactors at least some variables, e.g., pH and dissolved oxygen tension, may be kept constant.

As mentioned in the previous section the dilution rate is zero for a batch reactor and the mass balances for the biomass and the limiting substrate therefore take the form

$$\frac{dx}{dt} = \mu x \qquad x(t=0) = x_0 \tag{3.47}$$

$$\frac{dc_s}{dt} = -r_s x \qquad c_s(t=0) = c_{s,0} \tag{3.48}$$

where x_0 indicates the initial biomass concentration, which is obtained immediately after inoculation, and $c_s,0$ is the initial substrate concentration. According to the mass balance, the biomass concentration will increase as indicated in Figure 3.3

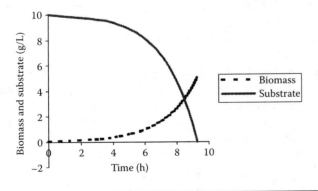

Figure 3.3
Batch fermentation described with Monod kinetics. The biomass concentration is found using Equation (c2) and the corresponding substrate concentration is found from Equation (c1). μ_{max} is 0.5 h^{-1}, K_s is 50 mg l^{-1} (a quite high value), and Y_{sx} is 0.5 g g^{-1}. The initial substrate concentration $c_s,0$ is 10 g l^{-1}. The substrate concentration decreases from 0.5 g l^{-1} to zero in less than 5 min, and this is the interesting substrate concentration range for estimation of K_s.

and the substrate concentration will decrease until its concentration reaches zero and growth stops. Because the substrate concentration is zero at the end of the cultivation, the overall yield of biomass on the substrate can be found from

$$Y_{sx}^{overall} = \frac{x_{final} - x_0}{c_{s,0}}$$ (3.49)

where x final is the biomass concentration at the end of the cultivation. Normally $X_0 \ll x_{final}$, and the overall yield coefficient can therefore be estimated from the final biomass concentration and the initial substrate concentration alone. Notice that the yield coefficient determined from the batch experiment is the overall yield coefficient and not Y_{sx} or $(Y_{xs}^{true})^{-1}$. The yield coefficient Ysx may well be time-dependent as it is the ratio between the specific growth rate and the substrate uptake rate [see Equation (3.16)]. However, if there is little variation in these rates during the batch culture (e.g., if there is a long exponential growth phase and only a very short declining growth phase) the overall yield coefficient may be very similar to the yield coefficient. The true yield coefficient, on the other hand, is difficult to determine from batch cultivation, because it requires information about the maintenance coefficients, which can hardly be determined from a batch experiment. However, in batch cultivation the specific growth rate is close to its maximum throughout most of the growth phase, and the substrate consumption due to maintenance is therefore negligible. According to Equation (3.17), the true yield coefficient is close to the observed yield coefficient determined from the final biomass concentration.

3.3.3 *The chemostat*

A typical operation of the continuous bioreactor is the so-called **chemostat**, where the added medium is designed such that there is a **single limiting substrate**. This allows for controlled variation in the specific growth rate of the biomass. The advantage of the continuous bioreactor is that a steady state can be achieved, which allows for precise experimental determination of specific rates. Furthermore, by varying the feed flow rate to the bioreactor the environmental conditions can be varied, and valuable information concerning the influence of the environmental conditions on the cellular physiology can be obtained. The continuous bioreactor is attractive for industrial applications because the productivity can be high. However, often the titer, i.e., the product concentration, is lower than can be obtained in the fed-batch reactor, and it is therefore a tradeoff between productivity and titer. Furthermore, it is rarely used in industrial processes because it is sensitive to contamination, e.g., via the feed stream, and to the appearance of spontaneously formed mutants that may out-compete the production strain. Other examples of continuous operation besides the chemostat are the **pH-stat**, where the feed flow is adjusted to maintain the pH constant in the bioreactor, and the **turbidostat**, where the feed flow is adjusted to maintain the biomass concentration at a constant level.

From the biomass mass balance (3.40), it can easily be seen that in a steady-state continuous reactor, the specific growth rate equals the dilution rate:

$$\mu = D \tag{3.50}$$

Thus, by varying the dilution rate (or the feed-flow rate) in a continuous culture, different specific growth rates can be obtained. This allows detailed physiological studies of the cells when they are grown at a predetermined specific growth rate (corresponding to a certain environment experienced by the cells). At steady state the substrate mass balance (3.38) gives

$$0 = -r_s x + D\left(c_s^f - c_s\right) \tag{3.51}$$

which, upon combination with Equation (3.50) and the definition of the yield coefficient, directly gives

$$x = Y_{sx}\left(c_s^f - c_s\right) \tag{3.52}$$

Thus, the yield coefficient can be determined from measurement of the biomass and the substrate concentrations in the bioreactor (it is assumed that the substrate concentration in the feed flow is known).

Besides the advantage for obtaining steady-state measurements the chemostat is well suited to study dynamic conditions, because it is possible to perform well-controlled transients. Thus, it is possible to study the cellular response to a sudden increase in the substrate concentration by adding a pulse of the limiting substrate to the reactor or to a sudden change in the dilution rate. These experiments both start and end with a steady state; so the initial and end conditions are well characterized, and this facilitates the interpretation of the cellular response. One type of transient experiment is especially suited to determining an important kinetic parameter – namely the maximum specific growth rate. By increasing the dilution rate to a value above μmax the cells will wash out from the bioreactor and the substrate concentration will increase (and eventually reach the same value as in the feed). After adaptation of the cells to the new conditions, they will attain their maximum specific growth rate and the dynamic mass balance for the biomass becomes

$$\frac{dx}{dt} = \left(\mu_{max} - D\right)x \tag{3.53}$$

or

$$\frac{x(t-t_0)}{x(t_0)} = \exp\left((\mu_{max} - D)(t-t_0)\right) \tag{3.54}$$

where t_0 is the time at which the cells have become adapted to the new conditions and grow at their maximum specific growth rate. Thus, the maximum specific growth rate is easily determined from a plot of the biomass concentration versus time on a semi-log plot.

3.3.4 *The fed-batch reactor*

This operation is probably the most common in industrial processes, because it allows for control of the environmental conditions, e.g., maintaining the glucose concentration at a certain level, as well as enabling formation of very high titers (up to several hundred grams per L of some metabolites), which is important for subsequent downstream processing. For a fed-batch reactor the mass balances for biomass and substrate are given by Equations (3.38) and (3.40). Normally, the feed concentration is very high, i.e., the feed is a very concentrated solution, and the feed flow is low, giving a low dilution rate.

For the fed-batch reactor the dilution rate is given by

$$D = \frac{1}{V}\frac{dV}{dt} \tag{3.55}$$

To keep D constant, there needs to be an exponentially increasing feed flow to the bioreactor, which is normally practically impossible as it may lead to oxygen limitations. The feed flow is therefore adjusted/increased until limitations in the oxygen supply set in, at which point the feed flow is kept constant. This will give a decreasing specific growth rate. However, because the biomass concentration usually increases, the volumetric uptake rate of substrates (including oxygen) may be kept approximately constant. From the above it is quite clear that there may be many different feeding strategies in a fed-batch process, and optimization of the operation is a complex problem that is difficult to solve empirically. Even when a very good process model is available, calculation of the optimal feeding strategy is a complex optimization problem. In an empirical search for the optimal feeding policy, the two most obvious criteria are

- Keep the concentration of the limiting substrate constant.
- Keep the volumetric growth rate of the biomass (or uptake of a given substrate) constant.

A constant concentration of the limiting substrate is often applied if the substrate inhibits product formation, and the chosen concentration is therefore dependent on the degree of inhibition and the desire to maintain a certain growth of the cells. A constant volumetric growth rate (or uptake of a given substrate) is applied if there are limitations in the supply of oxygen or in heat removal.

Fed-batch cultures were used in the production of baker's yeast as early as 1915. The method was introduced by Dansk Gæringsindustri and is therefore sometimes referred to as the Danish method. It was recognized that an excess of malt in the medium would lead to a higher growth rate, resulting in an oxygen demand in excess of what could be met in the fermentors. This resulted in the development of respiratory catabolism of the yeast, leading to ethanol formation at the expense of biomass production. The yeast was allowed to grow in an initially weak medium to which additional medium was added at a rate less than the maximum rate at which the organism could use it. In modern fed-batch processes for yeast production, the feed of molasses is under strict control, based on the automatic measurement of traces of ethanol in the exhaust gas of the bioreactor. Although such systems may result in low growth rates, the biomass yield is generally close to the maximum obtainable, and this is especially important in the production of baker's yeast, where there is much focus on the yield. Apart from the production of baker's yeast, the fed-batch process is used today for the production of secondary metabolites (where penicillin is the most prominent group of compounds), industrial enzymes, and many other products derived from cultivation processes.

3.4 Kinetic models

Kinetic modeling expresses the verbally or mathematically expressed correlation between rates and reactant/product concentrations that, when inserted into the mass balances derived in Section 3.3, permits a prediction of the degree of conversion of substrates and the yield of individual products at other operating conditions. If the rate expressions are correctly set up, it may be possible to express the course of an entire fermentation experiment based on initial values for the components of the state vector, e.g., concentration of substrates. This leads to simulations, which may finally result in an optimal design of the equipment or an optimal mode of operation for a given system. The basis of kinetic modeling is to express functional relationships between the forward reaction rates $\mathbf{v}$ of the reactions considered in the model and the concentrations of the substrates, metabolic products, biomass constituents, intracellular metabolites and/or the biomass concentration:

$$v_i = f_i(c_s, c_p, X_{macro}, X_{met}, x) \qquad (3.56)$$

If, during the cultivation, the biomass composition remains constant, then the rates of the internal reactions must necessarily be proportional. This is referred to as **balanced growth.** In this case the growth process can be described in terms of a single variable that defines the state of the biomass. This variable is quite naturally chosen as the biomass concentration x (g DW l^{-1}). This is the basis of the so-called **unstructured models** that have proved adequate during fifty years of practical application to design cultivation processes (especially steady state or batch cultivations), to install suitable control devices, and to estimate which process conditions are likely to give the best return on the investment in process equipment. However, these unstructured models generally have poor predictive strength and as such are of little value in fundamental studies of cellular function.

3.4.1 *The degree of model complexity*

A typical discussion on the complexity of mathematical modeling of biochemical systems may be initiated by asking the question of whether a mechanistic model or an empirical model should be applied instead. To illustrate this consider the fractional saturation y of a protein at a ligand concentration cl. This may be described by the Hill equation (Hill, 1910):

$$y = \frac{c_l^{h}}{c_l^{h} + K} \tag{3.57}$$

where h and K are empirical parameters. Alternatively, the fractional saturation may be described by the equation of Monod *et al.* (1965):

$$y = \frac{\left[La\left(1 + \dfrac{ac_l}{K_R}\right)^3 + \left(1 + \dfrac{c_l}{K_R}\right)^3\right]\dfrac{c_l}{K_R}}{L\left(1 + \dfrac{ac_l}{K_R}\right)^4 + \left(1 + \dfrac{c_l}{K_R}\right)^4} \tag{3.58}$$

where L, a, and KR are parameters. Both equations address the same experimental problem, but whereas Equation (3.57) is completely empirical with h and K as fitted parameters, Equation (3.58) is derived from a hypothesis for the mechanism; the parameters therefore have a direct physical interpretation. If the aim of the modeling is to understand the underlying mechanism of the process, Equation (3.57) can obviously not be applied because the kinetic parameters are completely empirical and give no (or little) information about the ligand binding to the protein. In this case Equation (3.58) should be applied, because by estimating the kinetic parameter the investigator is supplied with valuable information about the system and the parameters can be directly interpreted.

If, on the other hand, the aim of the modeling is to simulate the ligand binding to the protein, Equation (3.57) may be as good as (3.58) – Equation (3.58) may even be preferable because it is simpler in structure and has fewer parameters and it actually often gives a better fit to experimental data than Equation (3.57). Thus, the answer to which model is preferred depends on the aim of the modeling exercise. The same can be said about the unstructured growth models (Section 3.4.2), which are completely empirical, but which are valuable for extracting key kinetic parameters for growth. Furthermore, they are well suited to simple design problems and for teaching.

If the aim is to simulate dynamic growth conditions, one may turn to simple structured models (Section 4.4.3), e.g., the compartment models, which are also useful for illustration of structured modeling in the classroom. However, if the aim is to analyze a given system in further detail, it is necessary to include far more structure in the model. In this case one often describes only individual processes within the cell, e.g., a certain pathway or gene transcription from a certain promoter. Similarly, if the aim is to investigate the interaction between different cellular processes, e.g., the influence of plasmid copy number on chromosomal DNA replication, a single-cell model (Section 3.4.4) has to be applied.

Finally, if the aim is to look into population distributions, which in some cases may have an influence on growth or production kinetics, either a segregated or a morphologically structured model has to be applied (Section 3.5).

3.4.2 *Unstructured models*

Even when there are many substrates, one of these substrates is usually limiting, i.e., the rate of biomass production depends exclusively on the concentration of this substrate. At low concentrations c_s of this substrate μ is proportional to c_s, but for increasing values of c_s an upper value μmax for the specific growth rate is gradually reached. This is the verbal formulation of the Monod (1942) model:

$$\mu = \mu_{max} \frac{c_s}{c_s + K_s} \tag{3.59}$$

which has been shown to correlate fermentation data for many different micro-organisms. In the Monod model K_s is that value of the limiting substrate concentration at which the specific growth rate is half its maximum value. Roughly speaking, it divides the μ versus c_s plot into a low substrate concentration range where the specific growth rate is strongly (almost linearly) dependent on c_s, and a high substrate concentration range, where μ is independent of c_s.

Table 3.5 Compilation of K_s values for sugars

Species	Substrate	K_s (mg l^{-1})
Aerobacter aerogenes	Glucose	8
Escherichia coli	Glucose	4
Klebsiella aerogenes	Glucose	9
	Glycerol	9
Klebsiella oxytoca	Glucose	10
	Arabinose	50
	Fructose	10
Lactococcus cremoris	Glucose	2
	Lactose	10
	Fructose	3

Values are taken from Nielsen and Villadsen (1994).

When glucose is the limiting substrate, the value of K_s is normally in the micromolar range (corresponding to the mg l^{-1} range), and it is therefore experimentally difficult to determine. Some of the K_s values reported in the literature are compiled in Table 3.5. It should be stressed that the K_s value in the Monod model does not represent the saturation constant for substrate uptake, but an overall saturation constant for the entire growth process.

Some of the most characteristic features of microbial growth are represented quite well by the Monod model:

- The constant specific growth rate at high substrate concentration
- The first-order dependence of the specific growth rate on substrate concentration at low substrate concentrations

In fact, one may argue that the two features that make the Monod model work so well in fitting experimental data are deeply rooted in any naturally occurring conversion process: the size of the machinery that converts substrate must have an upper value, and all chemical reactions will end up as first-order processes when the reactant concentration tends to zero. The satisfactory fit of the Monod model to many experimental data should never be misconstrued to mean that Equation (3.59) is a mechanism of fermentation processes. The Langmuir rate expression of heterogeneous catalysis and the Michaelis–Menten rate expression in enzymatic catalysis are formally identical to Equation (3.59), but the denominator constant has a direct physical interpretation in both cases (the equilibrium constant for dissociation of a catalytic site-reactant complex) whereas K_s in Equation (3.59) is no more than an empirical parameter used to fit the average substrate influence on all cellular reactions pooled into the single reaction by which substrate is converted to biomass.

In the Monod model it is assumed that the yield of biomass from the limiting substrate is constant, i.e., there is proportionality between the specific growth rate and the specific substrate uptake rate. The Monod model is, however, normally used together with a maintenance consumption of substrate, i.e., the specific substrate uptake is described by the linear relation (3.18). The Monod model including maintenance is probably the most widely accepted model for microbial growth, and it is well suited for analysis of steady-state data from a chemostat (see Example 3c). Often the model is combined with the Luedeking and Piret (1959) model for metabolite production in which the specific rate of product formation is given by Equation (3.17). The Luedeking and Piret model was derived on the basis of an analysis of lactic acid fermentation and is in principle only valid for metabolic products formed as a direct consequence of the growth process, i.e., metabolites of primary metabolism. However, the model may in some cases be applied to other products, e.g., secondary metabolites, but this should not be done automatically.

Example 3c: The Monod model

Despite its simplicity, the Monod model is very useful for extracting key growth parameters, and it generally fits simple batch fermentations with one exponential growth phase and steady-state chemostat cultures (but rarely with the same parameters). We first consider a batch process, where substrate consumption due to maintenance can usually be neglected. In this case there is an analytical solution to the mass balances for the concentrations of substrate and the limiting substrate (Nielsen and Villadsen, 1994):

$$c_s = c_{s,0} - Y_{xs}(x - x_0) \tag{3c1}$$

$$\mu_{max}t = \left(1 + \frac{K_s}{c_{s,0} + Y_{xs}x_0}\right)\ln\left(\frac{x}{x_0}\right)$$
$$- \frac{K_s}{c_{s,0} + Y_{xs}x_0}\ln\left(1 + \frac{Y_{sx}(x_0 - x)}{c_{s,0}}\right) \tag{3c2}$$

Using this analytical solution, it is in principle possible to estimate the two kinetic parameters in the Monod model, but since K_s generally is very low it is in practice not possible to estimate this parameter from a batch cultivation (see Figure 3.3).

For a steady–state, continuous culture, the mass balance for the biomass, together with the Monod model, gives

$$D = \mu_{max}\frac{c_s}{c_s + K_s} \tag{3c3}$$

or

$$c_s = \frac{DK_s}{\mu_{max} - D} \tag{3c4}$$

Thus, the concentration of the limiting substrate increases with the dilution rate. When substrate concentration becomes equal to the substrate concentration in the feed, the dilution rate attains its maximum value, which is often called the critical dilution rate:

$$D_{crit} = \mu_{max} \frac{c_s^f}{c_s^f + K_s} \tag{3c5}$$

When the dilution rate becomes equal to or larger than this value, the biomass is washed out of the bioreactor. Equation (3b4) clearly shows that the steady-state chemostat is well suited to studying the influence of the substrate concentration on the cellular function, e.g., product formation, because by changing the dilution rate it is possible to change the substrate concentration as the only variable. Furthermore, it is possible to study the influence of different limiting substrates on the cellular physiology, e.g., glucose and ammonia.

Besides quantification of the Monod parameters, the chemostat is well suited to determine the maintenance coefficient. Because the dilution rate equals the specific growth rate, the yield coefficient is given by

$$Y_{sx} = \frac{D}{Y_{xs}^{true} D + m_s} \tag{3c6}$$

or, if we use Equation (3.52),

$$x = \frac{D}{Y_{xs}^{true} D + m_s} \left(c_s^f - c_s\right) \approx \frac{D}{Y_{xs}^{true} D + m_s} c_s^f \tag{3c7}$$

because $c_s^f \gg c_s$ except for dilution rates close to the critical dilution rate. Equation (3c7) shows that the biomass concentration decreases at low specific growth rates, where the substrate consumption for maintenance is significant compared with that for growth. At high specific growth rates (high dilution rates) maintenance is negligible and the yield coefficient becomes equal to the true yield coefficient (Figure 3.4). By rearrangement of Equation (3c7), a linear relationship between the specific substrate uptake rate (given by and the dilution rate is found, and using this, the true yield coefficient and the maintenance coefficient can easily be estimated using linear regression.

It is unlikely that the Monod model can be used to fit all kinds of fermentation data. Many authors have tried to improve on the Monod model, but generally these empirical models are of little value. However, in some cases growth is limited either

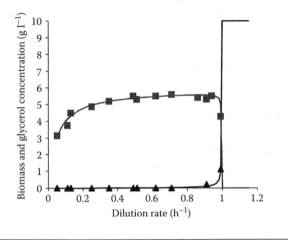

Figure 3.4

Growth of *Aerobacter aerogenes* in a chemostat with glycerol as the limiting substrate where the biomass concentration (■) decreases for increasing dilution rate due to the maintenance metabolism, and when the dilution rate approaches the critical value the biomass concentration decreases rapidly. The glycerol concentration (▲) increases slowly at low dilution rates, but when the dilution rate approaches the critical value it increases rapidly. The lines are model simulations using the Monod model with maintenance, and with the parameter values: cfs = 10 g L⁻¹; μmax = 1.0 h⁻¹; Ks = 0.01 g L⁻¹; ms = 0.08 g (g DW h)⁻¹; Y_{xs}^{true} = 1.70 g (g DW)⁻¹.

by substrate concentration or by the presence of a metabolic product, which acts as an inhibitor. In order to account for this, the Monod model is often extended with additional terms. Thus, for inhibition by high concentrations of the limiting substrate:

$$\mu = \mu_{max} \frac{c_s}{c_s^2 / K_i + c_s + K_s} \tag{3.60}$$

and for inhibition by a metabolic product

$$\mu = \mu_{max} \frac{c_s}{c_s + K_s} \frac{1}{1 + p / K_i} \tag{3.61}$$

Equations (3.60) and (3.61) may be a useful way of including product or substrate inhibition in a simple model, and often these expressions are also applied in connection with structured models. Extension of the Monod model with additional terms or factors should, however, be carried out with some restraint because the result may be a model with a large number of parameters but of little value outside the range in which the experiments were made.

3.4.3 Compartment models

Simple structured models are in one sense improvements to the
unstructured models, because some basic mechanisms of the
cellular behavior are at least qualitatively incorporated. Thus,
the structured models may have some predictive strength, i.e.,
they may describe the growth process at different operating conditions with the same set of parameters. But one should bear in
mind that "true" mechanisms of the metabolic processes are of
course not considered in simple structured models even if the
number of parameters is quite large.

In structured models all the biomass components are lumped
into a few key variables, i.e., the vectors $\mathbf{X}_{macro}$ and $\mathbf{X}_{met}$, which
are hopefully representative of the state of the cell. The micro-
bial activity thus becomes not only a function of the biotic variables, which may change with very small time constants, but
also of the cellular composition, and consequently of the "history" of the cells, i.e., the environmental conditions the cells
have experienced in the past.

The biomass can be structured in a number of ways. For
example, in simple structured models only a few cellular components are considered, whereas in highly structured models up
to twenty intracellular components are considered (Nielsen and
Villadsen, 1992). As discussed in Section 3.4.1, the choice of a
particular structure depends on the aim of the modeling exercise, but often one starts with a simple structured model onto
which more and more structures are added as new experiments
are added to the database. Even in highly structured models
many of the cellular components included in the model represent "pools" of different enzymes, metabolites, or other cellular components. The cellular reactions considered in structured
models are therefore empirical in nature because they do not
represent the conversion between "true" components. Consequently, it is permissible to write the kinetics for the individual
reactions in terms of reasonable empirical expressions, with a
form judged to fit the experimental data with a small number
of parameters. Thus, Monod-type expressions are often used
because they summarize some fundamental features of most
cellular reactions, i.e., being first order at low substrate concentration and zero order at high substrate concentration. Despite
their empirical nature, structured models are normally based on
some well-known cell mechanisms, and they therefore have the
ability to simulate certain features of experiments quite well.

The first structured models appeared in the late 1960s from
the group of Fredrickson and Tsuchiya at the University of Minnesota (Ramkrishna *et al.*, 1966, 1967; Williams, 1967), who
also were the first to formulate microbial models within a general mathematical framework similar to that used to describe
reaction networks in classical catalytic processes (Tsuchiya
et al., 1966; Fredrickson *et al.*, 1967; Fredrickson, 1976). Since

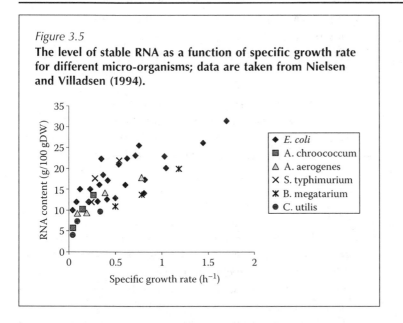

Figure 3.5
The level of stable RNA as a function of specific growth rate for different micro-organisms; data are taken from Nielsen and Villadsen (1994).

this pioneering work, many other simple structured models have been presented (Harder and Roels, 1982; Nielsen and Villadsen, 1992).

In these simple structured models, the biomass is divided into a few compartments. These compartments must be chosen with care, and cell components with similar functions should be placed in the same compartment, e.g., all membrane material and otherwise rather inactive components in one compartment, and all active material in another compartment. With the central role of the protein synthesizing system (PSS) in cellular metabolism, this is often a key component in compartment models. Besides a few enzymes, the PSS consists of ribosomes, which contain approximately 60% ribosomal RNA and 40% ribosomal protein. Because rRNA makes up more than 80% of the total stable RNA in the cell, the level of the ribosomes is easily identified through measurements of the RNA concentration in the biomass. As seen in Figure 3.5, the RNA content of many different micro-organisms increases linearly with the specific growth rate. Thus, the level of the PSS is well correlated with the specific growth rate. It is therefore a good representative of the activity of the cell, and this is the basis of most of the simple structured models (see Example 4d).

Example 3d: A two-compartment model

Nielsen *et al.* (1991a, b) presented a two-compartment model for the lactic acid bacterium *Lactococcus cremoris*. The model is a direct descendent of the model created by Williams (1967) with similar definitions for the following two compartments:

- Active (A) compartment contains the PSS and small building blocks.
- Structural and genetic (G) compartment contains the rest of the cell material.

The model considers both glucose and a complex nitrogen source (peptone and yeast extract), but in the following presentation we discuss the model with only one limiting substrate (glucose). The model considers two reactions for which the stoichiometry is

$$\gamma_{11}X_A - s = 0 \tag{3d1}$$

$$\gamma_{22}X_G - X_A = 0 \tag{3d2}$$

In the first reaction glucose is converted into small building blocks in the A compartment, and these are further converted into ribosomes. The stoichiometric coefficient γ_{11} can be considered as a yield coefficient because metabolic products (lactic acid, carbon dioxide, etc.) are not included in the stoichiometry. In the second reaction, building blocks present in the A compartment are converted into macromolecular components of the G compartment. In this process some by-products may be formed and the stoichiometric coefficient γ_{22} is therefore slightly less than 1. The kinetics of the two reactions have the same form, i.e.,

$$v_i = k_i \frac{c_s}{c_s + K_{s,i}} X_A \quad ; \quad i = 1, 2 \tag{3d3}$$

From Equation (3.13) the specific growth rate for the biomass is found to be

$$\mu = \begin{pmatrix} 1 & 1 \end{pmatrix} \begin{pmatrix} \gamma_{11} & -1 \\ 0 & \gamma_{22} \end{pmatrix} \begin{pmatrix} v_1 \\ v_2 \end{pmatrix} = \gamma_{11}v_1 - (1 - \gamma_{22})v_2 \tag{3d4}$$

or with the kinetic expression for $v1$ and $v2$ inserted:

$$\mu = \left(\gamma_{11}k_1 \frac{c_s}{c_s + K_{s,1}} - (1 - \gamma_{22})k_2 \frac{c_s}{c_s + K_{s,2}} \right) X_A \tag{3d5}$$

Thus, the specific growth rate is proportional to the size of the active compartment. The substrate concentration c_s influences the specific growth rate both directly and indirectly by determining the size of the active compartment. The influence of the substrate concentration on the synthesis of the active compartment can be evaluated through the ratio $r1/r2$:

$$\frac{r_1}{r_2} = \frac{k_1}{k_2} \frac{c_s + K_{s,2}}{c_s + K_{s,1}} \tag{3d6}$$

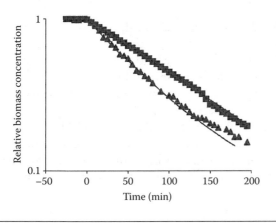

Figure 3.6
Measurement of biomass concentration, two transient experiments, of *L. cremoris* growing in glucose-limited chemostat; the dilution rate was shifted from an initial value of 0.10 h^{-1} (▲) or 0.50 h^{-1} (■) to 0.99 h^{-1}, respectively. The biomass concentration is normalized by the steady-state biomass concentration before the step change, which was made at time zero. The lines are model simulations. The data are taken from Nielsen et al. (1991b).

If $K_{s,1}$ is larger than $K_{s,2}$, the formation of XA is favored at high substrate concentration, and it is thus possible to explain the increase in the active compartment with the specific growth rate. Consequently, when the substrate concentration increases rapidly, there are two effects on the specific growth rate:

- A rapid increase in the specific growth rate, which is a result of mobilization of excess capacity in the cellular synthesis machinery.
- A slow increase in the specific growth rate, which is a result of a slow build-up of the active part of the cell, i.e., additional cellular synthesis machinery has to be formed in order for the cells to grow faster.

This is illustrated in Figure 3.6, which shows the biomass concentration in two independent wash-out experiments. In both cases the dilution rate was shifted to a value (0.99 h^{-1}) above the critical dilution rate (0.55 h^{-1}), but in one experiment the dilution rate before the shift was low (0.1 h^{-1}) and in the other experiment it was high (0.5 h^{-1}). The wash-out profile is seen to be very different, with a much faster wash-out when there was a shift from a low dilution rate. When the dilution rate is shifted to 0.99 h^{-1} the glucose concentration increases rapidly to a value much higher than $K_{s,1}$ and $K_{s,2}$, and this allows growth at the maximum rate. However, when the cells come from a low dilution rate, the size of the active compartment is not sufficiently large to allow rapid growth, and XA therefore has to be built up before the maximum specific growth rate is attained. On the other hand, if the cells come from a high dilution rate, XA is already high and the cells immediately attain their maximum specific growth rate. It is observed that the model is able to correctly describe the two experiments (all parameters were estimated from steady-state experiments), and the model correctly incorporates information about the previous history of the cells.

The model also includes the formation of lactic acid; the kinetics were described using a rate equation similar to Equation (3d3). Thus, the lactic acid formation increases when the activity of the cells increases, and so it is ensured that there is a close coupling between formation of this primary metabolite and growth of the cells.

It is interesting to note that even though the model does not include a specific maintenance reaction, it can actually describe a decrease in the yield coefficient of biomass on glucose at low specific growth rates. The yield coefficient is given by

$$Y_{sx} = \gamma_{21}\left(1 - \left(1 - \gamma_{22}\right)\frac{k_2}{k_1}\frac{c_s + K_{s,1}}{c_s + K_{s,2}}\right) \tag{3d7}$$

Because $K_{s,1}$ is larger than $K_{s,2}$ the last term within the parentheses decreases for increasing specific growth rates, and the yield coefficient will therefore also increase for increasing substrate concentration.

3.4.4 Single-cell models

Single-cell models are in principle an extension of the compartment models, but with the description of many different cellular functions. Furthermore, these models depart from the description of a population and focus on the description of single cells. This allows consideration of characteristic features of the cell, and it is therefore possible to study different aspects of cell function:

- Cell geometry can be accounted for explicitly and so it is possible to examine its potential effects on nutrient transport.
- Temporal events during the cell cycle can be included in the model, and the effect of these events on the overall cell growth can be studied.
- Spatial arrangements of intracellular events can be considered, even though this would lead to significant model complexity.

To set up single-cell models, it is necessary to have a detailed knowledge of the cell, and single-cell models have therefore only been described for well-studied cellular systems such as *Escherichia coli*, *Saccharomyces cerevisiae*, *Bacillus subtilis*, and human erythrocytes. The most comprehensive single-cell model is the so-called Cornell model set up by Shuler and co-workers (Shuler *et al.*, 1979), which contains twenty intracellular components. This model predicts a number of observations made with *E. coli*, and it has formed the basis for setting up several other models (Nielsen and Villadsen, 1992). Thus, Peretti and Bailey (1987) extended the model to describe plasmid replication and gene expression from a plasmid inserted into a host cell. This allowed study of host–plasmid interactions, especially the effect of copy number, promoter strength, and ribosome binding site strength on the metabolic activity of the host cell and on the plasmid gene expression.

3.4.5 *Molecular mechanistic models*

Despite the level of detail, the single-cell models are normally based on an empirical description of different cellular events, e.g., gene transcription and translation. This is a necessity because the complexity of the model would become very high if all these individual events were to be described with detailed models that include mechanistic information. In many cases it is, however, interesting to study these events separately, and for models where mechanistic information is included, they have to be used. These models are normally set up at the molecular level, and they can therefore be referred to as molecular mechanistic models. Many different models of this type can be found in the literature, but most fall in one of two categories:

- Gene transcription models
- Pathway models

Gene transcription models aim at quantifying gene transcription based on knowledge of the promoter function. The *lac*-promoter of *E. coli* is one of the best studied promoter systems of all, and so this system has been modeled most extensively. Furthermore, this promoter (or its derivatives) is often used in connection with the production of heterologous proteins by this bacterium, because it is an inducible promoter. In a series of papers, Lee and Bailey (1984a–d) presented an elegant piece of modeling of this system, and through combination with a model for plasmid replication they could investigate, for example, the role of point mutations in the promoter on gene transcription. This promoter system is quite complex, with both activator and repressor proteins, and empirical investigation can therefore be laborious; the detailed mathematical model is a valuable tool to guide the experimental work.

In pathway models the individual enzymatic reactions of a given pathway are described with enzyme kinetic models, and it is therefore possible to simulate the metabolite pool levels and the fluxes through different branches of the pathway. These models have mainly concentrated on glycolysis in *S. cerevisiae* (Gallazo and Bailey, 1990; Rizzi *et al.*, 1997), because much information about enzyme regulation is available for this pathway. However, complete models are also available for other pathways, e.g., the penicillin biosynthetic pathway (Pissarra *et al.*, 1996). These pathway models are experimentally verified by comparing modeling simulations with measurements of intracellular metabolite pool levels — something that has only been possible with sufficient precision in the last couple of years, because it requires rapid quenching of the cellular activity and sensitive measurement techniques.

Pathway models are very useful in studies of metabolic fluxes, because they allow quantification of the control of flux by the individual enzymes in the pathway. This can be done by calculation of sensitivity coefficients (or the so-called flux control coefficients; see Box 3.4), which quantify the relative importance of the individual enzymes in the control of flux through the pathway (Stephanopoulos *et al.*, 1998).

A general criticism of the application of kinetic models for complete pathways is that despite the level of detail included, they cannot possibly include all possible interactions in the system and therefore only represent one model of the system. The

Box 3.4   **Quantification of flux control**

In a study of flux control in a biochemical pathway the concept of metabolic control analysis is very useful (Stephanopoulos et al., 1998). Here the flux control of the individual enzymatic reactions on the steady-state flux *J* through the pathway is quantified by the so-called flux control coefficients (FCC):

$$C_i^J = \frac{v_i}{J}\frac{dJ}{dv_i}$$

where v_i is the rate of the *i*th reaction. If the enzyme concentration of the *i*th enzyme is increased, its rate will normally increase, and a higher flux through the pathway may be the result. However, it is likely that due to allosteric regulation (or other regulation phenomena) there may be a very small effect of increasing the enzyme concentration. This is exactly what is quantified by the FCC, i.e., the relative increase in the steady-state flux upon a relative increase in the enzyme activity. Clearly a step with a high FCC has a large control of the flux through the pathway. If a kinetic model is available for the pathway, the flux control coefficients for each step can easily be calculated using model simulations. The FCCs can also be determined experimentally by changing the enzyme concentration (or activity) genetically, by titration with the individual enzymes, or by adding specific enzyme inhibitors (Stephanopoulos et al., 1998).

robustness of the model is extremely important especially if the kinetic model is to be used for predictions, and unfortunately most biochemical models, even very detailed models, are only valid at operating conditions close to those where the parameters have been estimated, i.e., the predictive strength is limited. For analysis of complex systems it is, however, not necessary that the model gives a quantitatively correct description of all the variables, because even models that give a qualitatively correct description of the most important interactions in the system may be valuable in studies of flux control.

3.5 Population models

Normally it is assumed that the population of cells is homogeneous, i.e., all cells behave identically. Although this assumption is certainly crude if a small number of cells is considered, it gives a very good picture of certain properties of the cell population because there are billions of cells per ml medium (see Box 3.5).

Furthermore, the kinetics is often linear in the cellular properties, e.g., in the concentration of a certain enzyme, and the overall population kinetics can therefore be described as a function of the average property of the cells (Nielsen et al., 2003). There are, however, situations where cell property distributions influence the overall culture performance and here it is necessary to consider the cellular property distribution, and this is done in the so-called **segregated models**. In the following we will discuss two approaches to segregated modeling.

Box 3.5   **Deterministic versus stochastic modeling**

In a description of cellular kinetics macroscopic balances are normally used, i.e., the rates of the cellular reactions are functions of average concentrations of the intracellular components. However, living cells are extremely small systems with only a few molecules of certain key components, and it does not really make sense to talk about "the DNA concentration in the cell," for example, because the number of macromolecules in a cell is always small compared with Avogadro's number. Many cellular processes are therefore stochastic in nature and the deterministic description often applied is in principle not correct. However, the application of a macroscopic (or deterministic) description is convenient and it represents a typical engineering approximation for describing the kinetics in an average cell

in a population of cells. This approximation is reasonable for large populations because the standard deviation from the average "behavior" in a population with elements is related to the standard deviation for an individual cell through

$$\sigma_{pop} = \frac{\sigma}{\sqrt{e}}$$

Thus, with a population of 10^9 cells ml^{-1}, which is a typical cell concentration during a cultivation process, it can be seen that the standard deviation for the population is very small. There are, however, systems where small populations occur, e.g., at dilution rates close to the maximum in a chemostat, and here one may have to apply a stochastic model.

3.5.1 *Morphologically structured models*

The simplest approach to model distribution in the cellular property is by the so-called morphologically structured models (Nielsen and Villadsen, 1992; Nielsen, 1993). Here the cells are divided into a finite number Q of cell states Z (or morphological forms), and conversion between the different cell states is described by a sequence of empirical metamorphosis reactions. Ideally these metamorphosis reactions can be described as a set of intracellular reactions, but the mechanisms behind most morphological conversions are largely unknown. Thus, it is not known why filamentous fungi differentiate to cells with a completely different morphology from that of their origin. It is therefore not possible to set up detailed mechanistic models describing these changes in morphology; empirical metamorphosis reactions are therefore introduced. The stoichiometry of the metamorphosis reactions is given by analogy with Equation (3.4):

$$\Delta Z = 0 \tag{3.62}$$

where Δ is a stoichiometric matrix. Z_q represents both the qth morphological form and the fractional concentration (g q^{th}, morphological form (g DW^{-1}). With the metamorphosis reactions one morphological form is spontaneously converted to other forms. This is of course an extreme simplification because the conversion between morphological forms is the sum of many small changes in the intracellular composition of the cell. With the stoichiometry in equation (3.62) it is assumed that the metamorphosis reactions do not involve any change in the total mass, and the sum of all stoichiometric coefficients in each reaction is therefore taken to be zero. The forward reaction rates of the metamorphosis reactions are collected in the vector **u**. Each morphological form may convert substrates to biomass components and metabolic products. These reactions may be described by an intracellularly structured model, but in order to reduce the model complexity a simple unstructured model is used for description of the growth and product formation of each cell type, e.g., the specific growth rate of the qth morphological form is described by the Monod model. The specific growth rate of the total biomass is given as a weighted sum of the specific growth rates of the different morphological forms:

$$\mu = \sum_{i=1}^{Q} \mu_i Z_i \tag{3.63}$$

The rate of formation of each morphological form is determined both by the metamorphosis reactions and by the growth-associated reactions for each form (for derivation of mass balances for the morphological forms, see Nielsen and Villadsen, 1994). The concept of morphologically structured models is well suited to describing the growth and differentiation of filamentous micro-organisms (Nielsen, 1993), but it may also be used to describe other microbial systems where a cellular differentiation has an impact on the overall culture performance.

3.5.2 Population balance equations

The first example of a heterogeneous description of cellular populations was presented in 1963 by Fredrickson and Tsuchiya. In their model, single-cell growth kinetics was combined with a set of stochastic functions describing cell division and cell death. The model represents the first application of a completely segregated description of a cell population. In the model the cell population is described by a number density function $f(X,t)$, where $f(X,t)dX$ is the number of cells with property X being in the interval X to $X + dX$. The dynamic balance for $f(X,t)$ is given by

$$\frac{\partial f(X,t)}{\partial t} + \frac{\partial}{\partial X}\left(f(X,t)v(X,t)\right) =$$

$$2\int_X^\infty b(X^*,t)p(X^*,X,t)f(X^*,t)dX^* \qquad (3.64)$$

$$-b(X,t)f(X,t) - Df(X,t)$$

where v is the net rate of formation of the cell property, X. $b(X,t)$ is the breakage function, i.e., the rate of cell division for cells with property X, and $p(X^*,X,t)$ is the partitioning function, i.e., the probability that a cell with property X is formed upon division of a cell with property X^*. Through the functions p and b a stochastic element can be introduced into the model, but these functions can also be completely descriptive. The balance equation (3.64) was applied in the original work of Fredrickson and Tsuchiya, but in a later paper a general framework for segregated population models was presented (Fredrickson *et al.*, 1967). Segregated models represent the complete description of a cell population and they take into account that all cells in a population are not identical. However, complete cellular segregation is rarely applied in cell culture models for two main reasons:

- For large populations, the average properties will normally represent the overall population kinetics quite well.
- The mathematical complexity of Equation (3.64) is quite substantial, especially if more than a single-cell property is considered, i.e., the number density function becomes multidimensional.

If the kinetics for product formation is not zero or first order in a given cell property, application of an average property model will, however, not give the same result as a segregated description. This is the case for production of a heterologous protein in plasmid-containing cells of *E. coli*, where the product formation kinetics is not first order in the plasmid copy number. A segregated model therefore has to be applied to give a good description of the product formation kinetics (Seo and Bailey, 1985). The simplest segregated models are when the cellular property is described by a single variable, e.g., cell age, and in Example 3e, the age distribution of an exponentially growing culture is derived from the general balance (3.64).

Example 3e: Age distribution model

The simplest segregated population models are those where the cellular property is taken to be described solely by the cell age *a*. In this case the rate of increase in the cellular property $v(a,t)$ is equal to 1. Furthermore, if it is assumed that cell division occurs only at a certain cell age $a = td$, the two first terms on the right-hand side of Equation (3.64) become equal to zero. At steady state the balance therefore becomes

$$\frac{d\phi(a)}{da} = -D\phi(a) \tag{3e1}$$

where ϕ is a normalized distribution function:

$$\phi(a) = \frac{f(a)}{n} \tag{3e2}$$

with *n* being the total cell number [given as the zero moment of the number density function *f(a)*]. The solution to Equation (3e1) is

$$\phi(a) = \phi(0)e^{-Da} \tag{3e3}$$

Due to the normalization the zeroth moment of $\phi(a)$ is 1, i.e.,

$$\int_{o}^{t_d} \phi(a)da = 1 \tag{3e4}$$

which leads to

$$\phi(a) = \frac{D}{1 - e^{-Dt_d}} e^{-Da} \tag{3e5}$$

The cell balance relating to cell division (the so-called renewal equation) is given by

$$\phi(0) = 2\phi(t_d) \tag{3e6}$$

which together with Equation (3e3) directly gives Equation (3.2). Furthermore, when Equation (3.2) is inserted in Equation (3e5), we have the simpler expression:

$$\phi(a) = 2De^{-Da} \tag{3e7}$$

Thus, the fraction of cells with a given age decreases exponentially with age, and the decrease is determined by the specific growth rate of the culture (equal to the dilution rate at steady state). The average cell age is given as the first moment of $\phi(a)$:

$$\langle a \rangle = \int_0^{t_d} a\phi(a)\,da = \frac{1 - \ln 2}{D} \tag{3e8}$$

Consequently, the average age of the cells decreases for increasing specific growth rates.

Summary

- Understanding microbial growth kinetics is an essential requirement for the design and successful operation of industrial fermentation processes and for obtaining quantitative information about the function of microbial cells.
- The primary objective of this chapter is to introduce the reader to the basic principles of the wide ranging aspects of microbial growth kinetics and dynamics; from the basic principles to the more advanced concept of modeling.
- Based on the information given in this chapter, the reader should be able to design fermentation processes for the production of biomass and microbially derived products.
- Furthermore, the reader should be able to set up simple mathematical models describing microbial growth as well as evaluate more complex mathematical models.

References

Benthin, S., U. Schulze, J. Nielsen, J. Villadsen (1994). Growth energetics of *Lactococcus cremoris* FDI during energy, carbon and nitrogen limitation in steady state and transient cultures, *Chem. Eng. Sci.* **49**, 589–609.

Christensen, L. H., C. M. Henriksen, J. Nielsen, J. Villadsen, M. Egel-Mitani (1995). Continuous cultivation of *P. chrysogenum*. Growth on glucose and penicillin production, *J. Biotechnol.*, **42**, 95–107.

Fredrickson, A.G. (1976). Formulation of structured growth models, *Biotechnol. Bioeng.*, **18**, 1481–1486.

Fredrickson, A. G., H. M. Tsuchiya (1963). Continuous propagation of micro-organisms, *AIChE J.*, **9**, 459–468.

Fredrickson, A. G., D. Ramkrishna, H. M. Tsuchiya (1967). Statistics and dynamics of procaryotic cell populations, *Math. Biosci.*, **1**, 327–374.

Galazzo, J. L., J. E. Bailey (1990). Fermentation pathway kinetics and metabolic ßux control in suspended and immobilized *Saccharomyces cerevisiae*, *Enzym. Microb. Technol.*, **12**, 162–172.

Harder, A., J. A. Roels. (1982). Application of simple structured models in bioengineering, *Adv. Biochem. Eng.*, **21**, 55–107.

Herbert, D. (1959). Some principles of continuous culture. *Recent Prog. Microbiol.*, **7**, 381–396.

Hill, A. V. (1910). The possible effects of the aggregation of the molecules of haemoglobin on its dissociation curves, *J. Physiol. Lond.*, **40**, 4–7.

Ingraham, J. L., O. Maaloe, F. C. Neidhardt (1983). *Growth of the Bacterial Cell*, Sunderland, MA: Sinauer Associates.

Lee, S. B., J. E. Bailey (1994a). A mathematical model for λdv plasmid replication: analysis of wild-type plasmid, *Gene*, **11**, 151–165.

Lee, S. B., J. E.. Bailey (1994b). A mathematical model for λdv plasmid replication: Analysis of copy number mutants, *Gene*, **11**, 166–177.

Lee, S. B., J. E. Bailey (1994c). Genetically structured models for *lac* promoter-operator function in the *Escherichia coli* chromosome and in multicopy plasmids: *lac* operator function, *Biotechnol. Bioeng.*, **26**, 1372–1382.

Lee, S. B., J. E. Bailey (1994d). Genetically structured models for *lac* promoter-operator function in the *Escherichia coli* chromosome and in multicopy plasmids: *lac* promoter function, *Biotechnol. Bioeng.*, **26**, 1383–1389.

Luedeking, R., E. L. Piret (1959). A kinetic study of the lactic acid fermentation. Batch Process at controlled pH. *J. Biochem, Microbiol. Technol. Eng.*, **1**, 393–412.

Meyenburg, K. von (1969). Katabolit-Repression und der Sprossungszyklus von *Saccharomyces cerevisiae*. Dissertation, ETH, Zürich.

Monod, J. (1942). Recherches *sur* la Croissance des Cultures Bacteriennes, Paris: Hermann and Cie.

Monod, J., J. Wyman, J.-P. Changeux (1965). On the nature of allosteric transitions: A plausible model, *J. Molec. Biol.*, **12**, 88–118.

Neidhardt, F. C., J. L. Ingraham, M. Schaechter (1990). *Physiology of the Bacterial Cell. A Molecular Approach*, Sunderland, MA: Sinauer Associates.

Nielsen, J. (1993). A simple morphologically structured model describing the growth of Þlamentous micro-organisms, *Biotechnol, Bioeng.*, **41**, 715–727.

Nielsen, J., J. Villadsen (1992). Modeling of microbial kinetics, *Chem. Eng. Sci.*, **47**, 4225–4270.

Nielsen, J., J. Villadsen (1993). Bioreactors: Description and modelling, in Rehm, H.J. and Reed, G. (eds.) *Biotechnology*, Vol. 3, 2nd ed, Weinheim: VCR Verlag, pp. 77–104.

Nielsen, J., J. Villadsen (1994). *Bioreaction Engineering Principles*, New York: Plenum Press.

Nielsen, J., J. Villadsen, G. Lidén (2003). *Bioreaction Engineering Principles*. 2nd ed. New York. Kluwer Academic/Plenum Publishers.

Nielsen, J., K. Nikolajsen, J. Villadsen (1991a). Structured modeling of a microbial system 1. A theoretical study of the lactic acid fermentation, *Biotechnol. Bioeng.*, **38**, 1–10.

Nielsen, J., K. Nikolajsen, J. Villadsen (1991b). Structured modelling of a microbial system 2. Verification of a structured lactic acid fermentation model, *Biotechnol. Bioeng.*, **38**, 11–23.

Olsson, L., J. Nielsen (1997). On-line and in situ monitoring of biomass in submerged cultivations, *TIBTECH*, **15**, 517–522.

Oura, E. (1983). Biomass from carbohydrates, in Rehm, H.-J. and Reed, G. (eds.) *Biotechnology*, Vol. 3, 2nd ed, Weinheim: VCR Verlag, pp. 3–42.

Peretti, S. W., J. E. Bailey (1987). Simulations of host–plasmid interactions in *Escherichia coli*: Copy number, promoter strength, and ribosome binding site strength effects on metabolic activity and plasmid gene expression, *Biotechnol. Bioeng.*, **29**, 316–328.

Pirt, S. J. (1965). The maintenance energy of bacteria in growing cultures, *Proc. Roy. Soc. London, Ser. B.*, **163**, 224–231.

Pissarra, P. N., J. Nielsen, M. J. Bazin (1996). Pathway kinetics and metabolic control analysis of a high-yielding strain of *Penicillium chrysogenum* during fed-batch cultivations, *Biotechnol. Bioeng.*, **51**, 168–176.

Ramkrishna, D., A. G. Fredrickson, H. M. Tsuchiya (1966). Dynamics of microbial propagation: Models considering endogenous metabolism, *J. Gen. Appl. Microbiol.*, **12**, 311–327.

Ramkrishna, D., A. G. Fredrickson, H. M. Tsuchiya. (1967). Dynamics of microbial propagation: Models considering inhibitors and variable cell composition, *Biotechnol. Bioeng.*, **9**, 129–170.

Rizzi, M., M. Baltes, U. Theobald, M. Reuss (1997). In vivo analysis of metabolic dynamics in *Saccharomyces cerevisiae*: II. Mathematical model, *Biotechnol. Bioeng.*, **55**, 592–608.

Roels, J. A. (1983). *Energetics and Kinetics in Biotechnology*, Amsterdam: Elsevier Biomedical Press.

Seo, J. H., J. E. Bailey (1985). A segregated model for plasmid content and product synthesis in unstable binary fission recombinant organisms, *Biotechnol. Bioeng.*, **27**, 156–165.

Shuler, M. L., S. K. Leung, C. C. Dick (1979). A mathematical model for the growth of a single bacterial cell, *Ann. NY Acad. Sci.*, **326**, 35–55.

Sonnleitner, B., A. Fiechter (1988). High performance bioreactors: A new generation, *Anal. Chim. Acta,* **213**, 199–205.

Stephanopoulos, G., J. Nielsen, A. Aristodou (1998). *Metabolic Engineering,* San Diego: Academic Press.

Tsuchiya, H. M., A. G. Fredrickson, R. Aris (1966). Dynamics of microbial cell populations, *Adv. Chem. Eng.*, **6**, 125–206.

Williams, F. M. (1967). Av model of cell growth dynamics, *J. Theor. Biol.*, **15**, 190–207.

4

Microbial Synthesis of Primary Metabolites: Current Advances and Future Prospects

A.L. Demain and Sergio Sanchez

4.1 Introduction

Primary metabolites are microbial products made during the exponential phase of growth whose synthesis is an integral part of the normal growth process. They include intermediates and end products of anabolic metabolism, which are used by the cell as building blocks for essential macromolecules (e.g., amino acids, nucleotides) or are converted to coenzymes (e.g., vitamins). On the other hand, primary metabolites of catabolic metabolism (e.g., citric acid, acetic acid, and ethanol) are not biosynthetic precursors but their production, which is related to energy production and substrate utilization, is essential for growth. Industrially, the most important primary metabolites are amino acids, nucleotides, vitamins, solvents, and organic acids. These are made by a diverse range of bacteria and fungi and have numerous applications in the food, chemical, pharmaceutical, and nutraceutical industries. Many of these metabolites are manufactured by microbial fermentation rather than chemical synthesis because the fermentations are economically competitive and produce biologically active isomers. Several other industrially important chemicals could be manufactured via microbial fermentations (e.g., glycerol and other polyhydroxy alcohols) but are presently synthesized cheaply as petroleum by-products. However, since the cost of petroleum has skyrocketed recently, there is renewed interest in the microbial production of ethanol, organic acids, and solvents.

FEEDBACK
INHIBITION
A mechanism in which
the end product of a
given pathways acts
as an inhibitor of an
earlier reaction in the
pathway. Generally, the
end product acts as a
noncompetitive inhibitor
of the first committed
step in the pathway.

4.2   Control of primary metabolism

Microbial metabolism is a conservative process that usually
does not expend energy or nutrients to make compounds already
available in the environment and does not overproduce components of intermediary metabolism. Coordination of metabolic
functions ensures that, at any given moment, only the necessary
enzymes, and the correct amount of each, are made. Once a
sufficient quantity of a material is made, the enzymes concerned
with its formation are no longer synthesized and the activities of
preformed enzymes are curbed by a number of specific regulatory control mechanisms such as feedback inhibition and covalent modifications.

4.2.1   *Induction*

While the majority of anabolic enzymes are subject to repression,
most catabolic enzymes are subject to induction. The latter is a
control mechanism by which a substrate (or a compound structurally or metabolically related to the substrate), "turns on" the
synthesis of the enzymes required for its uptake and initiation
of metabolism. Enzymes that are synthesized as a result of genes
being turned "on" in response to signal molecules (substrates)
are called inducible enzymes with the substance that activates
gene transcription being referred to as the inducer. Inducible
enzymes are produced only in response to the presence of their
substrate or related compound and, in a sense, are produced
only when needed. The inducer molecule renders the repressor
molecule unable to bind to the operator region, thus facilitating the binding of RNA polymerase to the promoter region and
in turn initiation of transcription and translation into protein.
Although most inducers are substrates of catabolic enzymes,
products can sometimes function as inducers. For example,
fatty acids induce lipase while galacturonic acid induces polyga-
lacturonase. In some cases, however, the synthesis of a particular enzyme may be induced by substrate analogues that are not
attacked by the enzyme, e.g., IPTG in the case of β-galactosidase; such analogues are referred to as gratuitous inducers.

COVALENT
MODIFICATION
A regulatory control
mechanism in which
covalent binding of a
certain chemical, e.g.,
phosphate, activates
or inactivates the
enzyme. For example,
the phosphorylation of
isocitrate dehydrogenase
in *E. coli*, a reversible
reaction, renders the
enzyme catalytically
inactive as it makes
the enzyme unable
to bind its substrate,
a consequence of the
negative charge carried
on phosphoserine113
and the conformational
changes associated
with it.

4.2.2   *Catabolite repression*

Like enzyme induction, carbon source regulation (more commonly known as carbon catabolite repression, CCR) is one of
the conservative mechanisms that safeguards against wasting a
cell's protein-synthesizing machinery and operates when more
than one utilizable substrate is present in the environment. The
cell produces enzymes to catabolize the assimilated carbon
source while synthesis of enzymes utilizing other substrates is
repressed until the primary substrate is exhausted. The repressed
enzymes are usually inducible.

4.2.3 *Nitrogen source regulation (NSR)*

Nitrogen can be assimilated from inorganic or organic sources. Its assimilation from inorganic sources requires reduction to ammonia, followed by incorporation into intracellular metabolites. The appropriate distribution of nitrogen among various pathways usually involves specific or local regulatory mechanisms, such as end product inhibition or end product-mediated transcriptional control. In addition, some global regulators control the expression of genes from several pathways and thereby coordinate metabolism. The ability to assimilate particular inorganic or organic nitrogen sources depends on the particular organism. Organic nitrogen sources are usually monomeric units of macromolecules (e.g., amino acids or nucleic acid bases) or compounds derived from them (e.g., agmatine or putrescine). Ammonia usually supports the fastest growth rate and is therefore considered the preferred nitrogen source for *E. coli*. The biochemical basis of this "ammonium preference" is explained by the repression of enzymes acting on the alternative nitrogenous substrates present in the culture medium. NSR is known by many other names such as nitrogen metabolite repression, nitrogen catabolite repression, and ammonia repression. Enzymes typically under such control are proteases, amidases, ureases, and those that degrade amino acids.

4.2.4 *Phosphorus source regulation*

In natural environments, inorganic phosphorus is commonly the major growth-limiting nutrient. Thus, biological systems have evolved a variety of responses to modulate their phosphorus requirement or to optimize its utilization. In *E. coli*, over thirty genes are part of the phosphate regulon (*pho* regulon) and are transcriptionally activated by phosphorylated PhoB when the level of phosphate in the medium drops below a certain threshold. PhoR and PhoB comprise a two-component signal transduction system in which PhoR catalyses the reversible phosphorylation/activation of PhoB in response to low and high levels of phosphate, respectively. PhoR autophosphorylates and transfers the phosphate to PhoB. The environmental concentration of phosphate is monitored by the periplasmic phosphate-binding protein PstS, which transmits the signal for excess phosphate across the cytoplasmic membrane via PstC, PstA, PstB, and PhoU to PhoR. Phosphorylated PhoB binds to the promoters of thirty-one genes containing *pho* boxes and interacts with RNA polymerase allowing initiation of mRNA synthesis.

Nucleases and phosphatases are usually repressed by phosphate in fungi. In addition, phosphate represses proteases, isocitrate lyase, fructose diphosphate aldolase, NADP isocitrate dehydrogenase, and malate dehydrogenase in *Neurospora*. Phosphate also suppresses the production of riboflavin by *Eremothecium*

REGULON
In eukaryotes, a genetic unit consisting of a noncontiguous group of genes under the control of a single regulator gene. In bacteria, regulons are global regulatory systems involved in the interplay of pleiotropic regulatory domains of several genes or operons.

PERIPLASMIC
The periplasmic space is the space seen between the plasma membrane and the outer membrane in the Gram-negative bacteria; a smaller periplasmic space may be observed in the Gram-positive bacteria.

PHO BOX
The phosphate (*pho*) box is a consensus sequence shared by the regulatory regions of the genes of the pho regulon. For example, in *E. coli* it consists of eighteen nucleotides formed by two direct repeats of seven nucleotides (C/T)TGTCAT separated by four adenines or thymines. The phosphorylated PhoP protein (PhoP~P) activates, in response to phosphate starvation, expression of the *pho* regulon genes.

LIVERPOOL JOHN MOORES UNIVERSITY
LEARNING SERVICES

CHELATOR
A substance that binds
particular ions, removing
them from solution, e.g.,
EDTA is a chelator of
divalent cations such as
Mg++.

PLEIOTROPIC
MUTATION
 A mutation that
affects several different
characters/phenotypes.

TRANSCRIPTIONAL
ACTIVATOR
 Any transcription factor
required for initiation
or up-regulation of
transcription.

DEREPRESSION
A process/mechanism
by which repression of
a single or a group of
genes can be overcome or
relieved.

ashbyii. Phosphate-derepressed mutants can be selected by growth with a phosphate ester (e.g., β-glycerol phosphate) as the sole source of carbon in the presence of high phosphate.

Of great interest is inorganic polyphosphate (poly P), a linear polymer that carries numerous orthophosphate (P_i) residues linked by high-energy phosphoanhydride bonds. Poly P is found in cells of all bacteria, archaea, fungi, protozoa, plants, and animals and is prominent in many organisms. It is produced by polyphosphate kinase (PPK), which catalyzes the reversible transfer of the terminal phosphate of ATP to form a long-chain polyphosphate. The *E. coli* gene (*ppk*) encoding PPK has been cloned, sequenced, and overexpressed (about 100-fold). The poly P plays a significant role in metabolism including acting as a substitute for ATP as an energy source, reservoir for inorganic phosphate, chelator of metal ions, and regulator for stress and survival. It is fair to say, however, that the polyphosphate roles with respect to nutritional stringencies, environmental stress and stationary phase adaptations are the most widely recognized.

4.2.5 Sulfur source regulation

Sulfatases are regulated by sulfate and sulfur amino acids. As the case for other nutrients, the use of sulfur is controlled by one or a few pleiotropic transcriptional regulatory proteins. Thus, in *E. coli*, sulfur metabolism is controlled by the CysB transcriptional activator. The cysteine regulon includes most of the genes required for synthesis of cysteine and genes for uptake of sulfur sources such as L-cystine, sulfate, thiosulfate, and taurine. Transcriptional activation of these genes requires CysB protein, N-acetyl-L-serine as an inducer and sulfur limitation. CysB's activity is regulated by an efflux pump specific for cysteine and related metabolites. CysB is also an autorepressor, i.e., capable of repressing its own expression at the transcriptional level. Similarly, carbohydrate metabolism/fermentations were found to be adversely affected by mutations in *cysB* genes. The aforementioned effects are exerted at the transcriptional level and are partially reversible by exogenous cAMP or sulfur-containing substrates, e.g., cysteine or djenkolate. Cysteine inhibits inducer synthesis, resulting in maximal repression of the sulfur regulon, while growth under sulfur limitation or with poor sulfur sources, such as glutathione, enhances expression (derepression) of the sulfur regulon.

In *Neurospora crassa*, sulfate uptake is an important point of regulation of sulfur metabolism as it is subject to sulfur (metabolite) repression in which excess sulfate turns "off" the expression of sulfate permease encoding genes. Also, structural genes coding for aryl sulfatase, choline sulfatase, sulfate permeases I and II, a high-affinity methionine permease, and an extracellular protease are turned "on" when sulfur becomes limiting.

4.2.6 Feedback regulation

The most important mechanism responsible for regulation of anabolic enzymes involved in the biosynthesis of amino acids, nucleotides, and vitamins is not induction or nutrient repression, but feedback regulation. This category of regulation functions at two levels: enzyme action (feedback inhibition) and enzyme synthesis (feedback repression and attenuation).

In feedback inhibition, the final metabolite of a pathway, when present in sufficient quantities, inhibits the action of the first enzyme of the pathway, thus preventing the synthesis of unwanted intermediates on the one hand and the wasting of energy on the other. Feedback repression involves the turning "off" of enzyme synthesis when the amount of the product has been made in sufficient quantities to satisfy the biosynthetic demands. The end product of the pathway acts as a co-repressor. The apo-repressor specified by the regulator gene is inactive in the absence of its co-repressor and as such is unable to bind to the operator region. However, in the presence of a co-repressor, the inactive apo-repressor is converted to an active repressor that binds to the operator region, thus preventing the binding of RNA polymerase to the promoter region, which in turn brings enzyme biosynthesis to a halt.

4.2.7 Additional types of regulation

Other types of regulation include stringent control and regulatory inactivation. The effector of stringent control is the alarmone guanosine 5'-diphosphate 3'-diphosphate (ppGpp). Regulatory inactivation refers to the selective inactivation of enzymes by two different mechanisms. In "modification inactivation," the enzyme remains intact but its physical state is changed or it is covalently modified. Covalent modifications include phosphorylation of a specific serine or threonine residue, nucleotidylation of a specific tyrosine residue, ADP-ribosylation of an arginine residue, methylation of a glutamate or aspartate carboxyl group, acetylation of an ε-amino group of a lysine residue or tyrosinolation of a protein's terminal carboxyl group. In "degradative inactivation," at least one peptide bond is broken; it may represent the first step in protein turnover. It is carried out by proteases, which are restricted from nonselective action by confinement in vacuoles or by protease inhibitors. Regulatory inactivation usually occurs after the exponential phase of growth, especially after exhaustion of a source of carbon or nitrogen. This inactivation serves to prevent futile cycles of metabolism,

ATTENUATION
A mechanism for controlling gene expression. Typically transcription is terminated after initiation but before a full-length mRNA is produced.

CO-REPRESSOR
The metabolite that when bound to the repressor (of a repressible operon) forms a functional unit that can bind to its operator and block transcription.

APO-REPRESSOR
A metabolite that is capable of forming a functional unit with an inactive repressor thus rendering it able to bind to the operator region and, in turn, preventing transcription.

STRINGENT CONTROL
A translational control mechanism in prokaryotes that represses tRNA and rRNA synthesis during amino acid starvation.

ALARMONE
Substance that alerts
the cells to the onset
of metabolic stress and
subsequently regulating
gene expression and/or
enzyme function.

Figure 4.1

**Overproduction of an intermediate of a linear primary
metabolic pathway. Feedback inhibition by the end product
interferes with the activity of enzyme a and feedback
repression interferes with formation of enzymes a and
b. By making a genetic block (mutation) at enzyme c, an
auxotrophic mutant is made which cannot grow unless E is
added to the medium. As long the amount of E present is not
excessive, there will be no feedback effects and C will be
overproduced.**

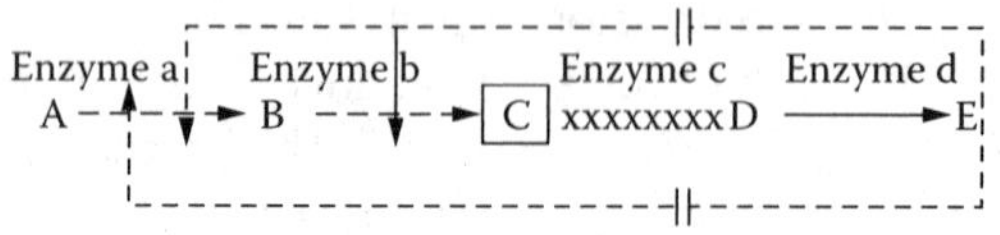

DEGRADATIVE
INACTIVATION
A limited degradation/
inactivation of the
catalytic subunit in a
given enzyme catalyzed
through the activity of
another.

to destroy enzymes no longer needed, and to divert carbon flow
at branch points from one branch to another.

4.3 Approaches to strain improvements

Organisms used today for industrial production of primary
metabolites have been developed by programs of intensive mutagenesis followed by screening and selection of overproducers.
Such efforts often start with organisms having some capacity
to make the desired product but that require multiple mutations
leading to deregulation in a particular biosynthetic pathway
before high productivity can be obtained. Auxotrophic mutants
are often very useful (Figures 4.1 and 4.2). The sequential mutations ensure that nutrients are channeled efficiently to the appropriate products without significant deviation to other pathways.
These mutations involve not only release of feedback controls
but also enhancement of the formation of pathway precursors
and intermediates. This approach to strain improvement has
been remarkably successful in producing organisms that make
industrially significant concentrations of primary metabolites.
However, some of the problems with this "brute force" approach
include (i) the necessity of screening large numbers of mutants
for the rare combination of traits sequentially obtained that lead
to overproduction, and (ii) the possibility that the vigor of the
producing strain may be substantially weakened following several rounds of mutagenesis.

FUTILE CYCLE
A combination of two
or more biochemical
reactions resulting
only in the hydrolysis
of ATP or other high-
energy compounds, thus
"wasting" energy.

AUXOTROPHIC
MUTANT
A mutant strain of
micro-organism that will
proliferate only when the
medium is supplemented
with a specific substance
not required by wild-type
organisms.

More recent approaches employ recombinant DNA technology to develop strains that are capable of overproducing primary metabolites. This rationale for strain construction relies
largely on the same principles of regulation discussed in the
previous sections, but aims at assembling the appropriate characteristics by means of *in vitro* recombinant DNA techniques.

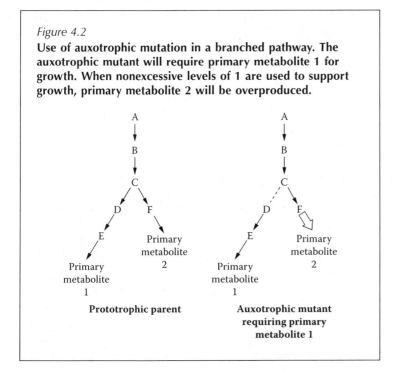

Figure 4.2
Use of auxotrophic mutation in a branched pathway. The auxotrophic mutant will require primary metabolite 1 for growth. When nonexcessive levels of 1 are used to support growth, primary metabolite 2 will be overproduced.

This is particularly valuable in organisms with complex regulatory systems, where deregulation would involve many genetic alterations.

Production of a particular primary metabolite by deregulated organisms may inevitably be limited by the inherent capacity of the particular organism to make the appropriate biosynthetic enzymes, i.e., even in the absence of repressive mechanisms, there may not be enough of the enzyme made to obtain high productivity. One way to overcome this is to increase the number of copies of structural genes coding for these enzymes by genetic engineering. Another way often used in combination with this strategy is to increase the frequency of transcription, which is related to the frequency of binding of RNA polymerase to the promoter region. The former can be achieved by cloning the biosynthetic genes *in vitro* into a plasmid that, when introduced into the cell through transformation, will replicate into multiple copies. Increasing the frequency of transcription involves the construction of a recombinant plasmid *in vitro* that contains the structural genes of the biosynthetic enzymes but lacks the regulatory sequences (promoter and operator) normally associated with them. Instead, the structural genes are cloned downstream of an efficient promoter, thus facilitating a higher level of expression. The ideal plasmid for metabolite synthesis would contain a regulatory region with a constitutive phenotype, preferably not subject to nutritional repression.

CONSTITUTIVE PHENOTYPE
A phenotype that is expressed all the time irrespective of growth conditions.

GENOME-BASED STRAIN RECONSTRUCTION
Involves identifying mutations by comparative genomic analysis, defining mutations beneficial for production, and assembling them in a single wild-type background.

MAGNETIZATION TRANSFER
A method for determining kinetics of chemical exchange by perturbing the magnetization of nuclei in a particular site or sites and following the rate at which magnetic equilibrium is restored. The most common perturbations are saturation and inversion, and the corresponding techniques are often called "saturation transfer" and "selective inversion-recovery."

MOLECULAR BREEDING
Identification and evaluation of useful traits using marker-assisted selection.

DNA SHUFFLING
A method for *in vitro* homologous recombination of pools of selected mutant genes by random fragmentation and polymerase chain reaction (PCR) reassembly.

SITE DIRECTED MUTAGENESIS
A method that introduces specific base-pair mutations into a gene.

Novel genetic technologies such as "genome-based strain reconstruction" achieve the construction of a superior strain that contains only mutations crucial to hyperproduction, but not other unknown mutations that accumulate by brute-force mutagenesis and screening. This approach has successfully been used to improve lysine production (see Section 4.4.1.2).

The directed improvement of product formation or cellular properties via modification of specific biochemical reactions or introduction of new ones with the use of recombinant DNA technology is known as "metabolic engineering." Analytical methods are combined to quantify fluxes and to control them with molecular biological techniques in order to implement suggested genetic modifications. Different means of analyzing flux are (i) kinetic-based models, (ii) control theories, (iii) tracer experiment, (iv) magnetization transfer, (v) metabolite balancing, (vi) enzyme analysis, and (vii) genetic analysis. The overall flux through a metabolic pathway depends on several steps, not just a single rate-limiting reaction (see chapter on flux control analysis).

A genome-wide transcript expression analysis called "massive parallel signature sequencing" has been successfully used to discover new targets for further improvement of riboflavin production by the fungus *Ashbya gossypii* (see Section 5.4.3.2). The development and combined application of the above technologies will help to develop "inverse metabolic engineering," which in turn will be used to construct certain phenotypes that are ideal for commercial purposes.

Molecular breeding techniques such as "DNA shuffling" come closer to mimicking natural recombination by allowing *in vitro* homologous recombination. These techniques not only recombine DNA fragments but also introduce point mutations at a very low but controlled rate. Unlike site directed mutagenesis, this method of pooling and recombining parts of similar genes from different species or strains has yielded remarkable improvements in enzymes in a very short amount of time. "Whole genome shuffling" is a novel technique for strain improvement combining the advantage of multi-parental crossing allowed by DNA shuffling with the recombination of entire genomes.

4.4   Production of primary metabolites

4.4.1   *Amino acids production*

Production of amino acids amounted to 2.3 million tons in 2002. Produced by fermentation are 1 million tons of L-glutamate, 600,000 tons of L-lysine-HCL, 20,000 tons of L-threonine, 13,000 tons of L-phenylalanine (including that made by chemical synthesis), 1,300 tons of L-glutamine, 1,200 tons of L-arginine, 600 tons of L-tryptophan (including enzymatic method), 500 tons of L-valine, 500 tons of L-leucine (including extraction), 400 tons of L-isoleucine (including extraction), 400 tons of L-histidine, 350 tons of L-proline, 300 tons of L-serine, and 165 tons of L-tyrosine. Ten thousand tons of L-aspartic acid and 500 tons of L-alanine are made enzymatically. DL-Methionine is made chemically at 400,000 tons per year. In 2004, the amino acid market was about 3.5 billion dollars.

Top fermentation titers reported in the literature are as follows: 88 g l^{-1} glutamic acid, 140 g l^{-1} L-lysine-HCL, 100 g l^{-1} L-threonine, 50 g l^{-1} L-phenylalanine, 96 g l^{-1} L-arginine, 58 g l^{-1} L-tryptophan, 99 g l^{-1} L-valine, 34 g l^{-1} L-leucine, 30 g l^{-1} L-isoleucine, 42 g l^{-1} L-histidine, 100 g l^{-1} L-proline, 65 g l^{-1} L-serine, and 75 g l^{-1} L-alanine. Genetic and metabolic engineering have made an impact by use of the following strategies: (i) amplification of the rate-limiting (controlling) enzyme of the pathway; (ii) amplification of the first enzyme after a branch point; (iii) cloning of a gene encoding an enzyme with more or less feedback regulation; (iv) introduction of a gene encoding an enzyme with a functional or energetic advantage as replacement for the normal enzyme; (v) amplification of the first enzyme leading from central metabolism to increase carbon flow into the pathway followed by sequential removal of bottlenecks caused by accumulation of intermediates. Transport mutations are also useful, i.e., mutations decreasing amino acid uptake often allow for improved excretion and lower intracellular feedback control. In cases where excretion is carrier-mediated, increase in activity of these carrier enzymes increases production of the amino acid.

Amino acids produced by microbial process are the L-forms. Such stereospecificity makes the process advantageous as compared to synthetic processes. Microbial strains employed for amino acid production are divided into four classes, i.e., wild-type strains, auxotrophic mutants, regulatory mutants, and auxotrophic regulatory mutants. Using bacterial mutants, all the essential amino acids except L-methionine can be produced by "direct fermentation" from cheap carbon sources such as carbohydrate materials or acetic acid.

Plasmid vector systems for cloning in *Corynebacterium glutamicum* have been established and amino acid production by *C. glutamicum* and related strains has been improved by gene cloning. Extensive research has been done on sequencing the genome of *C. glutamicum* and to investigate its genetic repertoire. The genome has been sequenced by several groups.

4.4.1.1 *Production of L-glutamic acid*

Monosodium glutamate (MSG) is a potent flavor enhancer, which was first made by fermentation in Japan in the late 1950s. Many organisms belonging to a wide range of taxonomically related genera including *Micrococcus*, *Corynebacterium*, *Brevibacterium*, and *Microbacterium* are capable of overproducing

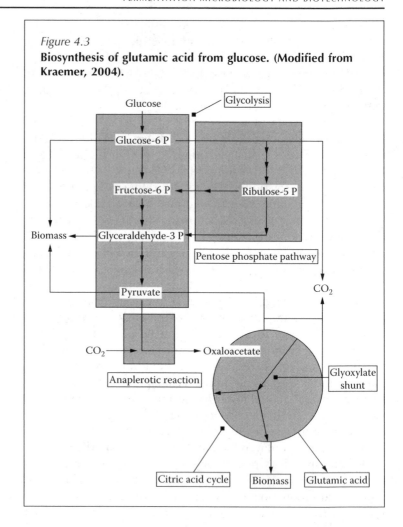

Figure 4.3

Biosynthesis of glutamic acid from glucose. (Modified from Kraemer, 2004).

glutamate. *Brevibacterium lactofermentum* and *Brevibacterium flavum* are now reclassified as subspecies of *C. glutamicum*. These organisms were shown to possess the Embden-Meyerhof Parnas glycolytic pathway (EMP), the hexose monophosphate pathway (HMP), the tricarboxylic acid (TCA) cycle, and the glyoxylate bypass (Figure 4.3). The TCA cycle, also widely known as the Krebs cycle, requires a continuous replenishment of oxaloacetate in order to replace the intermediates withdrawn for the synthesis of biomass and other amino acids. During growth on glucose and other glycolytic intermediates, the anaplerotic function is fulfilled by phosphoenolpyruvate carboxylase and pyruvate carboxylase.

Normally, glutamic acid overproduction would not be expected to occur due to feedback regulation. Glutamate feedback controls include repression of PEP carboxylase, citrate synthase, and NADP-glutamate dehydrogenase; the last-named enzyme is also inhibited by glutamate. However, by decreasing the effectiveness of the barrier to outward passage, glutamate can be pumped out

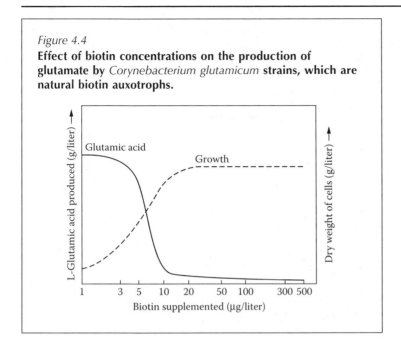

Figure 4.4

Effect of biotin concentrations on the production of glutamate by Corynebacterium glutamicum **strains, which are natural biotin auxotrophs.**

of the cell, thus allowing its biosynthesis to proceed unabated. The excretion of glutamate frees the glutamate pathway from feedback control until a very high level is accumulated; commercial L-glutamate titer is in excess of 80 g l^{-1}.

Glutamate excretion is intentionally influenced by manipulations of growth conditions; biotin limitation was the first means discovered to bring about glutamate overproduction in C. glutamicum; all glutamate overproducers are natural biotin auxotrophs (Figure 4.4). The finding that the addition of penicillin to cells grown in high biotin resulted in excretion of glutamic acid (Figure 4.5) led workers to postulate (i) that growth of the glutamate-overproducing bacterium in presence of nonlimiting levels of biotin results in a cell membrane permeability barrier that restricts the excretion of glutamate and (ii) that inhibition of cell wall biosynthesis by penicillin alters the permeability properties of the cell membrane and allows glutamate to flow out easily. The commonality in the various manipulations that were found to bring about high-level production of L-glutamic acid, i.e., limitation of biotin, addition of penicillin or fatty acid surfactants (e.g., tween 60) to exponentially growing cultures was recognized. Apparently, all of these manipulations result in a phospholipid-deficient cytoplasmic membrane, which favors active excretion of glutamate from the cell. This view was further substantiated by the discoveries that oleate limitation of an oleate auxotroph and glycerol limitation of a glycerol auxotroph also bring about glutamate excretion. Furthermore, glutamate-excreting cells were later found to have a very low level of cell lipids; especially phospholipids. In addition, it was found that

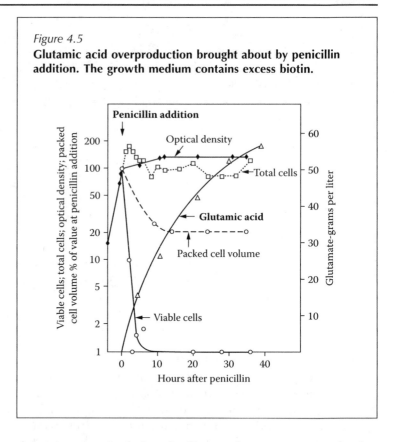

Figure 4.5

Glutamic acid overproduction brought about by penicillin addition. The growth medium contains excess biotin.

the various manipulations leading to glutamate overproduction cause increased permeability of the mycolic acid layer of the cell wall. The glutamate-overproducing bacteria are characterized by a special cell envelope containing mycolic acids which surrounds the entire cell as a structured layer and is thought to be involved in permeation of solutes. The mycolic acids esterified with arabinogalactan and the noncovalently bound mycolic acid derivatives form a second lipid layer of the cell; with the cytoplasmic membrane being the first. Overexpression or inactivation of enzymes that are involved in lipid synthesis alters the chemical and physical properties of the cytoplasmic membrane and changes glutamate efflux dramatically.

4.4.1.2 Production of L-lysine

The bulk of the cereals consumed in the world are deficient in the amino acid, L-lysine. This is an essential ingredient for the growth of animals and is an important part of a billion-dollar animal feed industry. Lysine supplementation converts cereals into balanced food or feed for animals including poultry, swine, and other livestock. In addition to animal feed, lysine is used in pharmaceuticals, dietary supplements, and cosmetics.

Figure 4.6
Biosynthetic pathways of L-lysine, L-threonine, and L-isoleucine productions.

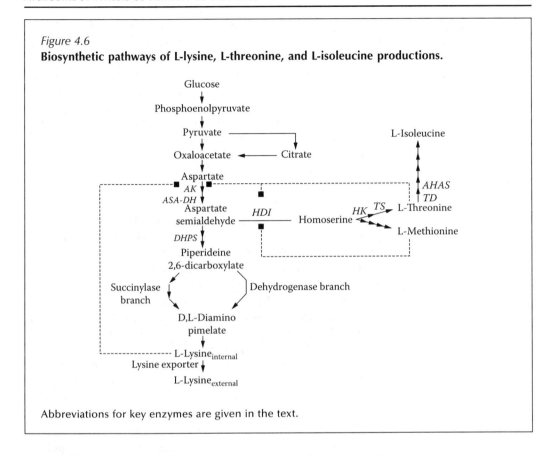

Abbreviations for key enzymes are given in the text.

Lysine is a member of the aspartate family of amino acids (Figure 4.6). It is made in bacteria by a branched pathway that also produces methionine, threonine, and isoleucine.

This pathway is controlled very tightly in organisms such as *E. coli*, which contains three aspartate kinases (AKs), each of which is regulated by a different end product. In addition, after each branch point, the initial enzymes are inhibited by their respective end product(s) and no overproduction usually occurs. However, *C. glutamicum*, the organism used for the commercial production of *L-lysine*, contains a single AK that is regulated via concerted feedback inhibition by threonine plus lysine. The relative contribution of carbon flux through the pentose phosphate pathway varies depending on the amino acid being produced, e.g., while it only contributes 20% of the total flux in the case of glutamate formation, it contributes 60–70% in the case of lysine production. This is evidently due to the high level of NADPH required for lysine formation. Use of rDNA technology has shown that the factors that significantly limit the overproduction of lysine are (i) the feedback inhibition of AK by lysine plus threonine (ii) the low level of dihydrodipicolinate synthase (DHPS), (iii) the low level of PEP carboxylase, and (iv) the low level of aspartase.

ANTIMETABOLITE
A drug that is very
similar to natural
chemicals in a normal
biochemical reaction
in cells but different
enough to interfere with
the normal division and
functions of cells.

Much work has been done on auxotrophic and regulatory
mutants of the glutamate-overproducing strains for the production of lysine. By genetic removal of homoserine dehydrogenase
(HDI), a glutamate-producing wild-type *Corynebacterium*
strain was converted into a lysine-overproducing mutant that
cannot grow unless methionine and threonine are added to the
medium. As long as the threonine supplement is kept low, the
intracellular concentration of threonine is limiting and feedback inhibition of AK is bypassed, leading to excretion of over
70 g l^{-1} of lysine in culture fluids. In some strains, addition of
methionine and isoleucine to the medium led to the increase
of lysine overproduction. Selection for S-2-aminoethylcysteine
(AEC; thialysine) resistance blocks feedback inhibition of AK.
Other antimetabolites useful for deregulation of AK include a
mixture of α-ketobutyrate and aspartate hydroxamate. Leucine
auxotrophy can also increase lysine production. L-Lysine titers
are known to be as high as 140 g l^{-1}.

Excretion of lysine by *C. glutamicum* is by active transport reaching a concentration of several hundred millimolar in the external
medium. Lysine, a cation, must be excreted against the membrane
potential (outside is positive) and excretion is carrier-mediated. The
system is dependent on electron motive force, not ATP.

Genome-based strain reconstruction has been used to improve
the lysine production rate of *C. glutamicum* by comparing a high
producing strain (production rate slightly less than 2 g l^{-1} h^{-1}) and
a wild-type strain. Comparison of sixteen genes from the production strain, encoding enzymes of the pathway from glucose
to lysine, revealed mutations in five of the genes. Introduction
of three of these mutations (*hom*, *lysC*, and *pyc* encoding HDI,
AK, and pyruvate carboxylase, respectively) into the wild type
created a new strain that produced 80 g l^{-1} in 27 hours, at a rate
of 3 g l^{-1} h^{-1}, the highest rate ever reported for a lysine fermentation. An additional increase (15%) in *L-lysine* production was
observed by introduction of a mutation in the 6-phosphoglu-
conate dehydrogenase gene (*gnd*). Enzymatic analysis revealed
that the mutant enzyme was less sensitive than the wild-type
enzyme to allosteric inhibition by intracellular metabolites. Iso-
tope-based metabolic flux analysis demonstrated that the *gnd*
mutation resulted in an 8% increase in carbon flux through the
pentose phosphate pathway during *L-lysine* production.

4.4.1.3 *Production of L-threonine*

This amino acid is the second major amino acid used for feeding
pigs and poultry. The pathway of threonine biosynthesis is similar in all micro-organisms (Figure 4.6). Starting from L-aspar-
tate, the pathway involves five steps catalyzed by five enzymes:
AK, aspartate-semialdehyde dehydrogenase (ASA-DH), HDI,
homoserine kinase (HK), and threonine synthetase (TS).

Production of L-threonine has been achieved with the use of several micro-organisms. In *Serratia marcescens,* construction of a high threonine producer was done by transductional crosses that combined several feedback control mutations into one organism. Three classes of mutants were obtained from the parental strain as the source of genetic material for transduction: (i) one strain in which both the threonine-regulated AK and HD were resistant to inhibition by threonine. It was selected on the basis of β-hydroxynorvaline resistance; (ii) a second strain, also selected for β-hydroxynorvaline resistance, in which HDI was resistant to both inhibition and repression and the threonine-regulated AK was constitutively synthesized; and (iii) a third strain that was resistant to thialysine, in which the lysine-regulated AK was resistant to feedback inhibition and repression. Since at least one of the three key enzymes in threonine synthesis was still subject to regulation in these strains, each produced only modest amounts of threonine (4.1 to 8.7 g l^{-1}). Recombination of the three mutations by transduction yielded a strain that produced higher levels of threonine (25 g l^{-1}), had AK and HDI activities that were resistant to feedback regulation by threonine and lysine, and that was also a methionine bradytroph (leaky auxotroph). Another six regulatory mutations derived by resistance to amino acid analogues were combined into a single strain of *S. marcescens* by transduction. These mutations led to desensitization and derepression of AKs I, II, and III and HDIs I and II. The resulting transductant produced 40 g l^{-1} of threonine, which was further improved to 63 g l^{-1} through overexpression of PEP carboxylase.

In *E. coli,* threonine production was increased to 76 g l^{-1} by conventional mutagenesis and selection/screening techniques. Of major importance were mutations to decrease both regulation of the pathway and degradation of the amino acid. An *E. coli* fed-batch process with methionine and phosphate feeding yielded 98 g l^{-1} L-threonine at 60 hours. Another *E. coli* strain was developed via mutation and genetic engineering and optimized by inactivation of threonine dehydratase (TD), resulting in a process yielding 100 g l^{-1} in 36 hours of fermentation.

Threonine excretion by *C. glutamicum* is mainly (>90%) effected by a carrier-mediated export mechanism dependent on membrane potential. Cloning in extra copies of threonine export genes into an *E. coli* strain producing threonine led to increased production. Also increased was resistance to toxic antimetabolites of threonine. Another means of increasing threonine production was reduction in the activity of serine hydroxytransferase, which breaks down threonine to glycine. In *C. glutamicum* ssp. *lactofermentum*, threonine production reached 58 g l^{-1} when a strain producing both threonine and lysine (isoleucine auxotroph resistant to thialysine, α-amino-β-hydroxyvaleric acid and S-methylcysteine sulfoxide) was transformed with a recombinant plasmid carrying its own *hom* (encoding HDI), *thrB* (encoding HK), and *thrC* (encoding TS) genes.

4.4.1.4 *Production of L-isoleucine*

Isoleucine is of commercial interest as a food and feed additive and for parenteral nutrition infusions. This branched-chain amino acid is currently produced both by extraction of protein hydrolysates and by fermentation with classically derived mutants of C. *glutamicum*. The biosynthesis of isoleucine by C. *glutamicum* involves eleven reaction steps, of which at least five are controlled with respect to activity or expression (Figure 4.6). L-isoleucine synthesis shares reactions with the lysine and methionine pathways. In addition, threonine is an intermediate in isoleucine formation, and the last four enzymes also carry out reactions involved in valine, leucine, and pantothenate biosynthesis. Therefore, it is not surprising that multiple regulatory steps identified in C. *glutamicum*, as in other bacteria, are required to ensure the balanced synthesis of all these metabolites for cellular demands. In C. *glutamicum*, flux control is exerted by repression of the *homthrB* and *ilvBNC* operons. The activities of AK, HDI, TD, and acetohydroxy acid synthase (AHAS) are controlled by allosteric transitions of the proteins to provide feedback control loops, and HK is inhibited in a competitive manner. Isoleucine increases the Km of TD from 21 to 78 mM whereas valine reduces it to 12 mM. The AHAS is 50% feedback inhibited by isoleucine plus valine plus leucine.

Isoleucine processes have been devised in various bacteria such as S. *marcescens,* C. *glutamicum* ssp. *flavum,* and C. *glutamicum*. In S. *marcescens*, resistance to isoleucine hydroxamate and α-aminobutyric acid led to derepressed TD and AHAS and production of 12 g l^{-1} of isoleucine. Further work involving transductional crosses into a threonine-overproducer yielded isoleucine at 25 g l^{-1}. The C. *glutamicum* ssp. *flavum* work employed resistance to α-amino-β-hydroxyvaleric acid and the resultant mutant produced 11 g l^{-1}. Mutation to D-ethionine resistance yielded a mutant producing 33.5 g l^{-1} isoleucine in a fermentation continuously fed with acetic acid.

A threonine-overproducing strain of C. *glutamicum* was sequentially mutated to resistance to thiaisoleucine, azaleucine, and α-aminobutyric acid; it produced 10 g l^{-1} of isoleucine. An improved strain was obtained by cloning multiple copies of *hom* (encoding HDI), and wild-type *ilvA* (encoding TD) into a lysine-overproducer, and by increasing HK (encoded by *thrB*); 15 g l^{-1} isoleucine was produced. Independently, cloning of three copies of the feedback-resistant HDI gene (*hom*) and multicopies of the deregulated TD gene (*ilvA*) in a deregulated lysine producer of C. *glutamicum* yielded an isoleucine producer (13 g l^{-1}) with no threonine production and reduced lysine production.

FED-BATCH CULTURE
A batch-culture to which nutrients are added/ supplied intermittently to promote the biosynthesis of a desired product.

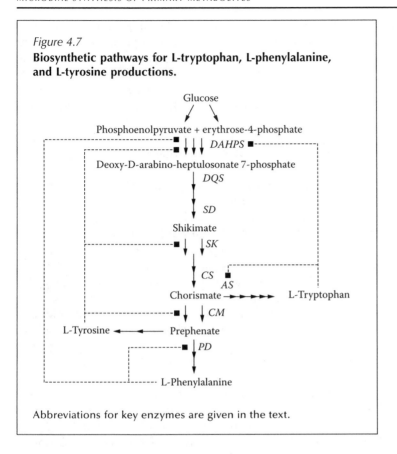

Figure 4.7
Biosynthetic pathways for L-tryptophan, L-phenylalanine, and L-tyrosine productions.

Abbreviations for key enzymes are given in the text.

Application of a closed-loop control fed-batch strategy raised production to 18 g l^{-1}, which was further amplified using metabolic engineering strategies to 30 g l^{-1} of isoleucine.

4.4.1.5 *Production of aromatic amino acids*

In C. *glutamicum* ssp. *flavum*, 3-deoxy-D-arabino-heptulosonate 7-phosphate synthase (DAHPS) is feedback inhibited concertedly by phenylalanine plus tyrosine and weakly repressed by tyrosine. Other enzymes of the common pathway (Figure 4.7) are not inhibited by phenylalanine, tyrosine, and tryptophan, but the following are repressed: shikimate dehydrogenase (SD), shikimate kinase (SK), and 5-enolpyruvylshikimate-3-phosphate synthase. Elimination of the uptake system for aromatic amino acids in C. *glutamicum* results in increased production of aromatic amino acids in deregulated strains.

A tryptophan process was improved from 8 g l^{-1} to 10 g l^{-1} by mutating the C. *glutamicum* ssp. *flavum* producer to azaserine resistance. Azaserine is an analogue of glutamine, the substrate of anthranilate synthase (AS). The mutant was two- to three-fold derepressed in DAHPS, dehydroquinate synthase (DQS), SD, SK, and chorismate synthase (CS). Another mutant, selected for its ability to resist sulfaguanidine, showed additional increases in DAHPS and DQS and tryptophan production. The reason that sulfaguanidine was chosen as the selective agent involves the next limiting step after derepression of DAHPS, i.e., conversion of the intermediate chorismate to anthranilate by AS. Chorismate can also be undesirably converted to p-aminobenzoic acid (PABA) and sulfonamides are PABA analogues. A sulfaguanidine-resistant mutant was obtained with *C. glutamicum* ssp. *flavum* and production increased from 10 g l^{-1} tryptophan to 19 g l^{-1}. The sulfaguanidine-resistant mutant was still repressed by tyrosine but showed higher enzyme levels at any particular level of tyrosine.

Gene cloning of the tryptophan branch and mutation to resistance to feedback inhibition yielded a *C. glutamicum* strain producing 43 g l^{-1} L-tryptophan. The genes cloned were those which encoded AS, anthranilate phosphoribosyl transferase, a deregulated DAHPS, and other genes of tryptophan biosynthesis. However, sugar utilization decreased at the late stage of the fermentation and plasmid stabilization required antibiotic addition. Sugar utilization stopped due to killing by accumulated indole. By cloning in the 3-phosphoglycerate dehydrogenase gene (to increase production of serine, which combines with indole to form more tryptophan) and by mutating the host cells to deficiency in this enzyme, both problems were solved. The new strain produced 50 g l^{-1} tryptophan with a productivity of 0.63 g l^{-1} h^{-1} and a yield from sucrose of 20%. Further genetic engineering to increase the activity of the pentose phosphate pathway increased production to 58 g l^{-1}.

A deregulated strain of *E. coli* in which feedback inhibition and repression controls were removed made 11 g l^{-1} phenylalanine in a fed-batch culture. Production was increased to 28.5 g l^{-1} when a plasmid was cloned into *E. coli* containing a feedback inhibition-resistant version of the CM-prephenate dehydratase (PD) gene, a feedback inhibition-resistant DAHPS and the $O_R P_R$ and $O_L O_L$ operator-promoter system of lambda phage. Further process development of genetically engineered *E. coli* strains brought phenylalanine titers up to 46 g l^{-1}. Independently, genetic engineering based on cloning *aroF* and feedback resistant *pheA* genes created an *E. coli* strain producing 50 g l^{-1}.

A *C. glutamicum* ssp. *lactofermentum* culture, obtained by selection with *m*-fluorophenylalanine produced 5 g l^{-1} phenylalanine, 7 g l^{-1} tyrosine, and 0.3 g l^{-1} anthranilate and contained desensitized DAHPS and PD. DAHPS in the wild type was inhibited cumulatively by phenylalanine and tyrosine, whereas PD was inhibited by phenylalanine. Cloning of the gene encoding PD from a desensitized mutant and the gene encoding desensitized DAHPS increased the enzyme activities and yielded a strain producing 18 g l^{-1} phenylalanine, 1 g l^{-1} tyrosine, and no anthranilate. Further cloning of a recombinant plasmid expressing desensitized DAHPS increased production to 26 g l⁻¹ phenylalanine. Similarly, *C. glutamicum* strains have been developed, producing up to 28 g l⁻¹ phenylalanine.

When SK was cloned into a tyrosine-producing *C. glutamicum* ssp. *lactofermentum* strain, tyrosine production increased from 17 g l⁻¹ to 22 g l⁻¹. Cloning of desensitized genes encoding DAHPS and CM from a deregulated phenylalanine-producing *C. glutamicum* strain into the deregulated tryptophan producer, *C. glutamicum* KY 10865 (CM-deficient strain, phenylalanine, and tyrosine double auxotroph with a desensitized AS) shifted production from 18 g l⁻¹ tryptophan to 26 g l⁻¹ tyrosine.

4.4.2 Production processes for purines and pyrimidines, their nucleosides and nucleotides

Commercial interest in nucleotide fermentations is due to the activity of two purine ribonucleoside 5′-monophosphates, namely guanylic acid (guanosine 5′-monophosphate; GMP) and inosinic acid (inosine 5′-monophosphate; IMP), as flavor enhancers. It is quite impressive that a 1:1 mixture of MSG with IMP or GMP gives flavor intensity thirty times stronger than MSG alone. Approximately 5000 tons of GMP and IMP are produced annually in Japan alone, with a world market of $350 million.

The purine residue of IMP is built up on a ribose ring in ten enzymatic reactions after ribose phosphate pyrophosphokinase (PRPP synthetase) catalyzes the conversion of α-D-ribose-5-phosphate (R5P) and ATP to 5-phosphoribosyl-α-pyrophosphate (PRPP) (Figure 4.8). Adenosine-5′-monophosphate (AMP) and GMP are synthesized from IMP. AMP formation involves participation of two enzymes, adenylosuccinate synthetase and adenylosuccinase. GMP synthesis requires the participation of IMP dehydrogenase and GPM synthetase. PRPP synthetase is feedback inhibited by AMP, GMP, and IMP; adenylosuccinate synthetase is inhibited by AMP; and IMP dehydrogenase is inhibited by xanthosine-5′-monphosphate (XMP) and GMP.

The genes encoding the enzymes of IMP biosynthesis in *B. subtilis* constitute the *pur* operon, whereas the genes encoding the GMP biosynthetic enzymes, *guaA* (GMP synthetase) and *guaB* (IMP dehydrogenase), and the *purA* gene encoding adenylosuccinate synthetase, all occur as single units. The *purB* gene encodes an enzyme involved in both IMP and AMP biosynthesis and is located in the *pur* operon. The levels of purine biosynthetic enzymes (except for GMP synthetase) are repressed in cells grown in the presence of purine compounds. Transcription of the *pur* operon is regulated negatively by adenine and guanine compounds including ATP, hypoxanthine, and guanine, which are corepressors. Feedback repression of purine nucleotide biosynthesis in *E. coli* is exerted by binding of the corepressor to the product of the *purR* gene. Hypoxanthine and guanine act

BIOCONVERSION
In industrial microbiology, use of micro-organisms to convert a substance to a chemically modified form.

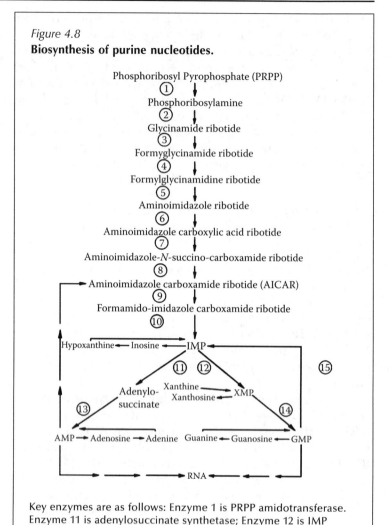

Figure 4.8
Biosynthesis of purine nucleotides.

Key enzymes are as follows: Enzyme 1 is PRPP amidotransferase. Enzyme 11 is adenylosuccinate synthetase; Enzyme 12 is IMP dehydrogenase; Enzyme 13 is adenylsuccinase; Enzyme 14 is XMP aminase.

cooperatively to change the conformation of *PurR*, thus enhancing its binding to DNA.

Techniques similar to those described above for amino acid fermentations have yielded IMP titers of 27 g l^{-1}. Since only low levels of GMP have been produced by direct fermentation, it is usually made by bioconversion of XMP. Genetic modification of *Corynebacterium ammoniagenes* involving transketolase (an enzyme of the non-oxidative branch of the pentose phosphate pathway) resulted in the accumulation of 39 g l^{-1} of XMP. This work demonstrates the need for high levels of pentose (ribose) for nucleotide and nucleoside biosynthesis and overproduction.

The key to effective accumulation of purines and their derivatives is the limitation of intracellular AMP and GMP. This limitation is best achieved by restricting purine supply during growth of the purine auxotrophs. Thus, adenine-requiring mutants lacking adenylosuccinate synthetase accumulate hypoxanthine or inosine that results from breakdown of intracellularly accumulated IMP. Certain adenine-auxotrophs of *B. subtilis* excrete over 10 g l^{-1} of inosine. These strains are still subject to GMP repression of enzymes of the common path. To minimize the severity of this regulation, the adenine auxotrophs are further mutated to eliminate IMP dehydrogenase. These adenine-xanthine double auxotrophs show a twofold increase in specific activity of some common-path enzymes and accumulate inosine up to 15 g l^{-1} under conditions of limiting adenine and xanthine (or guanosine). Further deregulation is achieved by selection of mutants resistant to purine analogues. Thus, mutants resistant to azaguanine with requirements for adenine and xanthine produce over 20 g l^{-1} inosine. Insertional inactivation of the IMP dehydrogenase gene in a *B. subtilis* strain yielded a culture producing inosine at 35 g l^{-1}.

Cloning of IMP dehydrogenase has been used to improve guanosine production in *B. subtilis*. The donor strain produced a low level of purine nucleosides but the distribution was in favor of guanosine (1 g l^{-1} of inosine versus 9 g l^{-1} guanosine). The recipient strain was auxotrophic for adenine, lacked GMP reductase and purine nucleoside phosphorylase, and was resistant to 8-azaguanine, adenine and adenosine; it produced 19 g l^{-1} inosine and 7 g l^{-1} guanosine. Cloning of IMP dehydrogenase from the first strain into the second strain yielded a recombinant that produced 5 g l^{-1} inosine and 20 g l^{-1} guanosine. Other *B. subtilis* mutants produce as much as 23 g l^{-1} of guanosine. Nucleosides such as inosine and guanosine are then converted to their active nucleotide derivatives chemically, microbiologically, or enzymatically.

The *de novo* pyrimidine biosynthetic pathway involves five enzymes and results in uridine-5′-monophosphate (UMP) production. Aspartate transcarbamoylase, the first pathway enzyme committed to pyrimidine biosynthesis, catalyzes the conversion of aspartate and carbamoylphosphate to carbamoylaspartate. The subsequent biosynthetic pathway enzymes are dihydroorotase, dihydroorotate dehydrogenase, orotate phosphoribosyltransferase, and orotidine-5′-monophosphate (OMP) decarboxylase. Uridine triphosphate (UTP) is produced from UMP by the sequential actions of two nucleoside kinases. Cytidine triphosphate (CTP) is formed by amination of UTP by CTP synthetase. The pyrimidine biosynthetic pathway is usually regulated at the level of gene expression. UMP kinases from *E. coli* and *B. subtilis* are activated by GTP and inhibited by UTP.

Selection for antimetabolite resistance has proven to be successful in development of pyrimidine nucleotide and nucleoside fermentations. Cytidine production by a *B. subtilis* cytidine deaminase-deficient mutant with resistance to fluorocytidine amounted to 10 g l^{-1}. Further mutation to 3-deazauracil resistance increased production to 14 g l^{-1}. By introducing a gene encoding a feedback resistant carbamyl phosphate synthase, cytidine production was raised to 18 g l^{-1}. Homoserine dehydrogenase (HSD) deficiency in *B. subtilis* increased cytidine production in a deregulated mutant from 9 g l^{-1} to 23 g l^{-1}. Increasing the glucose concentration raised production to 30 g l^{-1}. Uridine production by mutants of *B. subtilis* resistant to pyrimidine antimetabolites has reached 55 g l^{-1}.

4.4.3 *Production processes for vitamins*

More than half of vitamins produced commercially are fed to domestic animals. The vitamin market is several billion dollars per year. Microbes produce five vitamins commercially: vitamin B_{12} (cyanocobalamin), ascorbic acid (vitamin C), riboflavin (vitamin B_2), pantothenic acid (vitamin B_5), and biotin.

4.4.3.1 *Production of vitamin B12*

Vitamin B_{12} is produced commercially at about 10 tons per year. Fermentations have to be run under complete or partial anaerobiosis when using species of *Pseudomonas* or *Propionibacterium*. The major industrial organisms are *Pseudomonas denitrificans and Propionibacterium shermanii*. Conventional strain improvement has yielded *P. denitrificans* strains producing 150 mg l^{-1}. *Propionibacterium freudenreichii* can produce 206 mg l^{-1}.

Increasing the activity of S-adenosyl-L-methionine uroporphyrinogen III methyltransferase (SUMT) by cloning in a DNA fragment containing this gene (*cobA*) in *P. denitrificans* increased vitamin B_{12} production by 100%. SUMT is at the branch point of the heme and B_{12} biosynthetic pathways. Cloning of the gene *cob1* increased S-adenosyl-L-methionineprecorrin-2-methyltransferase (SP$_2$MT) and B_{12} production by 30%. SP$_2$MT is at the branch point of the siroheme and B_{12} pathways.

4.4.3.2 *Production of riboflavin*

Annual production of riboflavin is over 3000 tons/year. Riboflavin overproducers include two yeast-like molds, *E ashbyii* and *A. gossypii*, which synthesize riboflavin in concentrations greater than 20 g l^{-1}.

With *A. gossypii,* riboflavin production was found to be stimulated three- to four-fold by the addition of the precursors glycine and hypoxanthine. The level of production, which occurs after growth rate declines, is determined by the activity of the promoter of gene *RIB3*. This gene encodes 3, 4-dihydroxy-2-butanone-4-phosphate (DHBP) synthase, the first enzyme of the pathway. Mutation of *A. gossypii* to resistance to aminomethylphosphonic acid (a glycine antimetabolite) yielded improved producers. Isocitrate lyase (ICL) is important for use of fatty acids for riboflavin production. Itaconate, an inhibitor of ICL, eliminated the yellow color of *A. gossypii* colonies. A mutant strain, which was yellow on itaconate-containing agar, produced 15% more enzyme and 25-fold more riboflavin.

Processes using recombinant *B. subtilis* strains that produce at least 30 g l^{-1} riboflavin have been developed. In this species, riboflavin formation is regulated by feedback repression, not inhibition. An aporepressor encoded by *ribC*, whose effectors are riboflavin, FMN, and FAD, is responsible for this effect. Mutations of *ribC* lead to riboflavin overproduction. Sequential selection for resistance to 8-azaguanine, decoyinine, methionine sulfoxide, and roseoflavin plus cloning of multiple copies of the *riboflavin biosynthetic rib* operon yielded overproducing mutants. Further improvement was achieved when an extra copy of the *ribA* gene was introduced into the culture. This gene encodes both GTP cyclohydrolase II and 3, 4-dihydroxy-2-butanone 4-phosphate synthase, both of which act to commit precursors GTP and ribulose-5-phosphate to riboflavin biosynthesis.

A *Candida famata* (*Candida flareri*) strain produced 20 g l^{-1} in 200 hours. It was obtained by mutation and selection for resistance to 2-deoxyglucose (DOG), iron, tubercidin (a purine analogue) and depleted medium, plus protoplast fusion. The process depends on the addition of glycine and hypoxanthine. Selection for resistance to the adenine antimetabolite 4-aminopyrazolo (3, 4-d) pyrimidine improved production. Threonine showed a nine-fold stimulation in a strain with a cloned threonine aldolase, which converts threonine to glycine.

4.4.3.3 *Production of vitamin C*

Vitamin C has a global production of 100,000 tons/year. It is used for nutrition of humans and animals as well as a food antioxidant. The otherwise chemical Reichstein process utilizes one bioconversion reaction, the oxidation of D-sorbitol to L-sorbose by *Gluconobacter oxydans*, as the first step in ascorbic acid production. The biotransformation proceeds at the theoretical maximum, i.e., 200 g l^{-1} of D-sorbitol can be converted to 200 g l^{-1} of L-sorbose, when using a mutant of *G. oxydans* selected for resistance to a high sorbitol concentration. The bioconversion is used rather than a chemical reaction since the latter produces unwanted D-sorbose along with L-sorbose. An excellent fed-batch bioconversion process uses a starting concentration of 100 g l^{-1} of D-sorbitol and achieves production of 280 g l^{-1} of L-sorbose in 16 hours with a productivity of 17.6 g l^{-1} h^{-1}. The Reichstein process converts glucose to 2-keto-L-gulonic acid

---

ANTIOXIDANT
A molecule that is capable of reacting with free radicals and neutralizing them.

REICHSTEIN PROCESS
It is the use of a single bioconversion followed by a purely chemical route for vitamin C biosynthesis. The principal industrial process for the artificial synthesis of vitamin C still bears the name of the inventor.

(2-KLGA) in five steps with a yield of 50%. Then, 2-KLGA is chemically converted to L-ascorbic acid in two more steps.

Fermentation processes are competing with the Reichstein method. A mixed culture of *G. oxydans* (which converts L-sorbose to 2-KLGA) and *Gluconobacter suboxydans* (which converts D-sorbitol to L-sorbose) was able to convert 138 g l^{-1} of D-sorbitol to 112 g l^{-1} of 2-KLGA, with a molecular conversion yield of 75% in two days. A recombinant strain of *G. oxydans* containing genes encoding L-sorbose dehydrogenase and L-sorbosone dehydrogenase from *G. oxydans* was able to produce 2-KLGA effectively from D-sorbitol. Mutation to suppress the L-idonate pathway and improvement of the promoter led to production of 130 g l^{-1} 2-KLGA from 150 g l^{-1} D-sorbitol.

4.4.3.4 *Production of other vitamins*

Recombinant *E. coli*, transformed with genes encoding pantothenic acid (vitamin B$_5$) biosynthesis, and resistant to salicylic and/or other acids produce 65 g l^{-1} of D-pantothenic acid from glucose using β-alanine as precursor. Seven thousand tons per year are made chemically and microbiologically. Thiamine (vitamin B$_1$) is produced synthetically at 4000 tons per year. Pyridoxine (vitamin B$_6$) is made chemically at 2500 tons per year. The vitamin F (polyunsaturated fatty acids) processes of *Mortierella isabellina* or *Mucor circinelloides* yield 5 g l^{-1} of γ-linolenic acid.

4.4.4 *Production processes for organic acids*

Acetic, citric, gluconic, and lactic acids are the main primary metabolic organic acids with commercial application as chemicals, at least some of which are produced industrially by fermentation. Production has been improved by classical mutation and screening/selection techniques as well as by metabolic engineering.

4.4.4.1 *Production of acetic acid*

Over 7 million tons of acetic acid are made worldwide, over half of which is produced through microbial fermentations. Vinegar has been produced microbiologically as far back as 4000 B.C. Vinegar fermentation is best carried out with species of *Gluconacetobacter* and *Acetobacter*. A solution of ethanol is converted to acetic acid during which 90–98% of the ethanol is attacked yielding a solution of vinegar containing 12–17% acetic acid. Titers of acetic acid have reached 53 g l^{-1} with genetically engineered *E. coli*, 83 g l^{-1} by a *Clostridium thermoaceticum* mutant, and 97 g l^{-1} by an engineered strain of *Acetobacter aceti ssp. xylinum*.

4.4.4.2 Production of citric acid

Production of citric acid by *Aspergillus niger* and yeasts amounts to about one million tons per year. Annual sales reached $1.2 billion in the mid-1990s. The best strains of *A. niger* make over 200 g l^{-1} of citric acid. Keys to the fermentation are excess carbon source, low levels of pH and dissolved oxygen, and limited concentrations of certain trace metals and phosphate.

Glucose is converted to pyruvate by the Embden-Meyerhof-Parnas (EMP) pathway, then to acetyl coenzyme A, which enters the tricarboxylic acid (TCA) cycle by condensing with oxaloacetate to form citric acid. Citric acid production is stimulated by growing *A. niger* with a high sucrose concentration (10 to 20%). This is probably the result of the sugar's ability to cause the intracellular accumulation of fructose 2, 6-bisphosphate which activates glycolysis. Fructose 2, 6-bisphosphate is the product of phosphofructokinase II. Another key to successful citric acid fermentation by this fungus is a deficiency of Mn^{2+} ions. Also important is the restriction of the activity of isocitrate dehydrogenase, while maintaining an active citrate synthase. This prevents oxidative decarboxylation of isocitrate to α-ketoglutarate. Since the equilibrium of the reaction catalyzed by aconitase, which converts citric acid to isocitrate, is markedly in favor of citrate formation, citric acid accumulates. Two isocitrate dehydrogenases, mitochondrial and $NADP^+$ specific are inhibited by citrate. Since the cofactor of this enzyme is Mg^{2+} or Mn^{2+}, citrate's ability to chelate these bivalent ions restricts the enzymic activity of isocitrate dehydrogenase, which in turn allows citrate to accumulate.

In addition to Mn^{2+}ion, the metal deficiencies necessary for efficient citric acid production by different strains of *A. niger* include Fe^{2+} and Zn^{2+}. However, it is Mn^{2+} limitation that is paramount for the production of a high titer of citrate. The principal regulatory control site in the reactions from glucose to citrate is phosphofructokinase. This enzyme is inhibited by citrate, an event which would not be favorable for overproduction of citric acid. However, Mn^{2+} deficiency slows down growth, leading to degradation of intracellular nitrogenous macromolecules and a five-fold increase in the concentration of NH_4^+ in mycelia. The high ammonium concentration reverses citrate inhibition of phosphofructokinase, thus ensuring the continued conversion of glucose to citrate. Mutants whose phosphofructokinase I is partially desensitized to citrate inhibition are less dependent on low Mn^{2+} for high citric acid production. Citrate inhibition of phosphofructokinase is reversed by fructose 2, 6-diphosphate and AMP. The optimum pH for citric acid production by *A. niger* is 1.7–2.0. At pH values higher than 3.0, oxalic and gluconic acids are produced instead. Low pH inactivates glucose oxidase and prevents gluconate production. Mutants of *A. niger* with greater resistance to low pH are improved citric acid producers.

Other selective tools include resistance to high concentrations of citrate and sugars.

Species of the yeast genus *Candida* also excrete large amounts of citric acid and isocitric acid. The key event of the yeast citrate process appears to be a sharp drop in intracellular AMP following nitrogen depletion, inhibiting the AMP-requiring isocitrate dehydrogenase. *Candida guilliermondii* excretes large quantities of citric acid without the undesirable isocitric acid when cultured in the presence of metabolic inhibitors (e.g., sodium fluoroacetate, *n*-hexadecylcitric acid, and *trans*-aconitic acid). These inhibitors block the TCA cycle at the aconitase step. Mutation of *Candida lipolytica* to aconitase deficiency is also effective. The optimum pH for the yeast citrate process is above 5.0. Lower pH values lead to production of polyhydroxy compounds such as erythritol and arabitol. The yeast process utilizes hydrocarbons as substrate and citric acid yields as high as 225 g l^{-1} at a rate of 1.4 g l^{-1} h^{-1}.

High concentrations of citric acid are also produced by *Candida oleophila* from glucose. In chemostats, 200 g l^{-1} can be made and more than 230 g l^{-1} can be made in continuous repeated fed-batch fermentations. This compares to 150 to 180 g l^{-1} by *A. niger* in industrial batch or fed-batch fermentation in six to ten days. The key to the yeast fermentation is nitrogen limitation coupled with an excess of glucose. The citric acid is secreted by a specific energy-dependent transport system induced by intracellular nitrogen limitation. The transport system is selective for citrate over isocitrate. *Yarrowia lipolytica* produces up to 198 g l^{-1} citric acid in fed-batch fermentations on sunflower oil with a very low production of isocitric acid.

4.4.4.3 *Production of lactic acid*

The global market for lactic acid is about 250,000 tons per year. *Rhizopus oryzae* is favored for production since it makes stereochemically pure L (+)-lactic acid whereas lactobacilli produce mixed isomers; furthermore, lactobacilli require yeast extract. However, a mutant strain of *Lactobacillus lactis* has been developed which produces 195 g l^{-1} of L-lactic acid from 200 g l^{-1} of glucose. *R. oryzae* normally converts 60 to 80% of added glucose to lactate, the remainder going to ethanol. By increasing lactic dehydrogenase levels via cloning, more lactate and less ethanol are produced. Mutation of wild-type *R. oryzae* led to L (+)-lactic acid production of 131–136 g l^{-1}, a yield from glucose of 86–90% and a productivity of 3.6 g l^{-1} h^{-1}. This was a 75% improvement over the wild-type strain. The final strain was the result of a six-step mutation sequence.

A recombinant *E. coli* strain has been constructed that produces optically active pure D-lactic acid from glucose at virtually the theoretical maximum yield, e.g., two molecules from one molecule of glucose. The organism was engineered by eliminating genes of competing pathways encoding fumarate reductase, alcohol/aldehyde dehydrogenase, and pyruvate formate lyase, and by a mutation in the acetate kinase gene. Whole genome shuffling has been used to improve the acid tolerance of a commercial lactic acid-producing *Lactobacillus* sp.

Products in development are the nonchlorinated solvent, ethyl lactate, and the bioplastic, polylactide. Polylactide is made by converting corn starch to dextrose, fermenting dextrose to lactic acid, condensing lactic acid to lactide, and polymerizing lactide.

CHEMOSTAT
A continuous culture in which growth of microorganisms is restricted or limited by a certain nutrient.

4.4.4.4 *Production of pyruvic acid*

A recombinant strain of *E. coli* that is a lipoic acid auxotroph and defective in F_1ATPase produces 31 g l^{-1} of pyruvic acid from 50 g l^{-1} glucose. The lowering of the energy level in the cell by the F_1ATPase deletion increases glucose uptake and glycolysis rate, thereby leading to an increase in pyruvate production. An improved fermentation was developed using *Torulopsis glabrata* yielding a pyruvic acid concentration of 77g l^{-1}, a conversion of 0.80g g^{-1} glucose and a productivity of 0.91 g $l^{-1} h^{-1}$ in 85 hours. By directed evolution in a chemostat, a mutant *Saccharomyces cerevisiae* strain was obtained which produces from glucose 135 g l^{-1} pyruvic acid at a rate of 6 to 7 mmol per g biomass per hour during exponential growth with a yield of 0.54 g^{-1}.

4.4.4.5 *Production of other organic acids*

From 120 g l^{-1} of glucose, *Rhizopus arrhizus* produces 97 g l^{-1} fumaric acid. The molar yield from glucose is 145% and involves CO_2 fixation from pyruvate to oxaloacetate and the reductive reactions of the TCA cycle. Succinic acid is made chemically at 15,000 tons per year for commercial use as a (i) surfactant/detergent extender/foaming agent, (ii) ion chelator in electroplating to prevent metal corrosion and pitting, (iii) acidulant/pH modifier/flavoring agent/antimicrobial agent for food, and (iv) chemical in the production of pharmaceuticals. Market size is $400 million per year. Production by fermentation with *Actinobacillus succinogenes* amounts to 40 g l^{-1}, a productivity of 7 g $l^{-1} h^{-1}$, and a 76% yield from glucose. Bioconversion from fumarate yields 85 g l^{-1} succinate after 24 hours. Metabolic engineering of *E. coli* yielded a shikimic acid-overproducer making 84 g l^{-1} with a 0.33 molar yield from glucose. Production of gluconic acid amounted to 150 g l^{-1} from 150 g l^{-1} glucose plus corn steep liquor in 55 hours by *A. niger*. Production level is about 50,000 tons per year. Cloning of fumarase in *S. cerevisiae* remarkably improved the malic acid bioconversion from fumaric acid from 2 g l^{-1} to 125 g l^{-1}; conversion yield was nearly 90%.

4.4.5 *Production of ethanol and related compounds*

4.4.5.1 *Production of ethanol*

Ethanol is a primary metabolite produced by fermentation of sugar, or of a polysaccharide that can be depolymerized to a fermentable sugar. *S. cerevisiae* is used for the fermentation of hexoses, whereas *Kluyveromyces fragilis* or *Candida* species can be used if lactose or a pentose, respectively, is the substrate. Under optimum conditions, approximately 10–12% ethanol by volume is obtained within five days. At present, all beverage alcohol is made by fermentation. Industrial ethanol is mainly manufactured by fermentation, but some is still produced from ethylene by the petrochemical industry. Bacteria such as *Clostridia* and *Zymomonas* are being re-examined for ethanol production after years of neglect. *Clostridium thermocellum*, an anaerobic thermophile, can convert waste cellulose and crystalline cellulose directly to ethanol (see chapter on renewable resources conversion to fine chemicals). The available cellulosic feedstock in the USA could supply twenty billion gallons of ethanol in comparison to the three billion gallons currently made from corn. This would be enough to add 10% ethanol to all gasoline used in the USA. Other *Clostridia* produce acetate, lactate, acetone, and butanol and will be used to produce these chemicals when the global petroleum supplies begin to become depleted.

Ethyl alcohol is produced in Brazil from cane sugar at over four billion gallons per year and is used either as a 25% blend or as a pure fuel. Most new cars in that country use pure ethanol whereas the remainder utilizes a blend of 20 to 25% ethanol in gasoline. In the USA, over 3.4 billion gallons of ethanol were made from starchy crops (mainly corn) in 2004. It is chiefly added to gasoline to reduce CO_2 emissions by improving the overall oxidation and performance of gasoline.

Fuel ethanol produced from biomass would provide relief from air pollution caused by the use of gasoline and would not contribute to the greenhouse effect. *E. coli* has been converted into an excellent ethanol producer (43% yield, v/v) by cloning and expressing the alcohol dehydrogenase and pyruvate decarboxylase genes from *Zymomonas mobilis* and *Klebsiella oxytoca*. The recombinant strain was able to convert crystalline cellulose to ethanol in high yield when fungal cellulase was added. Other genetically engineered strains of *E. coli* can produce as much as 60 g l^{-1} of ethanol.

4.4.5.2 *Production of glycerol*

Glycerol has uses in the drug, food, cosmetics, paint, and many other industries. Production of glycerol is usually done by extraction of materials from the fat and oil industries, or by chemical synthesis from propylene, but good fermentations using *S. cerevisiae* and osmotolerant yeasts are available. Six hundred thousand tons of glycerol are produced annually. *S. cerevisiae* can produce up to 230 g l^{-1}. Osmotolerant *Candida glycerinogenes* can produce 137 g l^{-1} with yields of 63 to 65% and a productivity of 32 g l^{-1} h^{-1}. *Candida magnoliae* produces 170 g l^{-1} in fed-batch fermentation and in a similar type of process, *Pichia farinosa* can produce up to 300 g l^{-1}.

4.4.5.3   Production of 1, 3-propanediol

A strain of *Clostridium butyricum* converts glycerol to 1, 3-propanediol (PDO) at a yield of 0.55 g per g of glycerol consumed. In a two-stage continuous fermentation, a titer of 41 to 46 g l^{-1} was achieved with a maximum productivity of 3.4 g l^{-1} h^{-1}. At lower dilution rates, butyrate was produced, and at higher dilution rates, acetate was made. Recent metabolic engineering triumphs have included an *E. coli* culture that grows on glucose and produces PDO at 135 g l^{-1}, with a yield of 51% and a rate of 3.5g l^{-1} h^{-1}. To do this, eight new genes were introduced to convert dihydroxyacetone phosphate (DHAP) into PDO. These included yeast genes converting dihydroxyacetone to glycerol and *Klebsiella pneumoniae* genes converting glycerol to PDO. Production in the recombinant was improved by modifying eighteen *E. coli* genes, including regulatory genes. PDO is the monomer used to chemically synthesize polyurethanes and the polyester fiber Sorono^{Tm} by DuPont. This new bioplastic, polytrimethylene terephthalate (3GT polyester) is made by reacting terephthalic acid with PDO. PDO is also used as a polyglycol-like lubricant and as a solvent.

4.4.5.4   Production of erythritol

The noncariogenic, noncaloric, and diabetic-safe sweetener erythritol is made by fermentation. It has 70 to 80% the sweetness of sucrose. Osmotic pressure increase was found to raise volumetric and specific production, but to decrease growth. By growing cells first at a low glucose level, i.e., 100 g l^{-1} and then adding 200 g l^{-1} glucose at 2.5 days, erythritol titer was increased to 45 g l^{-1} as compared to single-stage fermentation with 300 g l^{-1} glucose, which yielded only 24 g l^{-1}. Production of erythritol by a *C. magnoliae* osmophilic mutant yielded a titer of 187 g l^{-1}, a rate of 2.8 g l^{-1} h^{-1}, and 41% conversion from glucose. Other processes have been carried out with *Aureobasidium* sp. (165 g l^{-1} from glucose with a 48% yield), and the osmophile *Trichosporon* sp. (188 g l^{-1} with a productivity of 1.18 g l^{-1} h^{-1} and 47% conversion). Erythritol can also be produced from sucrose by *Torula* sp. at 200 g l^{-1} in 120 h with a yield of 50% and a productivity of 1.67 g l^{-1} h^{-1}.

TRANSCRIPTOME
The full complement of activated genes, i.e., mRNA transcripts, in a particular tissue at a particular time.

PROTEOME
The full complement of proteins expressed by a cell at a particular time and under specific conditions.

METABOLOME
It is the study of changes in the concentrations and fluxes of endogenous metabolites in response to a given stimulus.

4.4.5.5 *Production of other compounds*

Dihydroxyacetone is used as a cosmetic tanning agent and surfactants. It is produced from glycerol by *Gluconobacter* species with a conversion rate of up to 90%. D-Mannitol is used in the food, chemical, and pharmaceutical industries. It is only poorly metabolized by humans, is about half as sweet as sucrose, and is considered a low-calorie sweetener. It is produced mainly by catalytic hydrogenation of glucose/fructose mixtures but 75% of the product is sorbitol, not mannitol. For this reason, fermentation processes are being considered. Recombinant *E. coli* produces up to 91 g l^{-1} mannitol and *Leuconostoc* sp. up to 98 g l^{-1}. Mannitol production reached 213 g l^{-1} from 250 g l^{-1} fructose after 110 hours by *C. magnoliae*. Sorbitol, also called D-glucitol, is 60% as sweet as sucrose and has use in the food, pharmaceutical, and other industries. Its worldwide production is 500,000 tons per year and it is made chemically by catalytic hydrogenation of D-glucose. Toluenized (permeabilized) cells of *Z. mobilis* produce 290 g l^{-1} of sorbitol and 283 g l^{-1} of gluconic acid from a glucose and fructose mixture in 16 hours with yields near 95% for both products. Xylitol is a naturally occurring sweetener with anticariogenic properties used in some diabetes patients. It can be produced chemically by chemical reduction of D-xylose. A mutant of *Candida tropicalis* produces 40 g l^{-1} l from D-xylose in over a 90% yield.

Early in the nineteenth century, the acetone-butanol fermentation process was a commercial operation but was later replaced by chemical synthesis from petroleum because of economic factors. These included the low concentration of butanol in the broth (1%) and the high cost of butanol recovery. *Clostridium beijerinckii* and *Clostridium acetobutylicum* are the organisms of choice for the fermentation. Research on this aspect of fermentation continued over many years, dealing with process engineering, mutation, and metabolic engineering. Butanol-resistant mutants were isolated for their ability to overproduce butanol and acetone. Further research involving biochemical engineering modifications increased the production of acetone, butanol, and ethanol (ABE) to 69 g l^{-1}. A mutant in the presence of added acetate produced 21 g l^{-v1} butanol and 10 g l^{-1} of acetone from glucose. Acetate both stimulates production and helps stabilize the culture, which is known to be highly unstable.

Summary

The microbial production of primary metabolites contributes significantly to the quality of life that we enjoy today. Fermentative production of these compounds is still an important goal of modern biotechnology. Through fermentation, micro-organisms growing on inexpensive carbon and nitrogen sources produce valuable products such as amino acids, nucleotides, organic acids, and vitamins, which can be added to food to enhance its flavor, or increase its nutritional value. The contribution of micro-organisms goes well beyond the food and health industries with the renewed interest in solvent fermentations. Micro-organisms have the potential to provide many petroleum-derived products as well as the ethanol necessary for liquid fuel. Additional applications of primary metabolites lie in their impact as precursors of many pharmaceutical compounds. The roles of primary metabolites and, in turn, microbial fermentations stand to grow in stature in the years ahead.

Overproduction of microbial metabolites is related to developmental phases of micro-organisms. Inducers, effectors, inhibitors, and various signal molecules play roles in different types of overproduction. Biosynthesis of enzymes catalyzing metabolic reactions in microbial cells is controlled by well-known positive and negative mechanisms, e.g., induction, nutritional regulation, feedback regulation, etc.

In the early years of fermentation processes, development of producing strains depended on classical strain breeding involving intensive rounds of random mutagenesis and screening and/or selection. More recently, methods of molecular genetics have been used for the overproduction of primary metabolic products. The development of modern tools of molecular biology enabled more rational approaches for strain improvement. Techniques of transcriptome, proteome and metabolome analysis, as well as metabolic flux analysis, have recently been introduced in order to identify new and important target genes and to quantify metabolic activities necessary for further strain improvement.

References and suggested reading

Demain, A. L. (2000). Small bugs, big business: The economic power of the microbe, *Biotechnol. Adv.* **18**, 499–514.

Demain, A. L., M. Newcomb, J. H. D. Wu (2005) Cellulase, clostridia, and ethanol, *Microbiol. Mol. Biol. Rev.*, **69**, 124–154.

Eggeling, L., H. Sahm (1999). L-Glutamate and L-lysine: traditional products with impetuous developments, *Appl. Microbiol. Biotechnol.*, **52**, 146–153.

Eggeling, L., H. Sahm (2001). The cell wall barrier in *Corynebacterium glutamicum* and amino acid efflux, *J. Biosci. Bioeng.*, **92**, 201–213.

El-Mansi, M. (2004). Flux to acetate and lactate excretions in industrial fermentations: physiological and biochemical implications, *J.Indust. Microbiol. Biotechnol.*, **31**, 295–300.

Han, L., S. Parekh (2004). Development of improved strains and optimization of fermentation processes, in Barredo, J. L. (ed.) *Microbial Processes and Products*, Totowa, New Jersey: Humana, pp. 1–23.

Jetten, M. S. M., A. J. Sinskey (1995). Recent advances in the physiology and genetics of amino acid-producing bacteria, *Crit. Rev. Biotechnol.*, **15**, 73–103.

Kraemer, R. (2004). Production of amino acids: physiological and genetic approaches. *Food Biotech.* **18**, (2), 1–46.

Kinoshita, S. (1987). Thom award address: Amino acid and nucleotide fermentations: from their genesis to the current state, *Devel. Indust. Microbiol.*, **28**, Suppl. 2, 1–12.

Magnuson, J. K., L. L. Lasure (2004). Organic acid production by filamentous fungi, in Tkacz, J. and Lange, L.(eds.) *Advances in Fungal Biotechnology for Industry, Agriculture, and Medicine*, New York: Kluwer Academic/Plenum, pp. 307–340.

Malumbre, M., J. F. Martin (1996). Molecular control mechanisms of lysine and threonine biosynthesis in amino acid-producing corynebacteria: redirecting carbon flow, *FEMS Microbiology Lett.*, **143**, 103–114.

Nielsen, J. (2001). Metabolic engineering, *Appl. Microbiol. Biotechnol.*, **55**, 263–283.

Ohnishi, J., S. Mitsuhashi, M. Hayashi, S. Ando, H. Yokoi, K. Ochiai, M. Ikeda (2002). A novel methodology employing *Corynebacterium glutamicum* genome information to generate a new L-lysine-producing mutant, *Appl. Microbiol. Biotechnol.*, **58**, 217–223.

Sahm, H., L. Eggeling, A. A. de Graaf (2000). Pathway analysis and metabolic engineering in *Corynebacterium glutamicum*, *Biol. Chem.* **381**, 899–910.

Sanchez, S., A. L. Demain (2002). Metabolic regulation of fermentation processes, *Enzyme Microb. Technol.*, **31**, 895–906.

Varela, C., E. Agosin, M. Baez, M. Klapa, G. Stephanopoulos (2003). Metabolic flux distribution in *Corynebacterium glutamicum* in response to osmotic stress, *Appl. Microbiol. Biotechnol.*, **60**, 547–555.

Vinci, V.A., G. Byng (1999). Strain improvement by nonrecombinant methods, in Demain, A.L. and Davies, J. (eds.) *Manual of Industrial Microbiology and Biotechnology*, 2nd ed., Washington: ASM Press, pp. 103–113.

5

Microbial Synthesis of Secondary Metabolites and Strain Improvement

Iain S. Hunter

5.1 Introduction

Although the use of micro-organisms in the production of commercially useful products is well established, the general public has yet to fully appreciate the microbial origin of some well-known commodities. For example, citric acid, an organic acid chemists found very difficult if not impossible to synthesize with the correct stereo-specificity, is now widely produced through fermentation by either *Aspergillus niger* or the yeast *Yarrowia lipolytica*. It was the expertise gained with such fermentations that allowed the rapid development of the antibiotics industry in the early 1940s (see below), with a consequential leap in quality of healthcare.

In our contemporary lifestyle, the microbial origins of many products are not well recognized. For example, xanthan gum, a product made by *Xanthomonas campestris*, is widely used in the manufacture of ice cream as it prevents the foams from collapsing with time. In a slightly modified form, xanthan gum gives paints their unique property of sticking to the brush without dripping but spreading smoothly and evenly from the brush to the surface.

In addition to natural products such as the chemicals mentioned above, recombinant strains of *Escherichia coli*, constructed by genetic engineering, are now used for the production of "non-natural products" to the organism such as human insulin. Another example is biological washing powders, which contain enzymes drawn from different species, particularly those belonging to the genus *Bacillus*. Microbial products thus have a significant impact on our everyday lives.

Significantly, the "new biotechnology" industry is critically dependent on microbial fermentations for the production of the

LOW-VALUE, HIGH-
VOLUME PRODUCT
A commercial term for
a product whose cost
is low (of the order
of £6/kg, or less) but
which is required in vast
quantities (many millions
of kg per year). Citric
acid and xanthan fall
into this category.

MEDIUM-VALUE,
MEDIUM-VOLUME
PRODUCT
A commercial term
for a product that falls
between the extremes
of a low-value, high-
volume and high-value,
low-volume product.
Typically, antibiotics fall
into this category, being
required on a scale of up
to 100,000 kg per year
at a cost of around
£60 per kg.

HIGH-VALUE, LOW-
VOLUME PRODUCT
A commercial term
for a product (usually
biologically active) that
is extremely potent
and is required only in
small (kg) quantities
throughout the world
per year. However, the
price charged for the
product may be high (up
to £60/mg). Recombinant
proteins, such a human
insulin, interferon, and
growth hormone fall into
this category.

extremely high-value recombinant therapeutic products, e.g.,
human insulin and human growth hormone, as well as the low-
value recombinant products, e.g., prochymosin, which provides
an ethically acceptable alternative to rennin (chymosin) in the
manufacture of cheese.

For each useful microbial product, it is necessary to develop a
strain that, in addition to demonstrating a high rate of production, fulfills a number of important criteria including:

- The ability to utilize effectively the cheapest available feed
  stock (substrate)
- The ability to complete fermentation in a relatively short
  time
- The excretion of fewer byproducts so that downstream
  processing costs are kept to a minimum.

Strain improvement programs must, therefore, be undertaken to
achieve the above targets. This chapter has two main themes:

1. The use of classical strain improvement strategies, which
   have been extremely successful in the past, and continue
   to provide a foundation to develop strains that have the
   potential to make elevated levels of new products, especially in the antibiotics industry and
2. The use of recombinant (cloning) techniques in the
   improvement of existing products and development of new
   therapeutics of extremely high value to mankind.

5.2 The economics and scale of microbial product fermentations

The type of fermentation used, as well as its size, duration, and
nutrient profile, will depend critically on the nature of the microbial product. For "low-value, high-volume" products, such as
citric acid and xanthan gum, high-capacity fermentors (often up
to 800 m^3 in volume) are generally used. However, the duration
of the fermentation process and cost of nutrients and "utilities"
(heating, cooling, and air) are the critical factors in the overall
profitability of this business.

"Medium-value, medium-volume" products, such as antibiotics, are typically made in fermentors that are considerably
smaller (100–200 m^3), and again the duration and utility/nutri-
ent cost are a significant factor.

"High-value, low-volume" products, such as the recombinant
therapeutic proteins, are made in small (approximately 400 L, i.e.,
0.4 m^3) fermentors for which the cost of the nutrients and utilities
is a minor factor in the overall feasibility and profitability.

For all but the high-value products, nutrient costs (especially of the primary carbon source) are critical. Depending on the vagaries of world commodity markets, complicated further by artificially imposed trade tariffs, the availability and, in turn, cost of nutrients can fluctuate at alarming rates. Flexibility in the choice of nutrients is therefore of paramount importance, and strain improvement programs must take this factor into consideration.

By contrast, for the high-value (recombinant) products, the emphasis of the strain improvement program focuses on the following aspects:

- The stability of the strain
- The level of expression
- Overall quality of product, rather than the cost of the fermentation process *per se*.

A fermentation process constitutes a business that aims to sell the product at an overall profit. This issue positions strain improvement programs at the interface between science and the commercial world, and requires a different set of criteria to judge whether a task is worthwhile or not. Strain improvement programs are positioned must first be vetted to establish the cost of the research and development (R&D) that will be necessary to achieve the stated goal, and to set the expected cost against the annual savings likely to ensue, should that piece of work deliver the expected gains in productivity. It often comes as a shock to researchers new to this field that projects that are highly innovative, but to which some risk of failure is attached, are not funded because the return on such an investment (set against the risk) is not high enough. For example, a strain improvement program for a "mature product" would be difficult to justify if the annual R&D cost could not be recouped in around three years through the projected increase in productivity. The annual cost of strain improvement programs must, therefore, be no more than three times the projected annual cost savings. Acclimatization to such rigorous reviews of research plans and draconian decision-taking constitutes a sharp learning curve for newly recruited research staff.

MATURE PRODUCT

A commercial term for a product that has been in the marketplace for some time and for which the cost of manufacture is very important. By contrast, a new product is either patented (in which case the company that owns the patent has sole rights to market it for up to 17 years) or has some novel property compared with its competitors. In either case the product can be sold at a premium price. However, as other competitive products appear on the market or, in the case of a patented product, when the patent runs out, the selling price comes under pressure and the profit associated with it is reduced. With such a "mature product," the cost of manufacture (amount of product made and efficiency of the overall process) becomes critical.

5.3 Different products need different fermentation processes

At the "low-value" end of the microbial products business, the margins on profitability are extremely tight. For citric acid, the titre must be greater than 100 gL^{-1}, with carbon conversion efficiency (amount of substrate converted to product on a per-gram basis) close to 100% for the process to become economic. Citric acid is produced concurrently with microbial growth, but a fine balance has to be struck between the amount of cells that are made in the fermentation (which consumes some of the carbon source that otherwise could be used to make the product) and the fact that a doubling of the cell mass may result in twice the volumetric rate of citric acid production (as each cell acts as its own "cell factory"), but may hinder the eventual purification of the product. Growth can be arrested by, typically, limiting the amount of nitrogen available to the culture, in which case citric acid is still produced for some time by the cells in stationary phase. Cheap carbon sources (e.g., unrefined molasses), fast production runs with minimal turnaround time, and a cheap and rapid means of extracting the product are most important, and the strain improvement program of this mature product will be focused on substrate flexibility and rapid production.

Medium-value products, such as antibiotics, are produced by microbes, which, when first isolated from the soil, usually make detectable, but vanishingly small, quantities (a few micrograms) of the bioactive substance. Most commercial fermentation processes will only become economically viable if the strain is eventually capable of making more than 15–20 gL^{-1} of the bioactive product using comparatively cheap carbon sources, such as rape seed oil, and low-grade sources of protein, such as Soya bean or fish meal. Antibiotic production (see below) usually occurs following the onset of the stationary phase, i.e., after growth has been arrested, but many antibiotics contain nitrogen as well as carbon, so growth is generally limited by the supply of phosphate. A steady supply of nutrients containing carbon and nitrogen, but not phosphate, is fed to the fermentation during the phase of antibiotic production.

Once a new microbial metabolite has been discovered, a campaign of "empirical strain improvement" is undertaken to boost the level of production to a titre at which the process becomes economically viable.

For "high-value" products, such as recombinant therapeutic proteins, the cost of the fermentation broth is not a major issue. To ensure consistency of the final product, expensive well-defined media (either Analar mineral medium or high-specification tryptone hydrolysates) are used along with a high purity carbon source (usually glucose). Following the onset of the stationary phase, the production of high-value products is triggered, generally in response to an external signal such as temperature shift or the addition of an exogenous gratuitous inducer. The levels of production may be relatively modest (less than 1 gL^{-1}) to meet commercial targets, but higher levels are always sought. Because the products are often formulated into injectable medicines, in which there is a fear of raising an immune response if the recombinant protein is not 100% pure or not folded into the correct tertiary structure. A major objective of strain improvement programs, therefore, is to address this important regulatory issue, rather than the cost of media and utilities.

5.4 Fed-batch culture; the paradigm for many efficient microbial processes

Simple batch culture is a fairly inefficient way to synthesize a microbial product, as a substantial proportion of the nutrients present in the fermentation is used to make the biomass and the opportunity then to use the biomass as a cell factory to make the product is limited to the time of the growth period. Although very high levels of productivity can often be achieved in continuous culture, this technique has the disadvantages that large volumes of medium (often expensive medium) are required, and the product is made in a dilute stream that has to be concentrated before final isolation can take place. Fed-batch culture is a "halfway" approach. The cells are grown up in batch culture and then the resident biomass, which is no longer growing, is dedicated to product formation by feeding nutrients, except for that chosen to limit growth. This type of culture technique was developed by the antibiotics industry.

To understand the nature of antibiotic fermentations, it is important to take account of the life cycle of the producing micro-organisms within the natural ecosystem. This example is from *Streptomyces*, the filamentous bacteria that make the majority (> 60%) of natural antibiotics, including streptomycin, the tetracyclines, and erythromycin. The life cycles of the filamentous eukaryotic organisms, which produce the natural penicillins, are broadly similar. The cycle begins with the spore, which may lie dormant in the soil for many years; when nutrients and water become available, the spore germinates, thus leading to the formation of mycelium (Figure 5.1). These organisms are said to be "mycelial" in nature, as they colonize soil particles (or agar medium in Petri dishes, if they are in the laboratory) by extending outwards in all directions (radial growth) in a fixed branched pattern; often called vegetative mycelium. Inevitably, at some point, nutrients or water become in short supply and it is at this point that they differentiate, ultimately to form spores again. Initially, the differentiation process involves the formation of aerial mycelium in which the biomass no longer extends out radially, but rises up and away from the plane of the radial growth and forms elaborate coiled structures that then septate after a while and the spores are formed again, thus completing the life cycle.

These enzymes digest the vegetative mycelium, in part, and the cellular building blocks that are released are used to construct the aerial mycelium. In this way, part of the vegetative mycelium is "sacrificed" to allow the aerial mycelium to be formed en route to sporulation. This process can be viewed as a survival strategy because, when the spores have formed, the organism's DNA is held in an inert state so that the life cycle can begin again when water and nutrients are plentiful. It has been suggested that the reason that these actinomycetes make antibiotics exactly at the same time as the differentiation step is to

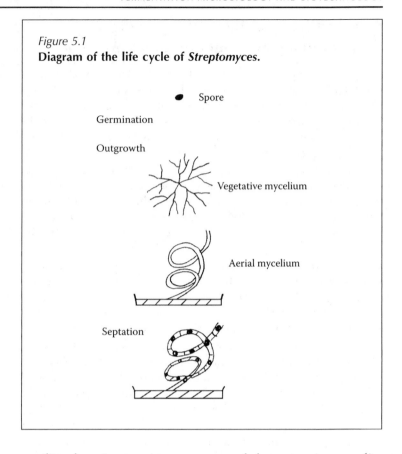

Figure 5.1
Diagram of the life cycle of *Streptomyces*.

sterilize the micro-environment around the vegetative mycelia, so that other microbial predators cannot take advantage of the available nutrients released following mycelial lyses (a view that is not held by all scientists in the field).

5.4.1 *Nutrient limitation and the onset of secondary metabolite formation*

Interestingly, antibiotics are made at the time of differentiation between the stage of vegetative mycelia and the aerial mycelial stage (Figures 5.1 and 5.2). This may seem a paradox, as the signal to undertake this differentiation step is lack of nutrients or water, which in turn begs the question of how a micro-organism, starved of nutrients, elaborates such a complex structure as the aerial mycelium on one hand and the triggering of antibiotic production on the other? The answer to this apparent paradox lies in the ability of the organism to coordinate the production and release of extracellular lytic enzymes (proteolytic, lipolytic, and hydrolytic) and the turning on of the differentiation switch.

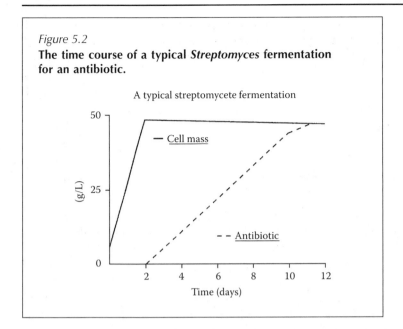

Figure 5.2
The time course of a typical *Streptomyces* fermentation for an antibiotic.

5.4.1.1 *The role of RelA in initiating transcription of antibiotic synthesis genes under nitrogen limitation*

Much attention has recently been focused on the role of RelA, the ribosome associated ppGpp synthetase, in initiating secondary metabolism in actinomycetes under nitrogen limitation, but not phosphate. RelA is apparently central to antibiotic production under conditions of nitrogen limitations in *Streptomyces coelicolor* A3 (Chakrabartty and Bibb, 1997; Bentley et al., 2002) and *Streptomyces clavuligerus* (Jin et al., 2004a, 2004b). The question of whether antibiotic production is triggered as a direct consequence of RelA or indirectly as a consequence of slow growth rate through inhibition of rRNA synthesis by ppGpp binding remains to be established (Bibb, 2005). Evidence supporting the direct participation of ppGpp in activating the transcription of genes involved in the biosynthesis of secondary metabolites, e.g., *act*II-*orf*4 genes involved in the biosynthesis of the antibiotic actinorhodin in *S. coelicolor*, has been given (Hesketh et al., 2001). However, the mechanism through which this is achieved has yet to be unraveled (Bibb, 2005).

5.4.1.2 *The role of phosphate in repressing transcription of antibiotic synthesis genes*

Unlike conditions of nitrogen limitation, the signal that triggers the synthesis of secondary metabolism under phosphate limitation is RelA (ppGpp)-independent (Chakrabartty and Bibb, 1997). Under these conditions, i.e., nitrogen limitation,

DIFFERENTIATION SWITCH
A genetic switch that changes the pattern of the genes that are expressed, in such a way that the morphology of the cell changes (differentiates) along with some of the functions that it performs. The step cannot be reversed easily, i.e., the cells do not return to their original shape and functionality in a single step.

the polyphosphate reserve is hydrolyzed to inorganic phosphate, which in turn represses biosynthetic enzymes, to satisfy growth requirements, The negative role portrayed for polyphosphate kinase (PPK), the enzyme responsible for the biosynthesis of polyphosphate polymer, was ascertained as loss of enzymic activity and accompanied by increased level of antibiotic formation in *S. lividans* (Chouayekh and Virolle, 2002). It is generally accepted that:

- Conditions leading to elevated levels of inorganic phosphate, lead to repression of enzymes of secondary metabolism and in turn diminish flux to antibiotic formation.
- Conditions leading to inorganic phosphate depravation stimulate secondary metabolism and in turn increase flux to antibiotic formation.

However, the question of how elevated levels of inorganic phosphate might interfere with the onset of secondary metabolism and morphological differentiation of actinomycetes awaits further investigations.

5.4.2 The role of quorum sensing and extracellular signals in the initiation of secondary metabolism and morphological differentiation in actinomycetes

It is generally agreed that γ-butyrolactones are produced and, in turn, implicated specifically in the production of secondary metabolites as well as morphological differentiation in several species of *Streptomyces* (Mochizuki et al., 2003), e.g., the production of streptomycin by *Streptomyces griseus* and the production of tylosin in *S. fradiae*.

In sharp contrast to γ-butyrolactones-binding proteins, which downregulate the transcription of secondary metabolite gene clusters, another molecule, namely SpbR, has been shown to stimulate the biosynthesis of pristinamycin (Folcher et al., 2001). Deletion of SpbR was accompanied not only by the abolition of antibiotic formation but also severely impaired growth on Agar media, thus lending further support to the positive role of SpbR in antibiotic production and cellular growth and differentiation in actinomycetes (Bibb, 2005).

Another factor, namely PI (2, 3-diamino-2, 3-bis (hydroxymethyl)-1, 4-butanediol), has recently been shown to be able to elicit the biosynthesis of pimaricin in *S. natalensis* (Recio et al., 2004), and its positive role was further substantiated when its addition to P1-defective mutants was accompanied by restoring the organism's ability to produce pimaricin.

5.4.3 Positive activators of antibiotic expression

In antibiotic fermentations, temporal expression of antibiotic biosynthesis is regulated tightly as part of the cellular differentiation pathway (Figure 5.2). Production of antibiotics is costly to the cell in terms of carbon and energy. Therefore, it is hardly surprising that tight control of expression has evolved. The genes for antibiotic biosynthesis are invariably clustered together, irrespective of whether the microbe is a prokaryote or eukaryote. In *Streptomyces*, regulation of expression of such gene clusters is controlled by a positive activator — a master gene that, when switched on, makes a protein that targets itself to the various promoters of the gene cluster and switches them on in concert (Figure 5.8).

In this way, the cell ensures that the full complement of enzymes necessary to make the antibiotic is produced at the same time, at the correct levels.

The great majority of antibiotic pathways are regulated in this way. For example, the *tyl*R system that controls tylosin biosynthesis has been characterized (Stratigopoulos et al., 2004) and a newly discovered regulator for actinorhodin biosynthesis in the model organism *S. coelicolor* (Uguru et al., 2005). Sometimes, the expression of the activator gene is integrated tightly with the developmental pathway, such as the *bld*A-dependent regulation of ActIIOrf4 in *S. coelicolor* and *bld*G-dependent regulation of *cca*R in *S. clavuligerus* (Bignell et al., 2005), in which CcaR upregulates the biosynthesis of both cephamycin C and clavulanic acid (Perez-Llarena et al., 1997).

Other factors that may act in a pleiotropic manner, e.g., AfsR of *S. coelicolor*, play a key role in integrating multiple signals that are transduced through phosphorylation cascades (Horinouchi, 2003). Another effector is AfsK, which on sensing appropriate signals autophosphorylates itself. Autophosphorylated (activated) AfsK phosphorylates the cytoplasmic AfsR, which, in turn, through its DNA binding activity, activates the transcription of *dfsS*, the product of which is apparently capable of stimulating the biosynthesis of a number of different antibiotics (Bibb, 2005).

MASTER GENE (POSITIVE ACTIVATOR)
A positive activator is a protein that is absolutely required to bind to the DNA upstream of the target gene to allow it to be transcribed, i.e., the target gene cannot be expressed without it. If a collection of genes has the same binding sequence upstream, then they are switched on simultaneously. Hence the positive activator acts as a "master switch" that allows the entire set of genes to be regulated coordinately.

5.5 Tactical issues for strain improvement programs

In attempting to harness the vast natural potential of the *Streptomyces* and related actinomycetes for the making of antibiotics on a large scale, the organisms have to be grown in large volumes of liquid cultures. Care has to be taken so that the essential features of the biology of antibiotic production process are not lost following the transfer (inoculation) of the organism from solid medium to the liquid medium in the fermentor. In a liquid medium, it is very unusual for aerial mycelia to be made, and sporulation is observed even less frequently. Despite this, it is possible to induce the cultures to make an antibiotic in a liquid culture.

The physiology of the fermentation process has to be adapted to suit the biology of the microbe. As one of the triggers for differentiation in the natural ecosystem is deprivation of nutrients, then the same strategy may also be applied to the fermentation process. All antibiotics are composed of carbon atoms and many contain nitrogen atoms. Therefore, it would be a poor tactic to attempt to trigger the onset of antibiotic production by limiting the cells' supply of either of these elements, because after antibiotic production had commenced, the cells would be starved of one of the most important chemical elements needed to biosynthesize the antibiotic structures. Fortunately, very few antibiotics contain phosphorus in their elemental composition, so the cells are most often limited by the supply of phosphate to trigger antibiotic production. If all other nutrients are supplied in sufficient quantities (but not in vast excess, see later), then antibiotic production will continue for many days (often 8–10 days) at a rate that is linear with time (Figure 5.2). Eventually, this rate starts to tail off. It makes economic sense to terminate the fermentation at this point, recover the maximal amount of product in the shortest possible time, and then prepare the vessel for another round of fermentation.

Fermentations for high-value recombinant therapeutic proteins are best undertaken by fed-batch cultures, which maximize the use of the biomass factory and deliver the product in the most concentrated form. However, the duration of the production phase for these high-value fermentations is considerably shorter than for the antibiotic fermentations, typically around 8 hours, or less.

In practice, the order in which the experimental strategies for improving strains are used depends on the nature of the desired end product and of the fermentation. In the initial stages of developing an antibiotic producing strain from a "soil isolate," random mutagenesis of a population of the producer microorganism is undertaken (see below) and the progeny of the mutagenic treatment are screened for higher levels of the antibiotic. Subsequently, more directed screens are used to further enhance the titre of the antibiotic.

By contrast, improvement of strains making recombinant therapeutic proteins starts with a very directed strategy and finishes with a more random, empirical approach. This is because, in the initial stages of strain improvement, there is a well-defined template of experimental improvements that can be followed. Subsequently, the performance of the strain can be improved further by the empirical approach.

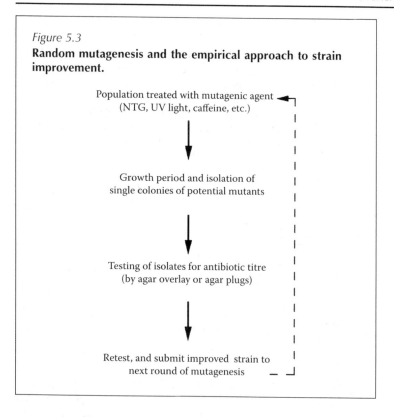

Figure 5.3

Random mutagenesis and the empirical approach to strain improvement.

Population treated with mutagenic agent
(NTG, UV light, caffeine, etc.)

Growth period and isolation of
single colonies of potential mutants

Testing of isolates for antibiotic titre
(by agar overlay or agar plugs)

Retest, and submit improved strain to
next round of mutagenesis

5.6 Strain improvement: The random, empirical approach

In this approach a population of micro-organisms is subjected to a mutagenic treatment (Figure 5.3), typically with the chemical carcinogen, nitrosoguanidine (NTG), but other mutagenic agents such as UV light or caffeine may be used. The treatment is tailored so that each cell, on average, has a single mutation induced. The mutagenized population is plated on agar, which will support the growth of the normal (wild-type) strain. Some mutations will be very deleterious to the growth of the organism; cells containing these will not grow up to become colonies. The mutant strains (colonies, each of which is derived from a single cell) that show little or no growth impairment are then tested randomly for their ability to overproduce antibiotics. Classically, this was achieved by overlaying the agar plate containing the surviving colonies after mutagenesis with soft agar containing another micro-organism that is sensitive to the antibiotic in question. Colonies producing more biologically active antibiotic will be surrounded with a larger clear zone of inhibition of the culture overlay. However, this technique does not take into account the fact that some colonies may be larger (i.e., have a greater diameter) than others. To circumvent this problem, a cylinder of the colony and the agar underneath is cut out (usually

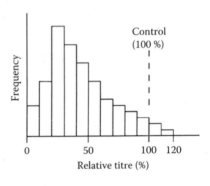

Figure 5.4

Antibiotic titres of individual survivors after random mutagenesis. The histogram shows the frequency distribution of titres of survivors after random mutagenesis, for a strain whose lineage has been developed for around 2 years. Notice that only a few isolates perform better than the control (some of this may be due to error in determining the titre) and that most survivors perform much worse than the control, as the mode is about 25%.

with a cork borer) and the "agar plug" placed on a fresh lawn of sensitive bacteria spread on a new agar plate. The cylinders are uniform and contain comparable amounts of cell material. This modification takes account of differences in sizes of different mutagenised colonies. The whole procedure is tedious but has to be undertaken in a painstaking manner. In recent times, the pharmaceutical industry has used robots to perform the tasks required in random screening. There has been a tendency to move away from screening for how much antibiotic is made by colonies on a plate, to liquid media-based cultures that reflect the situation in fermentors more accurately. It is not unusual for a screening robot to evaluate the performance of over a million mutants in a year — such "high throughput screening" is orders of magnitude more efficient than using people for these tasks, leaving the human input focused on the design and efficiency of the overall screening program.

Most of the progeny will make lower levels (Figure 5.4) of the microbial product, and only a few will have larger titres, often showing 10% (or less) improvement. The reason for this is that most mutations are deleterious to production and only a few mutations will result in slightly elevated levels of production. It is very unusual to isolate mutants that have large (> 20%) increases in titre.

These individual isolates are then tested in small-scale fermentations to confirm that they really are improved mutants. Most turn out not to have improved titres and are discarded.

Mutants that do show some promise will then be tested intensively at the laboratory and pilot plant levels.

Different companies have different strategies at this stage. Some may carry out extensive laboratory trials at around the 10 L level and scale up the mutants directly to the large production fermentors in one step. Other companies may adopt a more conservative posture and scale up the volume of fermentation in stages. For a process that is undertaken at 100,000 L production scale, the stages in scale up would typically be from 10 L to 500 L, then onward to 5000 L and finally to the large fermentors. The doctrine is that operation of these large fermentors is expensive — both in the cost of fermentation broth and in lost opportunity, as that large fermentor might otherwise have been used to make another valuable product. The view of those fermentation technologists who scale up in one step is that time is wasted solving problems associated with the intermediate stages and that a production fermentation that is modeled well at the bench scale should, by definition, predicate immediate translation of the new mutant and its process details to the production fermentor hall.

It is often overlooked that the initial improvement in titre, which is achieved with a new mutant at the production scale, can often be improved still further by process optimization through adjustments to the media and fermentation conditions that better suit the physiology of the new mutant. Process optimization invariably makes an equal, if not greater, contribution to overall increase in titre than the stepwise increase in fermentation performance achieved when the new mutant is adopted initially at the production scale.

5.7 Strain improvement: The power of recombination in "strain construction"

In the early stages of undertaking a random mutagenesis program for a new product (i.e., starting from a new soil isolate that makes a small amount of product), there are likely to be rapid, concurrent advances in the performances of several strains. These strains, separately, will have individual desirable properties. For example, in an antibiotic program, one strain may give a higher titre, while another may use less carbon source to make a level of antibiotic equivalent to that of the parent (and so be more economical). Yet another strain may show no better performance in terms of titre of product or the amount of nutrients consumed to make it, but may carry out the fermentation at a lower viscosity, which will make it easier for the product to be extracted from the fermentation broth and purified. Different "lineages" of the strain are developed very rapidly in these early days in the desire to improve the fermentation considerably. However, it soon becomes desirable to construct strains with a combination of several of the desirable properties already present in the individual lineages. To achieve this, desirable traits from each lineage must be "recombined" genetically into a single strain that possesses all of the desirable properties, or "traits."

Most microbial species are able to exchange genetic information with other members of their species by recombination. Construction of the genetic maps that appear in many textbooks is based on exploitation of these natural recombination processes. For *E. coli* and other enteric Gram-negative bacteria, protocols to undertake genetic recombination are well documented and depend on plasmid-based mobilization of the bacterial chromosome. However, the producers of many microbial metabolites, including the *Streptomyces*, have a life cycle that involves a sporulation stage (described above). With these species, the most effective method of constructing recombinants that have combinations of several desirable traits is to undertake "spore mating": to mix spores of the parents with the individual traits, germinate them together, allow them to go through an entire life cycle (Figure 5.1), and (from the spores formed at the end of this life cycle) to select those that have exchanged some genetic information (i.e., select those that show that they have undertaken some degree of recombination) and then to screen them to identify the individuals that have received the desired combination of traits (Figure 5.5).

This screening process is usually conducted empirically, in the same way as screening of strains after random mutagenesis. Unfortunately, a major drawback with this procedure can often arise: as strains with further improved titres are selected and the strain lineage becomes longer, the ability to sporulate is often lost. Although sporulation is associated with the differentiation process, antibiotic production can often become decoupled from sporulation in the advanced high-titre strains. The strains become asporogenous, which makes such a spore mating strategy impossible. For some strains, the problem can be circumvented by undertaking strain construction by recombination between the vegetative mycelia of different parents. However, not all strains are amenable to this approach. For them, a more involved strategy called protoplast fusion (Figure 5.6) has to be enacted. The most common pitfall with protoplast fusion is the time taken to develop the media conditions and culture techniques that allow the protoplasts to regenerate satisfactorily back into fully competent mycelium (see Figure 5.6). Often an extensive series of empirical range-finding experiments has to be undertaken to optimize the conditions for regeneration, and occasionally it proves impossible to develop a protocol for a particular species.

Recently, the power of recombination has been used to great advantage in combination with high-throughput screening in

Figure 5.5

Strain construction by recombination. In this example two lineages have been derived from the first soil isolate. One lineage shows a steady improvement in titre with each new strain, but this is accompanied by wasteful utilization of carbon source. The other lineage makes very little antibiotic, but does so in a very efficient way. By genetic recombination of these traits, the attributes of both strands of the culture lineage can be combined together in a single strain.

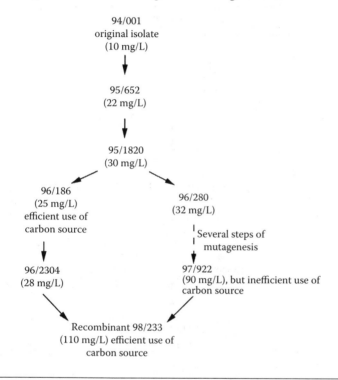

a "genome shuffling" strategy that lead to a rapid increase in tylosin titre in the industrially relevant *S. fradiae* (Zhang et al., 2002). Only two rounds of genome shuffling were needed to achieve results that had previously required at least 20 rounds of classical mutation and screening.

5.8 Directed screening for mutants with altered metabolism

Although the random mutagenesis and recombination approaches are fruitful in acquiring mutants, which give initial improvements in titres, in due course as the level of microbial product produced increases, it becomes more difficult (in parallel) to isolate further improved strains using the empirical,

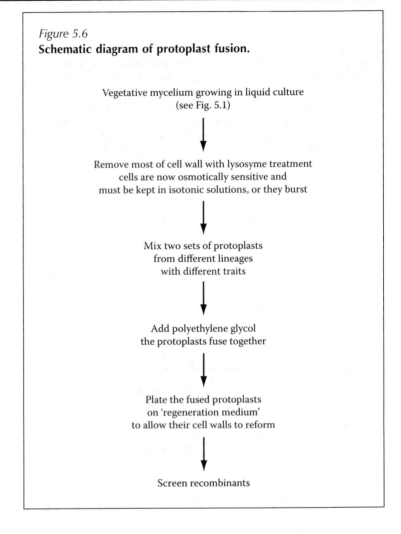

Figure 5.6
Schematic diagram of protoplast fusion.

Vegetative mycelium growing in liquid culture
(see Fig. 5.1)

↓

Remove most of cell wall with lysosyme treatment
cells are now osmotically sensitive and
must be kept in isotonic solutions, or they burst

↓

Mix two sets of protoplasts
from different lineages
with different traits

↓

Add polyethylene glycol
the protoplasts fuse together

↓

Plate the fused protoplasts
on 'regeneration medium'
to allow their cell walls to reform

↓

Screen recombinants

random approach. For example, in the development of an anti-biotic fermentation process, by this time the titre of product will have reached a few grams per litre and considerable background data will have been gathered on the physiology and biochemis-try of the fermentation process.

These data can be analyzed to diagnose whether wasteful metabolites are being made during the fermentation process (see Chapters 3, 6, and 7 for further details). For example, many *Streptomycete* fermentations often consume copious quantities of glucose and excrete pyruvate and α-ketoglutarate, giving low conversion yields of glucose into product. If the overutilization of glucose can be prevented by careful process control (e.g., lim-iting the supply of glucose), then the economics of the process will improve substantially. However, the same objective can be achieved by generating mutants that will use the glucose less quickly. Such a strategy is amenable to a "directed screening" approach.

In the case of uncontrolled uptake and wasteful metabolism of glucose, it is known that mutants resistant to 6-deoxyglucose (a toxic analogue of glucose) have reduced or impaired glucose uptake and subsequent metabolism. In this strategy, a mutagenic treatment is performed on a population of cells or spores, in the same way as described in Figure 5.3. However, instead of plating all of the survivors under nonselective conditions, the survivors are plated directly onto a nutrient agar to which the desired selective pressure can be applied. In this case, the survivors would be plated on media containing 6-deoxyglucose at a level that, under normal conditions, just prevents the growth of the micro-organism. The media would also contain a second carbon source (usually glycerol) on which the cultures can normally grow quite vigorously. Out of the millions of survivors of the initial mutagenic treatment, only those which are altered in some aspect of their metabolism of 6-deoxyglucose will survive and grow on such a selective medium. These isolates can then be screened conventionally to identify the individuals among the population that may have the potential in the fermentor to display a reduced level of glucose uptake and consequently a more balanced metabolism of the sugar, which does not involve the wasteful production of pyruvate and the α-oxo-organic acids.

Many survivors that have become resistant to 6-deoxyglucose will have mutated to confer complete exclusion of the toxic analogue from the cell, but in addition, they will no longer be able to utilize glucose at all and will have to be discarded. Strains that are totally incapable of glucose utilization are of no use to the fermentation industry, which is substantially based on cheap, glucose-rich carbon sources.

However, the minor class of analogue-resistant survivors, which have impaired, but still significant, glucose uptake, are the sought-after prizes in this "directed screen." The terminology "directed" is appropriate because the researcher defines the exact conditions under which such mutants should survive — to generate new culture isolates with improved fermentation performance through less wasteful metabolism of glucose.

Directed screening is an extremely important tactic that may be employed to great advantage after the basic details of the fermentation have been worked out and understood. The value of directed screening is the reduction in numbers of mutants to be screened (typically by around 10,000-fold) before an isolate with the desired characteristics is identified. Such a reduction in workload allows time to establish the metabolite production profile of each survivor and to evaluate the data in greater depth. Such is the power of the directed screen that rare spontaneous mutants may be isolated, rather than those generated by mutagenic agents. If 106 to 108 cells are plated on a single plate containing a toxic analogue, a few spontaneous mutants will always be isolated.

Figure 5.7

Diagram of the biosynthetic origin of isopenicillin N. This has three component amino acids, condensed together with peptide bonds. The origins of the amino acids are shown between the dotted lines.

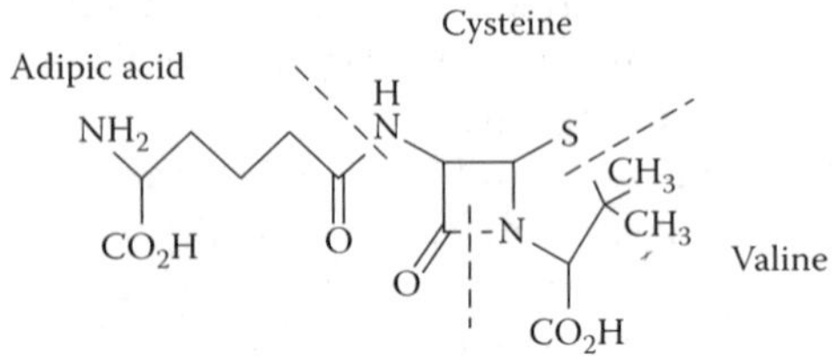

In addition to carbon utilization, the flux of nitrogen (fixation and metabolism of ammonia) and phosphorus (phosphate metabolism) can be altered by specific mutations, for which directed screens can be devised to select for potentially improved mutants. Fixation of ammonia takes place via the enzyme glutamine synthetase in virtually all microbes. The flux through this step is often altered in mutants that are resistant to bialaphos, a toxic compound that specifically inhibits glutamine synthetase. Mutants with altered phosphate metabolism may be obtained by directed screening for resistance to toxic analogues, such as arsenite or dimethyl arsenite.

Fluoroacetate is a classic metabolic inhibitor that poisons the tricarboxylic acid (TCA) cycle by being converted to fluorocitrate, a toxic analogue of citric acid. Mutants more resistant (or sometimes those more sensitive) to fluoroacetate often have a TCA cycle with altered properties. As the TCA cycle is a fundamental component of cellular aerobic metabolism (conditions under which most fermentations are conducted), then these mutants can have important properties and be fruitful sources of improved strains.

The building blocks for all microbial products, including the antibiotics, are common metabolic precursors used in other biosynthetic processes of the cell. For example, the three components of penicillin are two amino acids, cysteine and valine, together with adipic acid, which is a precursor of lysine. Thus, penicillin can be viewed biosynthetically as a simple but modified tripeptide (Figure 5.7) of three common amino acids. During microbial growth, cellular metabolism is painstakingly controlled to ensure that supplies of all twenty amino acids needed for growth are available in a balanced fashion.

In a fed-batch culture, such as an antibiotic fermentation, tightly regulated metabolism during the growth phase is followed by the production phase (Figure 5.2), during which the commercial aim is to produce a single product quickly and at high levels — to the exclusion of others. As this microbial product will probably be made from a few key metabolic intermediates (e.g., during production of penicillin, only a supply of the three amino acids will be in high demand), then metabolism must be altered to satisfy this increased demand, while minimizing the side reactions of wasteful metabolism. Directed screens can be devised that decouple the usual control strategies of the biosynthetic pathways (such as feedback inhibition and cross-pathway regulation), which normally keep the supply of all precursors just balanced to the needs of the growing cell.

By way of example, consider the supply of adipic acid (part of the lysine pathway) for the biosynthesis of penicillin (Figure 5.7). Normally, the lysine pathway is subject to end-product feedback inhibition. The toxic analogue of lysine, δ-(2-amino-ethyl)-I-cysteine, also inhibits the first step of lysine biosynthesis. Mutants resistant to this toxic analogue are no longer subject to end-product feedback inhibition of the early part of the biosynthetic pathway. They have an enhanced flux of precursor supply to adipic acid and often produce higher titres of penicillin.

Almost every metabolic pathway that supplies the precursors of microbial products is amenable to this type of directed screening strategy. Thus, the rate of "fuel supply" for biosynthesis of microbial products can be enhanced.

The last example of directed screening relates to the selection of mutants that produce elevated levels of the enzymes responsible for catalysis of the precursor building blocks for the biosynthesis of the backbone structures of antibiotics, such as the tetracyclines and erythromycin. They are polymers of acetyl-CoA and methylmalonyl-CoA respectively, and both are made by a process that is essentially the same as that for fatty acid biosynthesis. The antibiotic cerulenin targets the enzyme complexes responsible for fatty acid biosynthesis and acts to starve the growing cell of the fatty acids necessary for insertion into the membrane. Mutants resistant to toxic levels of cerulenin have circumvented this problem by making elevated levels of the fatty acid biosynthetic enzymes to "titrate out" the effect of the antibiotic. The close similarity between the enzymes of fatty acid biosynthesis and those which make the backbones of erythromycin and tetracycline allows cerulenin to inhibit the biosynthesis of these antibiotics. Mutants that can still make the antibiotic in the presence of cerulenin have elevated levels of the biosynthetic enzyme complexes. When cerulenin is removed, they retain the high level of biosynthetic capability, which, if supplied with enough of the metabolic precursor "fuel," results in higher titres of the antibiotic. The utility of the directed screening approach is that the survivors of the mutagenic treatment are invariably altered in some aspect of fatty acid or antibiotic biosynthesis.

Individual mutants made by directed screening approaches can be recombined together using genetic recombination, spore mating, or protoplast fusion, as described above, to combine several desired traits.

5.9 Recombinant DNA approaches to strain improvement for low- and medium-value products

The advent of recombinant DNA techniques, first devised for *E. coli* in the 1970s, has meant that new strategies can be applied to strain improvement for low- and medium-value microbial products. It has taken some time for recombinant techniques to be developed and applied to the commercial strains, such as the filamentous bacteria and filamentous fungi, which are the mainstay of this sector of the fermentation industry.

In addition, regulatory hurdles have to be crossed to gain approval to undertake fermentations at the production scale with these "genetically engineered micro-organisms." Gaining such approval is time-consuming and costly. Thus, there have to be good long-term economic reasons for adopting a recombinant DNA strategy for strain improvement.

By cloning this master gene regulator (see 5.4.3) and then expressing it at unnaturally high levels, the cellular complement of the entire biosynthetic machinery for production of an antibiotic may be enhanced. Of course, there has to be sufficient fuel (metabolic precursors and energy) to realize the full potential available from this boosted level of biosynthetic machinery. The roles of metabolic flux analysis and process optimization (see Chapters 6 and 7 for more details) in assuring this advantage are very important.

It is also possible to force expression of the master gene during the growth phase and so to produce antibiotics during growth. In some instances, this may be to the advantage of the overall process, but often it is better to mimic the situation in Nature: first to focus the design of the process on maximizing the acquisition of biomass, and then to turn that biomass to best advantage by using the cells as a factory (with the maximal level of installed biosynthetic capability) to make the product at a fast rate and to achieve a high titre. Therefore, controlled expression of the master gene so that it is switched on decisively at the end of growth is often preferred.

Molecular genetic analysis has also shown that some of the strains from improvement programs, isolated over the years by random mutagenesis and selection, have increased dosages of the biosynthetic genes. Thus, some of the best strains for penicillin production have multiple copies of the critical part of the biosynthetic gene cluster arranged in tandem arrays. This effect can also be achieved by gene cloning strategies.

Figure 5.8
Schematic diagram to illustrate how antibiotic gene clusters are controlled. The DNA encodes a number of genes clustered together on the chromosome. The genes (rectangles) are usually transcribed as polycistronic mRNAs (shown by the unbroken straight arrows). Transcription of the production genes (unshaded rectangles) must be dependent on transcription of the positive activator (master gene, shown by the shaded rectangle and the protein by the two circles) that migrates and binds to the DNA (curved arrows) to allow transcription of the production genes to take place. Without the activator protein, there is no expression of production genes.

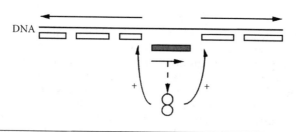

The biochemical pathways for antibiotic biosynthesis are long linear series of enzyme reactions. The flux through the entire pathway is governed by the pace of the slowest catalytic step. If the flow of metabolic intermediates through that step can be improved, then there is a good chance that the productivity of the overall process will be improved. Thus, for the antibiotic tylosin, produced by *Streptomyces fradiae*, it was established that the rate of the last step in the pathway, conversion of macrocin to tylosin by macrocin-O-methyltransferase (encoded by the gene *tyl*F), limited the overall productivity. As the strain improvement program had developed with time and the production strain lineage was reviewed, it was apparent that mutants with higher titres of tylosin also displayed extremely high levels of macrocin, which was excreted as a shunt metabolite (Figure 5.9) because conversion of macrocin was limiting. Strains that had the methyltransferase gene cloned and expressed at high levels showed improved titres of tylosin.

Often, process analysis shows that some metabolites on the main biosynthetic pathway are being diverted to other shunt products (Figure 5.10). This not only represents a waste of valuable carbon source, but also presents a problem for the ultimate purification of the desired product, as the second metabolite has to be purified away. Genetic manipulation can be used to specifically inactivate the gene for the enzyme that diverts the metabolite to the shunt metabolite. This precise "genetic surgery" enhances the flow through the pathway to the desired product.

Figure 5.9

Schematic diagram highlighting (a) the limitation of tylosin biosynthesis by Streptomyces fradiae and (b) the elevation of such limitation by the use of a recombinant strain.

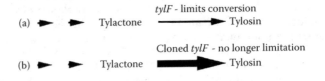

Figure 5.10

Schematic diagram highlighting the problem of shunt products and how this can be prevented. The metabolic pathway (substrate a to product d) is composed of three enzymes (1, 2, 3) with intermediates b and c. Because the carbon flow through steps 1 and 2 is greater than through step 3 (shown by the width of the arrows), the shunt metabolite, X, is formed through the action of enzyme 4. The presence of shunt metabolite, X, may complicate the recovery and purification of the desired product, d. By making a mutant devoid of enzyme 4, this is prevented.

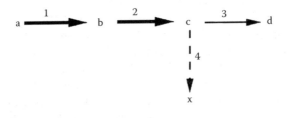

One last important contribution that genetic manipulation makes is in providing flexibility of utilization of a carbon source. If, for example, there is a plentiful supply of cheap lactose available (from the milk whey industry) but the strain does not use lactose naturally, then lactose utilization genes can be cloned from another species and introduced into it.

At times, the world marketplace has a glut of hydrolyzed sucrose. This cheap carbon source is problematic for fermentation because one of the hydrolysis products (glucose) is capable of inhibiting the utilization of the other (fructose). It is possible to overcome this problem by the cloning of the fructose utilization genes, which, in turn, ensures that the two monomeric carbon sources are used concurrently. The overall economics of such a process benefits from being able to use a cheap and plentiful carbon source in an efficient manner, without excess fructose carbon being left in the broth at the end of the fermentation.

5.10 Strain improvement for high-value recombinant products

Strain improvement programs for products derived from cloned genes follow a completely different strategy from those undertaken for their lower-value counterparts. The techniques used to clone a gene (or cDNA) usually place it precisely in a vector (most often, a plasmid), in a context that allows continuous high-level expression, or, more advisedly, expression of the gene is controlled and induced at a high level only after growth has ceased — in the same way as antibiotic fermentations are performed (Figure 5.2). A countless number of possibilities arise to tailor the gene within the vector, but this aspect involves only molecular biology and falls outside the remit of this chapter. The biology of the host strain is an equally critical factor in the overall performance of a fermentation process for a recombinant product, and strain improvement programs can have a significant impact on the economics of such a process.

In the early days of the "new biotechnology" industry, considerable difficulty was experienced in "scaling up" recombinant processes from the laboratory to the pilot plant (i.e., from a scale of around 100 ml to 400 L). This was because, at the larger volume, a greater proportion of cells had lost the recombinant plasmid from the cells. Plasmid-deficient cells do not make the recombinant protein, and this reduces the overall productivity. The root cause of the problem is the number of cell generations needed to attain significant growth of the culture at the larger volume, coupled to the inherent instability of engineered plasmids in cells. The growth rate of a cell carrying a plasmid is invariably slower than its counterpart that has shed its plasmid load. Therefore, plasmid-deficient cells will outgrow the plasmid-containing cells in a population, a phenomenon that becomes more significant as the number of generations in a culture increases: the larger the volume of the fully grown culture, the greater the proportion of plasmid-deficient cells in its population.

Resistance to an antibiotic, encoded by a plasmid-borne gene, is the usual selection strategy for the presence of a plasmid in a recombinant culture. The most commonly used selection is resistance to ampicillin, encoded by β-lactamase. Ampicillin-resistant cells survive because they are protected by the β-lactamase, which breaks down the chemical backbone of the antibiotic. Eventually, all of the ampicillin in the fermentation broth is broken down and the selective pressure is lost — the longer the fermentation (i.e., the more generations that take place), the more likely it will be that the antibiotic will become inactivated and that plasmid-deficient cells will form.

The frequency, at which plasmid-deficient cells are formed, in the absence of antibiotic, is biased because of their natural tendency to form oligomers within the cell. Consider a theoretical

Figure 5.11

Plasmid segregation — the complication of plasmid oligomers. In case (a) the plasmids are present as four monomers, and two segregate (on average) into each of the daughter cells. By contrast, in case (b), the plasmids are present as a tetramer. During growth, one daughter cell receives the tetramer, whereas the other does not and, in turn, becomes plasmid-deficient.

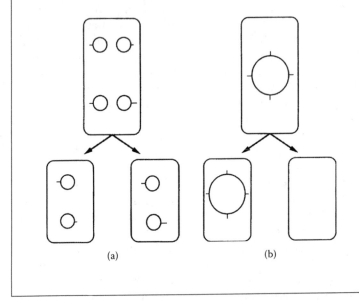

(a)   (b)

situation in which a cell has a plasmid copy number of four. This may be conceptualized as four separate plasmids, two of which segregate into each daughter cell at division (Figure 5.11a). However, because of oligomerization, an equally common situation is that there will be a single tetramer of plasmids within the parent cell (i.e., the copy number will still be four); when the daughter cells are formed, one will receive the plasmid, which consists of four monomers, and the other will not (Figure 5.11b). In the absence of selective pressure (i.e., after all of the antibiotic has been exhausted) both will survive, and the plasmid-deficient daughter will outgrow its plasmid-containing sibling because of the advantage in growth rate.

Engineered plasmids undergo such oligomerization events at alarming rates. However, natural plasmids do not undergo such "segregational instability," because they have a natural tendency to break back down from oligomers to monomers again (Balding et al., 2006). This ability has been lost during the process of conversion of natural plasmids into genetically engineered vectors. It was discovered that a small piece of DNA, called *cer*, was the missing factor in the engineered vectors. When *cer* was reintroduced into vectors from the natural plasmids, the segregational stability of the plasmid construct containing the cloned gene to be expressed was improved. However, operation of *cer* is dependent on the strain; this aspect of the biology of plasmid instability can be addressed by strain improvement programs. The cellular machinery that allows *cer* to operate is rather complex and involves at least three different genetic loci. If any of the three is defective or missing, the result is that plasmids (even those containing a fully competent *cer*) are unstable. Strain improvement of such cell lineages is best achieved using a targeted genetic approach, i.e., to use the power of genetic recombination to introduce the relevant machinery back into the chromosome of the strain, an extremely focused and targeted task.

Recombinant cultures are very prone to infection with bacterial viruses called bacteriophages. At a scale of fermentation above a liter, in a production environment, the recombinant cultures are at risk. There are two strain improvement strategies that can be brought into play. First, it is possible to mutate the strain to become resistant to viral infection. This approach works for known viruses, but there is always the risk that infection from a new source will take place. A second, more secure strategy is to introduce a restriction/modification system into the cell line. The modification enzyme will alter the host's DNA so that it is no longer a target for the resident restriction system. However, viral DNA that enters the cell will be recognized by the restriction system and degraded, thus preventing the infection. The strategy can be enhanced further by introduction of a second restriction/modification system as a "back-up," should some of the viral DNA escape restriction by the first enzyme system. Again, this is a much targeted approach to strain improvement.

Hosts for foreign gene expression often recognize the foreign protein as "not natural" and selectively degrade it, thus reducing the overall productivity. The enzyme system that undertakes this function, the *lon* protease, works in a similar way in ordinary cells, selectively degrading ordinary cellular proteins that have been made defectively with altered amino acids. By knocking out the *lon* protease gene, degradation of the foreign protein is prevented, albeit to the detriment of the host cell, which now has a slight growth rate disadvantage. Methods that generate *lon*-deficient cell lines are well established and extremely targeted.

In each of the examples in this section, there has been no need to use random mutagenesis and screening, as the blueprint to derive the improved strain is direct and straightforward. Subsequent to this phase, strain improvement programs use the random approach to fine-tune the genetic content of the production recombinant strain to gain further advantages in productivity. Thus, the order of events is the reverse of that for the lower-value counterparts.

Summary

Fermentation is an important source of many clinically useful drugs, and this trend is likely to increase in the future. The microbiologist, in collaboration with other scientists, has an important role to play in generating new microbial strains that produce higher levels of the desired metabolites. An impressive array of techniques can be used to achieve this goal, from random mutagenesis to the latest advances in genetic manipulation. Each has a role to play in the overall process, but the priorities of using different approaches will depend on the nature of the fermentation and the product being made.

References

Baltz, R. H. (1998). Genetic manipulation of antibiotic-producing *Streptomyces, Trends in Microbiol.* **6**, 76–83.

Bentley, S. D., K. F. Chater et al. (2002). Complete genome sequence of the model actinomycetes *Streptomyces coelicolor* A3. *Nature* **417**, 141–147.

Balding, C., Blaby, I., Summers, D. (2006). A mutational analysis of the ColE1-encoded cell cycle regulator Rcd confirms its role in plasmid stability. Plasmid **56**, 68–73.

Bibb, M. J. (2005). Regulation of secondary metabolism in *Streptomycetes. Current Opinion in Microbiology* **8**, 208–215.

Bignell, D. R. G., K. Tahlan, K. R. Colvin, S. E., Jensen, B. K. Leskiw (2005). Expression of *ccaR*, Encoding the positive activator of Cephamycin C and Clavulanic acid production in *Streptomyces clavuligerus* is dependent on *bldG. Antimic. Agents. Chemotherapy* **49**, 1529–1541.

Chakraburtty R., M. J. Bibb (1997). The ppGpp synthetase gene (relA) of *Streptomyces coelicolor* A3 plays a conditional role in antibiotic production and morphological differentiation. *J Bacteriol* **179**, 5854–5861.

Chater, K. F. (1998). Taking a genetic scalpel to the *Streptomyces* colony, *Microbiology* **144**, 727–738.

Chouayekh, H., M. J. Virolle (2002). The polyphosphate kinase plays a negative role in the control of antibiotic production in *Streptomyces lividans. Mol Microbiol,* **43**, 919–930.

Folcher, M., H. Gaillard, L. T. Nguyen, K. T. Nguyen, P. Lacroix, N. Bamas-Jacques, M., Rinkel, C. J. Thompson (2001). Pleiotropic functions of a *Streptomyces pristinaespiralis* autoregulator receptor in development, antibiotic biosynthesis, and expression of a superoxide dismutase. *J Biol Chem* **276**, 44297–44306.

Griffiths, A. J. F. et al. (1996). *An Introduction to Genetic Analysis*, 6th ed., New York: W.H. Freeman & Co., pp. 592–600.

Hardy, K. G., S. G. Oliver (1985). Conventional strain improvement, protoplast fusion and cloning, In Higgins, I. J., D. J. Best, J. Jones (eds.) *Biotechnology: Principles and Applications*, Oxford: Blackwell, pp. 257–282.

Hesketh, A., J. Sun, M. J. Bibb (2001). Induction of ppGpp synthesis in *Streptomyces coelicolor* A3 grown under conditions of nutritional sufficiency elicits actII-ORF4 transcription and actinorhodin biosynthesis. *Mol Microbiol* **39**, 136–144.

Hockney, R.C. (1994) Recent developments in heterologous protein production in *Escherichia coli*, *Trends in Biotechnol.*, **12**, 456–463.

Horinouchi, S. (2003). AfsR as an integrator of signals that are sensed by multiple serine/threonine kinases in *Streptomyces coelicolor* A3. *J Ind Microbiol Biotechnol* **30**, 462–467.

Jin, W., H. K. Kim, J. Y. Kim, S. G. Kang, S. H. Lee, K. J. Lee (2004a). Cephamycin C production is regulated by *rel*A and *rsh* genes in *Streptomyces clavuligerus* ATCC27064. *J Biotechnol* **114**, 81–87.

Jin, W., Y. G. Ryu, S. G. Kang, S. K. Kim, N. Saito, K. Ochi, S. H. Lee, K. J. Lee (2004b). Two relA/spoT homologous genes are involved in the morphological and physiological differentiation of *Streptomyces clavuligerus*. *Microbiology* **150**, 1485–1493.

S. Mochizuki, K. Hiratsu, M. Suwa, T. Ishii, F. Sugino, K. Yamada, H. Kinashi (2003). The large linear plasmid pSLA2-L of *Streptomyces rochei* has an unusually condensed gene organization for secondary metabolism. *Mol Microbiol* **48**, 1501–1510.

Perez-Llarena, F. J., P. Liras, A. Rodriguez-Garcia, J. F. Martin (1997). A regulatory gene (ccaR) required for cephamycin and clavulanic acid production in *Streptomyces clavuligerus*: amplification results in overproduction of both beta-lactam compounds. *J Bacteriol* **179**, 2053–2059.

Recio, E., A. Colinas, A. Rumbero, J. F. Aparicio, J. F. Martin (2004), PI actor, a novel type quorum-sensing inducer elicits pimaricin production in *Streptomyces natalensis*. *J Biol Chem* **279**, 41586–41593.

Segura, D., C. Santana, R. Gosh, L. Escalante, S. Sanchez (1997). Anthracyclines: Isolation of overproducing strains by the selection and genetic recombination of putative regulatory mutants of *Streptomyces peucetius* var. caesius, *Appl. Microbiol. Biotechnol.*, **48**, 615–620.

Stratigopoulos, G., N. Bate, E. Cundliffe (2004) Positive control of tylosin biosynthesis: Pivotal role of TylR. *Molec. Microbiol* **54**, 1326–1324.

Summers, D. K. (1991). The kinetics of plasmid loss, *Trends Biotechnol.*, **9**, 273–278.

Summers, D. K., D. J. Sherratt (1985). Bacterial plasmid stability, *Bioessays*, **2**, 209–211.

Uguru, G.C., K. E. Stephens, J. E. Stead, J. E. Towle, S. Baumberg, K. J. McDowall (2005). Transcriptional activation of the pathway-specific regulator of the actinorhodin biosynthetic genes in *Streptomyces coelicolor*. Molec. Microbiol. **58**, 131–150.

Zhang, Y., K. Perry, V. A. Vinci, K. Powell, W. P. C. Stemmer, S. P. del CardayreÂ (2002). Genome shuffling leads to rapid phenotypic improvement in bacteria. *Nature* **415**, 644–647.

6

Metabolic Analysis and Optimization of Microbial and Animal Cell Bioprocesses

David M. Mousdale

6.1 Metabolic analysis in the era of genomics, proteomics, and metabolomics

Of the thousands of genes sequenced and identified in the genomes of industrial micro-organisms, which are those key to controlling productivity?

Of the hundreds of proteins resolved on two-dimensional gels, which are those central to bioprocess metabolism and which are merely the inevitable consequences of the dynamics of growth during the course of fermentation in rapidly changing nutritional environments?

Of the hundreds of enzymes, proteins, and metabolites in interlocking metabolic pathways, in transport mechanisms across cellular and intracellular membranes, and in regulatory pathways that translate environmental stimuli into biological responses of cells, which can be modulated for ensuring the successful scaling up of novel bacterial strains synthesizing recombinant products or newly discovered bioactive materials?

The rise of the "omics" technologies and metabolic engineering together with the creation of databases for known genomes, proteins, and metabolic pathways (for example, the Enzymes and Metabolic Pathways project, www.empproject.com) offers a set of computation and information toolkits, which facilitates better understanding of cellular metabolism at the molecular level. Little of this multinational effort has of yet been translated into solving the problems of product formation (yield) and process efficiency with industrial fermentations. Partly this is because many industrial processes use organisms beyond the small and select group (*Escherichia coli*, *Saccharomyces cerevisiae*, etc.)

"CAN A BIOLOGIST FIX A RADIO?"
In this amusing but thought-provoking article (Lazebnik, 2002), the approach of the molecular biologist to dissecting the workings of a complex system (a radio) is shown to be limited by his (or her) lack of appreciation that only when all the components are connected does the "system" work (i.e., broadcast electromagnetic waves are translated into music). The typical radio is full of intriguing components of interesting shapes, sizes, and colors but, however important the study of the isolated parts are to the typical post-doctoral researcher, it is the integrated functioning of the various identifiable items of circuitry that actually matters.

that plays the role of a modern-day Rosetta Stone in unraveling the complexities of cellular metabolism. More generally, however, the quest for mathematically robust methodologies biases the choice of experimental system to those with chemically defined media and under steady-state conditions that are far removed from the realities of industrial fermentations.

The population of microbial or animal cells within fermentors is apparently highly suitable for the modern computational models of the systems biology approach. The problem is that the functions of many of the "components" coded in sequenced genomes are not clearly defined (Westerhoff, 2001). Empirical assumptions and estimates of model components are often explored, and the result is that a number of different mechanistic models can match small bodies of quantitative data with similar degrees of mathematical "fit" and the long-sought ability to predict cellular performance with *in silico* models is still a distant horizon.

The most important question is: what to measure? The application of fast-reaction sampling techniques and modern analytical methods has shown that intracellular pools of metabolites may be either small or large but can change significantly over a time frame of *seconds* (Oldiges et al., 2004). The accurate and successful "capturing" of the transient events that occur intracellularly is therefore a considerable technical feat. Even with the efforts of major research groups, however, these "metabolic snapshots" inevitably conclude that even large changes in the intracellular concentrations of individual metabolites are, by themselves, irrelevant to varying rates of cell growth and productivity (Raamsdonk et al., 2001).

With simple primary products such as carboxylic acids and amino acids a small number (five or less) of features in the functional biology of the producing cells are pivotal to developing a high-yielding commercial bioprocess (Karaffa and Kubicek, 2003; Ikeda, 2003). One (or more) of several hundred other possible biochemical events (and their changes in real time) may, however, determine the *commercial* viability of such a bioprocess against competition from other manufacturers. Nevertheless, the metabolic analysis of such primary products is relatively straightforward because:

- The pathways are short and well-defined.
- There may be a large body of *in vitro* enzymatic data for kinetics and for both inhibitions and activations.
- Transport proteins for product accumulation usually function well against large concentration and electrochemical gradients.
- Classical strain improvement over decades results in producing strains with high conversion efficiencies to desirable end products.

For secondary product producers, evolved over perhaps as much as fifty years of "black box" mutational selection, metabolic analysis must also include:

- Highly distorted patterns of metabolite accumulation but low product yield and/or low carbon conversion into the commercial product.
- The remnants of secondary pathways unrelated to that of the major product.
- Unrecognized products and pathways.
- The ability to utilize nutrients supplied by complex inputs such as cane sugar molasses, corn steep liquors, and yeast extracts.

Quantitative metabolic models can be constructed using measurements of inputs (glucose, ammonia, O_2, etc.) and outputs (cells, CO_2, accumulated primary metabolites, etc.). With a sufficiently large and reliable body of analyses, the data can be modeled with biochemical reaction matrices to investigate rate limitations to growth and product formation, how well carbon and nitrogen balances can be closed, and how much is undefined in the fermentation — frequently the most useful results are when models fail to predict or account for fermentation behavior, i.e., either metabolic "unknowns" accumulate or the relationships between the inputs and outputs are incomplete.

With computer programs of increasing sophistication the totality of "metabolism" can be seen as a matrix of major or minor pathways of fluctuating quantitative importance as growth rates increase and decrease and nutrients are fed or become exhausted (Almaas et al., 2004). Superimposed on the catabolic and anabolic superhighways is, in producers of antibiotics and other secondary metabolites, an array of often highly regulated biosynthetic pathways for which detailed biochemical information may be fragmentary. Reviewing the prospects for genetic engineering in antibiotic-producing *Streptomycetes* a distinguished molecular geneticist concluded (Chater, 1990): "These considerations lead to the hypothesis that many of the beneficial mutations obtained by random mutation and selection in current production strains improve the supply of metabolites and co-factors to the relevant pathway, but that few, if any, of the beneficial mutations cause an increase in the overall expression of pathway-specific genes."

In other words, secondary metabolism in micro-organisms can be viewed as a series of "optional" modules taking precursors from the network of primary pathways and that the mutational selections giving highly productive industrial strains may be the result of changes in *primary* metabolism. Metabolic analysis is therefore often quite literally a mapping procedure, discovering which of the many *possible* biochemical events known

---

CITRIC ACID: FEATURES OF THE BIOCHEMICAL "BLACK BOX"

Given the correct nutritional environment (high sugar concentration, low pH, low Mn(II), high aeration), *Aspergillus niger* can transform glucose to citric acid with an efficiency approaching the theoretical yield. Unregulated diffusion of glucose into the growth-restricted cells, unrestricted entry through the glycolytic pathway, an impaired ability to transform citric acid, a generally "shut down" state of cellular metabolism and the short-circuiting of energy conservation with an alternative oxidase system combine to channel carbon flow to acid formation.

PRIMARY AND SECONDARY METABOLISM

Primary metabolites are those compounds essential for the survival and well-being of the organism; secondary (or natural) products have no apparent utility but might be synthesized under conditions of stress or as agents of interspecies competition (Mann, 1987). Empirically, if a mutant incapable of producing the metabolite can grow as well as the producing parent in a medium no more nutritionally complex that that supporting the growth of the parent, then the metabolite is secondary. To some extent, the genes of secondary metabolite pathways are dispensable — or disposable (easily lost in strain degeneration) as industrial biotechnologists repeatedly discover! Such genetic instability includes both high-level DNA amplification and the spontaneous loss of large segments of DNA (Gravius et al., 1993).

or suspected to occur (Michal, 1999) are necessary and/or sufficient for production at commercially viable levels.

In the following sections, examples drawn from "classical" secondary metabolite productions including antibiotics, anthelmintics, carotenoids, enzymes, and (with animal cell systems) monoclonal antibodies will be used to illustrate the role of metabolic analysis in the development of industrial bioprocesses.

6.2   Secondary product fermentations

Antibiotics emphasize how complex biochemicals are synthesized from primary metabolites (sugars, carboxylic acids, amino acids, lipids, nucleosides, etc.) often by multifunctional enzymes that transform pathway intermediates so rapidly that intracellular pools never accumulate; such pathways are ideal for metabolic engineering to produce novel bioactives (Hutchinson, 1997). The "recruitment" of primary metabolites as biosynthetic precursors is, however, fundamentally the same for any secondary or genetically engineered (recombinant) product of fermentors or bioreactors, and the points of focus will trace relationships and patterns that inevitably repeat and reinforce across very different types of bioprocesses.

6.2.1   Clavulanic acid

Although itself only a weak antibiotic, clavulanic acid is a potent inhibitor of β-lactamases that degrade β-lactam antibiotics and is mixed with semisynthetic penicillins to give more potent mixtures for medical use. By the mid-1990s, clavulanic acid had worldwide sales of over $1 billion and was the second largest selling antibacterial agent (Elander, 2003). Clavulanic acid is one of a family of secondary metabolites produced by *Streptomyces clavuligerus* (Figure 6.1); cross-regulation of at least three different biosynthetic pathways has been recognized (Romero et al., 1984; de la Fuente et al., 2002).

Metabolic analysis was important in elucidating the unusual β-lactam biosynthetic pathway of the clavams, in particular in defining arginine as the amino acid precursor (Baggaley et al., 1997). The full characterization of all the biosynthetic and ancillary genes in the clavulanic acid gene cluster remains to be completed (Jensen et al., 2004), but a partial metabolic model was explored in chemostat cultures under varying nutritional limitations (Kirk et al., 2000).

The degradation of clavulanic acid in fermentations with *S. clavuligerus* has a strong influence on product titers and productivity (Roubos et al., 2002). Moreover, cell-associated decomposition of clavulanic acid accompanies its chemical instability (Mayer and Deckwer, 1996). This source of product loss might explain

Figure 6.1

Secondary metabolites produced by *Streptomyces clavuligerus*.

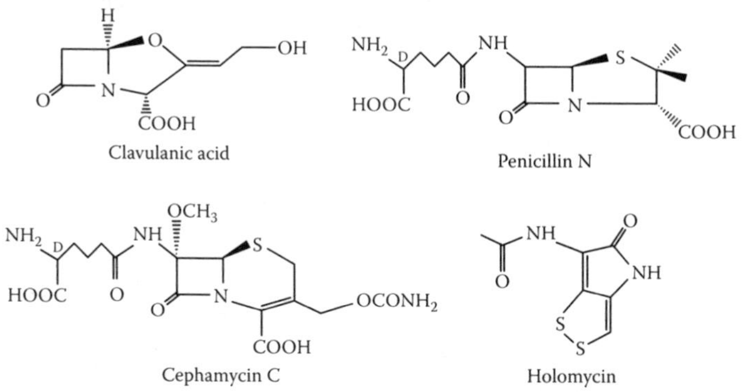

Clavulanic acid

Penicillin N

Cephamycin C

Holomycin

catastrophic failures to maintain clavulanic acid production noted with industrial-scale fermentations (Neves et al., 2001).

Using the published metabolic model as a starting point and including information from other published sources, fermentations for clavulanic acid using a complex medium with soybean meal as a nitrogen source can be modeled in greater depth to include the factors shown in Figure 6.2. In either batch or fed-batch mode, the fermentation will never be in a stationary state and therefore the numerical solutions for the mathematical model will change continuously — changes to both the process and the strain should, however, be interpretable in terms of the parameters of the metabolism and growth kinetics.

6.2.2 *Demethyl-chlortetracyclines*

Tetracyclines have a much longer history of development than clavulanic acid. In 2000, some fifty years after the discovery of this important group of antibiotics, researchers in Japan reported that the polyketide biosynthetic pathway for the 6-methyl family of tetracyclines produced the precursor for melanoid pigments — in fact, the melanin-like pigments (rather than the antibiotic) were the major products of the fermentation (Nakano et al., 2000).

Evidence from both microbial physiology and genetic mutants showed that the branch point occurs before the chlorination step in the lengthy polyketide pathway (Figure 6.3). Surprisingly, the enzyme responsible for a late step in the pathway, anhydrotetracycline (ATC) oxygenase, could use an earlier intermediate as a substrate and, in effect, redirect metabolism towards pigment formation.

MELANINS

Classically, melanin is the product of the action of the enzyme polyphenoloxidase on tyrosine but many chemically unstable products of aromatic metabolism can polymerize to yield the macromolecular "melanoid" pigments that are usually brown to black in coloration, sparingly soluble in water but highly soluble in alkali. "Bio-melanins" as microbial secondary products have been suggested as ingredients for suntan creams. Significantly, four of the twenty-five secondary metabolite gene clusters deduced from the whole-genome sequencing of the industrial avermectin (a macrolide antibiotic and antiparasitic) producer *Streptomyces avermitilis* code for enzymes of melanin pigments, a fifth for a polyketide-derived melanin and a sixth for an aromatic acid-derived pigment — all in addition to gene clusters for polyketides, carotenoids, siderophores, and non-ribosomally synthesized peptides (Omura et al., 2001).

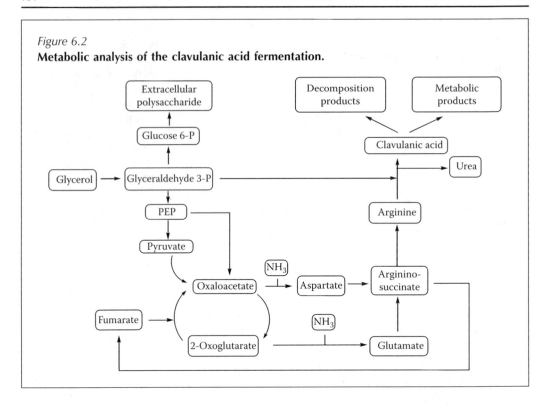

Figure 6.2
Metabolic analysis of the clavulanic acid fermentation.

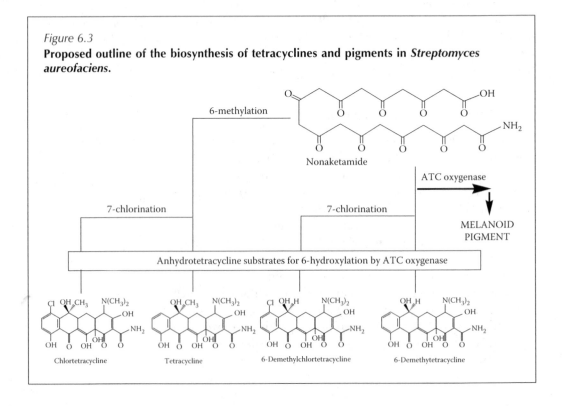

Figure 6.3
Proposed outline of the biosynthesis of tetracyclines and pigments in *Streptomyces aureofaciens*.

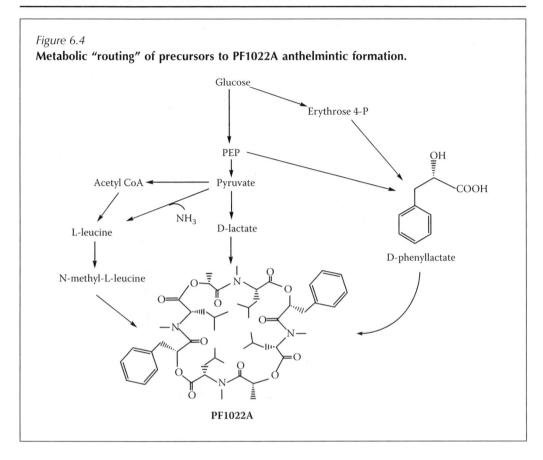

Figure 6.4
Metabolic "routing" of precursors to PF1022A anthelmintic formation.

Industrially, tetracyclines may have been manufactured for decades as "minor" fermentation products, more of the fluxes being used to elaborate pigments as "shunt" metabolites. This suggests that rational enzyme re-engineering or the use of gene-shuffling technologies could radically improve productivity in tetracycline producers.

6.2.3 *A novel cyclic octodepsipeptide*

Cyclic depsipeptides biologically active against nematodes of veterinary and agricultural importance synthesized by isolates of the *Rosellinia* fungus exemplify the problems posed by the development of fermentations for novel species and products (Yanai et al., 2004). The PF1022A depsipeptide anthelmintics typify the biochemical complexity of many secondary products, forming the precursors N-methyl-L-leucine, D-lactic acid, and D-phenyllactic acid from pools of amino acid (leucine), aromatics (L-phenyllactic acid), and glycolytic (L-lactic acid) intermediates as substrates for a nonribosomal peptide synthetase (Figure 6.4). Very little information has been published about the *Rosellinia* organism, but it can be deduced to use, among other inputs,

Figure 6.5
The major pathway of carotenoid biosynthesis in plants and micro-organisms.

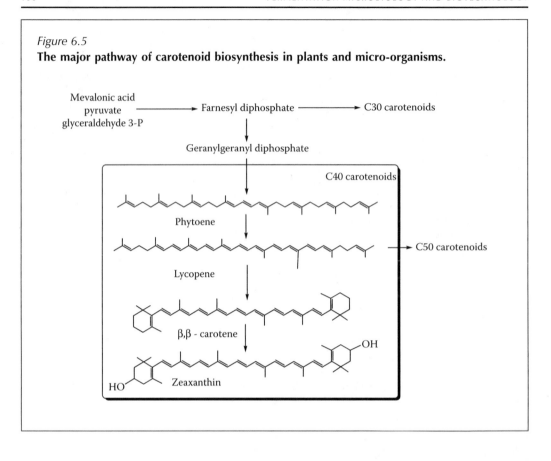

maltose as a carbon source and cottonseed protein as a nitrogen source. D-Phenyllactic acid accumulates in the culture filtrate during the fermentation; future targets for metabolic analysis could include how the supplies of the three separate precursors are balanced and the influence on this on the choice of carbon and nitrogen sources in the medium (Figure 6.4).

6.2.4 *Zeaxanthin*

Zeaxanthin and other carotenoids have been a focus of interest for biotechnological production for over twenty-five years as food coloring agents, antioxidants, vitamin A precursors, and animal feed supplements. *Flavobacterium* and *Erwinia* spp have been used as suitable micro-organisms but cyanobacteria and plasmid-based expression of carotenoid biosynthetic genes in *E. coli* have been explored (Figure 6.5).

Unlike hydrophilic antibiotics, which freely diffuse through an aqueous environment, carotenoids are essentially membrane constituents and are often highly water-insoluble and, although arguably themselves not secondary products, are guides to the behavior of the many antibiotics with poor water solubility. The metabolic "sink" for carotenoids may be the cell membranes

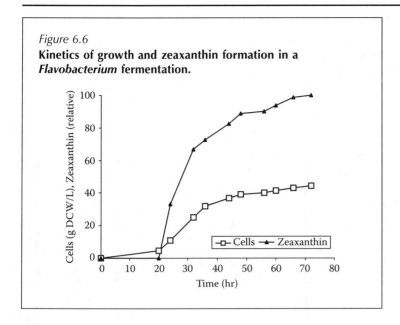

Figure 6.6

Kinetics of growth and zeaxanthin formation in a *Flavobacterium* fermentation.

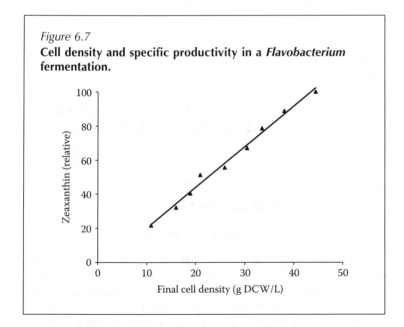

Figure 6.7

Cell density and specific productivity in a *Flavobacterium* fermentation.

in the producing organism, and this imposes the limitation of membrane surface area, i.e., growth.

In a *Flavobacterium*, for example, zeaxanthin production only persisted at a high rate for a few hours after the cessation of active growth (Figure 6.6). The "packing" of the carotenoid to fill up any available space (or volume) in the membrane is possible for only a limited time. Increased cell density can improve production but not specific productivity (Figure 6.7). One biochemical solution is to produce glucosides of zeaxanthin; the

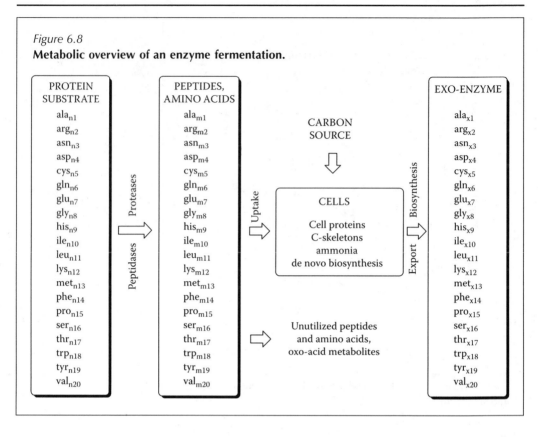

Figure 6.8
Metabolic overview of an enzyme fermentation.

diglucoside of zeaxanthin is 10–100 times more water-soluble than is the parent carotenoid (Hundle et al., 1992).

6.3   Microbial production of industrial enzymes

6.3.1   *Metabolic problems and perspectives*

Although the industrial production of enzymes has a long history, the use of enzymes as catalysts for biotransformations with the view of extending the versatility of organic chemistry is, however, a more recent development. Enzyme manufacturing has included the uses of both solid state and submerged fermentations. Simple bacteria have become favorite organisms, in particular *Bacillus* spp that can readily secrete highly active proteases and other degradative enzymes.

Whatever the organism used, however, the biochemistry of exo-enzyme formation is multifaceted (Figure 6.8). It is simplistic to view enzyme fermentations as simple conversions of input proteins (usually cost-effective sources of nitrogen, for example, monoculture crop proteins) to enzymes of higher value because the bioprocess poses several distinct questions:

- How do the cells degrade the protein substrate in the medium?
- How much of the nitrogen is biologically accessible to the producing cells?
- If peptidic nitrogen sources and ammonium salts are also supplied, do they inhibit protein use from the medium?
- How much *de novo* synthesis of amino acids is required by the producing cells to correct the mismatch between the amino acids in the inputs and those required for growth and exo-enzyme formation?
- Which is the best carbon source and how is it used?
- Which promoters or regulatory genes drive exo-enzyme secretion to high levels?
- How does the organism respond if the supply of immediate precursors (amino acids) fails?

6.3.2 *Protease fermentations*

Alkaline endopeptidases (including subtilisins) have been widely commercialized from *Bacillus* spp for use in industries as diverse as food, brewing, detergent ("bio" washing powder) manufacture, meat processing, and photography. For household detergents, alkaline proteases (optimally active at high pH) and stable at high temperature are widely used; genes encoding such alkaline proteases are widely distributed in micro-organisms and have been cloned and sequenced.

Spore-forming bacteria have an array of regulatory systems some of which act to suppress protease formation until the appropriate developmental stage is reached, often at the end of rapid or exponential growth (Pero and Sloma, 1993). This evolutionarily acquired feature acts to repress the metabolically expensive formation of exo-enzymes if the nutritional conditions (in particular, the availability of low molecular weight carbon and nitrogen sources) make their formation superfluous — an example of how micro-organisms avoid metabolic "burdens."

In chemostat experiments the regulation of protease formation is clearly seen in the dependence of specific protease production rate on specific growth rate (Figure 6.9, see Frankena et al., 1985). Maximal rates of protease formation are, therefore, confined to a narrow range of moderate growth rates; process optimization, therefore, centers on how growth rates can be best managed to maximize the time during which optimal growth rates are sustained while supplying the required precursors for protein synthesis.

If the protease gene(s) are contained on plasmids, a second effect of growth rate becomes apparent, i.e., the loss of nonessential plasmids during repeated cycles of cell division. Although high plasmid content can be ensured by application of a selection pressure (usually the presence of an antibiotic for which a

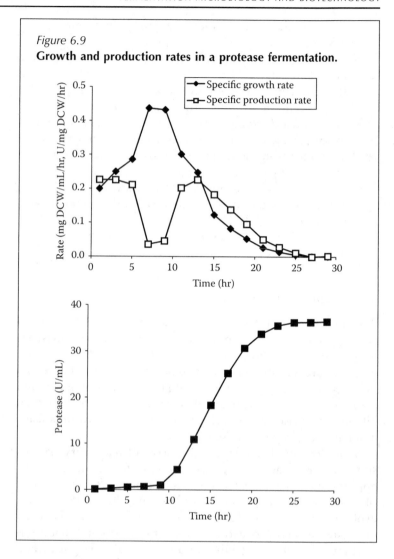

Figure 6.9

Growth and production rates in a protease fermentation.

resistance gene is carried on the plasmid), this is not feasible in large industrial fermentors (Kumar et al., 1991). Consequently, industrial strains have been generated by chromosomal insertion and under the control of constitutive promoters, although the relationship between protease production and gene copy number may be highly nonlinear because of the differing stabilities of the modified genomes with tandem and nontandem introduction of multiple gene copies (van der Laan et al., 1991; Jorgensen et al., 2000).

Within this framework of growth kinetics, the use of preformed amino acids is only fully apparent if analyses of total and peptidic amino acids are combined with measurements of protease activity, specific activity, and an exact knowledge of the primary amino acid sequence of the protease. For each individual amino acid — or, at least, those that survive acid

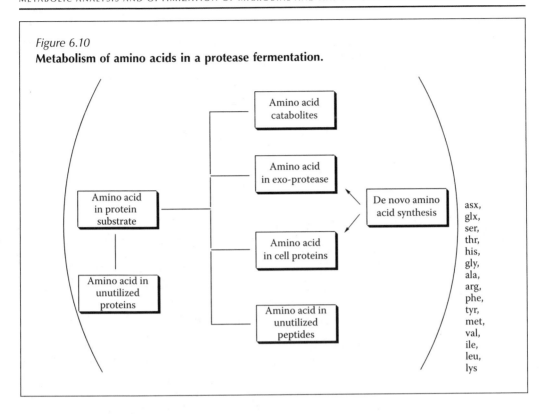

Figure 6.10
Metabolism of amino acids in a protease fermentation.

hydrolysis of proteins and peptides — the sequence of release, solubilization, anabolic use (polymerization into cellular proteins and exo-enzymes), and the catabolism of excess (as carbon and/or nitrogen sources) can be quantified (Figure 6.10). Such analysis can only account for some of the total amino acid use in the fermentation since tryptophan is entirely lost during acid hydrolysis while chemical protecting agents are required for cysteine, cystine, and methionine measurements; asparagine and glutamine are quantitatively converted to aspartate and glutamate, respectively, and only "lumped" values can be obtained.

Some amino acids can be used "excessively," i.e., over and above the requirements for growth and exo-protein production; amino acid and peptide uptake mechanisms are regulated independently of those for glucose and ammonia, and amino acid and peptides can be utilized as both carbon and nitrogen sources, depending on the metabolic requirements of the cell population. As a consequence, successive phases of the fermentation can exhibit competition between the demands of growth and exo-protein synthesis for amino acids and the partial catabolism of amino acids to yield oxo-acids and other degradation products that can be highly malodorous, including phenylacetic acid (formed by the degradation of phenylalanine) and short-chain fatty acids (from branched-chain fatty acids).

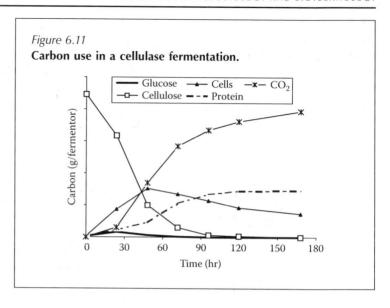

Figure 6.11

Carbon use in a cellulase fermentation.

6.3.3 *Cellulase fermentations*

In 1943 scientists in a United States Army laboratory isolated a mold (now known as *Trichoderma reesei*) from cellulose-based equipment deteriorating in storage. That discovery led to commercial production of cellulases, but the market was small (mostly in the food and drinks industries). However, interest in cellulases was spurred on by the possibility of new processes for the production of useful materials from the vast amounts of lignocellulosic material generated annually — approximately 10^{11} tons of cellulose in plant material per year — as well as from agricultural, horticultural, and urban wastes. Much interest has centered on the production of ethanol (as a fuel additive) from cellulosic biomass sources and, while the feasibility of this sustainable bio-industry has been demonstrated, economic pressures still inhibit its development in the face of low petroleum prices and competition from corn starch-based bioprocesses (for further details, see Chapter 9 on renewable resources conversion to fine chemicals).

"Cellulase" is a combination of three enzymic activities (exo- and endo-β-glucanases and a β-glucosidase) and over 10,000 species of fungi and bacteria are known to produce and secrete cellulases, although *T. reesei* remains the best-studied producer organism. While the biochemistry and molecular biology of cellulases have been extensively studied, the biotechnology of cellulase fermentations has only recently advanced from an empirical base (Lynd et al., 2002).

Despite cellulases being considered of low catalytic activity in comparison with other polysaccharide-degrading enzymes (for example, α-amylases), a quantitative study of carbon use in *T. reesei* showed that cellulose substrate use could be sufficiently fast so that a carbon starvation occurred in batch fermentations

(Sáez et al., 2002). Carbon use by the cellulase-hyperproducing strain was, however, inefficient with more than 75% of the output carbon accounted for as CO_2 (Figure 6.11 — redrawn from data in Sáez et al., 2002).

Since many industrial processes for cellulases are considered to be fed-batch and may use mixtures of carbon and nitrogen sources, a more ambitious metabolic analysis could consider:

- How glucose is metabolized to low molecular weight products (polyols, carboxylic acids, etc.).
- The growth-limiting nutrients in the medium.
- The degree of use of complex nitrogen inputs in the medium (protein, corn steep liquor) as well as ammonium salts.
- The possible selection of carbon-inefficient strains following random mutagenesis.

6.3.4 *Solid-state fermentations – the renaissance of an old technology?*

From a historical perspective, cultivations on solid substrates were the earliest forms of industrial fermentation, the first patent being granted in 1894. Solid-state fermentation (SSF) manufacture became in time suitable only for a small number of fermentations, usually for food or food-associated products, but continued scientific study (especially in India and Japan) has shown several benefits of this technology:

- High enzyme activities.
- Cost-effective use of agricultural waste substrates.
- A wider range of additional enzyme activities than found in submerged fermentations.

Gene expression is radically different under SSF or submerged fermentation conditions (te Biesebeke et al., 2005). In liquid culture, access to polymeric substrates is more rapid than in SSF due to the action of secreted enzymes which, in turn, release nutrients (glucose, amino acids, etc.), thus sustaining a faster growth rate. Under these circumstances, as discussed above for bacterial protease formation, the organism suppresses further exo-enzyme formation.

Detailed metabolic analysis is, therefore, feasible with SSF cultures. More and more SSF processes are being introduced for enzymes with a variety of modern, high-technology SSF fermentors with monitoring and control of temperature, O_2 content, and humidity and, because downstream processing is often much easier with SSF than with submerged fermentations, this manufacturing option is likely to be of increasing importance for speciality enzymes and applications where mixtures of different enzyme activities are required.

GLYCOSYLATED "CASSETTES" FOR IMMUNOGLOBULINS
Oligosaccharides can make up to 12% of the mass of mature immunoglobulin (IgG) antibodies and other proteins of mammalian origin. N-linked to an asparagine residue, the core structure of IgG's consists of N-acetylglucosamine, mannose, and fucose on to which varying combinations of other carbohydrates are attached (galactose, sialic acid, etc.). Quantitative and qualitative features of protein glycosylation are known to be influenced by growth (cell density) and nutrient levels in animal cell cultures, thus greatly affecting the isoelectric properties of the resulting molecules.

6.4 Animal cells and recombinant protein production in bioreactors

6.4.1 Production of biopharmaceuticals: Microbes or animal cells?

The production of monoclonal antibodies and other proteins from mammalian sources for diagnostic or therapeutic use requires the biosynthesis of proteins with precise molecular conformations (protein folding) and reproducible patterns of polypeptide glycosylation. While not essential for antigen-antibody binding, the pattern of glycosylation affects the stability of applied antibodies *in vivo* and the heterogeneity of the purified product. Fast-growing bacteria expressing heterologous gene products may not, therefore, be the ideal vehicles since product quality is of paramount importance and, despite long fermentation cycles and lower (sometimes very low) yields, eukaryotic cells have been used in modern bioprocesses at increasingly large scales of commercial production — and the market (estimated to be $10 billion per year by 2005) is expected to grow rapidly in the next 5–10 years.

6.4.2 Yeasts in stirred-tank fermentors

Yeasts can perform post-transcriptional protein modifications directed by foreign genes, and several different species have been used for the production of recombinant proteins.

Pichia pastoris has the added advantage of being able to grow to high cell densities and produce heterologous proteins using simple carbon sources (glycerol and methanol) with ammonia being the nitrogen source (usually added as ammonium hydroxide solution for pH control) in defined media free from animal-derived ingredients. The alcohol oxidase (*aox-1*) promoter from *P. pastoris* can be incorporated into methanol-inducible expression vectors for foreign genes. This results in a high production of the recombinant protein, and over one hundred different proteins have been produced in this system with correct protein folding and secretion into the medium. The lack of endotoxins in *P. pastoris* makes any produced protein eminently suitable for therapeutic use, and *P. pastoris* (unlike some other yeast) does not hyperglycosylate proteins, has no highly immunogenic cell wall oligosaccharides, and can accumulate the protein product to a high degree (80% of the total secreted protein), thus making downstream processing relatively straightforward.

An attempt to model fermentations with *P. pastoris* incorporated features of the known biochemistry of methanol use (Figure 6.12 — modified from Ren et al., 2003). Alcohol oxidase catalyzes the oxidation of methanol to formaldehyde, which then can be either catabolized (via formic acid) to CO_2 or condensed with xylulose 5-phosphate (a dihydroxyacetone synthase-catalyzed reaction) to give glyceraldehyde 3-phosphate.

Figure 6.12

Metabolic analysis of methanol feeding to a *Pichia pastoris* fermentation.

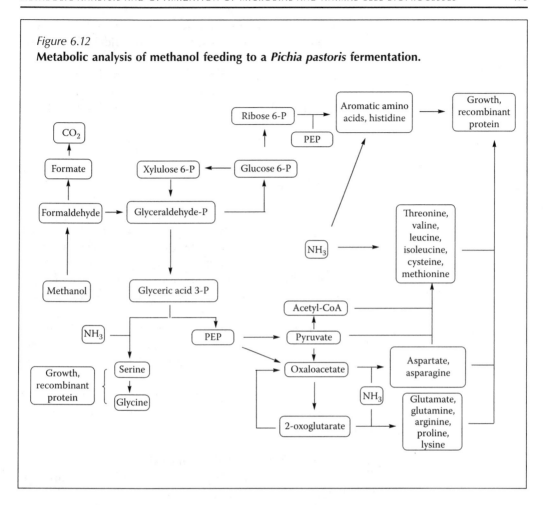

In the published model, only glyceraldehyde 3-phosphate and acetyl-CoA were directed to biomass formations as illustrated in Figure 6.12, which shows a more complete description of anabolic pathways. The biochemical model, combined with feed rates, measured concentrations and estimates for specific rates, etc., accounted for recombinant protein production in fed-batch mode but failed to predict the behavior in a methanol-overfed experiment, where, contradicting the belief that methanol in excess is toxic to *P. pastoris*, the growth rate increased greatly, the recombinant protein was degraded and the accumulated methanol rapidly utilized. An important area for future investigation is in understanding the metabolic responses of protein-secreting *P. pastoris* cells to methanol overfeeding.

One explanation for this unexpected breakdown of *P. pastoris* fermentation could be the known ability of *P. pastoris* to secrete proteases capable of degrading the secreted protein. This is worse in high-density fermentations and has only been partially remedied by adding amino acids, changing the pH, or using protease-deficient strains (Goodrick et al., 2001).

P. pastoris expression systems can also be based on glucose as a carbon source with constitutive glyceraldehyde 3-phosphate dehydrogenase promoters in fed-batch and continuous fermentations (Goodrick et al., 2001). Metabolic engineering of genes from citric acid-accumulating *Aspergillus niger* aims to test the metabolic effects of radically increasing the ability of *P. pastoris* to use glucose as a driving force for protein synthesis (www.antico.strath.ac.uk).

6.4.2 *Animal cells in bioreactors*

Monoclonal antibodies will have their biggest demand (by volume) in the therapy of diseases such as rheumatoid arthritis and multiple sclerosis. Current (2005) yields in animal cell processes for recombinant proteins approximate 1g/L. The estimated demand for cell culture-derived monoclonal antibodies of 5000 kg per year requires a total productive bioreactor volume of 5 million liters.

The major focus on improving the productivity of animal cell cultures has, therefore, been on gene expression. Process development has to be achieved within a narrow time window between proof of concept studies and the production of clinical-grade material under Good Manufacturing Practice-compliant conditions; obvious targets for yield enhancement commonly include:

- Selection of the highest yielding cell lines.
- Medium development by empirical experiments, statistical design, or the analysis of spent medium.
- Optimizations of hydrogen ion concentration (pH) and temperature.
- Investigation of the effects of osmotic concentration.
- Removal of CO_2.
- Use of larger bioreactors (10,000 L or larger in volume).
- Growth to higher cell densities in fed-batch processes.

Cell media evolved rapidly when the bovine spongiform encephalitis epidemic rendered animal-derived products undesirable as ingredients. Calf serum-containing media were replaced by either chemically defined recipes or media containing plant protein hydrolysates as the sole "magic ingredient" input, and this trend continues in the development of both commercial and in-house media with Chinese Hamster Ovary (CHO) and other widely used animal cells.

Glucose is the major carbon source and is metabolized via glycolysis, lactic acid synthesis, and the tricarboxylic acid (Figure 6.13). The other major metabolic fates of the supplied glucose are interconversion to other hexoses (for protein

Figure 6.13
Metabolic analysis of glucose and amino acids in an animal cell culture.

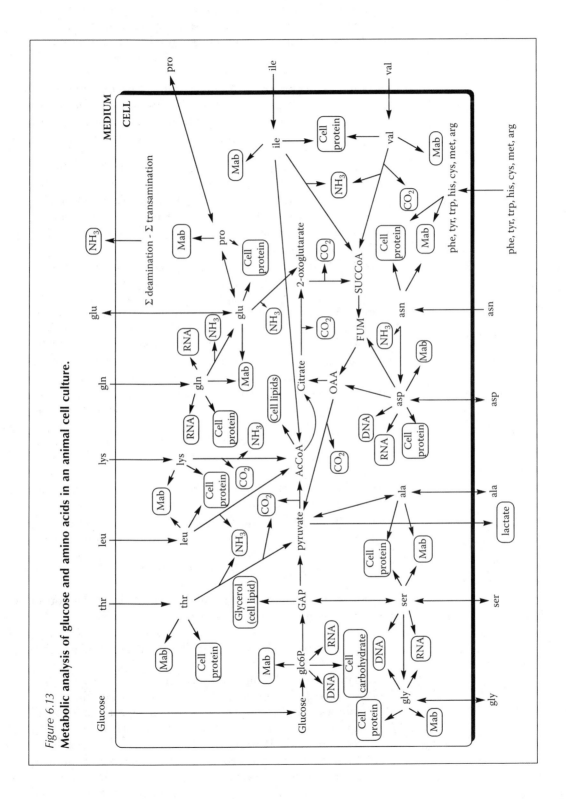

glycosylation), via the pentose phosphate pathway to ribose and deoxyribose for nucleic acid synthesis, and for cell lipids.

Glutamine is the major amino acid used as a nitrogen source, but other members of the traditionally defined group of "nonessential" amino acids are significant precursors of nucleic acid biosynthesis (Figure 6.13). At a recombinant protein titer of 1g/L and a total cell density of 1×10^7 cells/mL, calculations of amino acid use shows that 12–30% of the glutamine, alanine, glycine, and serine is used for recombinant protein production. Other reactions of intermediary metabolism can lead to the accumulation of the nonessential amino acids, for example, the use of serine as a C1 donor for nucleic acid synthesis inevitably produces glycine in a molar yield and alanine is often found as an "output" because the amination of pyruvic acid results in alanine in a similar way to pyruvate reduction producing lactic acid (Figure 6.13).

Of the amino acids essential for cell growth, several (valine, leucine, isoleucine, threonine, and lysine) can readily be deaminated (as nitrogen sources) and their carbon skeletons transformed, as in bacterial fermentations, to intermediates of the central metabolic pathways; these reactions are well-known in metabolic analyses of animal cell cultures (Nyberg et al., 1999b). Other essential amino acids (and including cystine, methionine, arginine, histidine, and tyrosine under this heading) are probably only used for cellular and recombinant protein synthesis. The minimum set of measured rates for glucose and amino acid use (or accumulation) and lactate and ammonia formation is therefore 23; a working metabolic rate model can be constructed if the following calculated rates are included:

- Forty rates of amino acid polymerization (cellular and recombinant protein).
- Eight rates of amino acid use for nucleic acid synthesis.
- Six rates of glucose and glycolytic intermediate use for cell constituents and recombinant protein.
- Thirty-four intracellular conversions.

To this working model of recombinant protein production in animal cell cultures three other factors can be added:

1. To minimize the chemical degradation of glutamine, dipeptides of glutamine (with either glycine or alanine) have been used as media ingredients.
2. Small peptides in general are known to be amino acid sources, for example, peptides from a peptone contributed significantly to amino acid use by cultured hybridoma cells (Nyberg et al., 1999a).
3. Glucose can be replaced as the carbon source by another, more slowly metabolizable sugar to maintain cell viability and recombinant protein production for longer and

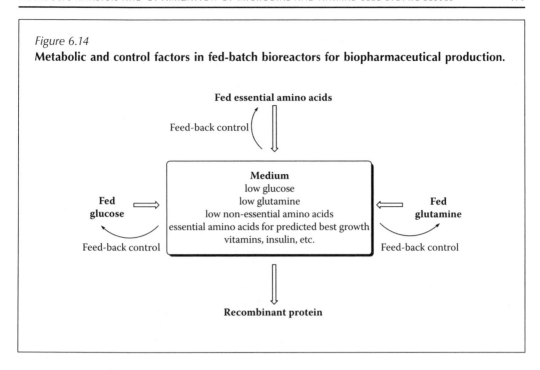

Figure 6.14
Metabolic and control factors in fed-batch bioreactors for biopharmaceutical production.

to reduce the accumulation of lactic acid, ammonia, and nonessential amino acids (Altamirano et al., 2001).

Metabolism in these variations of fed-batch processes can be analyzed in as great a detail as are microbial cultures of growth rates (for total and viable cells), estimates of cellular composition and recombinant protein concentrations are included in the data sets. Using this knowledge-based approach, bioreactors can be controlled more precisely through on-line monitoring of nutrient use rates to determine feed rates and the composition of the fed solutions to minimize the accumulation of lactate and ammonia and avoid depletion of any essential amino acid (Figure 6.14).

In fed-batch processes the maximum rate and specific rate of recombinant protein production can occur in the phase of net decrease in viable cell count. Figure 6.15 shows results redrawn and calculated from those published by a manufacturer of animal cell media (Gorfien et al., 2003). This production rate profile is quite unlike that found with secondary metabolite producers where lysis of the cell population is often associated with a drastic decline in productivity — indeed, the phenotype is neither that of growth-linked production or purely growth-independent production, both of which patterns are found with antibiotic-producing strains. Three factors may contribute to this feature of animal cell cultures:

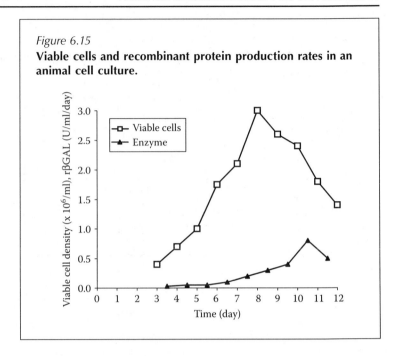

Figure 6.15
Viable cells and recombinant protein production rates in an animal cell culture.

1. Net loss of viable cells reduces competition between growth and recombinant protein synthesis (Figure 6.15).
2. Cell lysis liberates cellular materials (Nyberg et al., 1999b) that may contribute to the productive capacity of the surviving cells.
3. Regulatory genes may be activated to increase the rates of quantitative gene expression, mRNA translation, or polypeptide processing to secrete the mature protein.

6.5   Future prospects

Whole-organism genome sequencing of industrial microorganisms is gathering momentum and although these sequences will not rapidly reach the public domain, biochemical genomics and proteomics will gradually replace natural product chemistry as the preferred route to defining novel secondary pathways (Bormann, 2002). In this direction, it is interesting to see that the pathway for nikkomycin antibiotics, produced by *Streptomyces tendae*, involves no less than twenty genes deduced by:

1. Sequencing of the genomic region containing the *nik* cluster.
2. Comparison of the open reading frames with protein sequence databases and assignment of functionalities by homologies with the coding sequences of known enzymes.
3. Characterization of disruption mutants.
4. Heterologous expression and determination of enzymic activities.

Genome sequencing also revealed that the chromosomal DNA in *Streptomycetes* is large compared with "simpler" bacteria (*E. coli,* etc.) and codes for many secondary metabolite pathways, some of which are either "silent" or poorly expressed — unless, that is, the correct nutritional or environmental circumstances arise, for example, unpredictably in rapid growth/stationary phases during fed-batch fermentations and that might be adventitiously amplified during strain development.

Proteomics is, despite technical problems with assessing quantitative differences in expression profiles, being increasingly applied to analyzing the changes in abundance of biosynthetic enzymes in antibiotic producers. The enzymes of chlortetracycline biosynthesis in *S. aureofaciens* have been shown to be prominent intracellular proteins (Li et al., 2001). Future work will investigate how expression levels change (or have changed) during strain development and explore why, in the case of the tetracycline producer, very different kinetics of biosynthetic enzyme activities were found in the production phase and the implications of these findings for different theories of metabolic control and the regulation of enzyme activities.

None of the well-studied microbial "biological factories" is entirely problem-free for the production of enzymes or recombinant proteins. An ideal organism would grow rapidly to a high cell density and secrete the commercial protein at a continuously high rate (or, more realistically, for a prolonged time in comparison with the growth phase) in a stable culture independent of the effects of regulatory genes or developmental changes and limited only by the applied feed rates. Perfusion cultures have been a novel technology for achieving this with animal cells but, like "traditional" stirred-tank vessels, innovative approaches to accurately matching nutrient supply to productivity and maximizing cellular longevity will continue to be sought.

In the medium- to long-term view, recombinant proteins and antibiotics present radically different scenarios for future development. Hundreds of monoclonal antibodies are undergoing clinical trials; of these, many may not prove to be either effective or without side effects but enough will eventually require biomanufacturing to seriously challenge current and predicted production capacities. Few novel antibiotics are, on the other hand, being actively commercialized. The use of a relatively narrow range of medically important anti-infectives has been compromised by the evolution of resistant pathogens; the recognition of possible chronic contraindications to the use of major prescription drugs would rapidly undermine the struggle against natural and hospital-acquired diseases (Fallon, 2005). Many thousands of antibacterial and antifungal secondary metabolites have been discovered, a proportion of which probably had quite different modes of action from those of the well-established antibiotics such as penicillins but were not fully developed because they lacked any obvious clinical advantage over antibiotics then in clinical use. In parallel with investigation of secondary metabolites with anti-tumor, antiviral, and other actions, revisiting the enormous spectrum of antibiotic compounds elaborated by *Actinomycetes* and other organisms might be combined with fast-tracked strain development and process optimization to yield cost-effective alternatives in a pharmaceutical armory to which microbes will continue to adapt with their relentless metabolic ingenuity.

Summary

Metabolic analysis is the meeting point of biochemistry, cellular physiology, and process engineering. Examples selected from the landscape of microbial and animal cell fermentations for secondary metabolites, enzymes, and recombinant proteins demonstrate how kinetic events work in a framework of growth rates, feed rates, the expression of genes (and their changes in real time), and morphological transitions (including, at one extreme, cell death). For industrial purposes, the challenge is to define practical data sets that can be measured and used to devise control mechanisms for both growth and product formation.

Biochemistry and enzymology remain the key routes to discovering the unexpected features of metabolism in secondary product-forming bacteria, yeast, and fungi; the "unknowns" may be unrecognized products and pathways as well as the accumulation of side products that may have been amplified over generations of empirical strain selection. Significant features of both novel and established fermentations may include pathways to both divert precursors from the desired biosynthetic product and further metabolize the supposed end product of a long and complex exercise in "natural product chemistry" from primary metabolites.

Enzymes produced by bacteria, yeasts, and fungi illustrate the interactions between growth, developmental stages in a life cycle, and the metabolic utilization of components from media of great chemical complexity.

The metabolic science of animal cells used to manufacture biopharmaceuticals is much simpler but the simultaneous use of multiple carbon and nitrogen sources, the conflicting demands for growth and recombinant protein synthesis, and the unproductive use of some components from batched and fed media result in a surprising complexity of factors determining both product concentration and quality (the micro-heterogeneity of protein glycosylation).

References

Almaas, E., B. Kovács, T. Vicsek, Z. N. Oltvai, A. L. Barabási (2004). Global organization of metabolic fluxes in the bacterium *Escherichia coli*, *Nature* **427**, 839–843.

Altamirano, C., J. J. Cairó, F. Gòdia. (2001). Decoupling cell growth and product formation in Chinese Hamster Ovary cells through metabolic control, *Biotechnol. Bioeng.* **76**, 351–360.

Baggaley, K. H., A. G. Brown, C. J. Schofield (1997). Chemistry and biochemistry of clavulanic acid and other clavams, *Nat. Prod. Rep.* **140**, 309–333.

Bormann, C. (2002). Biosynthesis of the peptidyl nucleoside antibiotic nikkomycin in *Streptomyces tendae* Tü901 deduced from the analysis of the gene cluster and mutational studies, in Fierro, F. and Martín, J.F. (eds.) *Microbial Secondary Metabolites: Biosynthesis, Genetics and Regulation*, Trivandrum: Research Signpost, pp.43–61.

Chater, K. F. (1990). The improving prospects for yield increase by genetic engineering in antibiotic-producing Streptomycetes, *Bio/Technology* **8**, 115–121.

de la Fuente, A., L. M. Lorenzana, J. F., Martín, P. Liras (2002). Mutants of *Streptomyces clavuligerus* with disruptions in different genes for clavulanic acid biosynthesis produce large amounts of holomycin: Possible cross-regulation of two unrelated secondary metabolic pathways, *J. Bacteriol.* **184**, 6559–6565.

Elander, R. P. (2003). Industrial production of β-lactam antibiotics, *Appl. Microbiol. Biotechnol.* **61**, 385–392.

Fallon, J. (2005) Could one of the most widely prescribed antibiotics amoxicillin/clavulanate "augmentin" be a risk factor for autism? *Med. Hypotheses* **64**, 312–315.

Frankena, J., H. W. van Verseveld, A. H. Stouthamer (1985). A continuous culture study of the bioenergetic aspects of growth and production of exocellular protease in *Bacillus licheniformis*, *Appl. Microbiol. Biotechnol.* **22**, 169–176.

Goodrick, J. C., M. Xu, R. Finnegan, B. M. Schilling, S. Schiavi, H. Hoppe, N. C. Wan (2001). High-level expression and stabilization of recombinant human chitinase produced in a continuous constitutive *Pichia pastoris* expression system, *Biotechnol. Bioeng.* **74**, 492–497.

Gorfien, S. F., W. Paul, D. Judd, L. Tescione, D. W. Jayme (2003). Optimized nutrient additives for fed-batch cultures, *BioPharm International* **16 no. 4**, 34–40.

Gravius, B., T. Bezmalinović, D. Hranueli, J. Cullum (1993). Genetic instability and strain degeneration in *Streptomyces rimosus*, *Appl. Environ. Microbiol.* **59**, 2220–2228.

Hundle, B. S., D. A. O'Brien, M. Alberti, P. Beyer, J. E. Hearst (1992). Functional expression of zeaxanthin glucosyltransferase from *Erwinia herbicola* and a proposed uridine diphosphate binding site, *Proc. Natl. Acad. Sci. USA* **89**, 9321–9325.

Hutchinson, C. R. (1997) Antibiotics from genetically engineered micro-organisms, in Strohl W. R. (ed.) *Biotechnology of Antibiotics*, New York and Basel: Marcel Dekker, Inc., pp. 683–702.

Ikeda, M. (2003) Amino acid production processes, in Faurie, R. and Thommel, J. (eds.) *Advances in Biochemical Engineering/Biotechnology*, vol. 79, Microbial production of L-amino acids, Berlin: Springer-Verlag, pp. 1–35.

Jensen, S. E., A. S. Paradkar, R. H. Mosher, C. Anders, P. H. Beatty, M. J. Brumlik, A. Griffin, B. Barton (2004). Five additional genes are involved in clavulanic acid biosynthesis in *Streptomyces clavuligerus*, *Antimicrob. Agents Chemother.* **48**, 192–202.

Jorgensen, P. L., M. Tangney, P. E. Pedersen, S. Hastrup, B. Diderichsen, S. T. Jorgensen (2000). Cloning and sequencing of an alkaline protease gene from *Bacillus lentus* and amplification of the gene on the *B. lentus* chromosome by an improved technique, *Appl. Microbiol. Biotechnol.* **66**, 825–827.

Karaffa, L., C. P. Kubicek (2003). *Aspergillus niger* citric acid accumulation: do we understand this well working black box? *Appl. Microbiol. Biotechnol.* **61**, 189–196.

Kirk, S., C. A. Avignone-Rossa, M. E. Bushell (2000). Growth limiting substrate affects antibiotic production and associated metabolic fluxes in *Streptomyces clavuligerus*, *Biotechnol. Lett.* **22**, 1803–1809.

Kumar, P. K. R., H-E. Maschke, K. Friehs, K. Schügerl (1991). Strategies for improving plasmid stability in genetically modified bacteria in bioreactors, *TIBTECH* **9**, 279–283.

Lazebnik, Y. (2002). Can a biologist fix a radio? Or, what I learned while studying apoptosis, *Cancer Cell* **2**, 179–182.

Li, X. M., J. Novotná, J. Vohradský, J. Weiser (2001). Major proteins related to chlortetracycline biosynthesis in a *Streptomyces aureofaciens* production strain studied by quantitative proteomics, *Appl. Microbiol. Biotechnol.* **57**, 717–724.

Lynd, L. R., P. J. Weimer, W. H. van Zyl, I. S. Pretorius (2002). Microbial cellulose utilization: fundamentals and biotechnology, *Micro. Mol. Biol. Rev.* **66**, 506–577.

Mann, J. (1987). *Secondary Metabolism*, Oxford: Clarendon Press.

Mayer, A. F., W. D. Deckwer (1996). Simultaneous production and decomposition of clavulanic acid during *Streptomyces clavuligerus* cultivations, *Appl. Microbiol. Biotechnol.* **45**, 41–46.

Michal, G. (1999). *Biochemical Pathways. An Atlas of Biochemistry and Molecular Biology*, New York: John Wiley & Sons, Inc.

Nakano, T., K. Miyake, M. Ikeda, T. Mizukami, R. Katsumata (2000). Mechanism of the incidental production of a melanin-like pigment during 6-demethylchlortetracycline production in *Streptomyces aureofaciens*, *Appl. Environ. Microbiol.* **66**, 1400–1404.

Neves, A. A., L. M. Vieira, J. C. Menezes (2001). Effects of preculture on clavulanic acid fermentation, *Biotechnol. Bioeng.* **72**, 628–633.

Nyberg, G. B., R. R. Balcarcel, B. D. Follstad, G. Stephanopoulos, D. I. C. Wang (1999a). Metabolism of peptide amino acids by Chinese Hamster Ovary cells grown in a complex medium, *Biotechnol. Bioeng.* **62**, 324–335.

Nyberg, G. B., R. R. Balcarcel, B. D. Follstad, G. Stephanopoulos, D. I. C. Wang (1999b). Metabolic effects on recombinant Interferon-γ glycosylation in continuous culture of Chinese Hamster Ovary cells, *Biotechnol. Bioeng.* **62**, 336–347.

Oldiges, M., M. Kunze, D. Degenring, G. A. Sprenger, R. Takors (2004). Stimulation, monitoring and analysis of pathway dynamics by metabolic profiling in the aromatic amino acid pathway, *Biotechnol. Prog.* **20**, 1623–1633.

Omura, S., H. Ikeda, J. Ishikawa, A. Hanamoto, C. Takahashi, M. Shinose, Y. Takahashi, H. Horikawa, H. Nakazawa, T. Osonoe, H. Kikuchi, T. Shiba, Y. Sakaki, M. Hattori (2001). Genome sequence of an industrial micro-organism *Streptomyces avermitilis*: Deducing the ability of producing secondary metabolites, *Proc. Natl. Acad. Sci. USA* **98**, 12215–12220.

Pero, J., A. Sloma (1993). Proteases, in Sonensheim, A. L., Hoch J. A. and Losick, R. (eds.) Bacillus subtilis and Other Gram-Positive Bacteria: Biochemistry, Physiology, and Molecular Genetics, Washington, D.C: ASM Press, pp. 939–952.

Raamsdonk, L. M., B. Teusink, D. Broadhurst, N. Zhang, A. Hayes, M. C. Walsh, J. A. Berden, K. M. Brindle, D. B. Kell, J. J. Rowland, H. V. Westerhoff, K. van Dam, S. G. Oliver (2001). A functional genomics strategy that uses metabolome data to reveal the phenotype of silent mutations, *Nature Biotechnology* **19**, 45–50.

Ren, H. T., J. Q. Yuan, K. H. Bellgardt (2003). Macrokinetic model for methylotropic *Pichia pastoris* based on stoichiometric balance, *J. Biotech.* **106**, 53–68.

Romero, J., P. Liras, J. F. Martín (1984). Dissociation of cephamycin and clavulanic acid biosynthesis in *Streptomyces clavuligerus*, *Appl. Microbiol. Biotechnol.* **20**, 318–325.

Roubos, J. A., P. Krabben, W. T. A. M. de Laat, R. Babuška, J. J. Heijnen (2002). Clavulanic acid degradation in *Streptomyces clavuligerus* fed-batch cultivations, *Biotechnol. Prog.* **18**, 451–457.

Sáez, J. C., D. J. Schell, A. Tholodur, J. Farmer, J. Hamilton, J. A. Colucci, J. D. McMillan (2002). Carbon mass balance evaluation of cellulase production on soluble and insoluble substrates, *Biotechnol. Prog.* **18**, 1400–1407.

te Biesebeke, R., N. van Biezen, W. M. de Vos, C. A. M. J. J. van den Hondel, P. J. Punt (2005). Different control mechanisms regulate glucoamylase and protease gene transcription in *Aspergillus oryzae* in solid-state and submerged fermentation, *Appl. Microbiol. Biotechnol.* **67**, 75–82.

Westerhoff, H. V. (2001). The silicon cell, not dead but live! *Metabolic Engineering* **3**, 207–210.

van der Laan, J., G. Gerritse, L. J. S. M. Mulleners, R. A. C. van der Hoek, W. J. Quax (1991). Cloning, characterization, and multiple chromosomal integration of a *Bacillus* alkaline protease gene, *Appl. Environ. Microbiol.* **57**, 901–999.

Yanai, K., N. Sumida, K. Okakura, T. Moriya, M. Watanabe, T. Murakami (2004). *Para*-position derivatives of fungal anthelmintic cyclodepsipeptides engineered with *Streptomyces venezuelae* antibiotic resistance genes, *Nature Biotechnology* **22**, 848–855.

7

Flux Control Analysis: Basic Principles and Industrial Applications

E.M.T. El-Mansi and
Gregory Stephanopoulos

7.1 Introduction: traditional versus modern concepts

Industrial microbiologists, biochemists, and engineers are generally, but not entirely, of the view that flux through a given pathway is usually limited by one step. Such a step is termed the "rate-limiting" step or the "bottleneck" with the enzyme catalyzing such a step being referred to as the "pacemaker." However, the question of how such a step can be identified and quantified in a given pathway remained unanswered, largely due to the lack of an experimental procedure that describes how such a parameter, the rate-limiting step, can be identified and quantified.

Clearly, if a rate-limiting step exists in a given pathway, then increasing the activity of the enzyme catalyzing such a step will increase the overall flux through the pathway and, by the same token, varying the activity of any other enzyme will have no effect whatsoever on the overall flux of the pathway in question. Although a number of studies have been published in support of the concept of the rate-limiting step, the majority of studies that do not support it have not found their way into the public domain. However, a few studies have been published and the findings clearly demonstrate the inadequacy of such a concept. For example, a 3.5-fold (350%) increase in the activity of phosphofructokinase, the enzyme widely regarded as the rate-limiting step in glycolysis, had no appreciable effect on the flux through the glycolytic route of *Saccharomyces cerevisiae*.

By and large, the concept of the rate-limiting step does not appear to be a tenable proposition as it does not adequately

The rate-limiting step was defined as the slowest step in a given pathway. Such a definition resulted from the observation made in the mid-1960s that the reaction rate of a sequence of unsaturated enzymes (i.e., where the concentration of substrates is below the Km value for each of the enzymes involved) depended nonlinearly on the kinetic parameters of all the enzymes involved. However, no theoretical basis was given to validate or substantiate the existence of such a concept.

explain why flux and, in turn, yield could not be improved following the overexpression of enzymes that are considered to be rate-limiting. For example, pioneering work on the production of lysine by *Corynebacterium glutamicum* by Stephanopoulos and Vallino (1991) identified PEP carboxylase as rate-limiting in this process. Although their conclusion was based on well-founded reasons, further studies revealed that full inactivation of this enzyme had no effect on flux to lysine formation (Gubler et al., 1994). Furthermore, the enzymes that catalyze rate-limiting steps tend to be subject to feedback regulation and any changes to their properties will inevitably affect the overall velocity of flux through the pathway.

The erroneous nature of the concept of the rate-limiting step has also been emphasized in the elegant example given on tryptophan biosynthesis in the yeast *Saccharomyces cerevisiae* (Cornish-Bowden 1995; Cornish-Bowden et al., 1995). Biosynthesis of tryptophan in this organism involves the conversion of chorismate to anthranilate through the activity of anthranilate synthase (E1). The latter intermediate is then converted to tryptophan through the activities of four different enzymes namely, anthranilate phosphoribosyl transferase (E2), phosphoribosyl anthranilate isomerase (E3), indolglycerol phosphate synthase (E4), and tryptophan synthase (E5). In this study the aforementioned authors revealed that increasing the enzymic activity of any of the five enzymes involved singularly by a factor of up to 50-fold had no significant effect on the flux to tryptophan. On the other hand, however, increasing the concentration of all the five enzymes by a factor of 20-fold was accompanied by a significant increase in flux to tryptophan formation.

A number of techniques have been devised to assess the relative contribution of each enzyme to the overall velocity of a given pathway and although *in vitro* measurement of the maximal velocity of a particular enzyme is useful, the value obtained does not necessarily reflect the rate of catalysis *in vivo*, due to a lack of hard information regarding the intracellular level of substrates and effectors, not to mention the rapid "turnover" of metabolic pools, which renders such assessment difficult if not impossible. A new quantitative approach was therefore called for, not only to explain the many outstanding observations relating to flux control in industrial fermentations but also to provide a rational basis for the exploitation of the diverse array of metabolic pathways.

Although a number of approaches have been used as an alternative to the 'rate-limiting' step, including sensitivity analysis, an approach used to tackle similar problems in economics, the biochemical systems theory (Savageau et al., 1987), and the "metabolite balance technique" used for the calculation of carbon flux to acetate excretion (El-Mansi and Holms, 1989), amino acid production (Vallino and Stephanopoulos, 1993), and in general (Holms, 1996), it is Kacser's theory of metabolic control analysis (MCA) that has grown in stature since its inception in 1973 and proved, without undermining the intellectual capacity of other approaches, to be the ultimate approach. The controversy over the question of whether a given method can be used successfully to predict or determine the relative contribution of each enzyme to the overall flux in a given pathway was finally resolved when Kacser and Burns (1973) and Heinrich and Rapaport (1974) independently proposed the aforementioned theory of MCA. The fundamental difference between the rate-limiting step as a concept and that of the theory of MCA is that while the former is a qualitative parameter, the latter examines biological systems in a quantitative way that excludes bias, expectations, or preconceived ideas.

7.2   Flux control analysis: basic principles

Kacser's MCA theory facilitates the assessment of not only how perturbation of a particular enzymic activity affects metabolic flux, but also by how much. The response to changes in the concentration of a particular enzyme on flux varies over a wide range. For example, the response could be immediate, with a strong correlation between the increase in flux and the increase in enzymic activity, as is the case for adenylate kinase en route to ATP synthesis. In this case one might justifiably describe such an enzyme or a step as rate-limiting. The majority of enzymes, however, do not enjoy such a high profile and as such an increase in flux may or may not be brought about by the increase in enzymic activity.

Furthermore, the degree of control exerted by a particular enzyme on the overall flux of a given pathway is not purely dependent on the numerical value of its intracellular concentration but rather on whether the enzyme has the capacity for higher throughput, which can only be ascertained from the value of the enzyme's flux control coefficient, the first pillar in the theory of MCA.

7.2.1   *The flux control coefficient*

The flux control coefficient is a parameter that describes in quantitative terms the relative contribution of a particular enzyme to flux control in a given pathway. It is not an intrinsic property of the enzyme *per se* but rather a system property and so is subject to change as the environment changes. It is generally expressed as the fractional change in flux in response to a fractional change in the concentration of the enzyme in question, and its value ranges between 0 and 1.0.

The measurement of the flux control coefficient of a particular enzyme allows an accurate prediction of how flux through a

Figure 7.1
The enzymes and metabolites *en route* to acetate excretion.

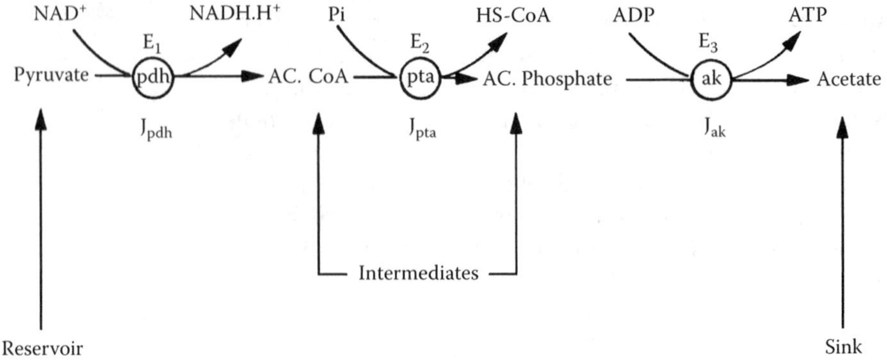

Key: pdh (E_1), pyruvate dehydrogenase
      pta (E_2), phosphotransacetylase
      ak (E_3), acetate kinase

given pathway might fluctuate in response to a specific change in the enzyme's catalytic activity or concentration. While a change in the concentration can be brought about by cloning and subsequent overexpression of the structural gene encoding the enzyme in question, changes in the catalytic activity of the enzyme without changing its concentration can be brought about through site-directed mutagenesis and protein engineering techniques. For example, consider pyruvate dehydrogenase (PDH) with the view of assessing its impact or influence on flux to acetate excretion in *E. coli* (Figure 7.1). The influence of PDH in that direction can be assessed from the enzyme's flux control coefficient, which can be calculated from the tangent to the curve of a log-log plot of flux (J) as a function of enzymic activity or concentration (E). Assuming that a small increase in the concentration of PDH (dEpdh) was accompanied by a small increase in the steady-state flux (J) of the enzyme acetate kinase (dJak), it follows, if we were to change the concentration of PDH very slightly, then the ratio dJak/dEpdh becomes equal to the slope of the tangent to the curve of Jak against Epdh as depicted in Figure 7.2. Analyzing the data in this way, however, is somewhat imperfect as the numerical value of enzyme concentration and units of enzymic activity will be different from one enzyme to another. This problem could be overcome if we were to relate the fractional changes in flux through acetate kinase to the fractional increase in the concentration of PDH, i.e., dJak/Jak and dEpdh/Epdh, and as such the flux control coefficient will assume a value between 0 and 1.0, which can then be expressed in terms of a percentage.

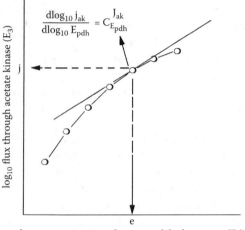

Figure 7.2

Determination of the flux control coefficient of pyruvate dehydrogenase (pdh) with respect to acetate excretion. The graph shows a typical pattern of variations in flux to acetate — measured as acetate kinase (Jak) — in response to changes in the concentration of pyruvate dehydrogenase (Epdh). Although $\log_{10}$ is used, natural logs can also be used providing that the data on the concentration of the enzyme and that on the flux are treated in the same way.

$$\frac{d\log_{10} j_{ak}}{d\log_{10} E_{pdh}} = C^{J_{ak}}_{E_{pdh}}$$

$\log_{10}$ flux through acetate kinase (E_3)

j

e

$\log_{10}$ concentration of pyruvate dehydrogenase (E_1)

However, it is possible for an enzyme to have a flux control coefficient with a negative value, as is the case at branch points where one metabolite has to be partitioned between two enzymes. In such a case the increase in flux through one branch is generally at the expense of the other, as exemplified in the case study for the partition of carbon flux at the junction of isocitrate (see Section 7.3). At this junction, any increase in the concentration of isocitrate dehydrogenase (ICDH) is concomitant with a decrease in flux through the competing enzyme, namely isocitrate lyase (ICL). It is possible, therefore, to describe ICDH as having a negative flux control coefficient on flux through ICL. While any increase in the concentration of ICDH is accompanied by a decrease in flux through ICL, the opposite is not true for reasons that will become apparent later on; for further details, see El-Mansi et al. (1994).

7.2.2 The summation theorem

The summation theorem, the second pillar of the MCA theory, states that the total sum of flux control coefficients of all enzymes in a given pathway adds up to 1.0. The summation theorem also shows that the flux control coefficient of an enzyme is a system property, because any increase in the concentration of a particular enzyme is accompanied by a decrease in its flux control coefficient. Such a decrease, according to the summation theorem, will have to be balanced by increasing the flux control coefficient of another enzyme — or more than one enzyme — within the same pathway so that the sum of all flux control coefficients remains constant, i.e., 1.0. For example, in a linear pathway consisting of enzymes with usual kinetic properties (i.e., where substrates stimulate and products inhibit reaction rate), the flux control coefficients for every enzyme must be 0 or higher, with a total sum of 1.0. If an enzyme were to show a flux control coefficient of 1.0 with all other enzymes showing flux control coefficients of 0, such an enzyme could justifiably be described as rate-limiting. The summation theorem also shows that this is not necessarily the case, because it is also possible for some or all of the enzymes to have values greater than 0 providing that the total does not exceed 1.0. In practice, we would expect a pathway flux to be influenced mainly by enzymes in that pathway, and to a much lesser extent by closely related pathways, and that distantly connected enzymes would have negligible influence or none at all. In other words, the flux control coefficients of hundreds or even thousands of enzymes that are not directly related or connected to the pathway in question will be zero despite the fact that flux control is shared among all enzymes.

Another consequence of the highly branched and intricate nature of cellular metabolism is that the central pathways provide biosynthetic precursors and energy for other pathways. So, as biosynthetic precursors are made, some are fed directly into the biosynthetic routes, which in turn diminishes flux through the central metabolic pathways. It follows, therefore, that biosynthetic enzymes are likely to have negative flux control coefficients with respect to flux through the central metabolic pathways. According to the summation theorem, if one or more enzymes possess a negative value of flux control coefficient, then it is possible to see some other enzymes displaying a flux control coefficient higher than the numerical value of 1.0. This is because if there are negative flux control coefficients, one or more flux control coefficients would have to be greater than 1.0 so that the total sum adds up to 1.0. This shows that the flux control coefficient is not an intrinsic property of the enzyme itself but rather a property of the whole system.

7.2.3 *Elasticity coefficient*

The flux control coefficient of an enzyme is influenced by the enzyme's ability to respond to changes in the concentration of its immediate substrate, as well as its ability to influence the concentrations of other metabolites in the pathway, a linkage which was first demonstrated by Heinrich and Rapaport (1974). The elasticity coefficient, the third pillar of the MCA theory, was therefore introduced to describe how flux is influenced by changes in the concentration of a given metabolite. In other words, elasticity is a parameter that describes, in quantitative terms, the sensitivity and responsiveness of an enzyme to a metabolite.

Unlike the flux control coefficient, elasticity is a property of individual enzymes and not of the pathway. The elasticity of an enzyme to a metabolite is defined by the slope of the curve of enzyme units (reaction rate) plotted as a function of metabolite concentrations, with the measurements taken at the metabolite concentration found *in vivo*. By analogy with the flux control coefficient (Figure 7.2), the value of the elasticity coefficient, which can be calculated from the slope, will depend upon the units used for the measurement of enzymic activities, which may vary from one enzyme to another. This can be avoided, as described earlier for the flux control coefficient, by calculating the elasticity coefficient directly from a log-log plot of catalytic activity versus metabolite concentration to give the fractional change in enzymic activity as a function of the fractional change in the concentration of the substrate. As highlighted in the case study presented in Section 7.3, elasticities have positive values for metabolites that stimulate enzymic activity (substrates, activators) and negative values for those that decrease reaction rate, such as products and inhibitors. Elasticity is, therefore, a parameter that describes, in quantitative terms, the sensitivity and responsiveness of an enzyme to a particular metabolite that could be a substrate, a product, or an effector.

7.2.4 *The connectivity theorem*

This theorem, the fourth pillar of the MCA theory, addresses the question of how the flux control coefficient of a given enzyme can be related to its kinetic properties. Such an inter-relationship is governed by the connectivity theorem, which states that the sum of all connectivity values in a given pathway is zero. The connectivity value for any given enzyme can be calculated by multiplying its flux control coefficient by its elasticity with respect to the metabolite in question. Naturally, enzymes not affected by the metabolite in question will have an elasticity of zero, and as such will make no contribution towards the final sum obtained. Further analysis of connectivity values has revealed that large elasticities are associated with small flux control coefficients, and vice versa. The mathematical equations relating the connectivity theorem to linear pathways, branch points, and cycles have been described and dealt with extensively elsewhere (Fell, 1997).

7.2.5 *Response coefficients*

Induction and repression of enzyme synthesis in response to internal or external environmental stimuli are widely distributed in nature and are very effective in "turning on" and "switching off" transcription. Covalent modification through reversible phosphorylation is another mechanism that regulates the activity of existing enzymes by rendering them active or inactive, as is the case for isocitrate dehydrogenase (ICDH) in *Escherichia coli* during adaptation to acetate (Koshland, 1987; Cozzone, 1988). In addition to degradation of mRNA and proteins, enzymes may also be the subject of allosteric control mechanisms, which change the enzyme's affinity towards its substrate and/or co-factor(s).

The response coefficient, the fifth pillar of the MCA theory, reflects the effectiveness of a particular effector on flux through a given pathway and is dependent on two factors, namely, the flux control coefficient of the target enzyme and the strength of the effector, given by its elasticity coefficient. Clearly, for an effector to have a significant effect on flux, each of the above parameters with respect to the target enzyme has to be of a value higher than zero.

Under circumstances where a particular effector may activate or inactivate more than one enzyme in a given pathway, the total response will be the sum of the individual responses from each enzyme affected (Hofmeyr and Cornish-Bowden, 1991). However, this is only true when the changes in the concentration of the effector are very small because of the nonlinear relationship of the kinetics in metabolic systems.

Now, let us consider how carbon flux is partitioned at the junction of isocitrate during growth of *Escherichia coli* on acetate.

7.3 Control of carbon flux at the junction of isocitrate in central metabolism during growth of *Escherichia coli* on acetate: a case study

During growth on acetate as sole source of carbon and energy, *E. coli* requires the operation of the anaplerotic sequence of the glyoxylate bypass for the provision of biosynthetic precursors (Kornberg, 1966). Under these conditions a new junction is generated at the level of isocitrate (Figure 7.3) where isocitrate lyase (ICL) of the glyoxylate bypass is in direct competition with the Krebs cycle enzyme isocitrate dehydrogenase (ICDH). Although ICDH has a much higher affinity for isocitrate, flux through ICL and thence the anaplerotic enzyme malate synthase (MS) is assured by virtue of high intracellular levels of isocitrate and the inactivation of a large fraction (75%) of ICDH (El-Mansi et al.,

Figure 7.3

A diagrammatic representation of the central metabolic pathways employed by *Escherichia coli* **for the metabolism of acetate, highlighting the direct competition between isocitrate lyase (ICL) of the glyoxylate bypass and the Krebs cycle enzyme isocitrate dehydrogenase (ICDH) for their common substrate. It also highlights the bifunctional role of isocitrate dehydrogenase kinase/phosphatase in the moiety-conserved cycle involved in the reversible inactivation of ICDH.**

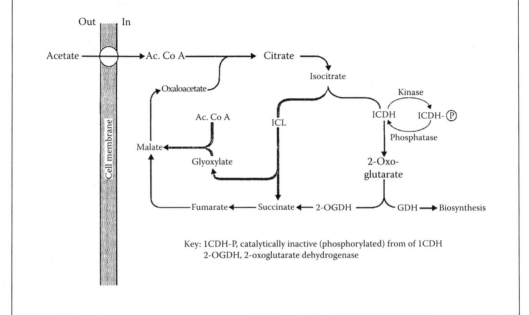

Key: 1CDH-P, catalytically inactive (phosphorylated) from of 1CDH
2-OGDH, 2-oxoglutarate dehydrogenase

1985; Cozzone and El-Mansi, 2005). Although the *in vivo* signal that triggers the "acetate switch" and, in turn, the expression of the glyoxylate bypass is yet to be determined, recent investigations have revealed that acetate *per se* can be safely ruled out as a possible signal (El-Mansi, 1998). Using radio-labeled isotopes and NMR spectroscopy, Walsh and Koshland (1984) have been able to quantitatively determine the rate of carbon flux through ICDH and ICL at the junction of isocitrate.

In order to assess the relative contribution of each of the above enzymes to the overall distribution of carbon flux among various enzymes of central metabolism, the computer software package MetaModel was used to calculate the steady-state fluxes and the concentration of various metabolites during growth of *Escherichia coli* on acetate. This computer package also enabled us to formulate the matrices of the elasticity coefficients and the control and response coefficients under different steady states. In the next section, we will discuss the data in the light of Kacser's MCA theory as well as the traditional concept of the rate-limiting step.

Figure 7.4

A skeleton model describing the central metabolic pathways of *Escherichia coli* **during growth on acetate; metabolites and enzymes are as described below.**

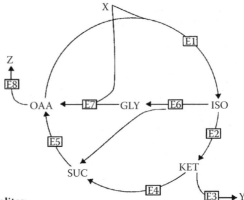

Metabolites:

X	= acetyl-CoA
ISO	= isocitrate
KET	= a-ketoglutarate
SUC	= succinate
OAA	= oxaloacetate
GLY	= glyoxylate
Y	= glutamate
Z	= gluconeogenesis and biosynthesis (sink for cell constituents)

ISO = isocitrate (reservoir excluding flux to fatty acids)

Y = glutamate (sink for cell constituents)

Enzymes:

E1 = citrate synthase/aconitase
E2 = isocitrate dehydrogenase
E3 = glutamate dehydrogenase
E4 = a-ketoglutarate dehydrogenase/succinate thiokinase
E5 = succinate dehydrogenase/fumarase/malate dehydrogenase
E6 = isocitrate lyase
E7 = malate synthase
E8 = PEP carboxykinase

7.3.1 *The model*

The complex enzyme system used by Walsh and Koshland (1984) for the central pathways was reduced to a skeleton model (Figure 7.4) in order to explore the consequences of controlled adjustment of ICL and ICDH enzymic activities on the partition of carbon flux among various enzymes of central metabolism. MetaModel 2.1, written and compiled by Cornish-Bowden and Hofmeyr (1991), was used to solve the steady state and to calculate the various coefficients. This computer program enabled the simulation of a whole series of variables, and any small change in any variable was detected, transmitted systematically, and all the changing fluxes, metabolites, and coefficients were calculated accordingly.

Table 7.1 Balance equations for the skeleton model proposed for the central metabolic pathways of *E. coli* during growth on acetate

Balance equation	Flux relationships		
$0 = dISO/dt = V1 - V2 - V6$	$>V1 = V2 + V6$		
$0 = dKET/dt = V2 - V4 - V3$	$>V2 = V3 + V4$	$>V3 = V2 - V4$	
$0 = dSUC/dt = V4 + V6 - V5$	$>V5 = V4 + V6$	$>V4 = V5 - V6$	
$0 = dGLY/dt = V6 - V7$	$>V6 = V7$		
$0 = dOAA/dt = V5 - V7 - V8 - V1$	$>V1 + V8 = V5 + V7$	$>V8 = V5 - V2$	

From this simple multi-enzyme system (Figure 7.4), differential equations were derived to describe the rate of change in the concentration of each metabolite as it is converted by the enzymes (Table 7.1). In a steady state, the rate of change for each metabolite equals zero, i.e., the net rate of formation equals the rate of consumption. It is clear from the skeleton model that we have eight different reactions, of which the velocities of conversion through enzymes 6, isocitrate lyase, and 7, malate synthase, are identical due to their linear relationship.

It is noteworthy that the model depicted above for the Krebs cycle and the glyoxylate bypass (Figure 7.4) does not constitute a moiety conserved cycle (Hofmeyr et al., 1986) because there are reversible sinks to the cycle as well as branching from within the cycles, thus giving rise to metabolite pools that are involved in three different reactions.

In formulating the reaction equations in our model, the following reversible Michaelis-Menten-type equation was used as a basis to describe the enzymic reactions:

$$v = \frac{V_f^* S/K_s - V_r^* P/K_p}{1 + S/K_s + P/K_p}$$

where V_f and V_r are the V_{max} values of the forward and the reverse reactions, respectively; S = substrate; P = product, and K_s and K_p are the Michaelis-Menten constants, i.e., K_m values for the relevant metabolites.

The differential equations that describe the rate of change of each substrate concentration in the central pathways following the entry of acetyl CoA are shown in Table 7.2. In this model we have taken into account the differing affinities of ICDH and ICL for their common substrate, isocitrate. Walsh and Koshland (1984) established that reversible inactivation of ICDH during growth on acetate allowed one-third of the flux through ICL (i.e., a ratio of 2.6: 1 in favor of ICDH). The concentration of E6 (ICL) in our model, therefore, was set at 0.388 while all other enzyme concentrations remained at 1.0. Furthermore, the intracellular concentrations of acetyl CoA, glutamate, and phosphoenolpyruvate (PEP) as external metabolites were fixed at concentrations of 56, 1.06, and 1.69 mM, respectively. These

LIVERPOOL
JOHN MOORES UNIVERSIT
AVRIL ROBARTS LRC
TITHEBARN STREET
LIVERPOOL L2 2ER
TEL. 0151 231 4022

Table 7.2   Reaction expressions and rate equations for the skeleton model proposed for the central pathways of *E. coli* during growth on acetate

---

1. X+OAA=ISO

    V/[E1] = (5.1*[X]*[OAA] - [ISO])/(1 + [X] + [OAA] + [ISO])

2. ISO = KET

    V/[E2] =(10*[ISO] - [KET])/(1 + [ISO] + [KET])

3. KET = Y

    V/[E3] = (7.5*[KET] - [Y])/(1 + [KET] + [Y])

4. KET = SUC

    V/[E4] = (10*[KET] - [SUC])/(1 + [KET] + [SUC])

5. SUC = OAA

    V/[E5] = (5*[SUC] - [OAA])/(1 + [SUC] + [OAA])

6. ISO=GLY+SUC

    V/[E6] = (10.6*[ISO] - [SUC]*[GLY])/(1 + [ISO] + [SUC] + [GLY])

7. GLY + X = OAA

    V/[E7] = (10*[X]*[GLY] - [OAA])/(1 + [X] + [GLY] + [OAA])

8. OAA = Z

    V/[E8] = (11*[OAA] - [Z])/(1 + [OAA] + [Z])

---

values represent the input of acetyl CoA and the outputs of glutamate and oxaloacetate required for the biosynthesis of one gram dry weight biomass of *E. coli*, as described previously (El-Mansi et al., 1994). The steady-state growth of *E. coli* on acetate was simulated and the data obtained (Table 7.3) are in good agreement with the *in vivo* data reported by Walsh and Koshland (1984).

The changes in the steady-state fluxes through ICDH and ICL as well as the intracellular concentration of isocitrate in response to changes in the concentration of ICDH (E2) are shown in Figure 7.5. As the rate of carbon flux through ICDH is diminished, the intracellular concentration of isocitrate rises

Table 7.3   Steady-state fluxes and pools as calculated by MetaModel for the skeleton model (Figure 7.4) representing the central metabolic pathways of *E.coli* during growth on acetate

Enzyme	Velocity	Metabolites (mM)	Velocity
[E1] = 1.0000	1.1117	[X] = 56.0000	Fixed
[E2] = 1.0000	0.8008	[OAA] = 0.2236	Variable
[E3] = 1.0000	0.0468	[ISO] = 0.1174	Variable
[E4] = 1.0000	0.7540	[KET] = 0.1551	Variable
[E5] = 1.0000	1.0469	[Y] = 1.0600	Fixed
[E6] = 0.3880	0.3109	[SUC] = 0.3880	Variable
[E7] = 1.0000	0.3109	[GLY] = 0.0322	Variable
[E8] = 1.0000	0.2642	[Z] = 1.6900	Fixed

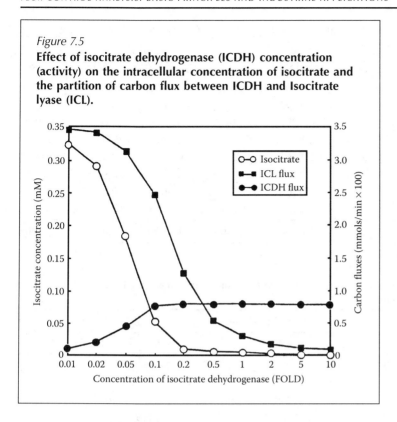

Figure 7.5

Effect of isocitrate dehydrogenase (ICDH) concentration (activity) on the intracellular concentration of isocitrate and the partition of carbon flux between ICDH and Isocitrate lyase (ICL).

to a level that sustains flux through ICL despite its low affinity. The data also show that any decrease in the concentration of ICDH results in an increase of carbon flux through ICL and, in turn, the anaplerotic enzyme malate synthase (MS). This in turn replenishes the central pathways with primary intermediates and biosynthetic precursors. During steady-state growth on acetate, the rate of carbon flux through ICL is 31 mmoles per min; this represents 33% of the total carbon processed at this junction.

According to the summation theorem, the flux through ICDH can be affected by other enzymes of the central pathways and the sum of the flux control coefficients of all enzymes on JICDH is 1.0. While the flux control coefficient is a measure of the relative change in flux through a particular enzyme in a given pathway, in response to a small change in its concentration, the elasticity coefficient on the other hand is a measure of the relative change in flux in response to a small change in the concentration of the substrate or a given co-factor. Tables 7.4 and 7.5 show the matrices of the flux control coefficients and the elasticity coefficients as calculated by MetaModel, respectively, for the enzymes of the central metabolic pathways during steady-state growth on acetate. The data outlined in Figure 7.6 clearly show that any increase in the concentration of ICDH above the level observed in acetate phenotype, i.e., 1.0, does not increase flux

Table 7.4 Flux control coefficients as calculated by MetaModel for the central metabolic pathways of *E.coli* during steady-state growth on acetate

Enzyme	J1	J2	J3	J4	J5	J6/J7	J8
E1	0.8599	0.8944	7.5300	0.4829	0.5670	0.7708	−0.4256
E2	−0.2399	−0.0164	−1.1192	0.0520	−0.2013	−0.8156	−0.7619
E3	−0.0257	−0.0265	0.2985	−0.0466	−0.0400	−0.0239	−0.0809
E4	0.1797	0.2359	−2.9514	0.4336	0.3172	0.0349	0.5635
E5	0.1996	0.1617	−1.5563	0.2682	0.2767	0.2971	0.6251
E6	0.2760	0.0224	1.3018	−0.0570	0.2309	0.9291	0.8631
E7	0.0083	0.0007	0.0400	−0.0017	0.0070	0.0285	0.0265
E8	−0.2579	−0.2722	−2.5432	−0.1314	−0.1575	−0.2208	0.1902
Sums	1.0000	1.0000	1.0000	1.0000	1.0000	1.0000	1.0000

Table 7.5 Elasticity matrices as calculated by MetaModel for the central metabolic pathways of *E.coli* during steady-state growth on acetate

	Metabolites							
Enzyme	X	OAA	ISO	KET	Y	SUC	GLY	Z
E1	0.0252	0.9979	−0.0039	0	0	0	0	0
E2	0	0	1.0600	−0.2742	0	0	0	0
E3	0	0	0	11.1642	−10.7127	0	0	0
E4	0	0	0	1.2329	0	−0.5849	0	0
E5	0	−0.2690	0	0	0	−0.8896	0	0
E6	0	0	0.9388	0	0	−0.2625	−0.0311	0
E7	0.0345	−0.0165	0	0	0	0	1.0120	0
E8	0	3.1190	0	0	0	0	0	−2.7707

through the enzyme itself nor does it increase the enzyme's flux control coefficient. On the other hand, however, any drop in the intracellular level of ICDH activity beyond that of acetate phenotype (1.0) appears to have a profound effect on carbon flux through the enzyme itself and the enzyme's flux control coefficient. Clearly, the reduction of ICDH activity to 20% or less (Figure 7.4) led to a sharp increase in the flux control coefficient of ICDH, mirrored in the rate of carbon flux through the enzyme itself. It follows therefore that flux through ICDH during growth on acetate is in excess of cellular demands and as such cannot be rate-limiting.

Modulation of ICL activity directly by systematically increasing the concentration of the enzyme revealed some interesting observations. From this simulation (Figure 7.7) the indications were that above a certain threshold of ICL concentration the two cycles work in concert and the partition of carbon flux between ICDH and ICL is no longer a problem. It is interesting that increasing the concentration of ICL resulted in an increase

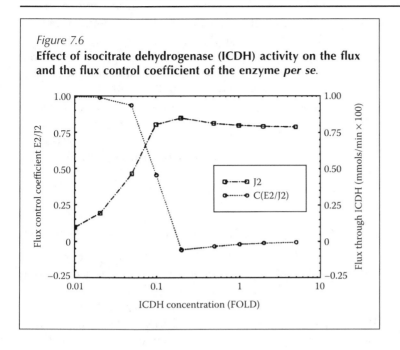

Figure 7.6

Effect of isocitrate dehydrogenase (ICDH) activity on the flux and the flux control coefficient of the enzyme *per se*.

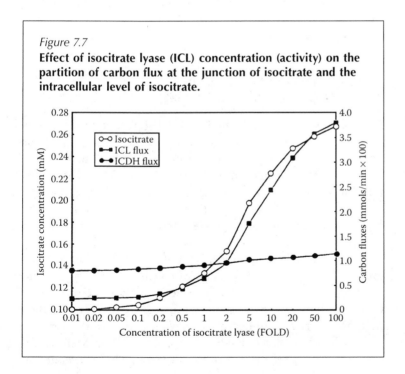

Figure 7.7

Effect of isocitrate lyase (ICL) concentration (activity) on the partition of carbon flux at the junction of isocitrate and the intracellular level of isocitrate.

of flux through ICDH as well as ICL and that this was inevitably at the expense of the intracellular concentration of isocitrate, which only rose to 0.271 mM. At an ICL concentration of 100, the rate of carbon flux through ICDH rose to 115 mmoles/min, while that through ICL rose to 378 mmoles /min. From the above data, we have been able to assess how perturbation of a given enzymic activity is transmitted throughout the various components of central metabolism. The data demonstrate that increasing the concentration of ICDH beyond the *in vivo* level was not accompanied by an increase in flux through ICDH *per se* or the central pathways as a whole. Furthermore, increasing the concentration of ICDH diminishes the intracellular pool of isocitrate, thus diminishing flux through the glyoxylate bypass.

7.4   Modeling using other computer programs

It is noteworthy that other metabolic modeling programs are also available and can be obtained free of charge; though some of which are no longer maintained by their respective authors for one reason or another, these include: Gepasi (Mendes, 1993, 1997; Mendes and Kell, 1998) and MMT (Metabolic Modeling Tool) (Hurlebaus, 2001; http://www.bioinfo.de/isb/gcb01/poster/hurlebaus.html), which unlike other programs uniquely contains a tool that provides a graphical output (a pathway overview) of a structured pathway overview and identifies parameter algorithms.

7.4.1 *Modeling of the partition of isocitrate flux with Gepasi*

Gepasi, version 3.30, a Windows-based application that is user-friendly was used in this study. The entry of reaction equations and rate laws (kinetics) is straightforward and easily achieved. A number of predefined rate laws are available, and it is also possible to create user-defined rate laws. Once the reaction rates and rate laws are entered, the starting concentrations of fixed and variable metabolites are entered. The entry of reaction equations, rate laws (kinetics), and starting concentrations of fixed and variable metabolites is similar to MetaModel. However, it should be noted that the Michaelis-Menten-type equations will need to be created as a new kinetic type in Gepasi. One important point to note when using programs other then MetaModel is the influence of enzyme concentration values. This influence is not immediately apparent in the MetaModel program, and so attention is drawn here to the full rate equation that is used by MetaModel. The full rate equation for reaction 1 in Table 7.2 reads as follows (note the multiplication factor of E1 concentration):

$$V/[E1] = E1*(5.1*X*OAA - ISO)/(1 + X + OAA + ISO)$$

Figure 7.8

Effect of ICL (E6) concentrations on flux through the enzyme itself (J6), flux through ICDH (J2) and isocitrate concentrations as calculated by Gepasi using the original rate equations listed in Table 7.2. Note that the data represented in this graph are composed of a combined scan of E6 (14 values) and E2 (10 values), thus giving 140 separate steady-state data points. The variations of E6 are shown along the abscissa, while the variations in E2 are shown as multiple contours of J6, J2, and isocitrate concentrations.

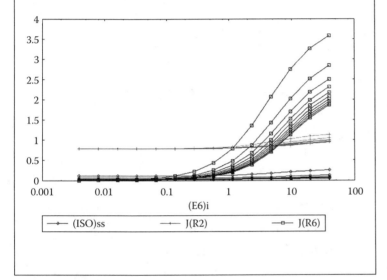

Once this basic configuration is achieved, the MCA steady state values can be calculated. Interestingly, the data obtained on the impact of ICL (E6) and ICDH (E2) on the partition of carbon flux at the junction of isocitrate as well as isocitrate concentrations using Gepasi is comparable to that of MetaModel with Gepasi giving the added advantage of plotting multifunction analysis as shown in Figures 7.8 and 7.9, respectively. Although the output in the two figures appears to be significantly different with respect to isocitrate concentration and flux through ICDH and ICL, the reaction equation input only differ in the sign used in the denominator; while plus was used in the Figure 7.8, multiplication was used in Figure 7.9. Although the question of whether which sign is correct is arguable, the *in vivo* analysis of isocitrate concentration is in good agreement with using plus rather than multiplication in the denominator of formula template. This change from straight addition of the substrate and product concentrations multiplication was originally observed in version 3 of MetaModel, but no explanation or rational was given for the change.

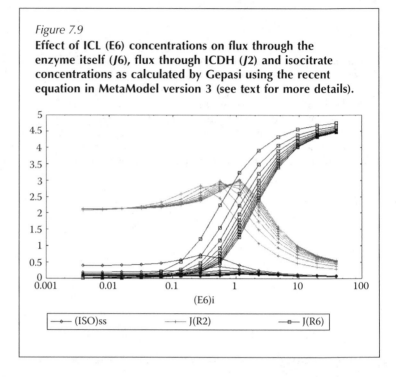

Figure 7.9
Effect of ICL (E6) concentrations on flux through the enzyme itself (J6), flux through ICDH (J2) and isocitrate concentrations as calculated by Gepasi using the recent equation in MetaModel version 3 (see text for more details).

7.5 Strategies for manipulating carbon fluxes en route to product formation in intermediary metabolism

7.5.1 *Validity of the concept of the "rate-limiting" step as an approach to increasing flux to product formation*

Industrialists, driven by a strong faith in genetic engineering, often argue that if one can identify the enzyme catalyzing the rate-limiting step in a given pathway, then the rest is simple. This approach, however, has proved to be, by and large, erroneous and examples illustrating this notion are given in the introductory section of this chapter. Alternative approaches, which may be beneficiary to the industrialists, include Kacser's theory of metabolic control analysis (MCA), which readily accounts for the failure of the rate-limiting step as a rationale for improving the efficiency of biotechnological processes. The MCA argues that the rate-limiting step is an oversimplification of metabolism and as such should be abandoned; instead, the control of metabolic flux in a given pathway is shared among all the enzymes involved, albeit to a varying degree of extent as described earlier in this chapter. Although the MCA overlooks peculiarities in metabolic network and ignores regulatory enzymes, the victim of MCA as Kacser's used to say, it is nevertheless a very

Figure 7.10

A model representing a branched biosynthetic route involving the production of two products: S_{4a} and S_{4b}. Note that the whole pathway can be divided into two main sections, i.e., supply and demand. While the supply involves enzymes E_1 to E_4, the demand section involves E5 only. Also note that each end product (S_{4a} and S_{4b}) inhibits the first enzyme required for its synthesis (E_{3a} and E_{3b} respectively) and that the branch metabolite (S_2) inhibits the first enzyme (E_1) in the whole pathway. It was assumed that the reactions obeyed Michaelis-Menten equation as described for isocitrate and that the pool concentration of X_0 was held constant at 10 with the steady-state concentrations of various intermediates; shown underneath each metabolite, were calculated for a value of 1 for K_{5a} using MetaModel as described earlier (Cornish-Bowden et al., 1995). The steady-state fluxes through the common pathway and the two branches are highlighted by shading, i.e., 1.5092 for the common pathway and 0.7621 and 0.7471 for the fluxes through branches a and b, respectively. This steady state was used as the starting point for examining the impact of different strategies on flux to S_{4a}. This diagram is based on the original model of Cornish-Bowden et al. (1995); reproduced with the kind permission of Elsevier.

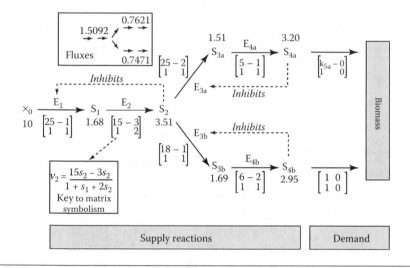

powerful tool for examining the general properties of metabolic networks.

7.5.2 Modulation of carbon flux en route to product formation

7.5.2.1 The Model

In the model pathway (Figure 7.10), the biosynthesis of two end products, S_{4a} and S_{4b} from a biosynthetic precursor, X_0, via a branch point metabolite, S_2 is highlighted. Also note that the precursor metabolite X_0 is external to the system, i.e., its concentration is fixed and, unlike all other metabolites within the pathway, does not depend on the properties of the eight enzymes involved in the skeleton model (Figure 7.10) (Cornish-Bowden et al., 1995). Although the numerical values assumed by the afore-mentioned authors for the eight kinetic equations are arbitrary, the essential features of the model are not and correspond to a typical case of sequential feedback inhibition; in which each of the two end products (S_{4a} and S_{4b}) inhibits the enzymes (E_{3a} and E_{3b}), catalyzing the first committed step of their formation, respectively, with the branch-point metabolite (S_2) inhibiting (E_1), the enzyme catalyzing the first step in the whole pathway (Figure 7.10). As is the case for the partition of carbon flux at the junction of isocitrate, all simulations were carried out using the latest available version of MetaModel (Cornish-Bowden and Hofmeyr, 1991).

7.5.2.1 Strategies for manipulating metabolic fluxes

A number of different strategies; namely, opposition, oblivion, evasion, suppression, and subversion, have been employed in an attempt to increase carbon flux to product formation; in this case S_{4a} and S_{4b} (Cornish-Bowden et al., 1995). The terms coined for the aforementioned strategies reflect the possible impact of each on cellular regulation (Cornish-Bowden, 1995). Opposition as a strategy is widely used and in which industrialists attempt to increase flux in a given pathway by increasing the relative con-centration of the enzyme considered to be rate-limiting; often the enzyme that is subject to feedback inhibition, e.g., E_1 and/or E_{3a} in the model system portrayed above (Figure 7.10). While obliv-ion strategy utilizes mathematical modeling of certain matri-ces to determine the degree of change needed to achieve certain increase in flux, evasion utilizes fluxes to calculate the changes needed in enzymic activity to achieve the desired increase in the concentration of end product without adversely affecting the concentrations of intracellular metabolites. Suppression as a strategy, however, makes full use of the primary regulatory function of feedback inhibition, i.e., to transfer control from the biosynthetic reactions (supply) to polymerization and assem-bly (demand) (Figure 7.10), and as such seeks to increase flux to product formation through increasing the demand made on central and intermediary metabolism. Suppression relies on the elimination of feedback loops while subversion relies on increas-ing the demands made on the biosynthetic enzymes.

With the exception of oblivion strategy, all other strategies have been examined *in silico* and their impact on flux to prod-uct formation assessed (Table 7.6). Suppression is an all-or-none strategy; as feedback loops are either present or absent and had only a very modest effect on flux. Elimination of the feedback loop to E_{3a} gave a 30% increase in flux, which was accompa-nied by 17-fold increase in the concentrations of the metabo-lites in branch a (Figure 7.10). On the other hand, elimination

Table 7.6 Impact of various strategies on metabolites concentrations and flux through branch 'a' in the model shown in Figure 7.10

| Strategy | Relative Activity of each enzyme | | | | | | | | Relative J_{5a} | Relative Metabolite Concentrations | | | |
	E_1	E_2	E_{3a}	E_{4a}	E_{5a}	E_{3b}	E_{4b}	E_{5b}		S_1	S_2	S_{3a}	S_{4a}
Wild-Type	1	1	1	1	1	1	1	1	1.00	1.00	1.00	1.00	1.00
Opposition [b]	5	1	1	1	1	1	1	1	1.02	1.50	1.51	1.11	1.11
Opposition [b]	1	1	5	1	1	1	1	1	1.08	1.02	0.99	1.49	1.47
Opposition [b]	5	1	5	1	1	1	1	1	1.10	1.52	1.49	1.66	1.63
Evasion	3.02	3.02	5	5	5	1	1	1	5.00	1.00	1.00	1.00	1.00
Suppression	*	1	1	1	1	1	1	1	1.17	49.5	47.7	2.64	2.56
Suppression	1	1	*	1	1	1	1	1	1.29	1.06	0.96	17.8	16.9
Suppression	*	1	*	1	1	1	1	1	1.31	47.4	42.7	795	780
Subversion	1	1	1	1	5	1	1	1	4.13	1.74	0.76	3.66	0.53

[a] In cases marked by asterisk (*), the activities of the enzymes were not changed except for omitting the feed back inhibition terms from the denominations of the rate equations.

[b] The three cases of opposition were also examined with much larger changes in activity, 100- instead of 5-fold in each case. Effects on the flux were only trivially different from those shown, but the effects on concentrations were much larger.

The factors of 5 and 1 for the enzymes reflect the desired increases in flux. In the common part of the pathway (see Fig. 7.10), the factor 3.02 was calculated from the fluxes in the starting steady-state as follows: (5J3a + J3b)/(J3a + J3b) = (5 x 0.7621 + 0.7471)/(0.7621 + 0.7471) = 3.02. This table is reproduced from Cornish-Bowden et al., (1995) with the kind permission of Elsevier.

of the feedback loop E_1 yielded a 17% increase in flux with some 50-fold increase in the concentrations of metabolites in the common part of the pathway (Figure 7.10). It is noteworthy to indicate that large increase in the concentrations of metabolites *in vivo* might adversely affect cell viability as well as flux to desired end product. In all other strategies, the manipulations were performed with the view of achieving five-fold increase in flux through branch a (Figure 7.10). Simulation analysis revealed that opposition as a strategy is largely ineffective as changes in the concentration of one or two enzymes alone was accompanied by a very limited increase in flux to product formation. On the other hand, evasion produced the desired increase in flux, five-fold, without adversely affecting the intracellular concentrations of metabolites (Table 7.6). Furthermore, although evasion is more effective than subversion as a strategy for increasing fluxes (Table 7.6), it is important. that we recognize that the latter strategy achieved an appreciable increase in flux as a result of manipulating only one enzymic activity rather than all enzymes and that the changes in the intracellular concentrations of metabolites associated with it are modest and as such may be favorable to the organism *in vivo*.

7.6 Conversion of feedstock to biomass and desirable end products

The conversion of a given feedstock to desirable end product by micro-organisms involves a wide range of metabolic activities depending on the nature of the feed-stock and the identity of end product. Unlike products of primary metabolism (see Chapter 4), synthesis of secondary metabolites (see Chapter 5) is usually sequential to cessation of growth at which point, more feedstock (input) is added intermittently (fed-batch culture) or continuously (continuous cultures, e.g., turbidostat, auxostat, pH-stat) to sustain the metabolic fluxes required for the biosynthesis of desired end product (output) as well as maintenance of cell viability. During the productive stage, the primary feedstock (input) is generally channeled through specific metabolic routes and providing such routes are known, careful measurements of all inputs and outputs in a given fermentation process provides the data from which the fluxes through all the pathways involved can be calculated (Holms, 1996). A metabolic chart for central and intermediary metabolism can then be constructed, that relates input to biosynthesis of desired end product. Invariably, flux analysis shows that the efficiency with which a given feedstock is converted to biomass is far from optimal due to excretion of intermediates (El-Mansi, 2004); usually those upstreams of bottlenecks, or due to the accumulation of intracellular polymers or both. Analysis of culture filtrates allows the identification of undesired fluxes, which become targets for metabolic intervention by the metabolic engineers (El-Mansi and Holms, 1989, El-Mansi, 2004). Flux analysis, therefore, highlights wasteful fluxes, thus enabling the metabolic engineer to develop a rationale for the diversion of wasted fluxes to product formation, a primary target for the discerning biotechnologist.

7.6.1 *Stoichiometric analysis*

Stoichiometric analysis of metabolic fluxes in a given process is strain-specific and requires precise information regarding the input (carbon, nitrogen, phosphorus, sulphur, and so on) required for the generation of biomass and desired end product. The output to biosynthesis and energy generation together with fluxes required to satisfy auxotrophic deficiencies are also required. Furthermore, a complete description of metabolic flux through various routes and branches in a given process requires knowledge of not only the various regulatory control mechanisms employed but also the kinetic properties of each enzyme together with the intracellular level of various intermediates. Such information is, of course, hardly available and recent efforts in an attempt to overcome or bypass those limitations were based on steady-state analysis and the metabolic requirements for cell growth. This approach has yielded valuable information on metabolic fluxes, their regulation and manipulation to increase the efficiency of carbon conversion to biomass and desirable end products. Consequently, a good deal of work has been devoted towards the understanding of the dynamics of metabolic pathways.

7.6.2 Formulation of metabolic flux models/charts

Formulation of metabolic flux models requires, in addition to maintenance requirements, which is strain-specific and can be experimentally determined, knowledge of yield coefficient and full awareness and understanding of the metabolic routes through which the primary carbon source is converted to biomass and desirable end product(s). Once the above details are established, the metabolic demands made on central and intermediary metabolism for biosynthesis, energy generation, and product formation can be calculated. Fortunately, the metabolic networks employed by *Escherichia coli* and eukaryotic organisms are well established. Furthermore, the metabolic demands on central metabolism for biosynthesis of new biomass of *E. coli* (Neidhardt et al., 1990) and murine hybridoma cells are well established. Once the information required has been gathered, an appropriate mathematical framework can be used to calculate the distribution of metabolic fluxes among various enzymes in a given network (Neidhardt, 1990; Holms, 1996).

7.6.3 Applications of flux distribution analysis

To illustrate the usefulness of stoichiometric analysis in biochemical and metabolic engineering, let us consider the classical example of phenylalanine production.

7.6.3.1 Optimization of phenylalanine production: A case study

In *Escherichia coli*, biosynthesis of phenylalanine and other aromatic amino acids is initiated by the condensation of phosphoenolpyruvate (PEP) and erythrose 4-phosphate (E-4P) to give 3-deoxy-D-arabinoheptulosonate 7-phosphate (DAHP), This condensation reaction is catalyzed through the activities of three different DAHP synthases, the products of *aroF, aroG,* and *aroH*, which are subject to feedback inhibition by tyrosine, phenylalanine, and tryptophan, respectively. The metabolic routes to phenylalanine are well established and assuming that no unknown by-products are being formed, the theoretical yield coefficient for the production of phenylalanine — or any given product — can be calculated (Foberg et al., 1988).

The overall reaction for the formation of phenylalanine can be expressed by the following equation:

2PEP + E4-P + 2NADPH + ATP + Glutamate
→ Phenylalanine + CO_2 + 2-Oxoglutarate
+ 2NADP⁺ + ADP + 4Pi   (7.1)

Assuming that E4-P is generated through the activities of transketolase and transaldolase of the pentose phosphate pathway (PPP) and that PEP, which is generated through glycolysis, is required for the uptake of glucose and its activation to glucose 6-phosphate, the overall reaction of phenylalanine production can be described as follows:

X-Glucose → 2 PEP + E4-P + X-Pyruvate → Phenylalanine   (7.2)

Taking the number of carbon atoms in each of the above molecules into consideration, the carbon balance of the above equation can be resolved as follows:

X (6) → 2 (3) + 1 (4) + X (3)

This, in turn, can be solved as follows:

X (6) → 6 + 4 + X (3)

X (6) – X (3) = 6 + 4

X 3 = 10; it follows

X = 10/3 = 3 and ⅓rd moles of glucose.

According to the above stoichiometry, 3 and ⅓rd moles of glucose (MW 180) are required for the production of one mole of phenylalanine (MW 165.2), thus giving a maximum theoretical yield coefficient of 0.275 g of phenylalanine per g of glucose.

Taken into consideration that the biosynthesis of phenylalanine (Eq. 7.1) requires 2 moles of NADPH; generated as a consequence of pyruvate oxidation in the Krebs cycle at the level of isocitrate dehydrogenase (ICDH), the overall reaction can be rewritten as follows:

X Glucose → 2 PEP + E4-P + 2 Pyruvate → Phenylalanine   (7.3)

Solving the above equation in the same way as for Eq. 7.2, the carbon balance under these circumstances is 2 and ⅔rd moles of glucose are required for the production of one mole phenylalanine and as such the overall stoichiometry will therefore be as follows:

2 and ⅔rd moles of glucose → one mole of phenylalanine
+ 7 moles of CO_2 (7.4)

Further consideration of the stoichiometry balance for the above equation reveals a shortfall of PEP supply by a factor of ⅔rd mole. Similarly, on close examination, one can see that 0.66 (⅔rd) mole of pyruvate is wasted in the form of CO_2 (Figure 7.11). Diversion

Figure 7.11

Diagrammatic representations highlighting the various reactions *en route* to phenylalanine formation in *E. coli*. Note the impact of PEP synthase expression on the diversion of pyruvate to PEP formation, rather than CO_2 production, and its consequences on the efficiency of carbon conversion to phenylalanine.

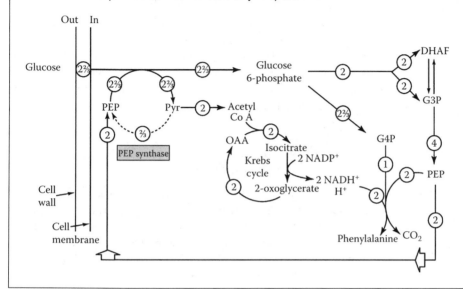

of pyruvate to PEP formation rather than CO_2 production becomes, therefore, a suitable target for metabolic intervention through the expression of PEP synthase (Figure 7.11). Since this enzyme is subject to catabolite repression by glucose in *E. coli*, the use of recombinant DNA technology to construct a mutant strain that is insensitive to catabolite repression or a recombinant strain that is capable of expressing PEP synthase constitutively affords a good example in which metabolic and genetic engineering work in concert.

Interestingly, an appreciable increase in flux to DAHP and, in turn, phenylalanine, as a result of overexpression of PEP synthase has been reported in the literature (e.g., Stephanopoulos et al., 1998), thus lending further support to the analysis portrayed above. The impact of overexpression of PEP synthase on the efficiency of carbon conversion to DAHP and, in turn, phenylalanine production can be assessed quantitatively as illustrated in the stoichiometric diagram shown in Figure 7.11. From the latter diagram (Figure 7.11), it can be seen that constitutive expression of PEP synthase resulted into the formation of a PEP regenerating cycle which, under these circumstances, proved to be advantageous — rather than futile — as evidenced by a 100% increase in flux to DAHP; the biosynthetic precursor for phenylalanine, to give a 0.787g of DAHP per g of glucose utilized. The production of phenylalanine is, therefore, a classic example in which stoichiometric flux analysis, metabolic and genetic engineering were integrated successfully.

7.6.3.2 *Kacser's universal method for increasing flux to metabolite production*

Kacser and Acerenza (1993) developed a simple method for calculating the degree of overexpression of a given enzyme in a particular pathway, so that a certain increase in productivity is achieved. What is significant about this method is that it does not perturb the steady-state fluxes in other pathways and can be applied to any metabolic pathway.

7.6.3.3 *Recent innovations to increasing flux to product formation*

Recently Xu et al. (2003) developed a new process for the production of phenylalanine using two different strains of *E. coli*. In their system, the coupling of transaminase activity from strain EP8–10, with that of aspartase from strain EA-1, led to rapid and efficient production of phenylalanine. This method has also the added advantage of being economical as aspartate could be replaced by ammonium fumarate, a relatively cheaper substrate than aspartate. This process is currently in operation on a pilot scale with an annual yield of 10 metric tons.

Summary

- It is proposed that qualitative terms used to describe an enzyme as rate-limiting, bottleneck, or pacemaker should be abandoned and replaced by the term "rate–controlling" and further qualified by the flux control coefficient to describe, in quantitative terms, the capacity of the enzyme with respect to flux control within a given pathway.
- The flux control coefficient of a particular enzyme is a system property, i.e., its value is not entirely independent of other enzymes in the pathway. The inter-relationship between the flux control coefficients in a given pathway is governed by the summation theorem, which dictates that the total sum of all flux control coefficients adds up to 1.0.
- In a steady state, the influence of a particular metabolite on flux through a given enzyme on the one hand and the whole pathway on the other can be determined from the enzyme's elasticity coefficient, a quantitative term that is directly related to the kinetic properties of the enzyme involved.
- The metabolic inter-relationship between the flux control coefficient and the elasticity coefficient is described by the connectivity theorem, which takes into account the kinetic properties of each of the enzymes involved.

- The action of external effectors on metabolic flux can be assessed by measuring the response coefficient, which, to a large extent, is dependent on the flux control coefficient and the elasticity coefficient of the enzyme with respect to the effector. For an effector to be able to influence the flux through a certain enzyme, the values of the aforementioned coefficients must be relatively high.

- Activating or increasing the catalytic activity of a single enzyme is not usually accompanied by a significant increase in flux (productivity) even with enzymes possessing a relatively large flux control coefficient. This is simply because the flux control shifts to other enzymes as the target enzyme is activated. This, in turn, implies that amplification of single enzymic activity is not a viable option for increasing productivity. This limitation, however, does not apply to the reduction of catalytic activity as reduction or inactivation is generally accompanied by a considerable drop in flux.

- A universal method has been developed to calculate the exact degree of overexpression required for increasing flux to product formation by a certain factor without perturbing the steady-state fluxes in other enzymes.

- Increasing the concentration of an effector that activates all enzymes in the pathway will be accompanied by an appreciable increase in flux and, in turn, yield. Furthermore, an effector that stimulates the activity of more than one enzyme in a given pathway may lead to increase in flux (productivity) particularly if those enzymes share a relatively high flux control coefficient.

- The case study presented in this chapter on the partition of carbon flux between ICDH and ICL revealed that during growth on acetate as sole source of carbon and energy, isocitrate dehydrogenase (ICDH) is not rate-controlling and that flux through isocitrate lyase (ICL) is essential not only to replenish central metabolism with biosynthetic precursors but also to sustain high intracellular level of isocitrate. Furthermore, above certain threshold concentration of ICL, the Krebs cycle and the glyoxylate bypass can work in concert without the need for the inactivation of ICDH.

- Among the five different strategies reported for the manipulation of carbon flux *in silico*, both evasion and subversion are the most suitable strategies for increasing fluxes to product formation without adversely affecting the intracellular concentrations of metabolites; though evasion has the added advantage of being applicable to all metabolic pathways as it does not make any assumption about the regulatory control systems employed *in vivo*.

- Subversion, however, affords a practical strategy as it only involves the manipulation of only one enzymic activity and as such may prove more of an attractive proposition for the industrialists. Subverting feedback inhibition by the desired end product can simply be achieved by generating a mutant that leaks the end product into the medium. Such a system is highly desirable as the recovery and downstream processing become more effective and less expensive.

- Formulation of flux distribution models/charts is relatively simple and only requires knowledge of the metabolic pathways employed, chemical composition of the cell, output to biosynthesis, and strain-specific requirements for maintenance and auxotrophy.
- Flux models, in association with the universal method of Kacser, allow the development of a rationale for strain improvement.

Acknowledgment

The authors wish to thank Gordon Lang, who was on an industrial placement at the school of life sciences, Napier University, Edinburgh, December 2005 to June 2006, for expert assistance in compiling matrices and modeling comparisons.

References

Cornish-Bowden, A., J. H. S. Hofmeyr (1991). MetaModel: A program for modeling and control analysis of metabolic pathways on the IBM PC and compatibles. *Comp. Appl. Biosci.*, 7, 89–93.

Cornish-Bowden, A. (1995). In *Biotechnology: A Comprehensive Treatise*, 2nd edition, H.-J. Rehm and G. Read (eds.), Vol 9, pp. 121–136.Weinheim: Springer-Verlag.

Cornish-Bowden, A,.J-H. S.Hofmeyr, M. L. Cardenas (1995). Stratigies for manipulating metabolic fluxes in biotechnology. *Bioorganic Chemistry*, 23, 439–449.

Cozzone, A. J. (1988). Protein phosphorylation in prokaryotes. *Ann. Rev. Microbiol.*, 42, 97–125.

Dng, O., C. Frieden (1997). New PC versions of the kinetic simulation and fitting programs KINSM and FITSIM. *Trends Biochem Scvi* 22, 8.

El-Mansi, E. M. T. (1998) Control of metabolic interconversion of isocitrate dehydrogenase between the catalytically active and inactive forms in *Escherichia coli, FEMS Microbiol. Lett.*, 166(2), 333–339.

El-Mansi, E. M. T., W. H. Holms (1989). Control of carbon flux to acetate excretion during growth of *Escherichia coli* in batch and continuous cultures. *J. Gen. Microbiol.*, 135, 2875–2883.

El-Mansi, E. M. T., G. C. Dawson, C. F. A. Bryce (1994). Steady-state modelling of metabolic flux between the tricarboxylic acid cycle and the glyoxylate bypass in *Escherichia coli, Comp. Appl. Biosci.*, 10, 295–299.

El-Mansi, E. M. T., H. G. Nimmo, W. H. Holms (1985). The role of isocitrate in control of the phosphorylation of isocitrate dehydrogenase in *Escherichia coli*, *FEBS Lett.*, **183**, 251–255.

Fell, D. (1997). *Understanding the Control of Metabolism*, London: Portland Press.

Foberg, C., T. Eliaeson, L. Haggstrom (1988). Correlation of theoretical and experimental yields of phenylalanine from non-growing cells of a recombinant *Escherichia coli* strain. *J. Biotechnology* 7, 319–331.

Gubler, M., S. M. Park, M. Jetten, G. Stephanopoulos, A. J. Sinskey (1994). Effect of phosphoenolpyruvate carb-oxylase deÞciency on metabolism and lysine production in *Corynebacterium glutamicum*, *Appl. Microbiol. Biotechnol.*, **40**, 857–863.

Heinrich, R., T. Rapaport (1974). A linear steady-state treatment of enzymatic chains. *Eur. J. Biochem.*, **42**, 89–95.

Hofmeyr, J. H. S., A. Cornish-Bowden (1991). Quantitative assessment of regulation in metabolic systems. *Eur. J. Biochem.*, **200**, 223–236.

Hofmeyr, J.-H. S., H. Kacser, K. J. Merwe (1986). Metabolic control analysis of moiety-conserved cycles. *Eur. J. Biochem.*, **155**, 631–641.

Holms, W. H. (1996). Flux analysis and control of the central metabolic pathways in *Escherichia coli*, *FEMS Microbiol. Rev.*, **19**, 85–116.

Hurlebaus, J. (2001) A pathway modeling tool for metabolic engineering. PhD thesis, University of Bonn, 2001.

Kacser, H., J. Burns (1973). The control of flux, *Symp. Soc. Exp. Biol.*, **27**, 65–104. (Reprinted in *Biochem. Soc. Trans.*, **23**, 341–366, 1995.)

Kacser, H., L. Acerenza (1993). A universal method for achieving increases in metabolite production. *Eur. J. Biochem.* **216**, 361–367.

Kornberg, H. L. (1966). The role and control of the glyoxylate cycle in *Escherichia coli*, *Biochem. J.*, **99**, 1–11.

Koshland, D. E. Jr (1987). Switches, thresholds and ultrasensitivity. *Trends Biochem. Sci.*, **12**, 225–229.

Mendes, P. (1993). GEPASI: A software package for modelling the dynamics, steady states and control of biochemical and other systems. *Comput. Appl. Biosci.* 9, 563–571.

Mendes, P. (1997). Biochemistry by numbers: simulation of biochemical pathways with Gepasi 3. *Trends Biochem. Sci.* 22, 361–363.

Mendes, P., D. B. Kell (1998). Non-linear optimization of biochemical pathways: applications to metabolic engineering and parameter estimation. *Bioinformatics* 14, 869–883.

Neidhardt, F. C., J. Ingraham, M. Schaechter (1990). Physiology of the bacterial cell: a molecular approach. Sinauer Associates, Sunderland, MA. **520**, 1–29.

Sauro, H. M. (1993). SCAMP: A general-purpose simulator and metabolic control analysis program. *Comp Appl Biosci.* **9**, 441–450.

Savageau, M. A., E. O. Voit, D. H. Irvine (1987). Biochemical systems theory and metabolic control theory: 1, Fundamental; similarities and differences. *Math. Biosci.*, **86**, 127–145.

Stephanopoulos, G., J. J. Vallino (1991). Network rigidity and metabolic engineering in metabolite overproduction. *Science*, **252**, 1675–1681.

Stephanopoulos G. N., A. A. Aristidou, J. Nielsen *Metabolic Engineering. Principles and Methodologies*, New York: Academic Press, 1998.

Vallino, J. J., G. Stephanopoulos (1993). Metabolic flux distributions in *Corynebacterium glutamicum* during growth and lysine overproduction. *Biotechnol. Bioeng.*, **41**, 633–646.

Walsh, K., D. E. Koshland (1984). Determination of flux through the branch point of two metabolic cycles: The tricarboxylic acid cycle and the glyoxylate shunt, *J. Biol. Chem.*, **259**, 9646–9654.

Xu, H., P. Wei, H. Zhou, F. Fan, P. Ouyang (2003). Efficient production of L-phenylalanine catalyzed by a coupled enzymatic system of transaminase and aspartase. *Enzyme and Microbial Technology* **33**, 537–543.

8

Enzyme and Co-factor Engineering and Their Applications in the Pharmaceutical and Fermentation Industries

George N. Bennett and K.-Y. San

8.1 Introduction

Over the last thirty years, the impact of molecular genetics on industrial biotechnology processes has become pervasive. Perspectives of the impact of molecular biology on industrial processes (Demain, 2001) and pharmaceuticals (Chartrain et al., 2000) have been published. A recent book covers areas related to the use of enzymes in catalysis and protein engineering (Bommarius, 2004). Other examples of useful recent reviews include some in the fields of polyketide antibiotics (Umeno et al., 2005), lipases (Akoh et al., 2004), and proteases (Bryan, 2000; Maurer, 2004).

This chapter is not meant as a comprehensive review of the field but will point out some general themes and use a few examples to illustrate certain points. In considering a process or metabolic engineering effort, there are now two main approaches to consider in identifying the appropriate enzyme for an ideal process. One approach makes use of the diversity in properties of enzymes found in nature and made available through the large-scale cloning and sequencing efforts for an appropriate choice. The other approach is based on structural features of the enzyme and by making alterations of a characterized enzyme through rational design or screening of random mutants, an appropriate enzyme variant can be found. In this chapter, we shall present several examples of the use of enzyme engineering for different applications and the accessory means of improving enzymatic conversions through manipulation of essential cofactors.

ENZYME ENGINEERING
The modification of an enzyme to improve a desired physical or catalytic property.

In the case of a single enzyme used in an industrial process *in vitro*, several criteria are important. These would include the catalytic characteristics of the enzyme such as specificity for substrates and products, specific activity or rate of the catalyzed process, and sensitivity to the presence and properties of inhibitors in the substrate feedstock. Other parameters of interest for a practical process involve the conditions that affect the activity and the lifetime of the catalyst. Such factors as the pH profile of activity of the enzyme, the temperature desired for the process, the stability of the enzyme under those conditions, and the presence of other substances in the substrate mixture that may affect the enzymatic activity (inhibitors) or stability (metals, reactive compounds, or redox compounds) need to be suitable for the process to be effective. In the case where several enzymes are operating in coordination to produce a product from a metabolic pathway in a whole cell system, the physical operating parameters are limited to the growth conditions of the cells, but the level of each enzymatic activity needs to be appropriately regulated so the whole pathway operates without imbalance and accumulation of toxic or undesired intermediate metabolites does not occur.

8.2   Types of major industrial enzymes and desired modifications

8.2.1   *Targets for enzyme engineering*

While many enzymes are known and can be engineered for improved function, the utility of the enzyme in the overall process must be an overriding factor and criteria for consideration have been discussed (Jacobsen and Finney, 1994). In identifying targets for enzyme engineering, the general application, economic benefit, and long-term potential must be evaluated. Considering the amount of research investment required for optimization of a particular trait, several factors need to be present to justify the focus on a particular enzyme. One is the long-term market value of the products and the cost advantage afforded by an enzyme with altered properties (if such an enzyme can be found or created). Second, there should be potential for the same process or enzyme be used in several areas or applied to multiple related products. Third, these variants should be identified through a suitable screening process. The ability to easily screen for the desired properties and the efficiency of a high-throughput system to find useful variants are practical aspects of successful enzyme engineering.

One general area where enzymes with engineered properties may become more important is the production of optically active amines as pharmaceutical precursors. Particular additional classes would include esterases, racemases, and the large field of oxidation-reduction reactions including those catalyzed by peroxidases, cytochrome P450s, and dehydrogenases. The use of enzymatic procedures that are stereospecific can enhance the overall synthetic process. Specific processes where the chemical reaction yields a mixture that is either hard to separate or resolve, or has a low yield, are attractive candidates for the introduction of enzymatic steps. With the great variety of enzymes available and the ability to alter specificity and sensitivity to various conditions, the field of developing better catalysts through protein engineering seems to offer great potential.

8.2.2 *Hydrolytic enzymes*

Most enzymes used in large quantity are hydrolytic enzymes. The addition of enzymes to laundry detergents has led to a major change in this industry. Protease enzymes are widely used as additives to remove stains from protein-rich foods or blood. The *Bacillus* alkaline serine-protease, subtilisin, is the most common enzyme used for this application. Desirable properties for such a use include stability and activity at high temperature and under the high pH associated with typical laundry conditions. Stability in the face of other harsh substances and chelating agents is also desirable. The amylases are another group of hydrolytic enzymes used in large quantity in starch liquefaction and to remove starchy foods from clothing in laundries. The alpha amylases are derived from *Bacillus* species and the beta amylases from *Bacillus* and plants. High-temperature activity is also usually desired as well as the ability to act on the crude substrate in the presence of other substances. In the processing of starch, glucoamylases from fungi are also important. One of the most large-scale uses of enzymes is glucose isomerase (xylose isomerase) in the production of high fructose corn syrup. The reaction conditions favored for industrial use are high temperature and low pH to avoid side reactions. A variety of enzymes from *Bacillus coagulans*, *Streptomyces rubiginusus*, and other bacterial species have been used for this purpose.

The manufacture of cheese uses a specific enzyme to hydrolyze casein in an early stage to curdle the milk protein. Recombinant chymosin (rennin) is used as the major source of the enzyme although enzymes from natural animal sources and fungi are also used. In this case, the specificity of the enzyme for a particular cleavage of the casein is important to minimize general protease degradation. The enzyme must also have high activity under the processing conditions needed.

Lipases are another group of hydrolytic enzymes used in a variety of ways including removing grease stains and clogs. Hydrolysis of fats in the food industry is a major use. They can also be used under conditions of low water content in an organic solvent to modify oils and fats for food use by transesterification of the acyl chains to generate a more desirable pattern in the product (e.g., the level of saturation or the chain length of the fatty acid). Another widely used hydrolytic enzyme is penicillin acylase. In the production of various β-lactam antibiotics, penicillin acylase is used to produce 6-amino penicillin derivatives. Addition of water to molecules in a specific fashion is a useful reaction of enzymes. Nitrile hydratase from *Pseudomonas chlororaphis* B23 has been used to form acrylamide from the nitrile in a large-scale industrial process that has environmental and economic advantages. Similarly, nicotinamide can be prepared from 3-cyanopyridine with an appropriate nitrile hydratase.

8.2.3 Specialty enzymes

In the pharmaceutical area, the interest has been in compounds where a specific chiral structure is needed in synthesis of the product. In addition to the specific synthetic pharmaceuticals, the nutritional area of natural amino acids and vitamins involve large volume processes that also require chiral production systems. Therefore, enzymes that can provide stereospecific reactions on complex molecules, particularly those such as sterols, alkaloids, etc., are employed in many synthetic processes. Another use of enzymes exhibiting high stereospecific action is in the separation or isomerization of a racemic mixture of an important compound produced by chemical synthesis. An example is the resolution of (R,S)-Naproxen 2,2,2-trifluoroethyl thioester to produce the desired (S)-Naproxen, an anti-inflammatory drug (Ng and Tsai, 2005). Another example of the use of enzymes to afford chiral products during synthesis is the production of Omapatrilat (Patel, 2001). The general aspects of formation of chiral products and specificity have been discussed (Jaeger and Eggert, 2004; Turner, 2003).

A large family of proteins that has received much attention from biochemists is the cytochrome P450 group. These redox-active enzymes act to carry out many specific oxidative reactions in eukaryotic organisms and can act on complex chemical compounds including toxic aromatics. These enzymes have been studied for their potential as industrial catalysts (Matsunaga and Shiro, 2004; Miles et al., 2000; Morant et al., 2003; Urlacher and Schmid, 2002; Urlacher et al., 2004) and their value as general oxygenases (Cirino and Arnold, 2002). The reactions of oxygenases and the need for coupling of the reaction to recycling of the reduced co-factor, often NADH, have been discussed (Li et al., 2002). The various applications of the readily available enzyme, horseradish peroxidase, in cancer therapy, cross-linking biomolecules and as a detection agent have been described (Veitch, 2004).

8.2.4 *Alteration of physical parameters of enzymes for process applications*

Temperature stability has been considered one of the major factors affecting use of industrial enzymes (Eijsink et al., 2004, 2005). The discovery and analysis of microbes adapted to high temperature have generated a natural source for enzymes active under high temperature, through analysis of the natural biodiversity of such environments (Wackett, 2004). The basis of thermostability has been investigated through theoretical and experimental studies (Kumar and Nussinov, 2001; Lehmann and Wyss, 2001; Scharnagl et al., 2005). These studies suggest that a dense hydrophobic core along with an increased density of intramolecular hydrogen bonds and ionic interactions are valuable for achieving high-temperature stability. Examples of high-temperature enzymes are thermostable glucoisomerases (Asboth and Naray-Szabo, 2000; Hartley et al., 2000) and glucoamylases (Hesselink et al., 2003; Sauer et al., 2000). The engineering of thermostable subtilisin (Zhao and Arnold, 1999) has illustrated approaches to improvement of temperature stability by use of molecular biology tools. The susceptibility to inactivation of enzymes by oxidation is also a problem that can be addressed, and in the case of subtilisin, specific mutant enzymes have been developed (Estell et al., 1985; Yang et al., 2000). Similarly, cold active enzymes are useful if the enzyme is to be easily inactivated after the reaction. The adaptation of enzymes to effective action at cold temperatures is linked to increased flexibility around the active site in enzymes from psychrophilic organisms (Feller, 2003; Georlette et al., 2004). These enzymes are also useful with substrates that may be themselves labile to high temperatures. Other examples that require cold conditions are those used in specific steps of food processing or cold laundry processes. In this case, protein engineering can be used but also the natural diversity of microbes can be taken advantage of in order to isolate enzymes that are cold-adapted. Such psychrophilic organisms have been isolated from polar, marine, and high-altitude environments. Examples of cold active enzymes have been cited (Cavicchioli et al., 2002; Georlette et al., 2004) and protein engineering of cold-adapted subtilisin has been described (Taguchi et al., 1998, 1999, 2000).

Other general examples of the modification of enzymes include efforts to degrade polymers at a desired pH for industrial applications. These include endoglucanase for cellulose degradation (Wang et al., 2005), and amylase for starch hydrolysis (Bessler et al., 2003). Compatibility of enzyme activity with conditions of high salt or organic solvents is also a challenge. These factors have been discussed (Castro and Knubovets, 2003). Again, the model of subtilisin has been used and its activity in the presence of dimethyl formamide has been improved (You and Arnold, 1996).

8.3   Summary of methods in enzyme engineering

8.3.1   Site-specific mutagenesis

One technique that has had great impact is that of site-specific mutagenesis. An overview is given in the review by (Brannigan and Wilkinson, 2002). The first uses of site-specific mutagenesis were to make a specific change in a particular amino acid in a protein in order to investigate the role of a particular functional group in the mechanism of action of the enzyme. Such a change could be made through molecular biology techniques by altering the gene encoding the protein. The codon within the gene that specifies that particular amino acid in the protein is changed, and the modified gene is expressed in a suitable organism to produce a quantity of the altered enzyme for biochemical analysis.

A widely used method for practical mutagenesis was developed by Kunkel (1985) and involved the use of a primer incorporating the desired nucleotide change and extending the primer annealed to a circular single-stranded template vector containing the gene to be mutated. In this case the vector was isolated from a source that allows a high level of incorporation of dU in the template. After extension of the primer by DNA polymerase, the double-stranded molecule is introduced into a host *Escherichia coli* strain. The strand made by extending the primer is preferentially replicated and the dU containing template strand is preferentially degraded by host enzymes yielding a high proportion of the progeny vector with the desired mutation (Kunkel, 1985; Kunkel et al., 1987; 1991). The method has been adapted in various ways for use with other plasmids (Jung et al., 1992). Recently, new methods based on PCR and standard kits have become available from many companies for rapid site-specific mutagenesis using commonly used versatile vectors and generating a high proportion of recombinant colonies bearing the desired mutant form.

8.3.2   Cassette mutagenesis

In order to introduce a number of mutations simultaneously, the technique of cassette mutagenesis has been used (Kegler-Ebo et al., 1996; Wells et al., 1985). This method allows a short segment of the gene to be cut out and replaced by a new chemically synthesized segment. The introduced segment can be made to contain a mixture of different nucleotides at a given position and so it can yield substitution of the original codon by various other triplets encoding a variety of different amino acids. When the modified DNA is introduced into a host strain, each molecule is replicated independently, and, upon plating of the mixture of transformed bacteria, each new variant can be isolated from a separate colony upon plating of the mixture of transformed

Figure 8.1

Schematic of site-specific mutagenesis. In the example shown, the specificity of isocitrate dehydrogenase from NADPH preferring to NADH preferring is illustrated and modified from the studies of Hurley et al. (1996). The lys-344, as a positively charged residue, can interact with the phosphate group of NADPH, aiding recognition and binding of NADPH. When mutagenesis alters the codon for amino acid 344 in the enzyme to a codon, GAT, specifying aspartic acid, a negatively charged amino acid residue, the interaction with a negatively charged phosphate from NADPH is disfavored, and a better interaction occurs with the unphosphorylated ribose hydroxyl groups that can form hydrogen bonds with the aspartic acid carboxylate group. Such an interaction would favor binding of NADH rather than NADPH at the active site. As with most enzymes there are a number of amino acids that contribute to specificity at the active site and more details are given in (Chen et al., 1995; Hurley et al., 1996; Yaoi et al., 1996) and other alterations affecting substrate specificity are described in a crystallographic study (Doyle et al., 2001).

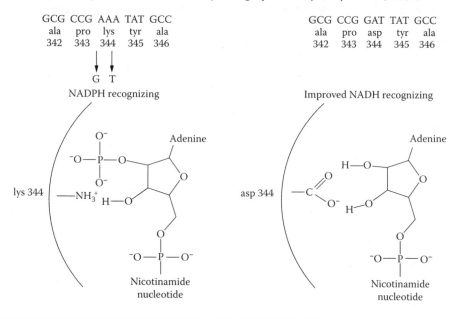

bacteria. In this way, a number of different amino acid residues at a particular position in the protein could be made and individually tested for their effect on enzyme properties. The use of segments with more extensively changed nucleotide sequences could allow the formation of protein variants with several amino acid changes within the section replaced giving a localized patch on the protein where there are many alterations.

As the technology for oligonucleotide synthesis improved, the possibility of synthesizing the entire gene became possible. With this approach, the gene encoding the desired protein could be designed to incorporate a number of useful restriction sites for replacement of different segments of the protein by cassette mutagenesis. This approach also allows extensive redesign of the protein. Some early uses of this approach are given in the articles (Rose et al., 1988; Shen et al., 1997).

Figure 8.2

Cassette mutagenesis. In this figure, the initial plasmid encodes the protein. The ATG indicates the codon for the first amino acid of the protein (N-terminus), and the TAA indicates the stop codon, which terminates synthesis at the C-terminus of the protein. The sites R₁, R₂, R₃, R₄ denote the positions of cleavage of the plasmid at a unique place by the restriction enzymes R₁, R₂, R₃, R₄, respectively. Cleavage by R₂ and R₃ then removes a defined segment, and this segment can be replaced by a synthetic segment having a mixture of nucleotides (N for a mixture of A, T, G, C) at a particular position within the segment. In the example shown, the nine positions specifying amino acids 344-345-346 are mixed and could then encode any amino acid at these positions. Replacement of the R₂-R₃ segment in the plasmid would generate a variety of plasmid molecules, with different possibilities for codons at positions 344, 345, 346 in this protein. After transformation and replication of individual molecules of the plasmid and plating to generate individual colonies, the protein from each colony could be isolated and examined and the nucleotide sequence of the specific R₂-R₃ segment determined.

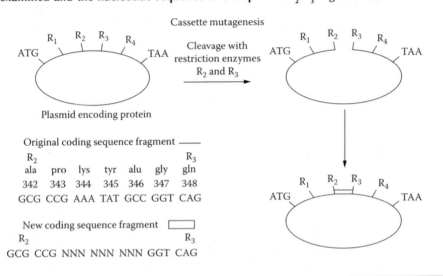

8.3.3 *3-D structure and specific mutations*

If one wished to enhance a particular property of an enzyme through site-specific mutagenesis, the three-dimensional structure of the enzyme would be studied and the three-dimensional structure of the enzyme with a bound substrate analogue would be examined to gain an understanding of the positioning of the substrate in the active site. Coupled networks of interacting functional groups in protein structures are often needed for catalysis, and this can make the identification of important residues to modify more complicated (Benkovic and Hammes-Schiffer, 2003). An example of the change of substrate specificity of the protease subtilisin illustrates this approach. When the binding cavity is reduced in size due to placement of a larger amino acid in substitution for the small glycine-127 within the substrate binding pocket, the enzyme exhibits a preference for smaller substrates that can fit into the active site of the enzyme

(Takagi et al., 1996). In general, expanding the size of the substrate binding site can lead to a broadening of the specificity as more, different substrates can fit near the catalytic residues of the enzymes; thus, the specificity may be broadened. However, the affinity of the enzyme for the substrate may decrease requiring a higher concentration of the substrate for effective catalysis.

Another specificity difference of industrial and metabolic interest is the co-factor requirement for reduction reactions. The pools of NADH and NADPH in the cell influence the reactions and NADH is generally preferred for *in vitro* systems due to its greater stability. Mutations that alter the specificity and convert the NADPH specificity to NADH preference of 2,5-diketo-D-gluconic acid reductase A, an enzyme used in some vitamin C synthetic processes, have been reported (Banta et al., 2002b). In a carbonyl reductase that is used for formation of optically active alcohols, NADH specificity has been engineered (Morikawa et al., 2005). In some cases, it may be preferable to have NADPH specificity to avoid interference with other enzymes or pathways and examples of the conversion of NADH specificity to that of NADPH have been reported for xylitol dehydrogenase (Watanabe et al., 2005) and lactate dehydrogenase (Holmberg et al., 1999).

8.3.4   *Random "directed evolution" methods*

The ability to make and screen a large number of random mutations within a localized region is referred to as "directed evolution." This approach can be used without knowledge of the three-dimensional structure of the enzyme and can inquire over a wider range of sequence and structure space to find a protein sequence with the appropriate properties. An important element in the strategy is being able to do genetic selections or efficient screens for identification of the improved variants of the enzyme among the large number of different mutant forms of the enzyme expressed in the transformed cell population (Jestin and Kaminski, 2004). The directed evolution strategy makes use of the techniques of error-prone PCR and DNA shuffling, which will be explained briefly below.

Overviews of the impact of these techniques on biomolecular engineering have been presented (Petrounia and Arnold, 2000; Ryu and Nam, 2000). Attention specifically focused on the directed evolution efforts with some industrial enzymes are presented in reviews (Cherry and Fidantsef, 2003; Hult and Berglund, 2003). As the use of this approach has grown and been applied to many enzymes, the general methods for directed evolution have been reviewed (Lutz and Patrick, 2004; Williams et al., 2004). Recent additional improvements and extensions include the use of data mining and bioinformatics to choose

DIRECTED EVOLUTION
Direct selection of specific variants of a given gene sequence with the desired properties.

ERROR-PRONE REPLICATION

During the process of replication, an incorrect nucleotide may be inserted and the natural rate of this kind of error during DNA replication may be less than one in a million. When a large variety of altered genes are needed for identification of variants with an improved property, conditions are varied so an error rate of one in a thousand may be generated.

COUPLED ENZYMES

In enzyme assays or in biosynthetic processes, two enzymes are often coupled with the second enzyme using the product of the reaction of the first enzyme. This can allow efficient processing of a starting compound along a biochemical pathway. Such coupling is needed in the recycling of a co-factor, for example, between oxidized and reduced forms.

THERAPEUTIC PROTEINS

Proteins that are administered to a patient for treatment of a medical condition.

appropriate related genes (Gustafsson et al., 2003) and the marriage of computational abilities with directed evolution techniques (Arnold, 2001).

Error-prone PCR takes advantage of the misincorporation of nucleotides into DNA during polymerization to yield a synthesized strand with random changes. The level of misincorporation can be altered by the use of different polymerases, metal ions, or addition of other substances that affect the structure of the polymerase or its ability to discriminate base pairing (Cirino et al., 2003; Pannekoek et al., 1993). Theoretical models have related the error rate to the overall distribution of base changes in the final population of molecules (Bessler et al., 2003).

DNA shuffling (Stemmer, 1994) uses fragments from closely related genes in a PCR misincorporation process to generate a high diversity of full-length gene product molecules made from randomly mutated segments derived from each of the parent genes. This technique allows a wider expanse of sequence space to be created and searched than the original error-prone PCR technique, and since there are a greater variety of altered forms of the enzyme in the mixture examined, there is a greater chance for an enzyme with the desired property to be found. Since a vast number of variants is generated, the challenge of developing a suitable protocol for searching through them with a sensitive selection regimen or screen is a serious consideration. The broader application of this technique to pharmaceutical and vaccine production (Locher et al., 2005; Patten et al., 1997) and improvement of industrial enzymes (Powell et al., 2001) have been reviewed.

While screening systems are not reviewed here, the usual practice would be to make use of a substrate or coupled reaction that can produce a colored compound. The color can serve as a signal of the appropriate enzyme activity under the conditions of the assay. The conditions of assay and detection can be modified to include physical parameters that are desired such as pH, temperature, or the presence of organic solvents or other inhibitors. The use of high-throughput detection systems in combination with a multiwell plate format allows interfacing sample recording with identification of modified protein candidates with enhanced activity. These candidates can be investigated in more detail by additional biochemical experiments.

8.4 Modification of pharmaceutical properties of protein agents

In the case of therapeutic proteins, protein engineering is used to produce a form with an appropriate half-life in circulation, reduced immunogenicity, and stability of therapeutic action (Shanafelt, 2005). Such modifications can alter the solubility,

Figure 8.3

Error-prone PCR. In error-prone PCR, the template is annealed with primers that allow extension by DNA polymerase and incorporation of the dNTPs in the incubation mixture. The condition of the extension and the enzyme used will define how frequently a mistake or misincorporation (for example, A for G) is made. Denaturation of the strand at high temperature followed by reannealing of the primers can then allow the strands with mistakes to be copied as well as the original strands, and further errors by misincorporation are introduced, giving a product that may have many variations within the product molecules. Conditions can be used that will give an appropriate level of altered molecules (e.g., 1 mutation per 100 or 1000 residues on average). Since the molecules can be cloned and replicated independently within a plasmid, again individual colonies with separate variants can be isolated and screened for enzyme properties (e.g., high-temperature activity) and the specific DNA sequence of interesting variants can be analyzed.

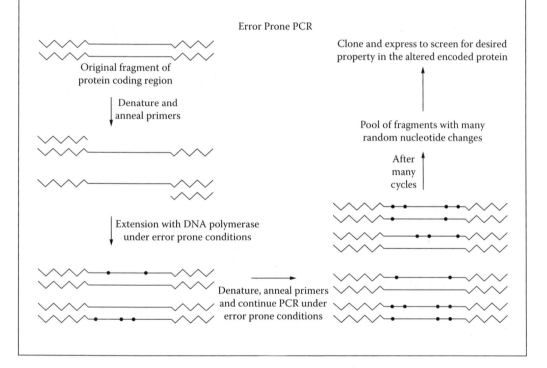

susceptibility to degradation by proteases, and reduction of the tendency of the enzyme to lose essential co-factors. Modifications to surface-exposed or terminal residues of the protein can affect these properties. One way to address these biological considerations is by connecting the protein to an agent with desirable properties for blood circulation. An example would be the introduction of site-specific modifications that will allow reactions such as pegylation or glycosylation to occur on suitable sites in the protein. These modifications allow the pharmaceutical properties of the protein to be enhanced (Graddis et al., 2002). The specific favorable characteristics of the

Figure 8.4

DNA shuffling. In the DNA shuffling procedure, several gene fragments encoding similar proteins are used. The segments must have a high enough sequence similarity to anneal well in heterologous combinations so the polymerase can extend the various duplexes formed. The repeated denaturation and reannealing give a chance for different sections of the molecule to be represented by any one of the original genes. The combination of various sections and the individual base pair mutations introduced during the repeated process generates a great variety of altered sequences. These DNA fragments can be cloned, and the individual variant genes encoding protein are expressed in individual colonies, and screening schemes or selections impressed on individual colonies are used to identify variants with desired properties. The DNA encoding the variant gene can be isolated and sequenced to identify the changes producing the effect.

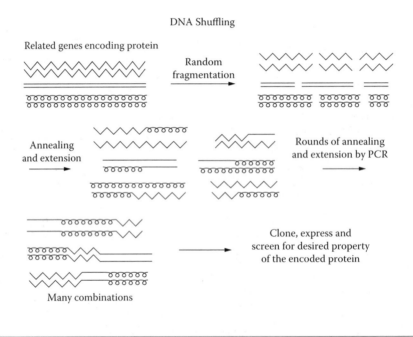

addition of polyethylene glycol (pegylation) to drugs has been reviewed (Harris et al., 2001; Molineux, 2002). The addition of the soluble polymer prevents clearance from the bloodstream by the kidney, reduces enzymatic degradation, and reduces immunogenic problems. Specific proteins where this has been advantageous include interferon alpha-2a. The enhanced pharmacokinetic properties of the modified protein can allow for more convenient and appropriate dosing schedules for patients.

Certain drugs can be better agents if they are modified to enhance uptake or pharmacokinetic properties. For example, the acylation of one of the two hydroxyl groups of lobucavir with valine is beneficial. The use of specific lipases in the synthesis of key intermediates has been described (Hanson et al., 2000).

8.5 Modification of enzymes for *in vivo* biosynthetic processes

The application of enzyme engineering to processes going on inside cells producing useful small molecules, or for whole cell biocatalysts or immobilized enzymes in bioprocess steps used in the production of pharmaceutical intermediates has been an area of expansion in recent years (Huisman and Gray, 2002). This is illustrated by the application to antibiotic production specifically involving beta-lactams (Sio and Quax, 2004). In an example, beta-lactam acylases have been used for reactions beyond their usual hydrolytic action on penicillin G or cephalosporin C, i.e., for new applications in the synthesis of novel semisynthetic beta-lactams. Knowledge of the 3-D structure of these enzymes has allowed defined site-specific mutagenesis in the application of enzyme engineering. Goals in this area include performance of immobilized enzymes and suitability for use with specialized reaction conditions that integrate well with other steps of downstream processing. Since the chiral nature of products are important (in many cases only one stereoisomer of a compound is pharmaceutically active), the resolution of racemic mixtures by selective hydrolysis of acyl groups has been a useful feature of these enzymes. In these processes, the racemic mixture is typically treated with an enzyme that can only hydrolyze one stereoisomer, and after the reaction, the unacylated form is easily separated from the acylated form by usual chemical isolation procedures.

Another group of antibiotics is the polyketides, and there has been considerable work on altering the complex enzymes involved in the biosynthesis of these compounds. Several articles describe work on these systems (Kealey, 2003; Khosla and Harbury, 2001; McDaniel et al., 2005; Pfeifer and Khosla, 2001; Umeno et al., 2005) and a general overview covers many features (Staunton and Weissman, 2001). In these articles, the modularity of the biosynthetic pathway enzymes and the formation of large multifunctional proteins and multiprotein complexes to biosynthesize the complex structure of the final polyketide antibiotic are emphasized. The modification of the individual functional domains and switching of these modules among the gene clusters specifying the biosynthesis of different antibiotics have given rise to a considerable industry that employs advanced genetic techniques to engineer restructured proteins. These are used to explore the synthesis of classes of novel polyketide compounds as potential pharmaceutical agents.

Sterols are an important group of compounds with many uses as pharmaceuticals. Yeast has been engineered via introduction and expression of genes to make pregnenolone and progesterone (Duport et al., 1998) or hydrocortisone (Szczebara et al., 2003). The broad potential for engineering of yeast for compounds of this type has been reviewed (Veen and Lang, 2004). Plants have been engineered for improved production of phytosterols and related compounds (Seo et al., 2005). A number of other organisms are used to make specific chiral modifications of the basic sterol structure and are used for chemical conversions or assays (Itagaki et al., 1990; MacLachlan et al., 2000). Protein engineering of specific amino acids at key positions in particular enzymes have altered the activities of steroid dehydrogenases involved in steroid hormone metabolism (Jez and Penning, 1998).

Carotenoids are a group of compounds containing a number of conjugated double bonds and are of interest for their color, nutritional, and pharmaceutical attributes. They are made by a variety of micro-organisms and the genes encoding the enzymes of the biosynthetic pathways are known. Recent work has involved the expression of biosynthetic enzymes in different hosts and in unique combinations (Mijts et al., 2004). The possibility of evolving novel pathways and forming a new series of useful carotenoids illustrates the use of a combinatorial approach in genetic engineering of organisms for formation of new compounds (Sandmann, 2002; 2003). The formation of new pathway products including new families of carotenoids has been reviewed (Mijts and Schmidt-Dannert, 2003; Umeno et al., 2005).

The above classes of compounds utilize NADH or CoA cofactors extensively in their biosyntheses and so form a connection between the engineering of the enzyme proper of the biosynthetic pathway and the next topic that of manipulations related to the co-factor supply.

8.6 Co-factor engineering

The term "co-factor engineering" was used by Duine (1991) in a general article that brought additional attention to the essential role of the co-factor in the efficiency of the reaction. It had long been known that the level of biotin could greatly affect the production of glutamate by *Corynebacterium glutamicum* (Kimura, 2003). A related experience was noted in the early production of citric acid with respect to iron, which is normally a constituent of certain enzymes and Fe-S or heme proteins.

Specific oxidation reactions and their bioprocess considerations have been reviewed (Buhler and Schmid, 2004). The co-factor most studied is the reductant used in many reactions, i.e., NADH or NADPH. While reductions are very important and the chiral nature of the product is often crucial, there is less industrial use of specific oxidoreduction due to the problem of regenerating the co-factor. *In vitro*, the co-factor can be regenerated by another enzyme that can form the reduced co-factor by oxidation of a readily available substrate that will not interfere with the desired reaction. An enzyme often used *in vitro* systems to form NADH is formate dehydrogenase. This enzyme can use formate to produce the reduced form of the co-factor NADH and CO_2. The CO_2 is easily removed and does not interfere with many redox reactions. *In vitro* application of formate dehydrogenase (FDH) for recycling of NADH has been frequently used and is described (Wandrey, 2004). This and other systems for regeneration of reduced cofactors have been reviewed (Wichmann and Vasic-Racki, 2005). Another system for NADH recycling has been proposed, that of phosphite oxidation, and the enzyme has been engineered for a generation of NADH (Woodyer et al., 2003). Since NADH is more stable and generally available in higher quantities, it is preferred for processes rather than NADPH. Thus, for *in vitro* processes, it is often considered useful to convert a NADPH-utilizing enzyme to one that can use NADH (see above example in Figure 8.1). This is also true for some processes that are done *in vivo* where the limitation on availability of NADPH must be addressed, usually via genetic means or through the use of particular growth conditions that will affect the pathways that produce NADPH. Several examples of engineering the specificity of an enzyme for NADH or NADPH have been reported and mentioned earlier in this chapter.

8.6.1 *NADH vs. NADPH specificity of enzymes*

The enzyme isocitrate dehydrogenase has been engineered based on its 3-D structure by the introduction of seven mutations that change the specificity from favoring NADPH by 7000-fold to an enzyme that prefers NADH by 200-fold (Hurley et al., 1996). In general, a major difference between the NADP and NAD utilizing enzymes is the presence of a carboxylate side chain that chelates the diol group at the ribose near the adenine of NAD, whereas in enzymes that utilize NADP, there is frequently a highly positively charged arginine side chain oriented toward the adenine that interacts with the negatively charged ribose 3'-phosphomonoester (Carugo and Argos, 1997). See the example of mutagenesis illustrated in Figure 8.1 and described in the literature (Chen et al., 1995; Hurley et al., 1996; Yaoi et al., 1996).

In the important group of cytochrome P450 enzymes, the NADPH specificity of an enzyme was altered so the enzyme could use NADH for recycling (Dohr et al., 2001). *Corynebacterium* 2, 5-diketo-D-gluconic acid (2, 5-DKG) reductase is an enzyme used in the synthesis of vitamin C. Efforts to improve the enzyme so it could also use NADH generated the F22Y/K232G/R238H/A272G quadruple mutant, which exhibits high activity with NADH while retaining a high level of activity with NADPH (Banta et al., 2002c). The effects of this altered system for vitamin C production have been modeled and it was suggested to provide cost savings (Banta et al., 2002a).

While *in vitro* regeneration of NADH has been addressed for some time (Wichmann and Vasic-Racki, 2005), alteration of *in vivo* metabolism by use of co-factor manipulation has been a more recent development. A few articles have addressed this general issue. An overview of the use of metabolic engineering in biosynthesis of antibiotics and the role of co-factor supply and the connection to primary metabolism have been presented (Gunnarsson et al., 2004). The engineering of recombinant yeast for chiral reductions has been reviewed (Johanson et al., 2005). Enzymatic reactions involved in the formation of flavor components vanillin, gamma-decalactone, carboxylic acids, C6 aldehydes and alcohols ('green notes'), esters, and 2-phenyleth-anol, including the need for co-factors and efficient recycling have been discussed (Schrader et al., 2004).

8.6.2 *Manipulation of NADH* in vivo

Expression of NADH oxidase *in vivo* was used to change the pattern of metabolic products formed in *Lactococcus lactis* (Lopez de Felipe et al., 1998). In such a strain, preference was for the production of more oxidized final products since there was less NADH available for conversion of intermediates to reduced final products. Several co-factor engineering alterations have been mentioned in the article (San et al., 2002) including a system for using formate dehydrogenase (FDH) to recycle NADH *in vivo* and the manipulation of CoA levels in the formation of products derived from acetyl-CoA. The use of the *Candida boidinii* FDH system (Allen and Holbrook, 1995; Sakai et al., 1997) *in vivo* was further investigated. The coupling of the FDH to L-amino acid dehydrogenases could produce optically active amino acids from alpha-keto acids in the presence of ammonium formate and a whole cell biocatalyst (Galkin et al., 1997). Enhancement of ethanol production during growth on glucose was found in cells expressing *C. boidinii* FDH grown under anaerobic conditions. In fact, some ethanol could be formed under aerobic conditions when formate was added to an FDH-expressing strain (Berrios-Rivera et al., 2002a; Berrios-Rivera et al., 2002b). The high availability of NADH in an organism could provide potential for reducing other molecules either present in the cell or added to the media. The coupling of FDH with a transhydrogenase to give higher availability of NADPH for chiral alcohol production has been reported (Weckbecker and Hummel, 2004). Engineering of the enzyme by mutagenesis to aid stabilization of the enzyme to oxidation has also improved the characteristics of the FDH (Slusarczyk et al., 2000) and the temperature stability of a similar enzyme from *Pseudomonas sp 101* (Rojkova et al., 1999). A number of FDHs from other species have been recently cloned and characterized. A comparison of FDH and a glucose dehydrogenase (GDH) from *Kluyveromyces aestuarii* suggested

Figure 8.5

Coupling of *in vivo* activity of formate dehydrogenase to other redox reactions. Through either internally produced or externally added formate, the reducing power of formate is used to produce NADH by the action of formate dehydrogenase. The NADH formed can be used by another enzyme in the cell, either an existing enzyme such as alcohol dehydrogenase or a heterologous enzyme from another organism capable of reducing a desired substrate. A variety of oxidoreductases can be produced in a host organism by introduction of appropriate genes by recombinant DNA techniques. The reduced product formed can be derived from a substrate normally produced by the growing cell from metabolism of a carbon source or it could be an added substrate molecule that is not otherwise metabolized by the organism. For those reduced products requiring formation by an NADPH-preferring enzyme, the system can be coupled to NADPH generation via a transhydrogenase.

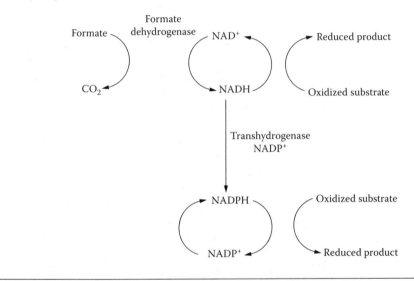

the GDH enzyme had a higher ability to regenerate NADH for coupling with the reduction of 4-chloroacetoacetate to form (S)-4-chloro-3-hydroxybutanoate, an important pharmaceutical intermediate (Yamamoto et al., 2004). The use of glucose dehydrogenase to supply reduced co-factor for reactions *in vivo* such as the production of chiral alcohols has been presented (Kataoka et al., 2003).

Mannitol production by an enzymatic reduction of fructose can be achieved by using a cloned mannitol dehydrogenase (Liu et al., 2005). The coupling of the mannitol dehydrogenase to FDH allowed the regeneration of NADH by formate while providing reductant for reduction of the fructose (Kaup et al., 2003; 2004). The extension of the system to the formation of other chiral compounds such as conversion of methyl acetoacetate to chiral hydroxy acid derivatives [methyl (R)-3-hydroxy butanoate] was developed by using the *fdh* gene from *Mycobacterium vaccae* N10. The gene encoding the NAD⁺-dependent formate dehydrogenase was co-expressed in *Escherichia coli* with an

CHIRAL COMPOUND
Many substances have different spatial forms but with the same overall covalent bonds.

alcohol dehydrogenase from *Lactobacillus brevis* that is able to catalyze a highly regioselective and enantioselective reduction (Ernst et al., 2005).

Cofactor regeneration was used to better produce a mixture of NADPH and NADH through incorporation of a soluble transhydrogenase to provide co-factors for the production of hydromorphone (Boonstra et al., 2000). In the utilization of xylose by *Saccharomyces cerevisiae* for ethanol production, an imbalance of NADPH was considered. Modifications of the ammonia pathway from the NADPH-dependent glutamate dehydrogenase to an NADH-dependent glutamate dehydrogenase (GDH2), or GLT1 and GLN1, which encode the glutamine synthetase-glutamate synthase complex (GS-GOGAT) were tested (Roca et al., 2003). The use of GDH2 increased the ethanol yield and reduced xylitol formation. This approach seems promising for improving batch culture performance. Overexpression of the GS-GOGAT complex improved ethanol yield modestly only under special continuous culture conditions. An improvement of ethanol production was found when GDP1, which codes for a fungal NADP+-dependent D-glyceraldehyde-3-phosphate dehydrogenase (NADP-GAPDH), was used to reduce the co-factor imbalance (Verho et al., 2003). The conversion of ketoisophorone (2,6,6-trimethyl-2-cyclohexen-1,4-dione) to (6R)-levodione (2,2,6-trimethylcyclohexane-1,4-dione) was stimulated by the overexpression of old yellow enzyme (Oye) from *Candida macedoniensis* in *Escherichia coli*. This enzyme is a well-characterized oxidoreductase. It was reported that the use of *Escherichia coli* BL21 (DE3) cells co-expressing *oye* and *gdh* as a whole cell catalyst was advantageous for the practical synthesis of (6R)-levodione (Kataoka et al., 2004).

8.6.3   *CoA compounds*

8.6.3.1   *CoA levels*

Manipulation of the acetyl-CoA levels for improved formation of esters derived from condensation of an alcohol with the acetyl-CoA has been reported.. (Vadali et al., 2004a; 2004b; 2004c). The high levels of CoA were produced by overexpression of a pantothenate kinase and increased acetyl-CoA was achieved by overexpression of pyruvate dehydrogenase. In the experiments described, the acetyl-CoA was condensed with an alcohol by an acetyl-CoA-alcohol transferase to form an ester. This reaction could serve as a useful reporter of available acetyl-CoA without disrupting other areas of metabolism. The recycling and balance of CoA utilizing pathways can serve as a possible means to influence the flux to other sinks for this intermediate and affect the overall pattern of metabolites. Although feedback systems generally control these CoA levels, they can be manipulated to some extent, which can contribute to enhanced production of acetyl-CoA derived compounds. Alteration of the acetyl-CoA levels could have wide application as many pathways for production of larger molecules rely on the sequential addition of acyl-CoA groups.

One area where the level of acyl-CoA compounds is important is in the production of polyhydroxybutyrate and its various derivatives. In this process, hydroxyacyl-CoA compounds are polymerized into a large polymer. Different physical properties within the polymer are obtained depending on the composition of monomers incorporated. Therefore, co-polymerization of hydroxybutyryl-CoA with other acyl-CoA compounds has been investigated extensively (Steinbuchel and Schlegel, 1991; van der Walle et al., 2001). To obtain high levels of the nonstandard acyl-CoA substrates, methods for uptake and activation of the corresponding acids have been employed. These approaches have included use of the *prpE* gene (Aldor and Keasling, 2001; Valentin et al., 2000), expression of a combination of a butyrate kinase and phosphotransbutyrylase (Liu et al., 2003), and development of means to take advantage of the connection in formation of acyl-CoA compounds between the pathways of fatty acid degradation and those acyl-CoA compounds needed for desired product formation (Nomura et al., 2004; Ren et al., 2000). The employment of CoA-transferases to bring other acids into the polymerization process has also been reported, for example, with 4-hydroxybutyrate-containing polymers (Song et al., 2005).

A large group of compounds built from acyl-CoA precursors are the polyketides, a group of bioactive structures [e.g., avermectin (Ikeda et al., 2001)]. In these reactions, malonyl-CoA is a major precursor. Malonyl-CoA is formed from acetyl-CoA by addition of a carboxyl group and is a normal component of fatty acid synthesis. In the case of polyketide synthesis, the group is transferred to an acyl carrier protein for incorporation into the chain as the longer chain polyketide is synthesized. The acyl carrier proteins require activation by addition of a 4'-phosphopantotheinyl group from CoA by specific enzymes, PPTases (Mofid et al., 2004). The analysis and overexpression of acetyl-CoA carboxylase from various sources have led to increased levels of malonyl-CoA (Davis et al., 2000; Gande et al., 2004). Manipulation of the acyl-CoA pools or use of selective acyl-transferases can lead to alterations of the polyketide product (Liou and Khosla, 2003). Providing enzymes such as CoA ligases for forming methylmalonyl-CoA or propionyl-CoA has increased the pool of the corresponding acyl-CoAs and improved synthesis of the erythromycin precursor, 6-deoxyerythronolide B (Murli et al., 2003). The use of CoA ligases with broad specificities offer a further route for manipulation of the precursor pools (Arora et al., 2005; Pohl et al., 2001).

Figure 8.6

Formation and uses of acetyl-CoA. The CoA level can be raised by overexpression of a key enzyme in the biosynthetic pathway of Coenzyme A from pantothenate. The CoA can then be more effectively converted to acetyl-CoA or other acyl-CoAs by pyruvate dehydrogenases or CoA ligases. The acetyl-CoA is then more available for conversion to esters as shown previously (Vadali et al., 2004a; 2004b; 2004c) or to possibly serve as a precursor to other compounds.

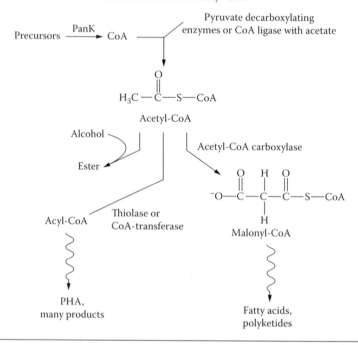

Formation and uses of acetyl-CoA

Summary

- The impact of molecular biology on production of pharmaceutical and industrial products has been immense. Many individual enzymes have been targeted for improvement based on their economic potential and methodological feasibility.

- Enzymatic properties, as well as physical and pharmokinetic properties, have been improved by a variety of mutagenesis strategies either *via* designed site-specific mutagenesis or through random methods and high-throughput screening programs.

- The possibility of manipulating co-factors required for specific reactions further expands the repertoire of tools available for metabolic engineering.

- The coupling of these experimental methods with better understanding of global cellular metabolism *via* advanced analytical techniques and mathematical treatments bodes well for further developments in the applications of enzymes and microbial biocatalysts in industrial and pharmaceutical bioprocesses.

References

Akoh, C. C., G. C. Lee, J. F. Shaw (2004). Protein engineering and applications of *Candida rugosa* lipase isoforms. *Lipids* **39**, 513–526.

Aldor, I., J. D. Keasling (2001). Metabolic engineering of poly(3-hydroxybutyrate-co-3-hydroxyvalerate) composition in recombinant *Salmonella enterica serovar typhimurium*. *Biotechnol. Bioeng.* **76**, 108–114.

Allen, S. J., J. J. Holbrook (1995). Isolation, sequence and overexpression of the gene encoding NAD-dependent formate dehydrogenase from the methylotrophic yeast *Candida methylica*. *Gene* **162**, 99–104.

Arnold, F. H. (2001). Combinatorial and computational challenges for biocatalyst design. *Nature* **409**, 253–257.

Arora, P., A.Vats, P. Saxena, D. Mohanty, R. S. Gokhale (2005). Promiscuous fatty acyl CoA ligases produce acyl-CoA and acyl-SNAC precursors for polyketide biosynthesis. *J. Am. Chem. Soc.* **127**, 9388–9389.

Asboth, B., G. Naray-Szabo (2000). Mechanism of action of D-xylose isomerase. *Curr Protein Pept. Sci.* **1**, 237–254.

Banta, S., M. Boston, A. Jarnagin, S. Anderson (2002a). Mathematical modeling of in vitro enzymatic production of 2-Keto-L-gluconic acid using NAD(H) or NADP(H) as cofactors. *Metab. Eng.* **4**, 273–284.

Banta, S., B. A. Swanson, S. Wu, A. Jarnagin, S. Anderson (2002b). Alteration of the specificity of the cofactor-binding pocket of *Corynebacterium* 2,5-diketo-D-gluconic acid reductase A. *Protein Eng.* **15**, 131–140.

Banta, S., B. A. Swanson, S. Wu, A. Jarnagin, S. Anderson (2002c). Optimizing an artificial metabolic pathway: Engineering the cofactor specificity of *Corynebacterium* 2,5-diketo-D-gluconic acid reductase for use in vitamin C biosynthesis. *Biochemistry* **41**, 6226–6236.

Benkovic, S. J., S. Hammes-Schiffer (2003). A perspective on enzyme catalysis. *Science* **301**, 1196–1202.

Berrios-Rivera, S. J., G. N. Bennett, K. Y. San (2002a). The effect of increasing NADH availability on the redistribution of metabolic fluxes in *Escherichia coli* chemostat cultures. *Metab. Eng.* **4**, 230–237.

Berrios-Rivera, S. J., G. N. Bennett, K. Y. San (2002b). Metabolic engineering of *Escherichia coli*: increase of NADH availability by overexpressing an NAD(+)-dependent formate dehydrogenase. *Metab. Eng.* **4**, 217–229.

Bessler, C., J. Schmitt, K. H. Maurer, R. D. Schmid (2003). Directed evolution of a bacterial alpha-amylase: toward enhanced pH-performance and higher specific activity. *Protein Sci.* **12**, 2141–2149.

Bommarius, A. S., B. R. Riebel (2004). *Biocatalysis*. Weinheim: Wiley-VCH Verlag GmbH & Co.

Boonstra, B., D. A. Rathbone, C. E. French, E. H.. Walker, N. C. Bruce (2000). Cofactor regeneration by a soluble pyridine nucleotide transhydrogenase for biological production of hydromorphone. *Appl. Environ. Microbiol.* **66**, 5161–5166.

Brannigan, J. A., A. J. Wilkinson (2002). Protein engineering 20 years on. *Nat. Rev. Mol. Cell Biol.* **3**, 964–970.

Bryan, P. N. (2000). Protein engineering of subtilisin. *Biochim. Biophys. Acta* **1543**, 203–222.

Buhler, B., A. Schmid (2004). Process implementation aspects for biocatalytic hydrocarbon oxyfunctionalization. *J. Biotechnol.* **113**, 183–210.

Carugo, O., P. Argos (1997). NADP-dependent enzymes. I: Conserved stereochemistry of cofactor binding. *Proteins* **28**, 10–28.

Castro, G. R., T. Knubovets (2003). Homogeneous biocatalysis in organic solvents and water-organic mixtures. *Crit. Rev. Biotechnol.* **23**, 195–231.

Cavicchioli, R., K. S. Siddiqui, D. Andrews, K. R. Sowers (2002). Low-temperature extremophiles and their applications. *Curr. Opin. Biotechnol.* **13**, 253–261.

Chartrain, M., P. M. Salmon, D. K. Robinson, B. C. Buckland (2000). Metabolic engineering and directed evolution for the production of pharmaceuticals. *Curr. Opin. Biotechnol.* **11**, 209–214.

Chen, R., A. Greer, A. M. Dean (1995). A highly active decarboxylating dehydrogenase with rationally inverted coenzyme specificity. *Proc. Natl. Acad. Sci. USA* **92**, 11666–11670.

Cherry, J. R., A. L. Fidantsef (2003). Directed evolution of industrial enzymes: An update. *Curr. Opin. Biotechnol.* **14**, 438–443.

Cirino, P. C. F. H. Arnold (2002). Protein engineering of oxygenases for biocatalysis. *Curr. Opin. Chem. Biol.* **6**, 130–135.

Cirino, P. C., K. M. Mayer, D. Umeno (2003). Generating mutant libraries using error-prone PCR. *Methods Mol. Biol.* **231**, 3–9.

Davis, M. S., J. Solbiati, J. E. Cronan Jr. (2000). Overproduction of acetyl-CoA carboxylase activity increases the rate of fatty acid biosynthesis in *Escherichia coli. J. Biol. Chem.* **275**, 28593–28598.

Demain, A. L. (2001). Molecular genetics and industrial microbiology—30 years of marriage. *J. Ind. Microbiol. Biotechnol.* **27**, 352–356.

Dohr, O., M. J. Paine, T. Friedberg, G. C. Roberts, C. R. Wolf (2001). Engineering of a functional human NADH-dependent cytochrome P450 system. *Proc. Nat. Acad. Sci. USA* **98**, 81–86.

Doyle, S. A., P. T. Beernink, D. E. Koshland Jr. (2001). Structural basis for a change in substrate specificity: crystal structure of S113E isocitrate dehydrogenase in a complex with isopropylmalate, Mg2+, and NADP. *Biochemistry* **40**, 4234–4241.

Duine, J. A. (1991). Cofactor engineering. *Trends Biotechnol.* **9**, 343–346.

Duport, C., R. Spagnoli, E. Degryse, D. Pompon (1998). Self-sufficient biosynthesis of pregnenolone and progesterone in engineered yeast. *Nat. Biotechnol.* **16**, 186–189.

Eijsink, V. G.,A. Bjork, S. Gaseidnes, R. Sirevag, B. Synstad, B. van den Burg, G. Vriend (2004). Rational engineering of enzyme stability. *J. Biotechnol.* **113**, 105–120.

Eijsink, V. G., S. Gaseidnes, T. V. Borchert, B. van den Burg (2005). Directed evolution of enzyme stability. *Biomol. Eng.* **22**, 21–30.

Ernst, M., B. Kaup, M. Muller, S. Bringer-Meyer H. Sahm (2005). Enantioselective reduction of carbonyl compounds by whole-cell biotransformation, combining a formate dehydrogenase and a (R)-specific alcohol dehydrogenase. *Appl. Microbiol. Biotechnol.* **66**, 629–634.

Estell, D. A., T. P. Graycar, J. A. Wells (1985). Engineering an enzyme by site-directed mutagenesis to be resistant to chemical oxidation. *J. Biol. Chem.* **260**, 6518–6521.

Feller, G. (2003). Molecular adaptations to cold in psychrophilic enzymes. *Cell. Mol. Life Sci.* **60**, 648–662.

Galkin, A., L. Kulakova, T. Yoshimura, K. Soda, N. Esaki (1997). Synthesis of optically active amino acids from alpha-keto acids with *Escherichia coli* cells expressing heterologous genes. *Appl. Environ. Microbiol.* **63**, 4651–4656.

Gande, R., K. J. Gibson, A. K. Brown, K. Krumbach, L. G. Dover, H. Sahm, S. Shioyama, T. Oikawa, G. S. Besra, L. Eggeling (2004). Acyl-CoA carboxylases (accD2 and accD3), together with a unique polyketide synthase (Cg-pks), are key to mycolic acid biosynthesis in *Corynebacterianeae* such as *Corynebacterium glutamicum* and *Mycobacterium tuberculosis*. *J. Biol. Chem.* **279**, 44847–44857.

Georlette, D., V. Blaise, T. Collins, S. D'Amico, E. Gratia, A. Hoyoux, J. C. Marx, G. Sonan, G. Feller, C. Gerday (2004). Some like it cold: Biocatalysis at low temperatures. *FEMS Microbiol. Rev.* **28**, 25–42.

Graddis, T. J., R. L. Remmele, Jr., J. T. McGrew (2002). Designing proteins that work using recombinant technologies. *Curr. Pharm. Biotechnol.* **3**, 285–297.

Gunnarsson, N., A. Eliasson, J. Nielsen (2004). Control of fluxes towards antibiotics and the role of primary metabolism in production of antibiotics. *Adv. Biochem. Eng. Biotechnol.* **88**, 137–178.

Gustafsson, C., S. Govindarajan, J. Minshull (2003). Putting engineering back into protein engineering: Bioinformatic approaches to catalyst design. *Curr. Opin. Biotechnol.* **14**, 366–370.

Hanson, R. L., Z. Shi, D. B. Brzozowski, A. Banerjee, T. P. Kissick, J. Singh, A. J. Pullockaran, J. T. North, J. Fan, J. Howell, S. C. Durand, M. A. Montana, D. R. Kronenthal, R. H. Mueller, R. N. Patel (2000). Regioselective enzymatic aminoacylation of lobucavir to give an intermediate for lobucavir prodrug. *Bioorg. Med. Chem.* **8**, 2681–2687.

Harris, J. M., N. E. Martin, M. Modi (2001). Pegylation: A novel process for modifying pharmacokinetics. *Clin. Pharmacokinet.* **40**, 539–551.

Hartley, B. S., N. Hanlon, R. J. Jackson, M. Rangarajan (2000). Glucose isomerase: Insights into protein engineering for increased thermostability. *Biochim. Biophys. Acta* **1543**, 294–335.

Hesselink, R. P., A. J. Wagenmakers, M. R. Drost, G. J. Van der Vusse (2003). Lysosomal dysfunction in muscle with special reference to glycogen storage disease type II. *Biochim. Biophys. Acta* **1637**, 164–170.

Holmberg, N., U. Ryde, L. Bulow (1999). Redesign of the coenzyme specificity in L-lactate dehydrogenase from *Bacillus stearothermophilus* using site-directed mutagenesis and media engineering. *Protein Eng.* **12**, 851–856.

Huisman, G. W., D. Gray (2002). Towards novel processes for the fine-chemical and pharmaceutical industries. *Curr. Opin. Biotechnol.* **13**, 352–358.

Hult, K., P. Berglund (2003). Engineered enzymes for improved organic synthesis. *Curr. Opin. Biotechnol.* **14**, 395–400.

Hurley, J. H., R. Chen, A. M. Dean (1996). Determinants of cofactor specificity in isocitrate dehydrogenase: Structure of an engineered $NADP^+ \rightarrow NAD^+$ specificity-reversal mutant. *Biochemistry* **35**, 5670–5678.

Ikeda, H., T. Nonomiya, S. Omura (2001). Organization of biosynthetic gene cluster for avermectin in *Streptomyces avermitilis*: Analysis of enzymatic domains in four polyketide synthases. *J. Ind. Microbiol. Biotechnol.* **27**, 170–176.

Itagaki, E., H. Matushita, T. Hatta (1990). Steroid transhydrogenase activity of 3-ketosteroid-delta 1-dehydrogenase from *Nocardia corallina*. *J. Biochem. (Tokyo)* **108**, 122–127.

Jacobsen, E. N., N. S. Finney (1994). Synthetic and biological catalysts in chemical synthesis: how to assess practical utility. *Chem. Biol.* **1**, 85–90.

Jaeger, K. E., T. Eggert (2004). Enantioselective biocatalysis optimized by directed evolution. *Curr. Opin. Biotechnol.* **15**, 305–313.

Jestin, J. L., P. A. Kaminski (2004). Directed enzyme evolution and selections for catalysis based on product formation. *J. Biotechnol.* **113**, 85–103.

Jez, J. M., T. M. Penning (1998). Engineering steroid 5 beta-reductase activity into rat liver 3 alpha-hydroxysteroid dehydrogenase. *Biochemistry* 37, 9695–9703.

Johanson, T., M. Katz, M. F. Gorwa-Grauslund (2005). Strain engineering for stereoselective bioreduction of dicarbonyl compounds by yeast reductases. *FEMS Yeast Res.* 5, 513–525.

Jung, R.,M. P. Scott, L. O. Oliveira, N. C. Nielsen (1992). A simple and efficient method for the oligodeoxyribonucle-otide-directed mutagenesis of double-stranded plasmid DNA. *Gene* 121, 17–24.

Kataoka, M., Kita, K., Wada, M., Yasohara, Y., Hasegawa, J. and Shimizu, S. (2003) Novel bioreduction system for the production of chiral alcohols. *Appl. Microbiol. Biotechnol.* 62, 437-445.

Kataoka, M., A. Kotaka, R. Thiwthong, M. Wada, S. Naka-mori, S. Shimizu (2004). Cloning and overexpression of the old yellow enzyme gene of *Candida macedoniensis*, and its application to the production of a chiral compound. *J. Biotechnol.* 114, 1–9.

Kaup, B., S. Bringer-Meyer, H. Sahm (2003). Metabolic engineering of *Escherichia coli*: Construction of an efficient biocatalyst for D-mannitol formation in a whole-cell biotransformation. *Commun. Agric. Appl. Biol. Sci.* 68, 235–240.

Kaup, B., S. Bringer-Meyer, H. Sahm (2004). Metabolic engineering of *Escherichia coli*: Construction of an efficient biocatalyst for D-mannitol formation in a whole-cell biotransformation. *Appl. Microbiol. Biotechnol.* 64, 333–339.

Kealey, J. T. (2003). Creating polyketide diversity through genetic engineering. *Front Biosci.* 8, c1–13.

Kegler-Ebo, D. M., G. W. Polack, D. DiMaio (1996). Use of codon cassette mutagenesis for saturation mutagenesis. *Methods Mol. Biol.* 57, 297–310.

Khosla, C., P. B. Harbury (2001). Modular enzymes. *Nature* 409, 247–252.

Kimura, E. (2003). Metabolic engineering of glutamate production. *Adv. Biochem. Eng. Biotechnol.* 79, 37–57.

Kumar, S., R. Nussinov (2001). How do thermophilic proteins deal with heat? *Cell Mol. Life Sci.* 58, 1216–1233.

Kunkel, T. A. (1985). Rapid and efficient site-specific mutagenesis without phenotypic selection. *Proc. Nat. Acad. Sci. USA* 82, 488–492.

Kunkel, T. A., K. Bebenek, J. McClary (1991). Efficient site-directed mutagenesis using uracil-containing DNA. *Methods Enzymol.* 204, 125–139.

Kunkel, T. A., J. D. Roberts, R. A. Zakour (1987). Rapid and efficient site-specific mutagenesis without phenotypic selection. *Methods Enzymol.* 154, 367–382.

Lehmann, M., M. Wyss (2001). Engineering proteins for thermostability: The use of sequence alignments versus rational design and directed evolution. *Curr. Opin. Biotechnol.* **12**, 371–375.

Li, Z., J. B. van Beilen, W. A. Duetz, A. Schmid, A. de Raadt, H. Griengl, B. Witholt (2002). Oxidative biotransformations using oxygenases. *Curr. Opin. Chem. Biol.* **6**, 136–144.

Liou, G. F., C. Khosla (2003). Building-block selectivity of polyketide synthases. *Curr. Opin. Chem. Biol.* **7**, 279–284.

Liu, S., B. Saha, M. Cotta (2005). Cloning, expression, purification, and analysis of mannitol dehydrogenase gene *mtlK* from *Lactobacillus brevis*. *Appl. Biochem. Biotechnol.* **121–124**, 391–401.

Liu, S. J., T. Lutke-Eversloh, A. Steinbuchel (2003). Biosynthesis of poly (3-mercaptopropionate) and poly (3-mercaptopropionate-co-3-hydroxybutyrate) with recombinant *Escherichia coli*. *Sheng Wu Gong Cheng Xue Bao* **19**, 195–199.

Locher, C. P., M. Paidhungat, R. G. Whalen, J. Punnonen (2005). DNA shuffling and screening strategies for improving vaccine efficacy. *DNA Cell Biol.* **24**, 256–263.

Lopez de Felipe, F., M. Kleerebezem, W. M. de Vos, J. Hugenholtz (1998). Cofactor engineering: A novel approach to metabolic engineering in *Lactococcus lactis* by controlled expression of NADH oxidase. *J. Bacteriol.* **180**, 3804–3808.

Lutz, S., W. M. Patrick (2004). Novel methods for directed evolution of enzymes: Quality, not quantity. *Curr. Opin. Biotechnol.* **15**, 291–297.

MacLachlan, J., A. T. Wotherspoon, R. O. Ansell, C. J. Brooks (2000). Cholesterol oxidase: sources, physical properties and analytical applications. *J. Steroid Biochem. Mol. Biol.* **72**, 169–195.

Matsunaga, I., Y. Shiro (2004). Peroxide-utilizing biocatalysts: Structural and functional diversity of heme-containing enzymes. *Curr. Opin. Chem. Biol.* **8**, 127–132.

Maurer, K. H. (2004). Detergent proteases. *Curr. Opin. Biotechnol.* **15**, 330–334.

McDaniel, R., M. Welch, C. R. Hutchinson (2005). Genetic approaches to polyketide antibiotics. 1. *Chem. Rev.* **105**, 543–558.

Mijts, B. N., P. C. Lee, C. Schmidt-Dannert (2004). Engineering carotenoid biosynthetic pathways. *Methods Enzymol.* **388**, 315–329.

Mijts, B. N., C. Schmidt-Dannert (2003). Engineering of secondary metabolite pathways. *Curr. Opin. Biotechnol.* **14**, 597–602.

Miles, C. S., T. W. Ost, M. A. Noble, A. W. Munro, S. K. Chapman (2000). Protein engineering of cytochromes P-450. *Biochim. Biophys. Acta* **1543**, 383–407.

Mofid, M. R., R. Finking, L. O. Essen, M. A. Marahiel (2004). Structure-based mutational analysis of the 4'-phosphopantetheinyl transferases Sfp from *Bacillus subtilis*: Carrier protein recognition and reaction mechanism. *Biochemistry* **43**, 4128–4136.

Molineux, G. (2002). Pegylation: Engineering improved pharmaceuticals for enhanced therapy. *Cancer Treat. Rev.* **28 Suppl A**, 13–16.

Morant, M., S. Bak, B. L. Moller, D. Werck-Reichhart.(2003). Plant cytochromes P450: tools for pharmacology, plant protection and phytoremediation. *Curr. Opin. Biotechnol.* **14**, 151–162.

Morikawa, S., T. Nakai, Y. Yasohara, H. Nanba, N. Kizaki, J. Hasegawa (2005). Highly active mutants of carbonyl reductase S1 with inverted coenzyme specificity and production of optically active alcohols. *Biosci. Biotechnol. Biochem.* **69**, 544–552.

Murli, S., J. Kennedy, L. C. Dayem, J. R. Carney, J. T. Kealey (2003). Metabolic engineering of *Escherichia coli* for improved 6-deoxyerythronolide B production. *J. Ind. Microbiol. Biotechnol.* **30**, 500–509.

Ng, I. S., S. W. Tsai (2005). Hydrolytic resolution of (R,S)-naproxen 2,2,2-trifluoroethyl thioester by *Carica papaya* lipase in water-saturated organic solvents. *Biotechnol. Bioeng.* **89**, 88–95.

Nomura, C. T., T. Tanaka, Z. Gan, K. Kuwabara, H. Abe, K. Takase, K. Taguchi, Y. Doi. (2004). Effective enhancement of short-chain-length-medium-chain-length polyhydroxyalkanoate copolymer production by coexpression of genetically engineered 3-ketoacyl-acyl-carrier-protein synthase III (*fabH*) and polyhydroxyalkanoate synthesis genes. *Biomacromolecules* **5**, 1457–1464.

Pannekoek, H., M. van Meijer, R. R. Schleef, D. J. Loskutoff, C. F. Barbas 3rd. (1993). Functional display of human plasminogen-activator inhibitor 1 (PAI-1) on phages: Novel perspectives for structure-function analysis by error-prone DNA synthesis. *Gene* **128**, 135–140.

Patel, R. N. (2001). Enzymatic synthesis of chiral intermediates for Omapatrilat, an antihypertensive drug. *Biomol. Eng.* **17**, 167–182.

Patten, P. A., R. J. Howard, W. P. Stemmer (1997). Applications of DNA shuffling to pharmaceuticals and vaccines. *Curr. Opin. Biotechnol.* **8**, 724–733.

Petrounia, I. P., F. H. Arnold (2000). Designed evolution of enzymatic properties. *Curr. Opin. Biotechnol.* **11**, 325–330.

Pfeifer, B. A., C. Khosla (2001). Biosynthesis of polyketides in heterologous hosts. *Microbiol. Mol. Biol. Rev.* **65**, 106–118.

Pohl, N. L., M. Hans, H. Y. Lee, Y. S. Kim, D. E. Cane, C. Khosla (2001). Remarkably broad substrate tolerance of malonyl-CoA synthetase, an enzyme capable of intracellular synthesis of polyketide precursors. *J. Am. Chem. Soc.* **123**, 5822–5823.

Powell, K. A., S. W. Ramer, S. B. Del Cardayre, W. P. Stemmer, M. B. Tobin, P. F. Longchamp, G. W. Huisman (2001). Directed evolution and biocatalysis. *Angew. Chem. Intl. Ed. Engl.* **40**, 3948–3959.

Ren, Q., N. Sierro, B. Witholt, B. Kessler (2000). FabG, an NADPH-dependent 3-ketoacyl reductase of *Pseudomonas aeruginosa*, provides precursors for medium-chain-length poly-3-hydroxyalkanoate biosynthesis in *Escherichia coli*. *J. Bacteriol.* **182**, 2978–2981.

Roca, C., J. Nielsen, L. Olsson (2003). Metabolic engineering of ammonium assimilation in xylose-fermenting *Saccharomyces cerevisiae* improves ethanol production. *Appl. Environ. Microbiol.* **69**, 4732–4736.

Rojkova, A. M., A. G. Galkin, L. B. Kulakova, A. E. Serov, P. A. Savitsky, V. V. Fedorchuk, V. I. Tishkov (1999). Bacterial formate dehydrogenase. Increasing the enzyme thermal stability by hydrophobization of alpha-helices. *FEBS Lett.* **445**, 183–188.

Rose, D. R., J. Phipps, J. Michniewicz, G. I. Birnbaum, F. R. Ahmed, A. Muir, W. F. Anderson, S. Narang (1988). Crystal structure of T4-lysozyme generated from synthetic coding DNA expressed in *Escherichia coli*. *Protein Eng.* **2**, 277–282.

Ryu, D. D., D. H. Nam (2000). Recent progress in biomolecular engineering. *Biotechnol. Prog.* **16**, 2–16.

Sakai, Y., A. P. Murdanoto, T. Konishi, A. Iwamatsu, N. Kato (1997). Regulation of the formate dehydrogenase gene, FDH1, in the methylotrophic yeast *Candida boidinii* and growth characteristics of an FDH1-disrupted strain on methanol, methylamine, and choline. *J. Bacteriol.* **179**, 4480–4485.

San, K. Y., G. N. Bennett, S. J. Berrios-Rivera, R. V. Vadali, Y. T. Yang, E. Horton, F. B. Rudolph, B. Sariyar, K. Blackwood (2002). Metabolic engineering through cofactor manipulation and its effects on metabolic flux redistribution in *Escherichia coli*. *Metab. Eng.* **4**, 182–192.

Sandmann, G. (2002). Combinatorial biosynthesis of carotenoids in a heterologous host: a powerful approach for the biosynthesis of novel structures. *Chembiochem.* **3**, 629–635.

Sandmann, G. (2003). Novel carotenoids genetically engineered in a heterologous host. *Chem. Biol.* **10**, 478–479.

Sauer, J., B. W. Sigurskjold, U. Christensen, T. P. Frandsen, E. Mirgorodskaya, M. Harrison, P. Roepstorff, B. Svensson (2000). Glucoamylase: Structure/function relationships, and protein engineering. *Biochim. Biophys. Acta* **1543**, 275–293.

Scharnagl, C., M. Reif, J. Friedrich (2005). Stability of proteins: Temperature, pressure and the role of the solvent. *Biochim. Biophys. Acta* **1749**, 187–213.

Schrader, J., M. M. Etschmann, D. Sell, J. M. Hilmer, J. Rabenhorst (2004). Applied biocatalysis for the synthesis of natural flavour compounds—current industrial processes and future prospects. *Biotechnol. Lett.* **26**, 463–472.

Seo, J. W., J. H. Jeong, C. G. Shin, S. C. Lo, S. S. Han, K. W. Yu, E. Harada, J. Y. Han, Y. E. Choi (2005). Overexpression of squalene synthase in *Eleutherococcus senticosus* increases phytosterol and triterpene accumulation. *Phytochemistry* **66**, 869–877.

Shanafelt, A. B. (2005). Medicinally useful proteins — enhancing the probability of technical success in the clinic. *Expert Opin. Biol. Ther.* **5**, 149–151.

Shen, T. J., Ho, N. T., Zou, M., Sun, D. P., Cottam, P. F., Simplaceanu, V., Tam, M. F., Bell, D. A., Jr. and Ho, C. (1997) Production of human normal adult and fetal hemoglobins in *Escherichia coli*. *Protein Eng.* **10**, 1085-1097.

Sio, C. F. and Quax, W. J. (2004) Improved beta-lactam acylases and their use as industrial biocatalysts. *Curr. Opin. Biotechnol.* **15**, 349-355.

Slusarczyk, H., Felber, S., Kula, M. R. and Pohl, M. (2000) Stabilization of NAD-dependent formate dehydrogenase from *Candida boidinii* by site-directed mutagenesis of cysteine residues. *Eur. J. Biochem.* **267**, 1280-1289.

Song, S. S., Ma, H., Gao, Z. X., Jia, Z. H. and Zhang, X. (2005) Construction of recombinant *Escherichia coli* strains producing poly (4-hydroxybutyric acid) homopolyester from glucose. *Wei Sheng Wu Xue Bao* **45**, 382-386.

Staunton, J. and Weissman, K. J. (2001) Polyketide biosynthesis: a millennium review. *Nat. Prod. Rep.* **18**, 380-416.

Steinbuchel, A. and Schlegel, H. G. (1991) Physiology and molecular genetics of poly(beta-hydroxy-alkanoic acid) synthesis in *Alcaligenes eutrophus*. *Mol. Microbiol.* **5**, 535-542.

Stemmer, W. P. (1994) Rapid evolution of a protein in vitro by DNA shuffling. *Nature* **370**, 389-391.

Szczebara, F. M., Chandelier, C., Villeret, C., Masurel, A., Bourot, S., Duport, C., Blanchard, S., Groisillier, A., Testet, E., Costaglioli, P., Cauet, G., Degryse, E., Balbuena, D., Winter, J., Achstetter, T., Spagnoli, R., Pompon, D. and Dumas, B. (2003) Total biosynthesis of hydrocortisone from a simple carbon source in yeast. *Nat. Biotechnol.* **21**, 143-149.

Taguchi, S., Komada, S. and Momose, H. (2000) The complete amino acid substitutions at position 131 that are positively involved in cold adaptation of subtilisin BPN'. *Appl. Environ. Microbiol.* **66**, 1410-1415.

Taguchi, S., Ozaki, A. and Momose, H. (1998) Engineering of a cold-adapted protease by sequential random mutagenesis and a screening system. *Appl. Environ. Microbiol.* **64**, 492-495.

Taguchi, S., Ozaki, A., Nonaka, T., Mitsui, Y. and Momose, H. (1999) A cold-adapted protease engineered by experimental evolution system. *J. Biochem. (Tokyo)* **126**, 689-693.

Takagi, H., Maeda, T., Ohtsu, I., Tsai, Y. C. and Nakamori, S. (1996) Restriction of substrate specificity of subtilisin E by introduction of a side chain into a conserved glycine residue. *FEBS Lett.* **395**, 127-132.

Turner, N. J. (2003) Controlling chirality. *Curr. Opin. Biotechnol.* **14**, 401-406.

Umeno, D., Tobias, A. V. and Arnold, F. H. (2005) Diversifying carotenoid biosynthetic pathways by directed evolution. *Microbiol. Mol. Biol. Rev.* **69**, 51-78.

Urlacher, V. and Schmid, R. D. (2002) Biotransformations using prokaryotic P450 monooxygenases. *Curr. Opin. Biotechnol.* **13**, 557-564.

Urlacher, V. B., Lutz-Wahl, S. and Schmid, R. D. (2004) Microbial P450 enzymes in biotechnology. *Appl. Microbiol. Biotechnol.* **64**, 317-325.

Vadali, R. V., Bennett, G. N. and San, K. Y. (2004a) Applicability of CoA/acetyl-CoA manipulation system to enhance isoamyl acetate production in *Escherichia coli. Metab. Eng.* **6**, 294-299.

Vadali, R. V., Bennett, G. N. and San, K. Y. (2004b) Cofactor engineering of intracellular CoA/acetyl-CoA and its effect on metabolic flux redistribution in *Escherichia coli. Metab. Eng.* **6**, 133-139.

Vadali, R. V., Bennett, G. N. and San, K. Y. (2004c) Enhanced isoamyl acetate production upon manipulation of the acetyl-CoA node in *Escherichia coli. Biotechnol. Prog.* **20**, 692-697.

Valentin, H. E., Mitsky, T. A., Mahadeo, D. A., Tran, M. and Gruys, K. J. (2000) Application of a propionyl coenzyme A synthetase for poly(3-hydroxypropionate-co-3-hydroxybutyrate) accumulation in recombinant *Escherichia coli. Appl. Environ. Microbiol.* **66**, 5253-5258.

van der Walle, G. A., de Koning, G. J., Weusthuis, R. A. and Eggink, G. (2001) Properties, modifications and applications of biopolyesters. *Adv. Biochem. Eng. Biotechnol.* **71**, 263-291.

Veen, M. and Lang, C. (2004) Production of lipid compounds in the yeast *Saccharomyces cerevisiae. Appl. Microbiol. Biotechnol.* **63**, 635-646.

Veitch, N. C. (2004) Horseradish peroxidase: a modern view of a classic enzyme. *Phytochemistry* **65**, 249-259.

Verho, R., Londesborough, J., Penttila, M. and Richard, P. (2003) Engineering redox cofactor regeneration for improved pentose fermentation in *Saccharomyces cerevisiae. Appl. Environ. Microbiol.* **69**, 5892-5897.

Wackett, L. P. (2004) Novel biocatalysis by database mining. *Curr. Opin. Biotechnol.* **15**, 280-284.

Wandrey, C. (2004) Biochemical reaction engineering for redox reactions. *Chem. Rec.* **4**, 254-265.

Wang, T., Liu, X., Yu, Q., Zhang, X., Qu, Y. and Gao, P. (2005) Directed evolution for engineering pH profile of endoglucanase III from *Trichoderma reesei. Biomol. Eng.* **22**, 89-94.

Watanabe, S., Kodaki, T. and Makino, K. (2005) Complete reversal of coenzyme specificity of xylitol dehydrogenase and increase of thermostability by the introduction of structural zinc. *J. Biol. Chem.* **280**, 10340-10349.

Weckbecker, A. and Hummel, W. (2004) Improved synthesis of chiral alcohols with *Escherichia coli* cells co-expressing pyridine nucleotide transhydrogenase, $NADP^+$-dependent alcohol dehydrogenase and NAD^+-dependent formate dehydrogenase. *Biotechnol. Lett.* **26**, 1739-1744.

Wells, J. A., Vasser, M. and Powers, D. B. (1985) Cassette mutagenesis: an efficient method for generation of multiple mutations at defined sites. *Gene* **34**, 315-323.

Wichmann, R. and Vasic-Racki, D. (2005) Cofactor regeneration at the lab scale. *Adv. Biochem. Eng. Biotechnol.* **92**, 225-260.

Williams, G. J., Nelson, A. S. and Berry, A. (2004) Directed evolution of enzymes for biocatalysis and the life sciences. *Cell. Mol. Life Sci.* **61**, 3034-3046.

Woodyer, R., van der Donk, W. A. and Zhao, H. (2003) Relaxing the nicotinamide cofactor specificity of phosphite dehydrogenase by rational design. *Biochemistry* **42**, 11604-11614.

Yamamoto, H., Mitsuhashi, K., Kimoto, N., Matsuyama, A., Esaki, N. and Kobayashi, Y. (2004) A novel NADH-dependent carbonyl reductase from *Kluyveromyces aestuarii* and comparison of NADH-regeneration system for the synthesis of ethyl (S)-4-chloro-3-hydroxybutanoate. *Biosci. Biotechnol. Biochem.* **68**, 638-649.

Yang, Y., Jiang, L., Zhu, L., Wu, Y. and Yang, S. (2000) Thermal stable and oxidation-resistant variant of subtilisin E. *J. Biotechnol.* **81**, 113-118.

Yaoi, T., Miyazaki, K., Oshima, T., Komukai, Y. and Go, M. (1996) Conversion of the coenzyme specificity of isocitrate dehydrogenase by module replacement. *J. Biochem. (Tokyo)* **119**, 1014-1018.

You, L. and Arnold, F. H. (1996) Directed evolution of subtilisin E in *Bacillus subtilis* to enhance total activity in aqueous dimethylformamide. *Protein Eng.* **9**, 77-83.

Zhao, H. and Arnold, F. H. (1999) Directed evolution converts subtilisin E into a functional equivalent of thermitase. *Protein Eng.* **12**, 47–53.

9

Application of Metabolic Engineering to the Conversion of Renewable Resources to Fuels and Fine Chemicals: Current Advances and Future Prospects

Aristos A. Aristidou

There's enough alcohol in one year's yield of an acre of potatoes to drive the machinery necessary to cultivate the fields for one hundred years.

— *Henry Ford*

9.1 Introduction

Industrial biotechnology (IB), also referred to as "white biotechnology," is poised to revolutionize the way we produce our energy, chemicals, and materials for the 21st century. In a way reminiscent of the transformation of the chemical industry back in the 1940s from a business based on inorganic mineral feedstocks to one based on organic petroleum, IB can allow us to substitute fossilized organic matter, i.e., petroleum, for organic biomass that can be renewed on a much shorter time scale, in many instances on an annual basis. This new technology can allow the chemical industries to produce fuels such as bioethanol, materials such as polylactic acid (PLA), fine or bulk chemicals such as various organic acids, polyols, or esters, from almost anything that grows. Since it takes millions of years to convert biomass into coal and petroleum, fossil fuels are not renewable

Figure 9.1

Historic U.S. and Brazil ethanol production.

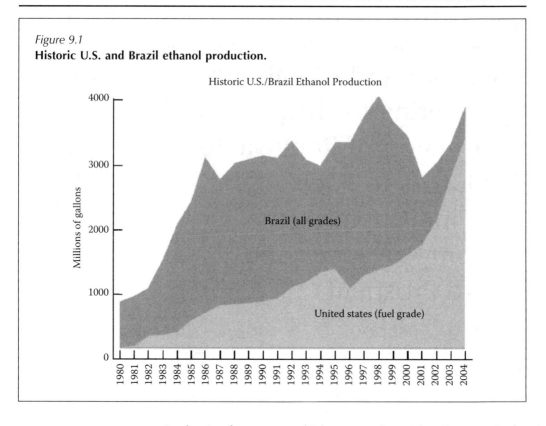

Historic U.S./Brazil Ethanol Production

in the timeframe over which we use them. Plant biomass is the only foreseeable sustainable source of organic fuels, chemicals. Success in this area can have a range of major implications on human society, including reducing the world's dependence on oil with all the ills associated with that, cleaner manufacturing technologies, reduction in CO_2 emissions and associated climate effects, as well as increasing economic development to rural areas. In a recent letter to the U.S. president, followed by a letter to the chairman of the U.S. Committee on Energy and Natural Resources, members of the Energy Future Coalition, which includes national security, labor, and energy policy experts, make strong recommendations on "a major new initiative" to significantly reduce U.S. consumption of foreign oil through "improved efficiency and the rapid substitution of advanced biomass, alcohol and other available alternative fuels."

Fuel ethanol — whose 2004 U.S. production of 3.4 billion gallons is more than double that produced in 2000 (Figure 9.1), and with a current world market of over 10 billion gallons (Table 9.1) — presents a great opportunity to demonstrate some of the potentials of IB. On one hand, it is an area where there is an established market and one that is projected to grow fairly rapidly, and on the other hand it is a yet suboptimal industry despite the fact that ethanol production has been part of human civilization for millennia. Current production of fuel ethanol

Table 9.1 World Ethanol Production, 2004

Country	Million Gallons	Country	Million Gallons
Brazil	3,989	Australia	33
United States	3,535	Japan	31
China	964	Pakistan	26
India	462	Sweden	26
France	219	Philippines	22
Russia	198	South Korea	22
South Africa	110	Guatemala	17
United Kingdom	106	Cuba	16
Saudi Arabia	79	Ecuador	12
Spain	79	Mexico	9
Thailand	74	Nicaragua	8
Germany	71	Mauritius	6
Ukraine	66	Zimbabwe	6
Canada	61	Kenya	3
Poland	53	Swaziland	3
Indonesia	44	Others	338
Argentina	42		
Italy	40	**Total**	**10,770**

From (Licht, 2005).

relies on the utilization of dextrose derived from corn starch (US) or sucrose derived from sugar cane (e.g., Brazil), whose relatively high value and high demand for other applications, coupled with the fact that sugar represents the major manufacturing cost for such commodity chemicals, results in costs that are not directly competitive with gasoline derived from petroleum. Longer term, the intent is to use biomass, i.e., the whole plant including cellulose, lignin, and hemicellulose instead of just the starchy part of the seed, hence, improves the economics of the raw material. Based on a number of economic studies, successful fermentation of the hemicellulosic sugars is crucial for achieving commercial success. Hemicellulose represents 33 to 45% of biomass before processing, and 75% in corn fiber from the corn wet milling process—one-third to three-fourths of the equivalent ethanol available. Furthermore, from a capacity perspective, utilization of hemicellulosic materials is likely to be essential in order to satisfy the long-term requirements of fuel ethanol, as well as for other chemicals derived from biomass sugars (Figure 9.2).

Many different biomass feedstocks can be used for the production of fuels and chemicals (Figure 9.3). These include various agricultural residues (corn stalks, wheat straws, potato or beet waste), wood residues (leftovers from harvested wood, unharvested dead and diseased trees), specifically grown crops (hybrid poplar, black locust, willow, silver maple, sugarcane, sugarbeet,

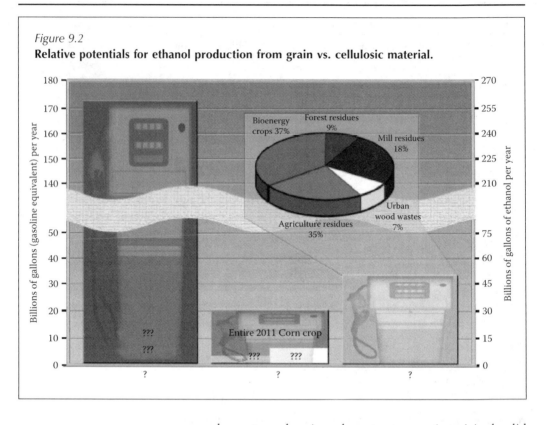

Figure 9.2
Relative potentials for ethanol production from grain vs. cellulosic material.

corn, and sweet sorghum), and waste streams (municipal solid waste, recycled paper, baggasse from sugar manufacture, corn fiber, sulfite waste). The chemical composition of biomass varies among species, but biomass consists of about 25% lignin and 75% carbohydrate polymers, mainly cellulose and hemicellulose (Lerouge et al., 1998). Cellulose is a high-molecular-weight linear glucose polysaccharide, with a degree of polymerization (DP) in the range of 200 to 2000 kDa (4000 to 8000 glucose molecules connected with β–1, 4 links). Cellulose is very strong and its links are broken by cellulase enzymes that are very uncommon in nature (Figure 9.4). The cellulases can be separated in two classes, endoglucanases (EG) and cellobiohydrolases (CBH) (Teeri et al., 1998). Cellobiohydrolases hydrolyze the cellulose chain from one end, whereas an endoglucanase hydrolyzes randomly along the cellulose chain. Hemicellulose, on the other hand, is a rather low-molecular-weight heteropolysaccharide (DP<200, typically β–1, 3 links), with a wide variation in both structure and composition. Commonly occurring hemicelluloses are xylans, arabino-xylan, gluco-mannan, galacto-glucomann, etc. In contrast to cellulose, which is crystalline, strong, and resistant to hydrolysis, hemicellulose has a random, amorphous structure with little strength. It is easily hydrolyzed by dilute acid or base, but nature provides an arsenal of hemicellulase enzymes for its hydrolysis (Kuhad et al., 1997). The cellulose

Figure 9.3

Examples of lignocellulosic biomass and relative macromolecular composition.

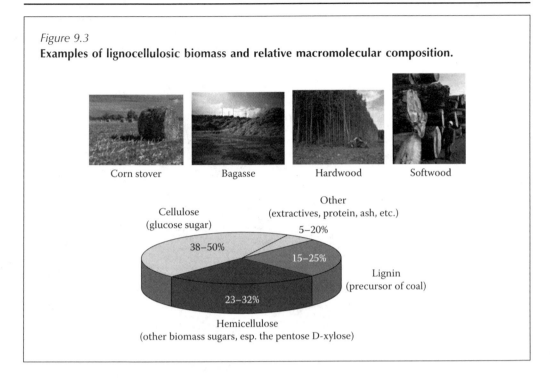

Corn stover   Bagasse   Hardwood   Softwood

Cellulose
(glucose sugar)

Other
(extractives, protein, ash, etc.)
5–20%

38–50%

15–25%

Lignin
(precursor of coal)

23–32%

Hemicellulose
(other biomass sugars, esp. the pentose D-xylose)

fraction of biomass is typically high (25 to 60%), whereas the hemicellulose fraction is typically in the range of 10 to 35%. The monomeric composition of lignocellulosic material can vary widely depending on the biomass source (see Table 9.2). In general, the carbohydrate fraction is made up primarily of the hexose sugar glucose (with small amounts of the hexoses galactose, mannose); however, the pentose fraction is rather significant: xylose 5 to 20% and arabinose 1 to 5%. Xylose is second only to glucose in natural abundance and it is the most copious sugar in the hemicellulose of hardwoods and crop residues. Analysis of the technological goals of the National Renewable Energy Laboratory (NREL) for cellulose ethanol conversion suggests that ethanol could compete favorably with other gasoline additives without the benefit of a federal subsidy if the goals were achieved. Enzymatic hydrolysis of cellulose appears to have the most potential for achieving the goals, but substantial reductions in the cost of producing cellulase enzymes and improvements in the fermentation of nonglucose sugars, especially xylose, to ethanol still are needed.

The conversion of both cellulose and hemicellulose for production of fuel ethanol is being studied intensively with a view to develop a technically and economically viable bioprocess. Ethanol is a versatile transportation fuel that offers high octane, high heat of vaporization, and other characteristics that allow it to achieve higher-efficiency use in optimized engines than gasoline. Ethanol is low in toxicity, volatility, and photochemical

Figure 9.4
Challenges in lignocellulose utilization.

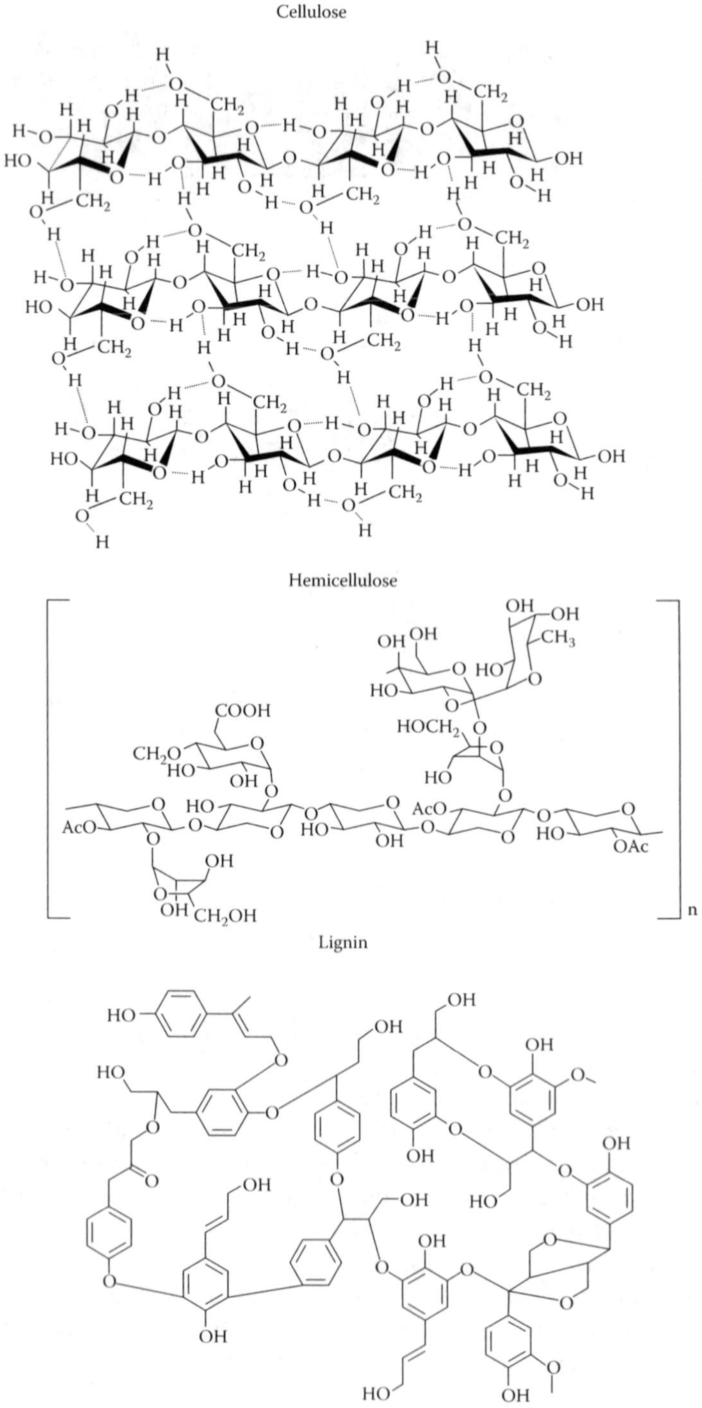

Cellulose

Hemicellulose

Lignin

Table 9.2 Composition of various biomass raw materials

	Corn Stover	Wheat Straw	Rice Straw	Rice hulls	Baggasse Fiber	News-print	Cotton gin trash	Douglas fir
Carbohydrate								
Glucose (C6)	39.0	36.6	41.0	36.1	38.1	64.4	20.0	50.0
Mannose (C6)	0.3	0.8	1.8	3.0	—	16.6	2.1	12.0
Galactose (C6)	0.8	2.4	0.4	0.1	1.1	—	0.1	1.3
Xylose (C5)	14.8	19.2	14.8	14.0	23.3	4.6	4.6	3.4
Arabinose (C5)	3.2	2.4	4.5	2.6	2.5	0.5	2.3	1.1
Total C6	40.1	39.8	43.2	39.2	39.2	81	22.2	63.3
Total C5	18	21.6	19.3	16.6	25.8	5.1	6.9	4.5
Noncarbohydrate								
Lignin	15.1	14.5	9.9	19.4	18.4	21.0	17.6	28.3
Ash	4.3	9.6	12.4	20.1	2.8	0.4	14.8	0.2
Protein	4.0	3.0	—	—	3.0	—	3.0	—

From (Lee, 1997).

reactivity, resulting in reduced ozone formation and smog compared to conventional fuels. Researchers at the National Renewable Energy Laboratory estimate that the United States potentially could convert 2.45 billion metric tons of biomass to 270 billion gallons of ethanol each year, which is approximately twice the current annual gasoline consumption in the United States. Bioethanol, also used as a hydrogen fuel source for fuel cells, could become a vital part of the long-term solution to climate change.

The important key technologies required for the successful biological conversion of lignocellulosic biomass to ethanol have been extensively reviewed (Lee, 1997; Chandrakant and Bisaria, 1998; Gong et al., 1999). Microbial conversion of the sugar residues present in wastepaper and yard trash from U.S. landfills alone could provide more than 400 billion liters of ethanol (Lynd et al., 1999), ten times the corn-derived ethanol burned annually as a 10% blend with gasoline (Keim and Venkatasubramanian, 1989). The biological process of ethanol fuel production utilizing lignocellulose as substrate requires (1) delignification to liberate cellulose and hemicellulose from their complex with lignin, (2) depolymerization of the carbohydrate polymers (cellulose and hemicellulose) to produce free sugars, and (3) fermentation of mixed hexose and pentose sugars to produce ethanol (Figure 9.5). The development of the feasible biological delignification process should be possible if lignin-degrading microorganisms, their ecophysiological requirements, and optimal bioreactor design are effectively coordinated. Some thermophilic anaerobes and recently developed recombinant bacteria have advantageous features for direct microbial conversion of cellulose to ethanol, i.e., the simultaneous depolymerization

Figure 9.5

Conversion of lignocellulose to ethanol. Crystalline cellulose, the largest (50%) and most difficult fraction, is hydrolyzed by a combination of acid and enzymatic processes. During these steps 95 to 98% of the xylose and glucose are recovered. These monosaccharides are subsequently converted to ethanol by appropriate micro-organisms.

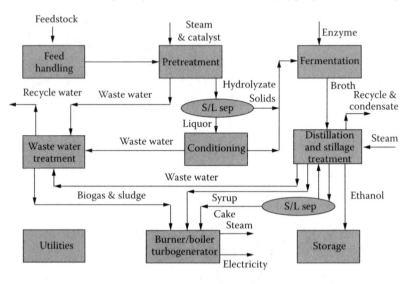

of cellulosic carbohydrate polymers with ethanol production. The new fermentation technology converting xylose to ethanol needs also to be developed to make the overall conversion process more cost-effective. The fermentation of glucose, the main constituent of cellulose hydrolysate, to ethanol can be carried out efficiently. Bioconversion of xylose, the main pentose sugar obtained on hydrolysis of hemicellulose, is essential for favorable process economics especially in the production of commodity chemicals such as ethanol. A lot of attention therefore has been focused on the utilization of both glucose and xylose to ethanol. The economics of the ethanol process is determined by the cost of sugar. The average biomass cost amounts to approximately $0.06 per kg of sugar or a contribution to the feedstock costs for ethanol production of as low as $0.10 per l.

Applied research in the area of biomass conversion to ethanol in the last 20 years has answered most of the major challenges on the road to commercialization but, as with any new technology, there is still room for performance improvement. Over the past decade, the total cost of ethanol has dropped significantly while the cost of gasoline has increased substantially to the level that these two are now very comparable. Further cost reductions in ethanol production can be accomplished through the use of lignocellulosics raw material (Figure 9.6), albeit a number of technical challenges still remain. As a number of studies have

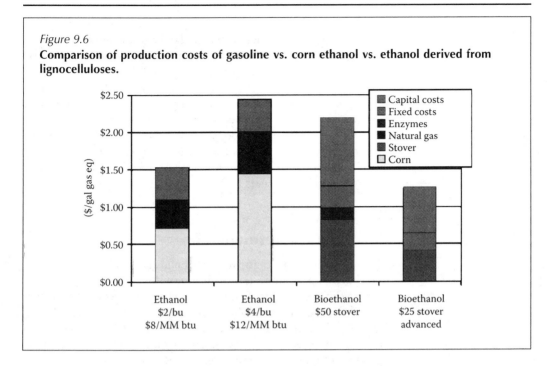

Figure 9.6

Comparison of production costs of gasoline vs. corn ethanol vs. ethanol derived from lignocelluloses.

indicated, efficient utilization of the hemicellulose component of lignocellulosic feedstocks offers an opportunity to reduce the cost of producing fuel ethanol by >25% (Hamelinck et al., 2005). This essentially requires the ability to convert all fermentable sugars (i.e., pentoses and hexoses) to product, which dictates the need to develop advanced hexose-/pentose-fermenting organisms. Given that no naturally occurring organism can satisfy all necessary specifications: high yield, high productivity, wide-substrate range, product tolerance, tolerance to inhibitors present in hydrolysates, biomass disposal cost, etc., a pragmatic approach is to utilize modern genetic engineering techniques aimed at improving organisms that are endowed with most of the desirable properties for such bioprocesses. Two such examples include the introduction of ethanol genes in the bacterium *Escherichia coli*, as well as the engineering of pentose-metabolizing pathways in natural ethanol producers such as the yeast *Saccharomyces cerevisiae* or the Gram-positive bacterium *Zymomonas mobilis*. *E. coli* offers a number of desirable attributes, such as its ability to effectively utilize both C6 and C5 carbon sources, fast growth on salts media, established genetic tools. etc. *S. cerevisiae* and *Z. mobilis*, albeit unable to utilize C5 sugars, are considered to be both good ethanol-producing microbes, with the former being one of the primary industrial organisms for ethanol production from C6 sugars.

Genetic improvements in the cultures have been made either to enlarge the range of substrate utilization or to channel metabolic intermediates specifically toward ethanol or other desirable products, such as organic acids. These contributions represent real significant advancements in the field and have also been adequately dealt with from the point of view of their impact on utilization of both cellulose and hemicellulose sugars to ethanol. The bioconversion process of lignocellulosics to ethanol could be successfully developed and optimized by aggressively applying the related novel science and technologies to solve the known key problems of conversion process. The efficient fermentation of xylose and other hemicellulose constituents may prove essential for the development of an economically viable process to produce fuels and chemicals from biomass.

9.2   Pentose fermentation

The fermentation organism must be able to ferment all monosaccharides present and, in addition, withstand potential inhibitors in the hydrolysate. Pentose-fermenting micro-organisms are found among bacteria, yeasts, and fungi, with the yeasts *Pichia stipitis*, *Candida shehatae*, and *Pachysolen tannophilus* being the most promising naturally occurring micro-organisms. Yeasts produce ethanol efficiently from hexoses by the pyruvate decarboxylase-alcohol dehydrogenase (PDC-ADH) system. However, during xylose fermentation the by-product xylitol accumulates, thereby reducing the yield of ethanol. Furthermore, yeasts are reported to ferment l-arabinose only very weakly. Only a handful of bacterial species are known that do possess the important PDC-ADH pathway to ethanol. Among these, *Zymomonas mobilis* has the most active PDC-ADH system; however, it is incapable of dissimilating pentose sugars.

Micro-organisms, in general, metabolize xylose to xylulose through two separate routes (Figure 9.7). The one-step pathway catalyzed by xylose isomerase (XI, EC 5.3.1.5) is typical in bacteria, whereas the two-step reaction involving xylose reductase (XR) and xylitol dehydrogenase (XDH) is usually found in yeast. Xylulose is subsequently phosphorylated with xylulokinase (XK) to xylulose-5P that can be further catabolized via the pentose phosphate pathway and the Embden-Meyerhof-Parnas (EMP) pathway or the Entner-Doudoroff (ED) pathway in organisms such as *Zymomonas mobilis*.

Empowered with the modern tools of genetic engineering, a number of groups have been pursuing the construction of yeast or bacterial organisms that can efficiently convert most of the sugars present in biomass derived hydrolysates to useful products. The approaches can be divided in two groups: (1) to engineer organisms with an expanded substrate spectrum, and (2) to engineer organisms with enhanced abilities of converting key intermediates of central carbon metabolism (e.g., pyruvate, the ultimate product of glycolysis) to useful compounds such as ethanol, lactate, or succinate. The former approach has

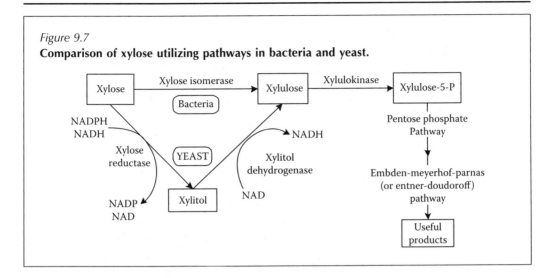

Figure 9.7
Comparison of xylose utilizing pathways in bacteria and yeast.

focused on good ethanologenic organisms, such as *Saccharomyces cerevisiae* or *Zymomonas mobilis*, with the aim of introducing the pathways for xylose or arabinose metabolism. The latter approach is to start with organisms that have a wide sugar substrate range (both C6 and C5), such as *Escherichia coli*, and introduce pathways for converting these sugars to various fermentation products, including ethanol, lactate, acetate, pyruvate, or succinate.

9.3   Genetically engineered bacteria

Initial studies were only partially successful in redirecting fermentative metabolism in *Erwinia chrysanthemi* (Tolan and Finn, 1987), *Klebsiella planticola* (Tolan and Finn, 1987), and *Escherichia coli* (Yomano et al., 1998; Ingram et al., 1999). The first generation of recombinant organisms amplified the PDC activity only and depended on endogenous levels of ADH activity to couple the further reduction of acetaldehyde to the oxidation of NADH (Figure 9.8). Since ethanol is just one of a number of fermentation products normally produced by these enteric bacteria, a deficiency in ADH activity together with NADH accumulation contributed to the formation of various unwanted by-products.

9.3.1   *Escherichia coli*

Most recent bacteria work in this area has focused on *Escherichia coli*. This is an attractive host organism for the conversion of renewable resources to ethanol and other useful products for a number of reasons: (1) it can grow efficiently on a wide range of carbon substrates that includes five-carbon sugars (has the

Figure 9.8

Competing pathways at the pyruvate branchpoint. EMP, Embden-Meyerhof-Parnas enzymes and intermediates; PDC, pyruvate decarboxylase; ADH, alcohol dehydrogenase; PFL, pyruvate formate-lyase; ACK/PTA, phosphotransacetylase and acetate kinase; ALDH, acetaldehyde dehydrogenase; FHL, formate hydrogen lyase; LDH, lactate dehydrogenase; PDH, pyruvate dehydrogenase.

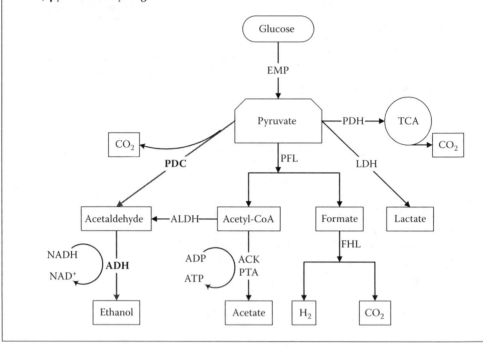

ability to ferment — besides glucose — all other sugar constituents of lignocellulosic material: xylose, mannose, arabinose, and galactose), (2) it can sustain high glycolytic fluxes (both aerobically and anaerobically), and (3) it has a reasonable ethanol tolerance (at least up to 50 g/l).

This was accomplished by assembling the *Zymomonas mobilis* genes coding for PDC and ADH (*pdc* and *adhB*) into an artificial operon to produce a portable genetic element for ethanol production (the so-called PET operon). In the recombinant *Escherichia coli*, both enzymes (PDC and ADH) required to divert pyruvate metabolism to ethanol were overexpressed to high levels. The combined effect of high PDC levels and its low apparent *Km* (Table 9.3) of this enzyme for pyruvate effectively is to divert carbon flow to ethanol even in the presence of native fermentation enzymes like lactate dehydrogenase. When the recombinant strain was grown on mixtures of sugars typically present in hemicellulose hydrolysates, sequential utilization was observed with glucose consumed first, followed by arabinose and xylose, to produce near-maximum theoretical yields of ethanol (Takahashi et al., 1994).

Table 9.3 Comparison of apparent Km values for pyruvate for selected *Escherichia coli* and *Zymomonas mobilis* pyruvate acting enzymes. PDH, pyruvate dehydrogenase; LDH, lactate dehydrogenase; PFL, pyruvate formate lyase; ALDH, aldehyde dehydrogenase; NADH-OX, NADH oxidase; PDC, pyruvate decarboxylase; ADH II, alcohol dehydrogenase II.

Organism	Enzyme	Km Pyruvate	Km NADH
Escherichia coli	PDH	0.4 mM	0.18 mM
	LDH	7.2 mM	> 0.5 mM
	PFL	2.0 mM	
	ALDH		50 μM
	NADH-OX		50 μM
Z. mobilis	PDC	0.4 mM	
	ADH II		12 μM

Under aerobic conditions, wild-type *Escherichia coli* metabolizes pyruvate through PDH and PFL (Km = 0.4 and 2.0 mM, respectively; Table 9.3), with main products CO_2 and acetate, the latter formed by the hydrolysis of excess acetyl-CoA. The apparent Km for the *Zymomonas mobilis* PDC is similar to that of PDH and lower than those of PFL and LDH, thereby facilitating acetaldehyde production. NAD^+ regeneration under aerobic conditions primarily results from biosynthesis and from the NADH oxidase coupled to the electron transport system. Again, because the apparent Km for *Zymomonas mobilis* ADH II is over fourfold lower than that for *Escherichia coli* NADH oxidase, the heterologous ADH II effectively competes for endogenous pools of NADH, allowing the reduction of acetaldehyde to ethanol. Under anaerobic conditions, wild-type *Escherichia coli* metabolize pyruvate primarily via LDH and PFL. As indicated again in Table 9.3, the apparent Km values for these two enzymes are eighteenfold and fivefold higher, respectively, than that for *Zymomonas mobilis* PDC. Furthermore, the apparent Km values for primary native enzymes involved in NAD^+ regeneration are also considerably higher in *Escherichia coli* than those of *Zymomonas mobilis* ADH. Thus, overexpressed ethanologenic *Zymomonas mobilis* enzymes in *Escherichia coli* can favorably compete with the native enzymes for pyruvate and redox co-substrates channeling pyruvate carbon to ethanol.

The University of Florida was awarded U.S. Patent No. 5,000,000 for the ingenious microbe created at its Institute of Food and Agricultural Sciences. Significant amounts of ethanol were produced in recombinant *Escherichia coli* containing the pet operon under both aerobic and anaerobic conditions (Table 9.4). Typical final ethanol concentrations are in excess of 50 g/l, with product yields on sugar approaching maximum theoretical, i.e., 0.5 g of ethanol/g of sugar (sugar → 2ethanol + $2CO_2$) (Ohta

Table 9.4 Comparison of fermentation products during aerobic and anaerobic growth of wild-type and recombinant *Escherichia coli*

Growth	Plasmid	Fermentation Product (mM)			
		Ethanol	Lactate	Acetate	Succinate
Aerobic					
	None	0	0.6	55	0.2
	PLO1308-10 (PET)	337	1.1	17	4.9
Anaerobic					
	None	0.4	22	7	0.9
	PLO1308-10 (PET)	482	10	1.2	5.0

From (Ingram and Conway, 1988).

et al., 1991). Ingram, who is director of the Florida Center for Renewable Chemicals and Fuels at UF, cited a recent report from the U.S. Department of Agriculture and DOE that indicates more than one billion tons of biomass can be produced on a sustainable basis each year. Converting this to fuel ethanol could replace half of all imported petroleum in the United States.

Published volumetric and specific ethanol productivities with xylose in simple batch fermentations are 0.6 g of ethanol per liter hour and 1.3 g of ethanol per g cell dry weigh h, respectively. Current sugar conversion to ethanol yields are in the 90–95% range. Elimination of pathways competing for pyruvate (e.g,. acetate kinase or alcohol-aldehyde dehydrogenase) was one approach to maximize yields. Inactivation of native *Escherichia coli* alcohol-aldehyde dehydrogenase (*adhE*), which uses acetyl-CoA as an electron acceptor, had no beneficial effect on growth, which was consistent with a minor role for this enzyme during ethanol production (Underwood et al., 2002). Further improvements have resulted in volumetric productivities exceeding 2 g of ethanol/l h. It is estimated that this organism can produce ethanol from biomass at a cost of approximately $1.30 per gallon (Luli and Ingram, 2005).

BC International in Alachua, FL., plans to build a 30-million-gallon biomass-to-ethanol plant in Jennings, LA, in 2006, presumably using the genetically engineered *Escherichia coli* as a production organism. BCI holds exclusive rights to use and license the engineering organism. According to the BBI Web site, they have successfully tested several cellulosic feedstocks, such as sugarcane, rice straw, rice hulls, softwood forest thinnings, pulp mill sludge, and baggasse. Waste from the sugarcane industry is believed to be the main plant feedstock for bioethanol production.

It is worth noting that there are various groups working on engineering *Escherichia coli* for the production of fermentation products other than ethanol from biomass. For example, the same group from U. of Florida has engineered *Escherichia coli* for the stereoselective production of either D-lactic or L-lactic acid, as well as pyruvate, acetate, or succinate.

The microbial production of L-(+)-lactic acid is rapidly expanding to allow increased production of polylactic acid (PLA), a renewable, biodegradable plastic. *Escherichia coli* W3110 was recently genetically engineered as a homofermentative producer of L-lactic acid as an alternative to lactic acid bacteria. One advantage that *E. coli* offers vs. lactic acid bacteria is its ability to grow and ferment on defined media. This can potentially eliminate the need for complex media components, such as oligopeptides and amino acids that are required by most lactic acid bacteria, hence reducing media and downstream costs. This engineered strain contains five chromosomal deletions (*focA-pflB frdBC adhE ackA ldhA*) and was constructed from a D-(−)-lactic acid producing strain, SZ63 (*focA-pflB frdBC adhE ackA*), by replacing part of the chromosomal *ldhA* coding region with *Pediococcus acidilactici ldhL* encoding for L-lactate dehydrogenase. The resulting strain produced L-lactic acid in M9 mineral salts medium containing glucose or xylose with a yield of 93 to 95%, a purity of 98%, and an optical purity greater than 99% (Zhou et al., 2003).

Along similar lines, a corresponding D-lactic acid *Escherichia coli* strain has also been constructed. D- and L-lactic acid can potentially be combined to generate PLA stereocomplexes with unique and desirable physical properties. These strains (SZ40, SZ58, and SZ63) require only mineral salts as nutrients and lack all plasmids and antibiotic resistance genes used during construction. Competing pathways were eliminated by chromosomal inactivation of genes encoding fumarate reductase (*frdABCD*), alcohol/aldehyde dehydrogenase (*adhE*), and pyruvate formate lyase (*pflB*). D-lactic acid production yield by these new strains approached the theoretical maximum of two moles per mole glucose. As above, the chemical purity of this D-lactic acid was close to 98% with respect to soluble organic compounds, and the optical purity exceeded 99%. Deleting the acetate kinase gene (*ackA*) further improved the cell yield and lactate productivity (Zhou et al., 2003).

Escherichia coli TC44, a derivative of W3110, was engineered for the production of pyruvate from glucose by combining the following genetic changes: $\Delta atpFH$ $\Delta adhE$ $\Delta sucA$, aimed at minimizing ATP yield, cell growth, and respiration. This was combined with gene deletions to eliminate acetate production: *poxB::FRT* $\Delta ackA$, and other by-products: $\Delta focA-pflB$ $\Delta frdBC$ $\Delta ldhA$ $\Delta adhE$. In mineral salts glucose medium, strain TC44 converted glucose to pyruvate with a yield of 0.75 g of pyruvate per g of glucose (77.9% of theoretical yield) at a rate of 1.2 g of pyruvate per liters per h, and a maximum pyruvate titer of ca. 0.75 M. According to the authors, the pyruvate production performance of strain TC44 medium was equivalent or better than previously reported for other biocatalysts including yeast or bacteria (Causey et al., 2004).

In a yet another approach, *Escherichia coli* W3110 was genetically engineered for the overproduction of acetate (Causey *et al.*, 2003). The resulting strain (TC36) converted 0.3 M glucose into ca. 0.6 M of acetate in less than 20 h. A maximum acetate titer of ca. 0.9 M was demonstrated. Strain TC36 was constructed by sequentially assembling deletions that inactivated oxidative phosphorylation *(ΔatpFH)*, disrupted the cyclic function of the tricarboxylic acid pathway *(ΔsucA)*, and eliminated native fermentation pathways *(ΔfocA-pflB ΔfrdBC ΔldhA ΔadhE)*. These mutations minimized the loss of substrate carbon and the oxygen requirement for redox balance. Although TC36 produces only four ATPs per glucose, this strain grows well in mineral salts medium and has no auxotrophic requirement.

Early attempts to engineer a succinate overproducing *Escherichia coli* involved the deletion of pyruvate-formate lyase and lactate dehydrogenase. Such organism can accumulate high amounts of succinate under anaerobic conditions (Vemuri et al., 2002). Recently, new genetic engineering approaches have been applied with an aim of generating strains that can produce succinate under aerobic conditions. Aerobic production offers the advantages over anaerobic fermentation in terms of faster biomass generation, carbon throughput, and product formation, albeit it introduces a significant capital and operating cost. Genetic manipulations were performed on two aerobic succinate producing systems to increase their succinate yield and productivity. One of the aerobic succinate production systems includes five gene deletions: *ΔsdhAB*, *Δicd*, *ΔiclR*, *ΔpoxB*, and *Δ(ackA-pta)*, resulting in a strain with a highly active glyoxylate cycle. A second variation of the above includes four of the five above mutations: *ΔsdhAB*, *ΔiclR*, *ΔpoxB*, and *Δ(ackA-pta)*, having two routes for succinate production. One is the glyoxylate cycle and the other is the oxidative branch of the TCA cycle. Furthermore, inactivation of *ptsG* and overexpression of a mutant sorghum *pepc* in these two production systems resulted in strains having succinate yield of 1.0 mol/mol glucose (close to maximum theoretical). Furthermore, the two-route production system with *ptsG* inactivation and *pepc* overexpression demonstrated substantially higher succinate productivity than the previous system (Lin et al., 2005).

9.3.2   *Klebsiella oxytoca*

In the early 1990s, Ohta et al. investigated the expression of the *pdc* and *adh* genes of *Zymomonas mobilis* in a related enteric bacterium, *Klebsiella oxytoca* (Ohta et al., 1991). The native organism has the capability to transport and metabolize cellobiose, thus, minimizing the need for extracellular additions of cellobiase. In *Klebsiella* strains, two additional fermentation pathways are present compared with *Escherichia coli* (Figure 9.8), converting pyruvate to succinate and butanediol. As in the case of *Escherichia coli*, it was possible to divert more than 90% of the carbon flow from sugar catabolism away from the native fermentative pathways and toward ethanol (strain P2). Overexpression of recombinant PDC alone produced only about twice the ethanol level of the parental strain. However, when both PDC and ADH were elevated in *Klebsiella oxytoca* M5A1, ethanol production was both very rapid and efficient: volumetric productivities >2.0 g per liter per h, yields 0.5 g of ethanol per g of sugar, and final ethanol of 45 g per liter for both glucose and xylose carbon sources were obtained.

The development of methods to reduce costs associated with the solubilization of cellulose is essential for the utilization of lignocellulose as a renewable feedstock for fuels and chemicals. One promising approach is the genetic engineering of ethanol-producing micro-organisms that also produce cellulase enzymes during fermentation. Recent efforts with this organism focused on enabling cellulose conversion to ethanol without addition of expensive cellulase enzymes. A derivative of *Klebsiella oxytoca* M5A1 containing chromosomally integrated genes for ethanol production from *Zymomonas mobilis* (*pdc, adhB*) and endoglucanase genes from Erwinia *chrysanthemi* (*celY, celZ*) produced over 20,000 U endoglucanase l-1 activity during fermentation. Since this organism has the native ability to metabolize cellobiose and cellotriose, this strain was able to ferment amorphous cellulose to ethanol without externally added cellulases with a yield of 58–76% of theoretical (Zhou and Ingram, 2001).

9.3.3 *Zymomonas mobilis*

Xylose also could be a useful carbon source for the ethanol producer *Zymomonas mobilis*. This is a bacterium that has been used as a natural fermentative agent in alcoholic beverage production and has been shown to have ethanol productivity superior to that of yeast. Overall, it demonstrates many of the desirable traits sought in an ideal biocatalyst for ethanol, such as high ethanol yield, selectivity and specific productivity, as well as low pH and high ethanol tolerance. In glucose medium, *Zymomonas mobilis* can achieve ethanol levels of at least 12% (w/v) at yields of up to 97% of the theoretical value. When compared to yeast, *Zymomonas mobilis* exhibits 5–10% higher yields and up to fivefold greater volumetric productivities. The notably high yield of this microbe is attributed to reduced biomass formation during fermentation, apparently limited by ATP availability.

As a matter of fact, *Zymomonas* is the only genus identified to date that exclusively utilizes the Entner-Doudoroff (ED)

LIVERPOOL JOHN MOORES UNIVERSITY
LEARNING SERVICES

pathway anaerobically. The stoichiometry of ethanol production in this recombinant organism can be summarized as follows [neglecting the NAD(P)H balances]:

$$3 \text{ Xylose} + 3 \text{ ADP} + 3 \text{ P}_i \rightarrow 5 \text{ Ethanol} + 5 \text{ CO}_2 + 3 \text{ ATP} + 3 \text{ H}_2\text{O}$$

Thus, the theoretical yield on ethanol is 0.51 g of ethanol/g of xylose (1.67 mol mol^{-1}). It is important to note that the metabolically engineered pathway yields only 1 mole of ATP from 1 mol of xylose, compared with 5/3 moles typically produced through a combination of the pentose phosphate and EMP pathways. When converting glucose to ethanol, this organism produces only one mol of ATP per mole of glucose through the ED pathway compared with two moles produced via the more common EMP pathway. The energy limitation is expected to result in a lower biomass formation, and, thus, a more efficient conversion of substrate to product.

Furthermore, glucose can readily cross the cell membrane of this organism by facilitated diffusion, efficiently be converted to ethanol by an overactive pyruvate decarboxylase/alcohol dehydrogenase system, and is generally recognized as a safe (GRAS) organism for use as an animal feed. As discussed earlier, the main drawback of this micro-organism is that it can only utilize glucose, fructose, and sucrose and thus is unable to ferment the widely available pentose sugars.

This led Zhang et al. to attempt to introduce a pathway for pentose metabolism in *Zymomonas mobilis* (Zhang et al., 1995). Early attempts by other groups using the xylose isomerase (*xylA*) and xylulokinase (*xylB*) genes (Figure 9.7) from either *Klebsiella* or *Xanthomonas* were met with limited success, despite the functional expression of these genes in *Zymomonas mobilis*. It soon became evident that such failures were due to the absence of detectable transketolase and transaldolase activities in *Zymomonas mobilis*, which are necessary to complete a functional pentose metabolic pathway (Figure 9.9). After the transketolase *Escherichia coli* gene was cloned and introduced in *Zymomonas mobilis*, a small conversion of xylose to CO$_2$ and ethanol occurred (Feldmann et al., 1992). The next step was to introduce the transaldolase reaction, as this strain accumulated significant amounts of sedoheptulose-7-P intracellularly. Sophisticated cloning techniques therefore were applied for the construction of a chimeric shuttle vector (pZB5) that carries two independent operons: the first encoding the *Escherichia coli xylA* and *xylB* genes and the second expressing transketolase (*tktA*) and transaldolase (*tal*) again from *Escherichia coli*. The two operons that included the four xylose assimilation and nonoxidative pentose phosphate pathway genes were expressed successfully in *Zymomonas mobilis* CP4. The recombinant strain was capable of fast growth on xylose as the sole carbon source, and moreover it efficiently converted glucose and xylose

Figure 9.9

Ethanol production from pentose sugars in metabolically engineered *Zymomonas mobilis*. XR, xylose reductase; XDH, xylulose dehydrogenase; TK, transketolase; TA, transaldolase.

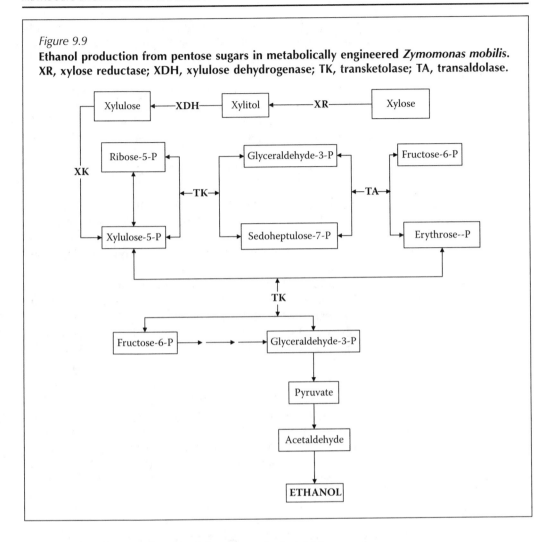

to ethanol with 86 and 94% of the theoretical yield from xylose and glucose, respectively.

In a subsequent article, the same laboratory reported the construction of a *Zymomonas mobilis* strain with a yet expanded substrate fermentation range to include the pentose sugar, l-arabinose, which is commonly found in agricultural residues and other lignocellulosic biomass (Deanda et al., 1996). Five genes, encoding l-arabinose isomerase (*araA*), l-ribulokinase (*araB*), l-ribulose-5-phosphate-4-epimerase (*araD*), transaldolase (*talB*), and transketolase (*tktA*), were isolated from *Escherichia coli* and introduced into *Zymomonas mobilis* under the control of constitutive promoters. The engineered strain grew on and produced ethanol from l-arabinose as a sole C source at 98% of the maximum theoretical ethanol yield, indicating that arabinose was metabolized almost exclusively to ethanol as the sole fermentation product. The authors indicate that this microorganism may be useful, along with the previously developed xylose-fermenting *Zymomonas mobilis* (Zhang *et al.*, 1995), in a mixed culture for efficient fermentation of the predominant hexose and pentose sugars in agricultural residues and other lignocellulosic feedstocks to ethanol.

Iogen Corporation of Ottawa, Canada, has recently built a 50 t/d biomass-to-ethanol demonstration plant adjacent to its enzyme production facility. Iogen, in collaboration with the University of Toronto, has recently tested the C6/C5 cofermentation performance characteristics of National Renewable Energy Laboratory's metabolically engineered *Zymomonas mobilis* using Iogen's biomass hydrolyzates (Lawford et al., 2001). In this study, the biomass feedstock was an agricultural waste, namely oat hulls, which was hydrolyzed in a proprietary two-stage process involving pretreatment with dilute sulfuric acid at 200–250°C, followed by cellulase hydrolysis. The oat hull hydrolyzate (OHH) contained glucose, xylose, and arabinose in a mass ratio of about 8:3:0.5. This work examined the growth and fermentation performance of xylose-utilizing, recombinant *Zymomonas mobilis* cultures CP4:pZB5, and a hardwood pre-hydrolyzate-adapted variant of 39676:pZB4L. In pH-stat batch fermentations with unconditioned 6% (w/v) glucose, 3% xylose, and 0.75% acetic acid, ZM4:pZB5 gave the best performance with a fermentation time of 30 h with a volumetric productivity of 1.4 g/L h. Based on the available glucose and xylose, the process ethanol yield for both strains was 0.47 g/g (92% max. theoretical). Acetic acid tolerance appeared to be a major determining factor in cofermentation performance. Acid tolerance may be one of the weakest attributes of this organism vs. other bacteria or yeast.

9.4   Genetically engineered yeast

Yeasts produce ethanol efficiently from hexoses by the pyruvate decarboxylase-alcohol dehydrogenase (PDC-ADH) system. The most commonly used ethanol producer, *Saccharomyces cerevisiae*, has the intrinsic limitation of not being able to ferment pentoses such as xylose or arabinose. Even though certain types of yeast such as *Pachysolen tannophilus*, *Pichia stipitis*, or *Candida shehatae* are xylose-fermenting, they have poor ethanol yields on pentoses (Bruinenberg et al., 1984) and low ethanol tolerance compared with the common glucose-fermenting yeasts, such as *Saccharomyces cerevisiae*.

9.4.1   *Saccharomyces cerevisiae*

Saccharomyces spp. are the safest and most effective microorganisms for fermenting sugars to ethanol and traditionally have been used in industry to ferment glucose (or hexose sugar)-

based agricultural products to ethanol. Cellulosic biomass, which includes agriculture residues, paper wastes, wood chips, etc., is an ideal inexpensive, renewable, abundantly available source of sugars for fermentation to ethanol, particularly ethanol used as a liquid fuel for transportation. However, *Saccharomyces* spp. are not suitable for fermenting sugars derived from cellulosic biomass since *Saccharomyces* spp., including *Saccharomyces cerevisiae*, are not able to ferment xylose to ethanol or to use this pentose sugar for aerobic growth. Even though *Saccharomyces* spp. are not able to metabolize xylose aerobically and anaerobically, there are other yeasts, such as *Pichia stipitis* and *Candida shehatae*, that are able to ferment xylose to ethanol and to use xylose for aerobic growth. However, these naturally occurring xylose-fermenting yeasts are not effective fermentative micro-organisms, and they also have a relatively low ethanol tolerance.

Many attempts to introduce the one-step pathway by cloning the gene coding for xylose isomerase from either *Escherichia coli* or *Bacillus subtilis* in *Saccharomyces cerevisiae* were unsuccessful due to the inactivity of the heterologous protein in the recombinant host cell (Sarthy et al., 1987). Subsequently, the *Thermus thermophilus xylA* gene encoding xylose (or glucose) isomerase was cloned and expressed in *Saccharomyces cerevisiae* under the control of the yeast *PGK1* promoter (Walfridsson et al., 1996). The recombinant xylose isomerase showed the highest activity at 85°C with a specific activity of 1.0 U per mg protein. This study also demonstrated a new functional, yet low throughput, metabolic pathway in *Saccharomyces cerevisiae* with ethanol formation during oxygen-limited xylose fermentation.

In most yeasts and fungi, xylose reductase and xylitol dehydrogenase are dependent on NADPH and NAD, respectively (Figure 9.10). However, examples of yeast xylose reductases exist that have dual coenzyme specificity, i.e., NADPH and NADH, such as those from *Pichia stipitis* and *Candida shehatae*. Such a type of enzyme has the advantage of preventing imbalances of the NAD/NADH redox system, especially under oxygen-limiting conditions (Granstrom et al., 2000).

The first step in yeast xylose metabolism is carried out by xylose (aldose) reductase. The gene coding for this enzyme has been given the designation *XYL1*. In most yeasts and fungi, this enzyme has a co-factor specificity for NADPH, but in *Pichia stipitis*, the enzyme shows 70% as much activity with NADH as with NADPH (Verduyn et al., 1985). The *Pichia stipitis* *XYL1* gene has been cloned by at least three groups independently (Amore et al., 1991; Hallborn et al., 1991; Takuma et al., 1991). Several xylose reductase genes from various organisms have been since cloned, including those from *Kluyveromyces lactis* (Billard et al., 1995), *Pachysolen tannophilus* (Bolen et al., 1996), and even *Saccharomyces cerevisiae* (Toivari et al., 2004). The relative affinity of various xylose reductases for

Figure 9.10

Anaerobic xylose utilization and co-factor regeneration in recombinant *S. cerevisiae*. EMP, Embden-Meyerhof-Parnas; PPP, pentose phosphate pathway; XR, xylose reductase; XDH, xylitol dehydrogenase; XK, xylulose kinase.

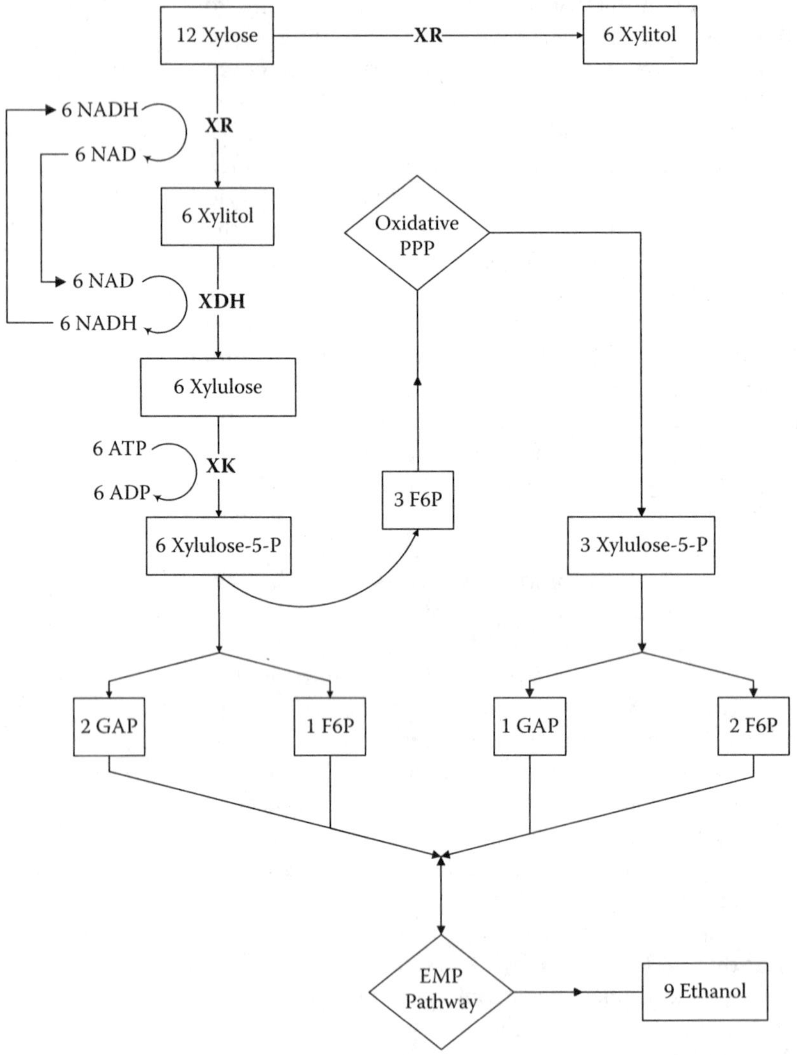

NADH and NADPH vary widely, however. For the most part, this enzyme has a preference for NADPH. One notable exception is the XR enzyme from *Candida boidinii* that was reported to have a higher activity with NADH than with NADPH (Vandeska et al., 1995).

The second step in xylose metabolism is coded for by xylitol dehydrogenase (*XYL2*), which unlike XR, this enzyme is almost always specific for NAD. Attempts have been made to modify the XR co-factor specificity (Metzger and Hollenberg, 1995;

Leitgeb et al., 2005). The mutation D207→G and the double mutation D207→G and D210→G within the binding domain (GXGXXG) increased the apparent Km for NAD ninefold and decreased the xylitol dehydrogenase activity to 47% and 35%, respectively, as compared to the unaltered enzyme. The introduction of the potential NADP-recognition sequence (GSRPVC) of the alcohol dehydrogenase from *Thermoanaerobium brockii* into the xylitol dehydrogenase allowed the mutant enzyme to use both NAD and NADP as co-factor with equal apparent Km values. The mutagenized *XYL2* gene could still mediate growth of *Saccharomyces cerevisiae* transformants on xylose minimal-medium plates when expressed together with the *XYL1*. More recently, the gene coding for a *Saccharomyces cerevisiae* xylitol dehydrogenase enzyme was also discovered (Aristidou et al., 2000; Toivari et al., 2004).

Several laboratories have attempted to engineer a xylose-fermenting *Saccharomyces cerevisiae* through the expression of *XYL1* or both *XYL1* and *XYL2*. Expression of *XYL1* alone has not proven sufficient to enable *S. cerevisiae* to ferment or even to grow on xylose, but in the presence of glucose, *Saccharomyces cerevisiae* strains expressing *XYL1* will produce primarily xylitol from xylose (Hallborn et al., 1991; Meinander et al., 1994). Production of xylitol appears to be a consequence of redox imbalance in the cell and is affected by glycerol production (Meinander et al., 1996).

Expression of both *XYL1* and *XYL2* has proven to be more successful, enabling *Saccharomyces cerevisiae* to grow aerobically on xylose as well as accumulate low levels of ethanol. Kotter and Ciriacy studied the xylose fermentation in *Saccharomyces cerevisiae* more extensively and compared the fermentative activities to *Pichia* stipitis (Amore et al., 1991). In the absence of respiration, *Saccharomyces cerevisiae* transformed with both *XYL1* and *XYL2* converts about half of the xylose present in the medium into xylitol and ethanol in roughly equimolar amounts. By comparison, *Pichia stipitis* produces only ethanol. They proposed, as had Hahn-Hägerdal et al., that in *Saccharomyces cerevisiae*, ethanol production is limited by cofactor imbalance. Additional limitations of xylose utilization in *Saccharomyces cerevisiae* were also attributed to the inefficient capacity of the nonoxidative PPP, as indicated by the accumulation of sedoheptulose-7-P (Senac and Hahn-Hägerdal, 1989; Senac and Hahn-Hägerdal, 1991).

Tantirungkij et al. took the approach one step further by subcloning *Pichia stipitis XYL1* into *Saccharomyces cerevisiae* under the control of the enolase promoter on a multicopy vector (Tantirungkij et al., 1993). This achieved two to three times the level of *XYL1* expression as was observed in *Pichia stipitis*. *XYL2* was also cloned and co-expressed in *Saccharomyces cerevisiae* at about twice the level achieved in induced *Pichia stipitis*. Despite these higher levels of expression, only low levels of ethanol (ca. 5 g/l) were observed under optimal conditions after 100 h. These researchers also selected mutants of *Saccharomyces cerevisiae* carrying *XYL1* and *XYL2* that exhibited rapid growth on xylose medium (Tantirungkij et al., 1994). The fastest growing strain showed a lower activity of XR, but a higher ratio of XDH to XR activity. Southern hybridization showed that the vector carrying the two genes had integrated into the genome resulting in increased stability of the cloned genes. The yield and production rate of ethanol increased 1.6- and 2.7-fold, respectively, but the maximum concentration of ethanol reported was only 7 g/l after 144 h.

The effect of the relative levels of expression of the *XYL1* and *XYL2* genes from *Pichia stipitis*, in *Saccharomyces cerevisiae* has also been investigated (Walfridsson et al., 1997). These two genes were placed in different directions under the control of the alcohol dehydrogenase I (*ADHI*) and phosphoglycerate kinase (*PGK*) promoters and inserted into the *Escherichia coli*-yeast shuttle plasmid YEp24. Different recombinant *Saccharomyces cerevisiae* strains were constructed with different specific activities of XR and XDH. The highest XR or XDH activities were obtained when the expressed gene was controlled by the *PGK* promoter and located downstream after the *ADHI* promoter-gene-terminator sequence. The XR/XDH ratio, i.e., the ratio of specific enzyme activities of XR and XDH, in these recombinant *Saccharomyces cerevisiae* strains varied from 17.5 to 0.06. In order to enhance xylose utilization, in the *XYL1*, *XYL2* containing *Saccharomyces cerevisiae* strains, the native *TKL1* gene encoding transketolase and the *TALI* gene encoding transaldolase were also overexpressed, which showed considerably good growth on the xylose plate. Fermentation of the recombinant *S. cerevisiae* strains containing *XYL1*, *XYL2*, *TKL1*, and *TAL1* were studied with mixtures of glucose and xylose. A strain with an XR:XDH ratio of 17.5 formed 0.82 g xylitol/g consumed xylose, whereas a strain with an XR:XDH ratio of 5.0 formed 0.58 g xylitol/g xylose. The strain with an XR:XDH ratio of 0.06, on the other hand, formed no xylitol and less glycerol and acetic acid compared with strains with the higher XR:XDH ratios. In addition, the strain with an XR:XDH ratio of 0.06 produced more ethanol than the other strains.

Ho and Chang (Ho and Chang, 1989) have reported the construction of a recombinant *Saccharomyces* strain expressing the genes for the three xylose metabolizing enzymes: xylose reductase gene, xylitol dehydrogenase gene, both from *Pichia stipitis*, and xylulokinase from *Saccharomyces cerevisiae*. Cloning of the *XYL3* gene from *Pachysolen tannophilus* was first reported in 1987 (Stevis et al., 1987), and cloning of *Saccharomyces cerevisiae XYL3* by complementation of a xylulokinase deficient mutant of *Escherichia coli* was first reported in 1988 (Rodriguez-Peña et al., 1998), and its role in xylose utilization by *Saccharomyces cerevisiae* was established soon thereafter

(Ho et al., 1998). Ho's group developed recombinant plasmids that can transform *Saccharomyces* spp. into xylose-fermenting yeasts. These plasmids, designated pLNH31, -32, -33, and -34, are 2μm-based high-copy-number yeast-*Escherichia coli* shuttle plasmids. In addition to the geneticin resistance and ampicillin resistance genes that serve as dominant selectable markers, these plasmids also contain three xylose-metabolizing genes: the *Pichia stipitis* genes for XR (*PsXYL1*) and XDH (*PsXYL2*), and the *Saccharomyces cerevisiae* gene for XK (*XYL3*). The parental yeast strain *Saccharomyces* 1400 is a fusion product of *Saccharomyces diastaticus* and *Saccharomyces uvarum*. It exhibits high ethanol and temperature tolerance and a high fermentation rate. Overexpression of *XYL3* in the *Saccharomyces* 1400 fusant along with *XYL1* and *XYL2* results in production of about 47 g/l of ethanol in 84% of theoretical yield from a 1:1 glucose/xylose mixture (Ho et al., 1998).

Researchers at the Finnish Technical Research Institute (VTT) have addressed the redox imbalance of the XR and XDH reactions by introducing artificial transhydrogenase cycles in xylose utilizing *Saccharomyces cerevisiae* (Aristidou et al., 1999). This work was based on the simultaneous expression of dehydrogenase enzymes having different co-factor specificities, e.g., the yeast glutamate dehydrogenase 1 and 2 — *GDH1*, *GDH2* — and also in combination with enzymes that can be driven by ATP, e.g., the malic enzyme, thus overcoming intrinsic limitations due to the physiological redox co-factor concentrations. Results from such genetically engineered organisms so far have been encouraging in terms of improving xylose utilization rates and ethanol productivities. In addition to *Saccharomyces cerevisiae*, these ideas were further successfully applied for the fission yeast *Schizosaccharomyces pombe* that also overexpresses the *Pichia* XR and XDR.

9.4.2 *Pichia stipitis*

This organism has also received significant attention in the past few years, in terms of developing and applying genetic engineering techniques to address metabolic imperfections, such as the oxygen requirements for efficient xylose utilization. Respiratory and fermentative pathways coexist to support growth and product formation in *Pichia stipitis*. This yeast grows rapidly without ethanol production under fully aerobic conditions, and it ferments glucose or xylose under oxygen-limited conditions, but it stops growing within one generation under anaerobic conditions.

Expression of *Saccharomyces cerevisiae URA1* (*ScURA1*) in *Pichia stipitis* enabled rapid anaerobic growth in minimal defined medium containing glucose when essential lipids were present. *ScURA1* encodes a dihydroorotate dehydrogenase that uses fumarate as an alternative electron acceptor to confer anaerobic growth. Initial *Pichia stipitis* transformants grew and produced 32 g/l ethanol from 78 g/l glucose. Cells produced even more ethanol faster following two anaerobic serial subcultures. Control strains without *ScURA1* were incapable of growing anaerobically and showed only limited fermentation. *Pichia stipitis* cells bearing *ScURA1* were viable in anaerobic xylose medium for long periods, and supplemental glucose allowed cell growth, but xylose alone could not support anaerobic growth even after serial anaerobic subculture on glucose. These data imply that *Pichia stipitis* can grow anaerobically using metabolic energy generated through fermentation but that it exhibits fundamental differences in cofactor selection and electron transport with glucose and xylose metabolism. This is the first report of genetic engineering to enable anaerobic growth of a eukaryote.

The *Pichia stipitis* xylose reductase gene (*XYL1*) was inserted into an autonomous plasmid that *Pichia stipitis* maintains in multicopy (Dahn et al., 1996). The plasmid pXOR with the *XYL1* insert or a control plasmid pJM6 without *XYL1* was introduced into *Pichia stipitis*. When grown on xylose under aerobic conditions, the strain with pXOR had up to 1.8-fold higher xylose reductase (XOR) activity than the control strain. Oxygen limitation led to higher XOR activity in both experimental and control strains grown on xylose. However, the XOR activities of the two strains grown on xylose were similar under oxygen limitation. When grown on glucose under aerobic or oxygen-limited conditions, the experimental strain had XOR activity up to 10 times higher than that of the control strain. Ethanol production was not improved, but rather it decreased with the introduction of pXOR compared to the control, and this was attributed to nonspecific effects of the plasmid.

Jeffries's group also studied the expression of the genes encoding group I alcohol dehydrogenases (*PsADH1* and *PsADH2*) in the xylose-fermenting yeast *Pichia stipitis* CBS 6054. The cells expressed PsADH1 approximately 10 times higher under oxygen-limited conditions than under fully aerobic conditions when cultivated on xylose. Transcripts of PsADH2 were not detectable under either aeration condition. The *PsADH1::lacZ* fusion was used to monitor *PsADH1* expression and found that expression increased as oxygen decreased. The level of *PsADH1* transcript was repressed about tenfold in cells grown in the presence of heme under oxygen-limited conditions. Concomitantly with the induction of PsADH1, PsCYC1 expression was repressed. These results indicate that oxygen availability regulates *PsADH1* expression and that regulation may be mediated by heme. The regulation of *PsADH2* expression was also examined in other genetic backgrounds. Disruption of *PsADH1* dramatically increased *PsADH2* expression on nonfermentable carbon sources under fully aerobic conditions, indicating that the expression of *PsADH2* is subject to feedback regulation under these conditions.

9.4.3 *Pichia pastoris*

A xylose reductase gene (*xyl1*) of *Candida guilliermondii* ATCC 20118 was cloned and characterized. The derived amino acid sequence of *Candida guilliermondii* xylose reductase was 70.4% homologous to that of *Pichia stipitis*. The gene was placed under the control of an alcohol oxidase promoter (*AOX1*) and integrated into the genome of the methylotrophic yeast *Pichia pastoris*. Methanol induced the expression of the xylose reductase and the expressed enzyme preferentially utilized NADPH as a co-factor. The authors speculated that the different co-factor specificity between *Pichia pastoris* and *Candida guilliermondii* xylose reductases might be due to the difference in the numbers of histidine residues and their locations between the two proteins. The recombinant was able to ferment xylose, and the maximum xylitol accumulation (7.8 g/l) was observed when the organism was grown under aerobic conditions.

9.4.4 Fungal xylose isomerase in yeast

Recently, a significant breakthrough in xylose conversion to ethanol or other fermentation products by yeast came about as a result of completely independent and parallel efforts at Nature-Works LLC (Minneapolis, Minnesota) and the Delft Technical University in the Netherlands. All previous efforts to overexpress XI in yeast focused on bacterial XI genes, and their success was limited. None of these bacterial XIs, though, resulted in ethanol titers, rates, or yields that could be considered as commercially relevant. The most notable of these attempts was the overexpression of the original or mutated XI gene from *Thermus thermophilus* (Walfridsson et al., 1996).

This breakthrough came about as a result of identifying the first fungal XI gene, isolated from the anaerobic fungus *Piromyces sp.* strain E2 (Xarhangi et al., 2003). This organism metabolizes xylose via xylose isomerase and D-xylulokinase as was shown by enzymatic and molecular analyses, which resembles the situation in bacteria. An early attempt to introduce the *Piromyces sp.* E2 XI gene into *S. cerevisiae* resulted in good XI activity. Slow growth on xylose, however, did not result in reported ethanol production (Kuyper et al., 2003). Subsequently, it's been reported that expression of this particular gene in yeast, including *Saccharomyces cerevisiae* (Kuyper et al., 2004, 2005, 2005) as well as nonconventional yeast, such as *Kluyveromyces sp.* or *Candida sp.* (Rajgarhia et al., 2004), in conjunction with other targeted-or-not genetic changes of the xylose pathway, resulted in engineered yeast able to convert xylose to ethanol at high rates and yields. Since the isolation of the *Piromyces* XI gene, additional homologous genes have also been isolated from other anaerobic fungi, such as *Cyllamyces aberensis*, and successfully expressed in yeast (Rajgarhia et al., 2004). Interestingly, XI isolated from such anaerobic yeasts turned out to be very homologous to the XI gene of the anaerobic Gram-positive bacterium *Bacteroides thetaiotaomicron*, which is a dominant member of human distal intestinal microbiota.

Based on public literature, a Saccharomyces strain expressing the *Piromyces* XI, together with some additional genetic modifications of the xylose pathway was reported to have good anaerobic growth and fermentation on xylose. Besides XI, the overexpressed enzymes were xylulokinase, ribulose 5-phosphate isomerase, ribulose 5-phosphate epimerase, transketolase, and transaldolase. Furthermore, the GRE3 gene encoding aldose reductase was deleted to further minimize xylitol production. During growth on xylose, xylulose formation was absent and xylitol production was negligible. The specific xylose consumption rate in anaerobic xylose cultures was 1.1 g xylose per g biomass per h (Kuyper et al., 2005). Further improvements were achieved through evolutionary engineering in xylose-limited chemostats followed by selection in anaerobic cultivation in automated sequencing-batch reactors on glucose-xylose mixtures. A final single-strain isolate, RWB 218, rapidly consumed glucose-xylose mixtures. anaerobically, in synthetic medium, with a specific rate of xylose consumption exceeding 0.9 g g-1 h-1, with a corresponding ethanol specific productivity of about 0.5 g/g h. When the kinetics of zero trans-influx of glucose and xylose of RWB 218 were compared to that of the initial strain, a twofold higher capacity (V_{max}) as well as an improved Km for xylose was apparent in the selected strain (Kuyper et al., 2005).

When put in perspective, the above is quite an accomplishment given that up to 2003 the maximum corresponding ethanol rate in other xylose-engineered yeast has been less than 0.1–0.15 g/ g h. This could potentially be the breakthrough necessary for the effective conversion of biomass sugars, in sugar hydrolysates, to useful products such as ethanol or organic acids. Nevertheless, a tremendous amount of further technological development will be necessary prior to having a technology that can be industrially implemented, and this includes aspects such as enhancing tolerance to hydrolysate inhibitors, genetic stability especially in continuous processes, as well as ability to utilize all sugars present in hydrolyzates that includes not only glucose and xylose, but also mannose, galactose, as well as arabinose.

9.5 Microbes producing ethanol from lignocellulose

It would be desirable if microbes producing ethanol from lignocellulose also had means to de-polymerize cellulose, hemicellulose, and associated carbohydrates. Many plant pathogenic bacteria (soft-rot bacteria), such as *Erwinia carotovora* and *Erwinia chrysanthemi*, have evolved sophisticated systems of hydrolases and lyases that aid the solubilization of lignocellulose and allow them to break up and penetrate plant tissue (Brencic and Winans, 2005). Genetic engineering of these bacteria for ethanol production represents an attractive alternative to the solubilization of lignocellulosic biomass by chemical or enzymatic means. *Erwinia carotovora* SR38 and *E. chrysanthemi* EC16 were genetically engineered with the PET operon and shown to produce ethanol and CO_2 efficiently as primary fermentation products from cellobiose and glucose (Beall and Ingram, 1993). Both ethanologenic *Erwinia* strains produced about 50 g per liter ethanol from 100 g per liter cellobiose in less than 48 h with a maximum volumetric productivity of 1.5 g per liter of ethanol per hour. This rate is over twice that reported for the cellobiose-utilizing yeast, *Brettanomyces custersii*, in batch culture.

Along similar lines, the incorporation of saccharifying traits to ethanol-producing micro-organism was also attempted. The gene encoding for the xylanase enzyme (*xynZ*) from the thermophilic bacterium *Clostridium thermocellum* was expressed at high cytoplasmic levels in ethanologenic strains of *Escherichia coli* KO11 and *Klebsiella oxytoca* M5A1(pLOI555) (Ohta et al., 1991). This is a temperature-stable enzyme that de-polymerizes xylan to its primary monomer (99%) xylose. In order to increase the amount of xylanase in the medium and facilitate xylan hydrolysis, a two-stage, cyclical process was employed for the fermentation of polymeric feedstocks to ethanol by a single, genetically engineered micro-organism. Cells containing xylanase were harvested and added to a xylan solution at 60°C, thereby lysing and releasing xylanase for saccharification. After cooling to 30°C, the hydrolysate was fermented to ethanol, in the meantime replenishing the supply of xylanase for the subsequent saccharification. *Klebsiella oxytoca* was found to be a superior strain for such an application, because, in addition to xylose (metabolizable by *Escherichia coli*), it can also consume xylobiose and xylotriose. Even though the maximum theoretical yield of M5A1(pLOI555) is in excess of 48 g per liter ethanol from 100 g per liter xylose, about one-third of that was achieved in this process because xylotetrose and longer oligomers remained unmetabolized by this strain. The yield appeared to be limited by the digestibility of commercial xylan rather than by the lack of sufficient xylanase activity or by ethanol toxicity.

Conclusions

- The application of metabolic engineering offers the potential to revolutionize the chemical industry. It is estimated that in less than 10 years, integrated biorefineries will play a role comparable to today's petrochemical manufacturers. Such biorefineries will use row crops, energy crops, and agricultural waste as inputs to extract oil and starch for food, protein for feed, lignin for combustion or chemical conversion, cellulose for conversion into fermentable sugars, as well as other by-products.

- Carbohydrates will be the central feedstock of the future vis-à-vis current hydrocarbons. Biomass sugars will then be used to produce bioethanol as a transportation fuel, but also for a whole host of new, basic building blocks otherwise known as "platform chemicals," including lactic, citric, acetic, or succinic acids, propylene glycol, sorbitol, glycerol, as well as novel chemicals such as 3-hydroxypropionic acid. These are referred to as "platforms" because the basic technology would generate the platform chemicals from which industry could make a wide range of chemicals. Although several obstacles remain to be addressed, the production of fuels and chemicals is advancing with a rapid pace.

- Significant efforts are currently focused on the application of metabolic engineering to biocatalyst formation with the view of increasing the efficiency of converting various sugars in the hydrolysates mixture to useful products using *Escherichia coli* and *Zymomonas mobilis* as well as the yeast *Saccharomyces cerevisiae*.

- With the rapid advances of genetic tools and genome sequencing we are bound to see an expansion in the list of organisms that are genetically engineered for such purposes, with perhaps more emphasis on less conventional species, such as extremophiles (e.g,. thermophilic bacteria), filamentous fungi, or even photosynthetic organisms.

- Looking at Brazil may provide a glimpse into the future of biorefineries. Here are some interesting facts regarding the bioethanol in Brazil:

  1. By law, all gasoline contains a minimum of 25 percent alcohol, yet ethanol is so popular it actually accounts for 40% of all vehicle fuel;
  2. By 2007, 100% of all new cars may be able to run on 100% ethanol;
  3. Sugar-cane-fed biorefineries will be capable of producing sufficient ethanol to allow the entire fleet, new and old cars alike, to do so;
  4. Ethanol is now being used in aviation;
  5. Brazil ended its ethanol subsidies in the mid-1990s. Nevertheless, with world oil prices hovering around

$55–60 a barrel, the price of ethanol today is only half that of gasoline;

6. Since its inception, Brazil's ethanol program has displaced imported oil worth $120 billion — this is comparable to a savings of almost $2 trillion for a U.S.-sized economy.

- Based on Brazil's experience and similar successes, one can only conclude that the future of integrated biorefineries is bright, and that the role of metabolic engineering is only limited by our imagination.

Summary

- Lignocellulosic materials (cellulose, hemicellulose, and lignin) are the most abundant renewable organic resource on earth and as such the development of new processes for their conversion to useful products has been recognized as an urgent need.
- The conversion of both cellulose, which is composed entirely of glucose, and hemicellulose, which in addition to glucose contains xylose and arabinose, for the production of fuel ethanol is being studied intensively with a view to developing a technically and economically viable bioprocess.
- While the fermentation of glucose can be carried out efficiently, the bioconversion of the pentose fraction (xylose, arabinose) released on hydrolysis of hemicellulose presents a greater challenge. A good deal of attention therefore has been focused on the developments of genetically modified strains that are capable of efficiently co-utilizing and, in turn, converting glucose and pentoses to useful compounds.
- This chapter deals mainly with the metabolic strategies employed in the development of efficient biocatalyst (bacteria and yeast) for the bioconversion of most hemicellulosic sugars to products of primary metabolism, e.g., ethanol (see Chapter 4 for more details).
- The metabolic engineering objectives have thus far been focused on increasing the efficiency of carbon conversion to desirable end products on one hand and expanding the substrate and product spectra on the other.

References

Amore, R., P. Kotter, C. Kuster, M. Ciriacy, C. P. Hollenberg (1991). Cloning and expression in *Saccharomyces cerevisiae* of the NAD(P)H-dependent xylose reductase-encoding gene (*XYL1*) from the xylose- assimilating yeast *Pichia stipitis*. *Gene* **109**, 89.

Aristidou, A., J. Londesborough, M. Penttilä, P. Richard, L. Ruohonen, H. Soderlund, A. Teleman, M. Toivari (1999). Transformed micro-organisms with improved properties. Patent Application PCT/FI99/00185, WO 99/46363, 93 pp.

Aristidou, A., P. Richard, L. Ruohonen, M. Toivari, J. Londes-borough, M. Penttilä (2000). Redox balance in fermenting yeast. *Monograph - European Brewing Convention* **28**, 161–170.

Beall, D. S., L. O. Ingram (1993). Genetic engineering of soft-rot bacteria for ethanol production from lignocellulose. *Journal of Industrial Microbiology* **11**, 151–155.

Billard, P., S. Ménart, R. Fleer, M. Bolotin-Fukuhara (1995). Isolation and characterization of the gene encoding xylose reductase from *Kluyveromyces lactis. Gene* **162**, 93.

Bolen, P. L., G. T. Hayman, H. S. Shepherd (1996). Sequence and analysis of an aldose (xylose) reductase gene from the xylose-fermenting yeast *Pachysolen tannophilus. Yeast* **12**, 1367.

Brencic, A., S. C. Winans (2005). Detection of and response to signals involved in host-microbe interactions by plant-associated bacteria *Microbiology and Molecular Biology Reviews* **69**, 155–194.

Bruinenberg, P. M., P. H. M. De Bot, J. P. Van Dijken, W. A. Scheffers (1984). NADH-linked aldose reductase: the key to anaerobic alcoholic fermentation of xylose by yeasts. *Applied Microbiology and Biotechnology* **19**, 256–260.

Causey, T. B., K. T. Shanmugam, L. P. Yomano, L. O. Ingram (2004). Engineering *Escherichia coli* for efficient conversion of glucose to pyruvate. *Proceedings of the National Academy of Sciences of the United States of America* **101**, 2235–2240.

Causey, T. B., S. Zhou, K. T. Shanmugam, L. O. Ingram (2003). Engineering the metabolism of *Escherichia coli* W3110 for the conversion of sugar to redox-neutral and oxidized products: homoacetate production. *Proceedings of the National Academy of Sciences of the United States of America* **100**, 825–832.

Chandrakant, P., V. S. Bisaria (1998). Simultaneous bioconversion of cellulose and hemicellulose to ethanol. *Critical Reviews in Biotechnology* **18**, 295–331.

Dahn, K., B. Davis, P. Pittman, W. Kenealy, T. Jeffries (1996). Increased xylose reductase activity in the xylose-fermenting yeast *Pichia stipitis* by overexpression of *XYL1. Applied Biochemistry and Biotechnology* **57-58**, 267–276.

Deanda, K., M. Zhang, C. Eddy, S. Picataggio (1996). Development of an arabinose-fermenting *Zymomonas mobilis* strain by metabolic pathway engineering. *Applied and Environmental Microbiology* **62**, 4465.

Feldmann, S., H. Sahm, G. A. Sprenger (1992). Pentose metabolism in *Zymomonas mobilis* wild type and recombinant strains. *Applied Microbiology and Biotechnology* **38**, 354.

Gong, C. S., N. J. Cao, J. Du, G. T. Tsao (1999). Ethanol production from renewable resources. *Advances in Biochemical Engineering and Biotechnology* **65**, 207.

Granstrom, T. B., A. A. Aristidou, J. Jokela, M. Leisola (2000). Growth characteristics and metabolic flux analysis of *Candida milleri*. *Biotechnology and Bioengineering* **70**, 197–207.

Hallborn, J., M. Walfridsson, U. Airaksinen, H. Ojamo, B. Hahn-Hagerdal, M. Penttilä, S. Keranen (1991). Xylitol production by recombinant *Saccharomyces cerevisiae*. *Biotechnology* **9**, 1090–1095.

Hamelinck, C. N., G. V. Hooijdonk, A. P. C. Faaij (2005). Ethanol from lignocellulosic biomass: Techno-economic performance in short-, middle- and long-term. *Biomass and Bioenergy* **28**, 384–410.

Ho, N. W., Z. Chen, A. P. Brainard (1998). Genetically engineered *Saccharomyces* yeast capable of effective cofermentation of glucose and xylose. *Applied and Environmental Microbiology* **64**, 1852.

Ho, N. W. Y., S. F. Chang (1989). Cloning of yeast xylulokinase gene by complementation of *E. coli* and yeast mutations. *Enzyme Microbiology and Technology* **11**, 417–421.

Ingram, L. O., H. C. Aldrich, A. C. C. Borges, T. B. Causey, A. Martinez, F. Morales, A. Saleh, S. A. Underwood, L. P. Yomano, S. W. York, J. Zaldivar, S. Zhou (1999). Enteric bacterial catalysts for fuel ethanol production. *Biotechnology Progress* **15**, 855–866.

Ingram, L. O., T. Conway (1988). Expression of different levels of ethanologenic enzymes from *Zymomonas mobilis* in recombinant strains of *Escherichia coli*. *Applied and Environmental Microbiology* **54**, 397.

Keim, C. R., K. Venkatasubramanian (1989). Economics of current biotechnological methods of producing ethanol. *Trends in Biotechnology* **7**, 22.

Kuhad, R. C., A. Singh, K. E. Eriksson (1997). Micro-organisms and enzymes involved in the degradation of plant fiber cell walls. *Advances in Biochemical Engineering/Biotechnology* **57**, 45.

Kuyper, M., H. R. Harhangi, A. K. Stave, A. A. Winkler, M. S. M. Jetten, W. T. A. M. De Laat, J. J. J. Den Ridder, H. J. M. Op den Camp, J. P. Van Dijken, J. T. Pronk (2003). High-level functional expression of a fungal xylose isomerase: The key to efficient ethanolic fermentation of xylose by *Saccharomyces cerevisiae*? *FEMS Yeast Research* **4**, 69–78.

Kuyper, M., M. M. P. Hartog, M. J. Toirkens, M. J. H. Almering, A. A. Winkler, J. P. van Dijken, J. T. Pronk (2005). Metabolic engineering of a xylose-isomerase-expressing *Saccharomyces cerevisiae* strain for rapid anaerobic xylose fermentation. *FEMS Yeast Research* **5**, 399–409.

Kuyper, M., M. J. Toirkens, J. A. Diderich, A. A. Winkler, J. P. van Dijken, J. T. Pronk (2005). Evolutionary engineering of mixed-sugar utilization by a xylose-fermenting *Saccharomyces cerevisiae* strain. *FEMS Yeast Research* 5, 925–934.

Kuyper, M., A. A. Winkler, J. P. van Dijken, J. T. Pronk (2004). Minimal metabolic engineering of *Saccharomyces cerevisiae* for efficient anaerobic xylose fermentation: A proof of principle. *FEMS Yeast Research* 4, 655–664.

Lawford, H. G., J. D. Rousseau, J. S. Tolan (2001). Comparative ethanol productivities of different *Zymomonas* recombinants fermenting oat hull hydrolysate. *Applied Biochemistry and Biotechnology* 91-93, 133–146.

Lee, J. (1997). Biological conversion of lignocellulosic biomass to ethanol. *Journal of Biotechnology* 56, 1.

Leitgeb, S., B. Petschacher, D. K. Wilson, B. Nidetzky (2005). Fine tuning of coenzyme specificity in family 2 aldo-keto reductases revealed by crystal structures of the Lys-274toArg mutant of *Candida tenuis* xylose reductase (*AKR2B5*) bound to NAD+ and NADP+. *FEBS Letters* 579, 763–767.

Lerouge, P., M. Cabanes-Macheteau, C. Rayon, A. C. Fischette-Lainé, V. Gomord, L. Faye (1998). N-glycoprotein biosynthesis in plants: recent developments and future trends. *Plant Molecular Biology* 38, 31.

Licht, F. O. (2005). Homegrown for the Homeland: Industry Outlook 2005. from http://www.ethanolrfa.org/outlook2005.html

Lin, H., G. N. Bennett, K.-Y. San (2005). Metabolic engineering of aerobic succinate production systems in *Escherichia coli* to improve process productivity and achieve the maximum theoretical succinate yield. *Metabolic Engineering* 7, 116–127.

Luli, G., L. Ingram (2005). UF/IFAS researcher's biomass-to-ethanol technology could help replace half of auto fuel in U.S. *University of Florida News*.

Lynd, L. R., C. E. Wyman, T. U. Gerngross (1999). Biocommodity engineering. *Biotechnology Progress* 15, 777–793.

Meinander, N., B. Hahn-Haegerdal, M. Linko, P. Linko, H. Ojamo (1994). Fed-batch xylitol production with recombinant *XYL1* expressing *Saccharomyces cerevisiae* using ethanol as a co-substrate. *Applied Microbiology and Biotechnology* 42, 334–339.

Meinander, N., G. Zacchi, B. Hahn-Hagerdal (1996). A heterologous reductase affects the redox balance of recombinant *Saccharomyces cerevisiae*. *Microbiology* 142, 165–172.

Metzger, M. H., C. P. Hollenberg (1995). Amino acid substitutions in the yeast *Pichia stipitis* xylitol dehydrogenase coenzyme-binding domain affect the coenzyme specificity. *European Journal of Biochemistry* 228, 50.

Ohta, K., D. S. Beall, J. P. Mejia, K. T. Shanmugam, L. O. Ingram (1991). Metabolic engineering of *Klebsiella oxytoca* M5A1 for ethanol production from xylose and glucose. *Applied and Environmental Microbiology* 57, 2810.

Rajgarhia, V., K. Koivuranta, M. Penttilä, M. Ilmen, P. Suominen, A. Aristidou, C. Miller, S. Olson, L. Ruohonen (2004). Genetically modified yeast species and fermentation processes using genetically modified yeast. Patent Application WO 2004099381, 148 pp.

Rodriguez-Peña, J. M., V. J. Id, J. Arroyo, C. Nombela (1998). The *YGR194c (XKS1)* gene encodes the xylulokinase from the budding yeast *Saccharomyces cerevisiae*. *FEMS Microbiology Letters* 162, 155.

Sarthy, A. V., B. L. McConaughy, Z. Lobo, J. A. Sundstrom, C. E. Furlong, B. D. Hall (1987). Expression of the *Escherichia coli* xylose isomerase gene in *Saccharomyces cerevisiae*. *Applied and Environmental Microbiology* 53, 1996–2000.

Senac, T., B. Hahn-Hägerdal (1989). Intermediary metabolite concentrations in xylulose- and glucose-fermenting *Saccharomyces cerevisiae* cells. *Applied and Environmental Microbiology* 56, 120.

Senac, T., B. Hahn-Hägerdal (1991). Effects of increased transaldolase activity on D-xylulose and D-glucose metabolism in *Saccharomyces cerevisiae* cell extracts. *Applied and Environmental Microbiology* 57, 1701.

Stevis, P. A., J. J. Hang, N. W. Y. Ho (1987). Cloning of the *Pachysolen tannophilus* xylulokinase gene by complementation in *Escherichia coli*. *Applied and Environmental Microbiology* 53, 2975–2977.

Takahashi, D. F., M. L. Carvalhal, F. Alterhum (1994). Ethanol production from pentoses and hexoses by recombinant *Escherichia coli*. *Biotechnology Letters* 16, 747.

Takuma, S., N. Nakashima, M. Tantirungkij, S. Kinoshita, H. Okada, T. Seki, T. Yoshida (1991). Isolation of xylose reductase gene of *Pichia stipitis* and its expression in *Saccharomyces cerevisiae*. *Applied Biochemistry and Biotechnology* 28-29, 327.

Tantirungkij, M., T. Izuishi, T. Seki, T. Yoshida (1994). Fed-batch fermentation of xylose by a fast growing mutant of xylose-assimilating recombinant *Saccharomyces cerevisiae*. *Applied Microbiology and Biotechnology* 41, 8–12.

Tantirungkij, M., N. Nakashima, T. Seki, T. Yoshida (1993). Construction of xylose-assimilating *Saccharomyces cerevisiae*. *Journal of Fermentation and Bioengineering* 75, 83.

Teeri, T. T., A. Koivula, M. Linder, G. Wohlfahrt, C. Divne, T. A. Jones (1998). *Trichoderma reesei* cellobiohydrolases: why so efficient on crystalline cellulose. *Biochemical Society Transactions* 26, 173.

Toivari, M. H., L. Salusjärvi, L. Ruohonen, M. Penttilä (2004). Endogenous xylose pathway in *Saccharomyces cerevisiae*. *Applied and Environmental Microbiology* **70**, 3681–3686.

Tolan, J. S., R. K. Finn (1987). Fermentation of D-xylose and L-arabinose to ethanol by *Erwinia chrysanthemi*. *Applied and Environmental Microbiology* **53**, 2033.

Tolan, J. S., R. K. Finn (1987). Fermentation of D-xylose to ethanol by genetically modified *Klebsiella planticola*. *Applied and Environmental Microbiology* **53**, 2039.

Underwood, S. A., S. Zhou, T. B. Causey, L. P. Yomano, K. T. Shanmugam, L. O. Ingram (2002). Genetic changes to optimize carbon partitioning between ethanol and biosynthesis in ethanologenic *Escherichia coli*. *Applied and Environmental Microbiology* **68**, 6263–6272.

Vandeska, E., S. Kuzmanova, T. W. Jeffries (1995). Xylitol formation and key enzyme activities in *Candida boidinii* under different oxygen transfer rates. *Journal of Fermentation and Bioengineering* **80**, 513.

Vemuri, G. N., M. A. Eiteman, E. Altman (2002). Succinate production in dual-phase *Escherichia coli* fermentations depends on the time of transition from aerobic to anaerobic conditions. *Journal of Industrial Microbiology and Biotechnology* **28**, 325–332.

Verduyn, C., R. Van Kleef, J. Frank, H. Schreuder, J. P. Van Dijken, W. A. Scheffers (1985). Properties of the NAD(P)H-dependent xylose reductase from the xylose-fermenting yeast *Pichia stipitis*. *Biochemical Journal* **226**, 669.

Walfridsson, M., M. Anderlund, X. Bao, B. Hahn-Hagerdal (1997). Expression of different levels of enzymes from the *Pichia stipitis XYL1* and *XYL2* genes in *Saccharomyces cerevisiae* and its effects on product formation during xylose utilisation. *Applied Microbiology and Biotechnology* **48**, 218–224.

Walfridsson, M., X. Bao, M. Anderlund, G. Lilius, L. Bulow, B. Hahn-Hägerdal (1996). Ethanolic fermentation of xylose with *Saccharomyces cerevisiae* harboring the *Thermus thermophilus xylA* gene, which expresses an active xylose (glucose) isomerase. *Applied and Environmental Microbiology* **62**, 4648.

Xarhangi, H. R., A. S. Akhmanova, R. Emmens, C. van der Drift, W. T. A. M. de Laat, J. P. van Dijken, M. S. M. Jetten, J. T. Pronk, H. J. M. Op den Camp (2003). Xylose metabolism in the anaerobic fungus *Piromyces sp.* strain E2 follows the bacterial pathway. *Archives of Microbiology* **180**, 134–141.

Yomano, L. P., S. W. York, L. O. Ingram (1998). Isolation and characterization of ethanol-tolerant mutants of *Escherichia coli* KO11 for fuel ethanol production. *Journal of Industrial Microbiology and Biotechnology* **20**, 132.

Zhang, M., C. Eddy, K. Deanda, M. Finkelstein, S. Picataggio (1995). Metabolic engineering of a pentose metabolism pathway in ethanologenic *Zymomonas mobilis*. *Science* **267**, 240.

Zhou, S., T. B. Causey, A. Hasona, K. T. Shanmugam, L. O. Ingram (2003). Production of optically pure D-lactic acid in mineral salts medium by metabolically engineered *Escherichia coli* W3110. *Applied and Environmental Microbiology* **69**, 399–407.

Zhou, S., L. O. Ingram (2001). Simultaneous saccharification and fermentation of amorphous cellulose to ethanol by recombinant *Klebsiella oxytoca* SZ21 without supplemental cellulase. *Biotechnology Letters* **23**, 1455–1462.

Zhou, S., K. T. Shanmugam, L. O. Ingram (2003). Functional replacement of the *Escherichia coli* D-(–)-lactate dehydrogenase gene (*ldhA*) with the L-(+)-lactate dehydrogenase gene (*ldhL*) from *Pediococcus acidilactici*. *Applied and Environmental Microbiology* **69**, 2237–2244.

10

Cell Immobilization and Its Applications in Biotechnology: Current Trends and Future Prospects

Ronnie Willaert

10.1 Introduction

The immobilization of whole cells can be defined as "the physical confinement or localization of intact cells to a certain region of space without loss of desired biological activity." When cells are encapsulated in an immobilized cell system, the term "bioencapsulation" or "microencapsulation" is used; the latter is used when cells are immobilized in microcapsules, i.e., micrometer-sized systems surrounded by a barrier membrane. Recently, nanoencapsulation of biological molecules such as proteins have also been successful.

Immobilizing individual enzymes, for simple reactions such as hydrolysis and isomerisation as well as multi-enzyme systems have been used as biocatalysts for the production of various chemicals through simple and conjugated reactions. Many applications have also been developed for single and multicellular organisms.

The suitability of a given system of immobilization is dictated by the type of application and the physical and biochemical characteristics of the immobilizing matrix/agent. Accordingly, requirements will be different for each particular case, but the following characteristics are generally desirable:

- Has a high cell mass-loading capacity.
- Affords easy access to nutrient media.
- Is a simple and "nontoxic" immobilization procedure.
- Affords high surface-to-volume ratio.

Immobilized cell system/aggregate is composed of three components, namely: the cells, the support material (or carrier or matrix), and the solution that fills the remainder of the space (interstitial solution), which is also known as microenvironment.

Figure 10.1

Classification of immobilized cell systems according to the physical localisation and the nature of the microenvironment (Willaert and Baron, 1996).

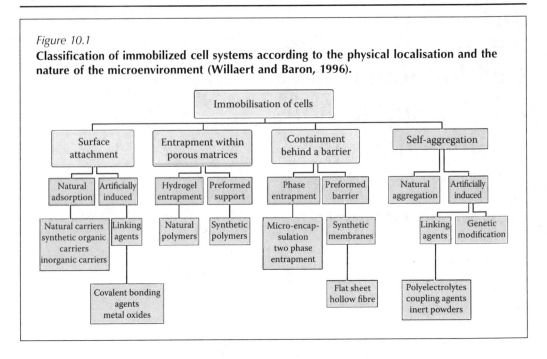

- Facilitates optimum mass transfer.
- Is sterilizable and reusable.
- Facilitates easy separation of cells and carrier from media.
- Is suitable for conventional reactor systems as well as cell suspension and anchorage-dependent cells.
- Should be biocompatible for animal cells.
- Contains immunoprotection barrier.
- Should be economically viable.

10.2   Immobilized cell systems

On the bases of physical localization and the nature of microenvironment, immobilized cell systems can be classified into four categories (Figure 10.1).

10.2.1   *Surface attachment of cells*

Although this is not suitable where cell-free effluent is desired, immobilization of cells by adsorption to a support material can be achieved naturally or induced artificially by using linking agents (metal oxides or covalent bonding agents such as glutaraldehyde or amminosilane) (Figure 10.2). A suitable adsorbent for spontaneous attachment should possess a high affinity towards the biocatalyst and cause minimal denaturation.

Figure 10.2

Cell-immobilization by adsorption/attachment to a surface: A. Adsorption of a monolayer, B. Adsorption of a biofilm, and C. Adsorption of an "artificial" biofilm, i.e., adsorption of a gel layer containing immobilized cells or cell aggregates.

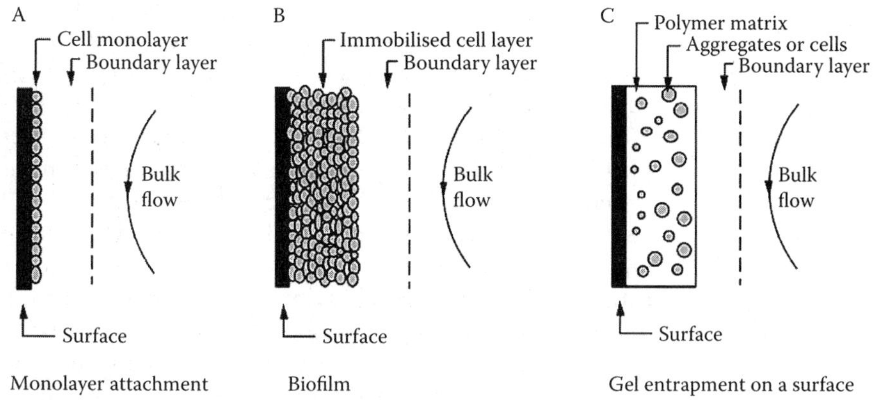

- The adsorption of cells to an organic or inorganic support material is achieved *via* the Van der Waals forces and ionic interactions.

The adhesion behavior of viable cells is influenced by the following factors:

- The physical and chemical properties of the adsorption matrix.
- The identity and the biochemical characteristics of the immobilized organism (especially the outer surface of the cell wall).
- The composition, the chemical and physical properties of the surrounding mobile phase.

The effects of different environmental and/or physiological conditions on the adhesion mechanisms of different bacteria have been intensively studied. Critical physicochemical parameters that govern the interactions of cells with surfaces can be quantified by several techniques (Mavituna, 2004). Since different types of forces are involved in cell-immobilization to the matrix surface, carrier surfaces can therefore be modified in order to affect a more efficient system of cell immobilization.

Recent genetic and molecular approaches have identified that cells develop surface-sensing responses following cell-support contact. Micro-organisms interact with surfaces through many specialised structures on their walls, e.g., pilus-associated adhesions, exopolymers (glycocalyx in the case of bacteria), and complex ligand interactions involving signaling molecules as

CONTACT
INHIBITION
Animal cells, that exhibit
contact inhibition, stop
growing when they make
contact since cell division
is gradually inhibited
by cell-cell contact as
the culture reaches
confluence.

well as quorum sensing mechanisms. It follows, therefore, that sensing either a biotic or an abiotic surface may turn on genetic switches, and this in turn may lead to changes in the organism's phenotype (Eberl et al., 1999; Loo et al., 2000).

Recently, the structural and developmental complexities of microbial biofilm formation investigated using new sophisticated techniques including proteomics revealed that biofilm formation in *Pseudomonas aeruginosa* proceeds as a regulated developmental sequence comprising five stages (Stoodley et al., 2002; Sauer et al., 2002). Stages one and two are generally identified by loose or transient association with the surface, followed by a robust adhesion. Stages three and four involve the aggregation of cells into microcolonies and subsequent growth and maturation, while stage five is characterized by a return to transient motility, where biofilm cells are sloughed or shed.

Biofilm structures can be flat or mushroom-shaped depending on the nutrient source, which seems to influence the interactions between localized clonal growth and the subsequent rearrangement of cells through type IV pilus-mediated gliding motility (Klausen et al., 2003).

Most animal cells from solid tissue grow as adherent monolayers, unless transformed into anchorage independent. Anchorage-dependent cells, however, are often diploid and exhibit contact inhibition. Following tissue disaggregation or subculturing, they will need to attach and spread out on the substrate before proliferating.

While cell adhesion is mediated by specific cell surface receptors, cell-substrate interactions are mediated primarily by integrins, receptors for matrix molecules such as fibronectin, entactin, laminin, and collagen, which bind them *via* a specific motif usually containing the arginine-glycine-aspartic acid (RDG) sequence (Yamada and Geiger, 1997). Each integrin comprises one α and one β subunit, both of which are highly polymorphic, thus generating considerable diversity among integrins.

The first step of the cell-surface interaction of anchorage dependent cells is attachment, in which the cells retain the round shape they possessed in suspension. The cells undergo conformational change, known as spreading, in which the cells increase their surface area prior to attachment to the surface. The kinetics of attachment and spreading has been determined by measuring the effective refractive index of the waveguide, the number of cells per unit area, and a parameter uniquely characterizing their shape, such as the area in contact with the surface.

10 2.2 *Entrapment within porous matrices*

Cell entrapment can be achieved through *in situ* immobilization in the presences of the porous matrix (i.e., gel entrapment), or by allowing the cells to move into a preformed porous matrix

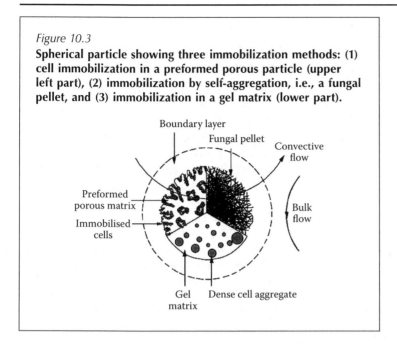

Figure 10.3

Spherical particle showing three immobilization methods: (1) cell immobilization in a preformed porous particle (upper left part), (2) immobilization by self-aggregation, i.e., a fungal pellet, and (3) immobilization in a gel matrix (lower part).

(Figure 10.3). Entrapped cells can reach high densities in the matrix and – compared to surface immobilization – cells are well protected from fluid shear. However, these dense cell packings may lead to mass transport limitations.

10.2.2.1 Hydrogel entrapment

Mainly for its simplicity and excellent cell containment, most of the research in the domain of immobilized cells has used gel entrapment.

A wide variety of natural (polysaccharides and proteins) and synthetic polymers can be gelled into hydrophilic matrices under mild conditions to allow cell entrapment with minimal loss of viability. Gel formation mechanisms of frequently used gels are shown in Table 10.1. The polymer-cell mixture can be formed in different shapes and sizes. The most common forms are small beads about 1 to 5 mm in diameter. Even though natural polymers dominate, synthetic polymers have recently been developed and applied for the immobilization of living cells. The synthetic polymers can be easily and artificially designed for adequate properties. The porosity of the gel as well as the ionic and hydrophobic or hydrophilic properties can be adjusted. Additionally, the mechanical strength and longevity of the gels formed from synthetic polymers are generally superior to those from natural polymers.

Gel entrapment has the disadvantage of limited mechanical stability. It has been frequently observed that the gel structure is easily destroyed by cell growth in the gel matrix and carbon

Table 10.1   Gel formation mechanisms for cell entrapment

Principle of gelation	Material
Ionotropic gelation	Alginate, chitosan
Thermal gelation	Agar, agarose, κ-carrageenan, collagen, gelatin, gellan gum, curdlan
Precipitation	Cellulose, cellulose triacetate
Polymerization with cross-linking reagent	Polyacrylamide, polymethacrylate, polyacrylamide-hydrazide
Polycondensation	Polyurethane, epoxy resin
Radical-mediated polymerization by irradiation with near ultraviolet light	Photo-cross-linkable resin prepolymers
Cross-linking through photo-dimerisation by irradiation with visible or UV light	Photosensitive resin prepolymers
Radiation polymerization	Poly(2-hydroxyethyl methacrylate/acrylate), Poly(vinyl alcohol), poly(ethylene glycol) diacrylate/ dimethacrylate
Gelation by iterative freezing and thawing or cross-linking with boric acid (and Ca-alginate)	Poly(vinyl alcohol)

A polyelectrolyte is a macromolecular substance, which on dissolving in water or another ionizing solvent, dissociates to give polyions (polycations or polyanions) – multiply charged ions – together with an equivalent amount of ions of small charge and opposite sign. Polyelectrolytes dissociating into polycations and polyanions, with no ions of small charge, are also conceivable. A polyelectrolyte can be a polyacid, a polybase, a polysalt, or a polyampholyte.

dioxide production. However, the gel structure can usually be reinforced, e.g., alginate gel was made stronger by the reaction of alginate with other molecules like polyethyleneimine, glutaraldehyde cross-linking, silica, genipin, poly(vinyl alcohol), or by partial drying of the gel. Another disadvantage of cell entrapment, compared to other immobilized cell systems, is oxygen limitations in the matrix. Examples of cell immobilization by attachment to surface and cell entrapments applications are given in Tables 10.2 and 10.3, respectively.

10.2.2.1.1   Hydrogels from natural polymers

Various natural hydrogels have been used to immobilize living cells. The most important gel materials will be discussed in more detail.

Alginates: Alginates constitute a family of unbranched copolymers of 1,4-linked β-D-mannuronic (M) and α-L-guluronic acid (G) of widely varying composition. The monomers are arranged in a pattern of blocks along the chain, with homopolymeric regions interspersed with regions of alternating structure (MG blocks). Divalent cation-induced gelling of alginates in solution reflects their specific ion binding capacity and the conformational change associated with it.

Entrapment of cells within spherical beads of Ca-alginate has become one of the most widely used methods for immobilizing living cells. The success of this method is mainly due to very mild conditions under which immobilization is performed, and

Table 10.2 Examples of cell immobilization by attachment to a surface

Material	Cell Type	Application	Reference
Bacteria			
Ion exchange resin	*Bacillus stearothermophilus*	Amylase production	Glassner et al., 1989
Coke	*Zymomonas mobilis*	Ethanol production	Dempsey, 1990
Seashell pieces	*Bacillus* sp. + *Aeromonas* sp. + *Alcaligenes* sp.	Decolourisation and degradation of triphenyl methane dyes	Sharma et al., 2004
Fungi			
Celite	*Penicillium chrysogenum*	Penicillin production	Lilly et al., 1990
Stainless steel fiber cloth	*Saccharomyces cerevisiae*	Beer production	Shen et al., 2004
Sugarcane bagasse	*Candida guilliermondii*	Xylitol production	Santos et al., 2005
Straw	*Agaricus* sp.	Laccacase production	Kaluskar et al., 1999
Structural fibrous network of papaya wood	*Aspergillus terreus*	Itaconic acid production	Iqbal and Saeed, 2005
Animal cells			
Surface modified polyethylene film	Midbrain cells	Neural differentiation	Nakaoka et al., 2003
Poly(lactide-co-glycolide), poly(D,L-lactide)	Chondrocyte	Growth on biodegradable scaffolds (tissue engineering)	Tsai and Wang, 2005
Pyrex glass, polysterene, glass beads	*Trichoplusia ni*	Recombinant protein production	Shuler et al., 1990

it is a fast, simple and cost-effective technique. The cell suspension is mixed with a sodium alginate solution, and the mixture dripped into a solution containing calcium ions. The droplets form gel spheres instantaneously, entrapping the cells in a three-dimensional lattice of ionically cross-linked alginate. A major disadvantage of the use of Ca-alginate beads is its sensitivity towards chelating agents such as phosphate, citrate, EDTA, and lactate, or antigelling cations such as Na^+ or Mg^{2+}. Various ways to overcome this limitation is to keep the beads in a medium containing a few millimolar free calcium ions and keep the Na^+/Ca^{2+} ratio low. Alginate beads can also be stabilized by replacing Ca^{2+} with other divalent ($Ba^{2+} > Sr^{2+} > Ca^{2+} >> Mg^{2+}$) or multivalent cations, e.g., Al^{3+} or Ti^{3+}. Other stabilization methods involve the use of polyelectrolytes, such as polyethyleneimine (PEI) and polypropyleneimine (PEI), glutaraldehyde, colloidal silica, propylene glycol ester of alginic acid with PEI,

Table 10.3   Some recent applications of cell entrapment in hydrogels

Material	Organism	Application	Reference
Bacterial cells			
Alginate	*Bifidobacterium longum*	Lactic acid production from whey	Shahbazi et al., 2005
	Lactococcus lactis	Cell growth and release	Klinkenberg et al., 2001
	Streptococcus thermophilus	Inoculation of milk	Champagne et al., 2000
Alginate-whey protein	*Lactococcus lactis*	Nisin production	Millette et al., 2004
NPAE[c]-alginate	*Lactobacillus rhamnosus*	Probiotics	Le-Tien et al., 2004
PVA[a] cryogel	*Bacillus agaradhaerens*	β-Cyclodextrin production	Martins et al., 2003
Fungi			
Ca-alginate	*Saccharomyces cerevisiae*	Ethanol production	Najafpour et al., 2004
Polyethylene-oxide	*Candida versatilis* and *Zygosaccharomyces rouxii*	Bioflavor for soy-sauce	van der Sluis et al., 2001
Mammalian cells			
Ba-alginate	Engineered NIH3T3 cells	Continuous release of interleukin 12 for cancer therapy	Zheng et al., 2003
Ca-alginate	Engineered HEK 293 EBNA	Angiostatin release for cancer therapy	Visted et al., 2003
	Rat bone marrow cells	Cell proliferation in a 3D scaffold	Wang et al., 2003
RGDS[b] chitosan	Rat osteosarcoma cells	Regeneration of bone-like tissue	Ho et al., 2005

[a] PVA = poly(vinyl alcohol); [b] RGDS = Arg-Gly-Asp-Ser; [c] NPAE = N-palmitoylaminoethyl

and potassium poly(vinyl alcohol) sulphate, and trimethylammonium glycol chitosan iodide.

Carrageenans: Carrageenans possess a backbone of alternating 1,3-linked β-D-galactose and 1,4-linked α-D-galactose. Differences in structure arise from the number and location of ester sulphate groups on these sugars and the extent to which the 1,4-linked residues exist as the 3,6-anhydro derivative. Gelation can be achieved by cooling or by contact with a solution containing gel-inducing reagents such as K^+, NH_4^+, Ca^{2+}, Cu^{2+}, Mg^{2+}, Fe^{3+}, amines, and water-miscible organic solvents. Both procedures are easy to perform and facilitate high viable cell content.

There are three main types of carrageenan, namely: lambda (λ)-, kappa (κ)-, and iota (ι), all of which are extracted from red seaweeds; with the κ-form considered to be the most suitable for cell immobilization. κ-Carrageenan gel can easily be produced in different shapes (bead, cube, membrane) according to the particular application. Bead formation can be accomplished by the dripping technique. The resonance nozzle technique has also been used to produce beads of a more spherical and uniform shape and to scale up the process (Buitelaar et al., 1990).

Agar and agarose: Agar and agarose are polysaccharides isolated from marine red algae. Two kinds of agar can be distinguished by their gelling temperatures and their methoxyl content. While the first type gels in the 40s (°C) and contains methyl ether groups, the second type gels in the low to mid-30s (°C) and is essentially devoid of methyl ether groups.

Agarose represents the basic gel-forming component of agar. Agarose is a linear polysaccharide composed of repeating agarobiose units consisting of alternating 1,3-linked β-D-galactopyranose and 1,4-linked 3,6-anhydro-α-L-galactopyranose, i.e., L-galactose anhydride. The other species present in agar include compounds derived from the β-D-galactose by substitution of anhydride in position 6, e.g., the 6-methyl ether and 6-sulphate, and the compounds obtained by substitution of anhydride in position 2 (the 2-sulphate derivative). Pentagonal pores are the essential feature of agarose supports. The pores are large enough to be readily penetrated by a protein with a molecular mass of several millions. The stability of the pores is dependent on the hydrogen bond formation between the strands of the triple helix of agarose chains. Disruption of these bonds results in dissolving the soluble monomeric agarose. Urea, guanidine hydrochloride, chaotropic agents, and certain detergents may disrupt the hydrogen bonds. The strength of the bonds and, therefore, the porosity and size of the beads, are altered by a change in ionic strength.

Agar or agarose (2–5% w/w) is dissolved in a suitable buffer/medium by heating followed by cooling to a temperature; 5–10°C above the gelling temperature, before mixing or adding the cells. The physical form can be cast as a sheet, bead, or cylinder. A mould can be used or a "gel-block" can be produced with subsequent mechanical disintegration into smaller particles. Spherical beads can also be produced by adding the molten preparation dropwise to ice-cold buffer, via emulsification in vegetable oil (Nilsson et al., 1987), or using the resonance nozzle immobilization technique (Buitelaar et al., 1990).

Chitin and chitosan: Chitin is a fibrous glucan derivative; 1,4-linked 2-deoxy-β-D-glucan, that is partially acetylated, is water insoluble, and contains amino groups. Chitosan is artificially deacetylated chitin that is water soluble with nitrogen contents higher than 7% (w/w). Chitosan can be formed ionotropically in a manner similar to that described earlier for alginate: gel formation will occur using a chitosan solution with a

COUNTERIONS
Ions of low relative
molecular mass, with a
charge opposite to that
of the colloidal ion, are
called counterions; if their
charge has the same sign
as that of the colloidal
ion, they are called
co-ions.

pH value lower than 6 to protonate the NH_2 groups and multi-valent anion counterions.

Cross-linking of chitosan with high-molecular-weight counterions results in capsules, while cross-linking with low-molecular-weight counterions results in globules in which the cells are entrapped. Both low-molecular-weight ions such as ferricyanide, ferrocyanide, and polyphosphates; and high-molecular-weight ions such as poly(aldehydocarbonic acid), poly(1-hydroxy-1-sulfonate-2-propene), and alginate can be used. With more hydrophobic counterions (like octyl sulphate, lauryl sulphate, hexadecyl sulphate, and cetylstearyl sulphate), it is also possible to produce hydrophobic gels. Bead formation in a cross-linking solution can also occur at a pH above 7.5, but in this case it is merely a precipitation of chitosan. At pH values greater than 7.5, chitosan is totally deprotonated and becomes water insoluble.

Combining chitosan with other polymers led to the development of a wide range of tissues such as bone, liver, neural tissue, vascular grafts, cartilage, and skin (Brown and Hoffman, 2002; Di Martino et al., 2005). Chitosan's ability to stabilize and deliver proteins has allowed the incorporation of growth factors into many of these structures, thus promoting angiogenesis in tissue-engineered structures.

Chitosan ionotropic gel beads are, unlike calcium alginate and potassium carrageenan, stable in phosphate buffered media, and their mechanical stability is comparable to that of Ca-alginate beads.

Pectins, pectates, and pectinates: Pectins are acidic, structural polysaccharides, which are present in the cell wall of plant cells. Although pectins are branched in their native form, when extracted they are predominantly linear polymers, based on a 1,4-linked α-D-galacturonate backbone, which is randomly interrupted by 1,2 linked L-rhamnose. Pectin is – like alginate – based on a diaxially linked backbone and forms gels with calcium ions. Gel formation studies indicate a very strong binding of these counterions. Due to the variability of the chemical structure of the commercially available pectins, gels can be formed in several ways. Polygalacturonic acid as the principal constituent is partly esterified with methoxyl groups. The free acid groups may be partly or fully neutralized with monovalent ions (i.e., Na^+, K^+, NH_4^+). The degree of methoxylation (DM) has an essential influence on the properties of pectin, especially on its solubility and its requirements for gelation, which is directly derived from the solubility. Pectins with a DM lower than 5% are called pectic acids, while those with higher DM value are called pectinic acids. Pectinic acids with a DM greater than 50% are called high-methoxyl (HM) pectins, and those with lower DM value are called low-methoxyl (LM) pectins.

The ionotropic gelation of pectate is simple, mild, and inexpensive. Dripping the polymer-cell solution into a cross-linking solution can easily produce beads with calcium or aluminium as counterions.

Calcium pectate as well as aluminium pectate beads are much less sensitive to small mono- and multivalent anions, i.e., citrate, phosphate, lactate, gluconate, and chloride, and Ca^{2+} (Al^{3+}) complexing agents, which diminish the stability of the beads. Stabilization and hardening of pectate beads can be performed using a treatment with polyethyleneimine followed by glutaraldehyde (Gemeiner et al., 1994).

Gellan gum: Gellan gum is a gel-forming polysaccharide secreted by the bacterium *Pseudomonas elodea* and is produced via aerobic fermentation. Gellan gum is a linear, anionic heteropolysaccharide with tetrasaccharide repeating units consisting of two β-D-glucose, one β-D-guluronic acid, and one α-L-rhamnose residue. The polymer contains approximately 1.5 acyl substituents per tetrasaccharide repeating unit. These substituents have been identified as an L-glyceric ester on C-2 of the 3-linked D-glucose and an acetic ester on C-6 of the same glucose residue. The presence of these substituents, in particular the bulky glycerate groups, hinders chain association and accounts for the change in gel texture brought about by de-esterification. The substituted form produces soft, elastic, and cohesive gels, whereas hard, firm, and brittle gels are obtained from the unsubstituted form. The rheological properties of unsubstituted gels are superior to those of other common polysaccharides such as agar, κ-carrageenan, and alginate at equimolar concentrations (Sanderson et al., 1989).

Gelation initially occurs by the formation of double helices followed by ion-induced association of these helices. Gel formation occurs when the fibrils associate in the presence of gel-promoting cations. Gelation depends upon the gum concentration, ionic strength, and the type of stabilizing cation (divalent are more effective than monovalent cations). A dispersion of gellan gum with a minimum amount of sequestrant and divalent cations will form a coherent, demouldable gel on cooling to ambient temperature. The gelation temperature increases from 35 to 55°C with increasing cation concentration.

Gellan gum is suitable for the immobilization of thermophilic bacteria because of its high setting temperature (over 50°C), which can be decreased by the addition of sequestrants such as citrate, metaphosphate, and EDTA, thus rendering it suitable for the immobilization of mesophilic bacteria (Camelin et al., 1993).

Hyaluronic acid: Hyaluronic acid is the largest glycosaminoglycan (GAG) found in nature. It can be readily isolated from abundant natural sources. It is well suited to be used in tissue engineering since it shows a minimal inflammatory or foreign body reaction upon implantation (Gutowska et al., 2001).

Collagen: Collagens are a family of highly characteristic fibrous proteins found in all multicellular animals. The central feature of all collagen molecules is their stiff, triple-stranded helical structure. Collagen chains are extremely rich in glycine and pro-line, both of which are important in the formation of the stable triple helix. Collagen is hydrophilic; it swells in the presence of water, is soluble at low pH, and is insoluble at high pH values.

The mechanism of cell immobilization in collagen involves the formation of multiple ionic interactions, hydrogen bonds and van der Waals forces between the cells and collagen. Prepara-tion of the collagen solution and mixing with cells has to be per-formed at low temperatures (4°C). Gelification is accomplished by raising the pH and ionic strength of the collagen solution and exposure to 37°C. Its natural ability to bind cells makes it a promising material for controlling cellular distribution within immunoisolated devices, and its enzymatic degradation can pro-vide appropriate degradation kinetics for tissue regeneration in micro- and macroporous scaffolds (Riddle and Mooney, 2004). Although collagen is widely used in cell immobilization, it is expensive to purify for use in tissue engineering.

Gelatine: Gelatine is a hydrolytic derivative of collagen. Gelifi-cation is accomplished by cooling the gelatine solution below a temperature of 30 to 35°C. The sol-gel transformation is revers-ible. The gel structure can be stabilized by adding organic, such as glutaraldehyde or formaldehyde, or inorganic (e.g., acetate or sulphate chromium salts) compounds (Sungur and Akbulut, 1994). The three-dimensional structure of gelatin is formed by secondary interactions between the polypeptide chains. These interactions are broken upon heating. The stabilization by alde-hydes is based on covalent bond formations between the gela-tine strands.

In a typical procedure, the cell suspension is mixed with an aqueous gelatin solution at 40°C, the suspension is cooled, and the gel lyophilized. Subsequently, the dry preparation is disin-tegrated into small particles. The glutaraldehyde treatment can be performed before the cooling of the suspension or after the lyophylization process. Uniform beads can be formed by dripping the hot suspension in a hydrophobic liquid (e.g., butyl acetate).

10.2.2.1.2   Hydrogels from synthetic polymers

The application of synthetic polymers for the immobilization of living cells has some interesting benefits compared to the use of natural polymers, since synthetic polymers of adequate properties can be easily and artificially designed. The porosity of the gel as well as the ionic and hydrophobic or hydrophilic properties can be easily adjusted. Additionally, the mechanical strength and longevity of the gels formed from synthetic polymers are generally superior to those from natural polymers.

Polyacrylamide: The first synthetic gel used to entrap living microbial cells was polyacrylamide. Entrapment is performed by the polymerization of an aqueous solution of acrylamide monomers in which the cells are suspended. Polymerization of polyacrylamide is a free radical process in which linear chains of polyacrylamide are cross-linked by inclusion of a bifunctional reagent (e.g., N,N'-methylene-bisacrylamide). The cross-linking degree is a function of the relative amounts of acrylamide and bifunctional reagent and determines the porosity and fragility of the gel. Initiation of the free radical polymerization can be performed by a chemical or a photochemical reaction. Persulphate and β-dimethylaminopropionitrile or N,N,N',N'-tetramethylenediamine (TEMED) are chemical catalysts, whereas sodium hydrosulphite, riboflavin, and TEMED act as photochemical polymerization initiators.

It is an easy immobilization technique, but the polymerization of the acrylamide monomers in the presence of viable cells usually results in a reduction of the viability of the entrapped cells due to the toxicity of the monomers (e.g., acrylamide and bisacrylamide) and the heat evolved during polymerization. The level of the "immobilization shock" of polyacrylamide-entrapped cells depends to a large extent on the initial physiological state of the population (Lusta et al., 1990).

Polyacrylamide-hydrazide: To eliminate the unfavorable influences caused by the acrylamide monomer, techniques have been developed that use prepolymerized linear polyacrylamides, partially substituted with acylhydrazide groups, for the entrapment of living cell with good retention of viability. The prepolymerized material is cross-linked in the presence of viable cells by the addition of controlled amounts of dialdehydes (glyoxal, glutaraldehyde, periodate-oxidized polyvinylalcohol). The porosity of these gels is affected by the cross-linking agent. The best results were obtained using glyoxal. The concentration of polymeric backbone also affects gel porosity. The mechanical stability of this gel is superior to gels made with similar concentrations of polymeric backbone from acrylamide-bisacrylamide copolymerization. This polyacrylamide-hydrazide (PAAH) gel is less brittle, is chemically stable, and does not undergo deformation as a result of changes in salinity or pH.

Methacrylates: The preparation of methacrylate gels is analogous to that of polyacrylamide gels. Methacrylate monomers such as methylacrylamide, hydroxyethylmethacrylate, or methylmethacrylate are polymerized in the presence of a cross-linking agent (e.g., tetraethyleneglycol dimethacrylate) to form a porous gel.

Poly(ethylene glycol) dimethacrylate has also been used as a cross-linking agent for the entrapment of biocatalysts by radical polymerization of acrylic acid and N,N-dimethylaminoethyl methacrylate. Living cells can also be immobilized in methacrylate by γ-ray irradiation at low temperatures (Carenza and Veronese, 1994).

Photo-cross-linkable resin prepolymers: Illumination with near-ultraviolet light of photo-cross-linkable resin prepolymers initiates radical polymerization of the prepolymers and completes gel formation within 3 to 5 minutes. Various types of prepolymers possessing photosensitive functional groups have been developed. Poly(ethylene glycol) dimethacrylate (PEGM) was synthesized from poly(ethylene glycol) (PEG) and methacrylate. ENT and ENTP were prepared from hydroxyethylacrylate, isophorone diisocyanate, and poly(ethylene glycol) or poly(propylene glycol), respectively. Each prepolymer has a linear skeleton of optional length, at both terminals of which are attached the photosensitive functional groups such as acryloyl or methacroyl. PEGM and ENT containing PEG as the main skeleton are water-soluble and give hydrophilic gels, while ENTP with poly(propylene glycol) as the main skeleton is water-insoluble and forms hydrophobic gels. Using PEG or poly(propylene glycol) of different molecular weight, prepolymers of different chain length can be prepared: from PEGM-1000 to PEGM-4000 (MW of main chain from ~1000 to 4000, respectively), ENT-1000 to ENT-6000, and ENTP-1000 to ENTP-4000. The chain length of the prepolymers correlates to the size of the network of gels formed from these prepolymers. Anionic and cationic prepolymers can also be prepared by introducing anionic and cationic functional group(s) to the main skeleton of the prepolymers.

The entrapment of cells can be achieved by illumination of a mixture consisting of a prepolymer, a photo-sensitizer (such as benzoin ethyl ether or benzoin isobutyl ether), and the cell suspension. A suitable buffer is used for the hydrophilic prepolymer and an adequate organic solvent for the hydrophobic prepolymer. Suitable mixtures of these two types of prepolymers can also be utilized. In some cases, a detergent is employed to mix a hydrophobic prepolymer with the suspension of biocatalysts.

Laser-induced photo-polymerization of poly(ethylene glycol) diacrylates and multiacrylates has been used to entrap living mammalian cell. These molecules of various molecular weights were synthesized by reaction of PEG with acryloyl chloride, using triethylamine as a proton acceptor. Tetrahydroxy PEG was used for the multiacrylate synthesis. The advantages of this entrapment method are that the laser light is not absorbed by the cells in the absence of an exogenous, cell-binding chromophore;

there is no significant heat of polymerization due to the nature, size, and dilution of the macromers used; the polymerization can proceed extremely rapidly in oxygen-containing aqueous environments at physiological pH; gels with the proper formulation are capable of being immunoprotective and can be used for cell therapy purposes.

Synthetic resin prepolymers have the following advantages as gel-forming starting materials:

- Entrapment procedures are very simple under very mild conditions;
- Prepolymers do not contain monomers that have adverse effects on the biocatalysts to be entrapped;
- The network structure of gels can be controlled by using prepolymers of optional chain length;
- Optional physicochemical properties of gels, such as hydrophobicity-hydrophilicity balance and ionic nature, can be changed by selecting suitable prepolymers, which had been synthesized in advance in the absence of biocatalysts.

Polyurethane: Urethane prepolymers have isocyanate groups at both terminals of the linear chain and are synthesized by heating for 1-2 hours at 80°C from toluene diisocyanate and polyether diols composed of poly(ethylene glycol) and poly(propylene glycol) or poly(ethylene glycol) alone. Prepolymers with a different hydrophilic or hydrophobic character can be obtained by changing the ratio of poly(ethylene glycol) and poly(propylene glycol) in the polyether diol moiety of the prepolymers. The chain length and the content of isocyanate group can also be changed.

Cells can be entrapped by the "self-cross-linking" gel. The prepolymers are water-miscible and when a liquid prepolymer is mixed with an aqueous cell suspension, the isocyanate functional groups at both terminals of the molecule react with each other only in the presence of water, forming urea linkages with liberation of carbon dioxide. Cells have also been entrapped in conventional polyurethanes, which were obtained by polycondensation of polyisocyanates. The polyurethane can be made in foam or in a gel structure depending on the type and concentration of the polycyanate used.

Poly(vinyl alcohol):   Poly(vinyl alcohol) (PVA) is a raw material of vinylon and is a low-cost material. PVA is nontoxic to micro-organisms and, consequently, can be used to entrap living cells. A PVA solution becomes gelatinous by freezing and the gel strength increases during iterations of freezing and thawing. Using this technique, a rubber-like, elastic hydrogel can be obtained without using any chemical reagent. The gel strength increases with iteration number of freezing-thawing until seven iterations. A decrease of activity due to the freezing and thawing can be prevented by adding cryoprotectants such as glycerol and skim milk to the PVA-cell solution. PVA cryogels can be used up to a temperature of 65°C and have also been used to immobilize thermophilic micro-organisms like *Clostridium thermocellum*, *C. thermosaccharolyticum,* and *C. thermoautotrophicum* to perform fermentations at 60°C (Varfolomeyev et al., 1990).

Elastic PVA gels with a high strength and durability can also be formed by cross-linking PVA with a boric acid solution to produce a monodiol-type PVA-boric acid gel lattice. This technique has two potential problems. First, the saturated boric acid solution used to cross-link the PVA is highly acidic and could cause difficulty in maintaining cell viability. Second, PVA is an extremely sticky material. As a result, PVA beads have a tendency to agglomerate, which can cause problems in fluidized-bed reactors. This latter problem can be solved by using a combination of PVA-boric acid and a small amount of calcium alginate (0.02%) (Wu and Wisecarver, 1992).

Chen and Lin (1994) developed a method based on the usage of phosphorylated PVA, where PVA was first cross-linked with boric acid for a short time to form a spherical structure, which was followed by solidification of the gel beads by esterification of PVA with phosphate. The short contact time with boric acid prevented severe damage to the entrapped micro-organisms.

Recently, PVA has been used to immobilise cells in Lentikats®, which are lens-shaped hydrogel particles (Wittlich et al., 2004). Due to their lenticular shape (diameter of 3–4 mm and a thickness of 200–400 µm), Lentikats® have the advantage of improved mass transport compared to beads with the same diameter. They are prepared by mixing a sol solution with the biocatalyst solution, and small droplets are floored on a suitable surface where the gelation takes place. These particles have also been produced on an industrial scale (production output between 2 and 50 kg/h) using a conveyer-belt system.

Photosensitive resin prepolymers: Photosensitive resin prepolymers are derivatives of poly(vinyl alcohol) (PVA) introduced by styrylpyridinium (SbQ) groups as photosensitive sites and are polymerized by photo-dimerization with irradiation of visible or ultraviolet light (PVA-SbQ gel). Hydrophilicity of the prepolymers can be controlled by changing the saponification degree of poly(vinyl alcohol).

Radiation polymers: Living cells can be entrapped by γ-irradiation of a wide variety of functional monomers or prepolymers at low temperature. Utilization of this method is limited because radiation equipment is required. Irradiation of cells at low temperatures is necessary to avoid the radiation damage. The rigidity of the polymer matrix can be increased and the porosity decreased, by increasing the monomer concentration.

Stimuli-sensitive hydrogels: Stimuli-sensitive polymer hydrogels, which swell or shrink in response to changes in he environmental conditions, have been extensively investigated and used as "smart" biomaterials and drug-delivery systems, e.g., copoly(N-isopropylacrylamide/acrylamide) (NIAAm/AAm), which is a "lower critical solution temperature" (LCST) hydrogel (Hoffman, 2004). This gel gradually shrinks as temperature is raised and then collapses when it is warmed through the LCST region. It expands and reswells as it is cooled below LCST. Cells containing gel beads have been prepared by inverse suspension polymerization. The conversion and activity of the immobilized cells may be enhanced by thermal cycling of the gel below its LCST, due to the reduced mass transfer resistance as the gel bead "squeezes out" and "draws in" substrate when it shrinks and swells during the thermal cycling. Pore sizes and their interconnections will change significantly as the gel is shrunk or swelled. This can significantly affect the diffusion rates of substrate in and product out of the gel.

10.2.2 2 *Preformed support materials*

Cell immobilization in preformed carriers involves passive/natural immobilization usually *in situ* in the bioreactor or the culture environment (Baron and Willaert, 2004; Mavituna, 2004). Most of the carriers are porous with a wide range of pore sizes to suit immobilization of various organisms or tissues. For passive immobilization, cells, flocs, mycelia, cell aggregates, or spores are inoculated into the sterilized medium containing empty preformed carriers. Depending on the cell and the carrier type, immobilization then takes place in a combination of filtration, adsorption, growth, and colonization processes. Furthermore, surfaces of the carriers can be modified by various pretreatments to enhance immobilization efficiency.

The cells are entrapped in a matrix that protects them from the shear field outside the particles. This is of particular importance for fragile cells such as mammalian cells. Unlike gel systems, porous supports can be inoculated directly from the bulk medium. As with the adsorption method, cells are not completely separated from the effluent in these systems. Mass transport of substrates and products can be achieved by molecular diffusion as well as convection, by proper particle design and organisation of external flow. Consequently, mass transport limitations are less severe under optimal conditions. When ideally, the colonized porous matrix retains some free space for flow, immobilization occurs partly by attachment to the internal surface, self-aggregation, and retention in dead-end pockets within the material. This is only possible when cell adhesion is not very strong and the application of high external flow rates reversibly removes cells from the matrix. When high cell densities are obtained, convection is no longer possible and the cell system behaves as dense cell agglomerates with strong diffusion limitations. Cell immobilization methods are simple and a high degree of cell viability is retained upon entrapment. The preformed matrix is chemically inert, resistant to microbial attack and incompressible. Often steam sterilization is possible and the matrix can be reused. Usually, the matrix takes up a significant volume fraction resulting in a lower immobilized cell density compared to other immobilization methods.

Various porous matrices have been described for living cell immobilization as shown in Table 10.4. The choice will usually depend on the cell type used and the kind of application. For example, the immobilization of microbial cells for the production of biochemicals in large packed-bed bioreactors requires a matrix with excellent mechanical characteristics to withstand the high-pressure drop in the reactor; tissue engineering porous matrices need to be an excellent scaffold for cell attachment and growth and must be characterized by excellent biocompatibility characteristics. Usually, guided by the following parameters, a choice has to be made from a number of suitable matrices:

- What matrix material is the cheapest?
- Which matrix is the most suitable for our purpose, and is it not patented?
- Is it reusable?

In tissue engineering, some applications require porous biodegradable scaffolds. The repair of large cartilage defects requires the use of a three-dimensional scaffold to provide a structure for cell proliferation and control the shape of the regenerated tissue. For example, cartilage implants based on chondrocytes and 3-D fibrous polyglycolic acid scaffolds closely resembled normal cartilage histologically, as well as with respect to cell density and tissue composition (Freed et al., 1994).

10.2.3   *Containment behind a barrier*

When cell separation from the effluent is required or when some high-molecular-weight or specific product (permselectivity) needs to be separated from the effluent, these systems are highly useful. The barrier can be either preformed (hollow fiber systems and flat membrane reactors) or formed around the cells to be immobilized (microcapsules and two-phase entrapment). The synthetic membranes are usually polymeric microfiltration or ultrafiltration membranes, although other types of membranes have been used such as ceramic, silicone rubber or ion exchange membranes. Mass transfer through the membrane is not only dependent on the pore size and structure but also on the hydrophobicity/hydrophilicity and charge. Transport can be

Table 10.4 Examples of cell immobilization in preformed porous support materials

Material	Micro-organism	Application/product	Reference
Natural organic polymers			
Bacteria			
Cotton towel	*Propionibacterium acidipropionici*	Propionic acid production	Suwannakham and Yang, 2005
Fungi			
Cellulose carrier	*Rhizopus niveus*	Wax esters production	Chen and Wang, 1997
Loffa sponge	*Saccharomyces cerevisiae*	Ethanol production	Ogbonna et al., 1996
Synthetic organic polymers			
Bacteria			
Silicone carrier (ImmobaSil®)	*Lactobacillus rhamnosus*	Exopolysaccharide production	Bergmaier et al., 2005
Fungi			
Polystyrene foam	*Phanerochaete chrysosporium*	Peroxidase production	Urek and Pazarlioglu, 2004
Polyurethane foam	*Agaricus* sp.	Laccacase production	Kaluskar et al., 1999
	Aspergillus niger	Citric acid production	Jianlong, 2000
	Yarrowia lipolytica	Oil degradation	Oh et al., 2000
Mammalian cell			
Polyester fibrous matrix	Hybridoma cells	Monoclonal antibody production	Yang et al., 2004
Polyethylene terephthalate	Human trophoblast	Tissue engineering	Ma et al., 2000
Anorganic materials			
Bacteria			
Kiezelguhr (Celite)	*Xanthomonas campestris*	Xanthan gum production	Nilsson et al., 1987
Fungi			
Ceramics	*Saccharomyces cerevisiae*	Ethanol production	Demuyakor and Ohta, 1992
Kiezelguhr (Celite)	*Penicillium chrysogenum*	Penicillin production	Gbewonyo et al., 1987
Porous glass	*Saccharomyces cerevisiae*	Beer maturation	Back et al., 1998
Metallics			
Bacteria			
Allumina pellets	*Zymomonas mobilis*	Ethanol production	Koutinas et al., 1988
Stainless steel knitted mesh	*Zymomonas mobilis*	Levan and ethanol production	Bekers et al., 2001
Fungi			
Stainless steel fiber cloth	*Saccharomyces cerevisiae*	Beer production	Shen et al., 2004
Stainless steel knitted mesh	*Phanerochaete chrysosporium*	Peroxidase production	Gerin et al., 1997

STARLING FLOW
In a hollow fiber reactor, some fluid will flow from the lumen into the extracapillary space (ECS) in the entrance half of the reactor, along the fibers in the ECS, and return to the lumen in the exit half of the reactor, because of the pressure gradient in the lumen (highest pressure at the entrance). This type of flow is called Starling (or toroidal) flow, in honor of the discoverer of this same fluid behavior in tissues surrounding blood capillaries.

by diffusion and/or by flow induced by application of a pressure difference over the membrane. Various mild micro-encapsulation methods have been developed to entrap living cells, and some are combined with gel immobilization within the microcapsule. The barrier can be as simple as the liquid/liquid phase interface between two immiscible fluids.

Entrapment behind preformed membranes represents a gentle immobilization method since no chemical agents or harsh conditions are employed. Cells are often immobilized by filtration of a cell suspension followed by some growth in the seeded reactor. Two-phase systems can be used in applications where substrates or products are partitioned separately (product inhibition, water-insoluble substrates). These systems also allow the recycling of the cell-containing phase, which is difficult with other immobilization methods. The small spheres involved in phase entrapment have a superior surface-to-volume ratio compared to flat sheets or hollow fibers and can be used in conventional bioreactors. Also membranes have been used to contact aqueous and organic streams in bioprocesses. The high membrane surface area in hollow fiber reactors is especially advantageous for the culture of anchorage-dependent mammalian cells, without the drawbacks occurring with immobilized growing micro-organisms (Sirkar and Kang, 2004). The maintenance of high-density cultures of animal and plant cells in membrane reactors resulted in long-term production with reduced cell growth. Membrane entrapment may be particularly helpful if aggregation is an advantage, as suggested for plant cells. In the biomedical engineering field, micro-encapsulation or macro-encapsulation leads to immunoprotection and artificial organs (Lysaght and Rein, 2004). In most cases, nutrients are supplied to, and products are removed from, the cell mass by diffusion. Consequently, mass transfer limitations can reduce the efficiency of these systems. In the case of hollow fiber systems, however, mass transfer can be governed by convective mass transport; otherwise known as "Starling flow."

In these systems, the cells are usually situated in the membrane porous support layer or the shell space and medium flows in the lumen. Alternatively, the cells can be contained in the lumen space in the case of macro-encapsulation (also known as "diffusion chambers") for encapsulated cell therapy such as diabetes, alleviation of chronic pain, treatment of neurodegenerative disorders, delivery of neurotrophic factors (Goosen, 1993). Here, the hollow fiber membrane allows the diffusion of low-molecular-weight molecules (lower than the molecular weight cut-off value) — like oxygen, glucose, nutrients, and waste product — but prevents the passage of larger molecules like antibodies.

MOLECULAR WEIGHT CUT-OFF
The molecular weight cut-off (MWCO) of a membrane is defined as the molecular weight at which the membrane rejects 90% of solute.

10.2.3.1 Micro-encapsulation

A typical micro-encapsulation process involves the formation of a spherical gel mold containing cells, on which is deposited a polymeric membrane. The internal gel matrix can also be lique- fied and allowed to diffuse out of the capsule, leaving behind the membrane and the contained cells. The type and porosity of the membrane and size of the microcapsules can be varied to accommodate many reactant-product systems. Capsule diame- ters from 20 µm to 2 mm are possible. The porosity of the mem- brane can be varied over a range of several orders of magnitude such that on the low end, molecules as small as glucose (180 Da) can be constrained to remain intracapsular while on the high end molecules as large as IgG (155,000 Da) can be made to be freely permeable. The polymer membrane should offer minimal resistance to the mass transfer of essential molecules as well as the toxic end products of cell metabolism in order for the encap- sulated mass to maintain normal physiological activity. For immunoisolation purposes, the capsule must be impermeable to host cells and soluble components of the immune system.

> IMMUNOISOLATION
> By using membranes, the free passage of immunoglobulins and complement proteins can be prevented from interacting with implanted "foreign" biological material.

The demand for specific properties of the capsules may vary very much depending on the system where it is to be used (Strand et al., 2004). Proliferating cells may need a stronger capsule mem- brane than nonproliferating cells, but the former may tolerate a tougher encapsulation procedure as viable cells will grow and replace the dead cells in the capsules. The criteria for wide indus- trial acceptance of immobilization technology may be different than for biomedical applications when it comes to safety, simplic- ity, long-term immobilization, activity, and price. Also, the free- dom to choose stabilizing surrounding solution may vary with the use of capsules, e.g., in continuous reactors and in implantation.

This immobilization technique is of particular interest for the immobilization of animal cells. Large process intensifica- tion over conventional cell suspension culture with low den- sity and low productivity can be achieved. Cell encapsulation and long-term continuous culture lead to significantly higher cell densities, which results in higher productivities. The high- culture densities provide a high degree of cell-cell contact and interaction, resulting in possible more favorable micro-envi- ronmental conditions. In addition, this technique can provide protection from shear for the sensitive animal cells and other sudden changes in the culture medium. It also permits direct aeration by air bubbles without risk of damaging the cells. The produced toxic metabolites, such as lactic acid and ammonium, will diffuse out of the capsule due to the concentration gradient, resulting in higher growth and product formation rates. Micro- encapsulation can provide simultaneous product separation and cell cultivation, resulting in concentration of high-molecular- weight metabolic products (e.g., monoclonal antibodies) within the capsule. The preconcentration of products within the capsule facilitates further purification steps. The immobilized cells are totally separated from the culture medium, which results in an easier and cost-effective downstream processing. This is not true for conventional gel immobilization growing cell systems, where cell release from the matrix is observed.

Micro-encapsulation of human cells or tissues is a recent technology to overcome biomedical problems, because the membrane may create an immunological barrier between the host and the transplanted cells. The immunoisolation of the encapsulated cells or tissue from the elements of the immune system prevents the rejection of the transplanted cells/tissue. Consequently, the necessity of immunosuppressive drugs in allo- and xenotransplantations can be avoided. The idea of using an ultrathin polymer membrane for the immunoprotection of transplanted cells was first implemented by T.M.S. Chang in the early 1960s (Chang, 1964). Despite its apparent promise in preclinical trials, it failed to live up to expectations and continues to elude the scientific community (Orive et al., 2004).

Micro-encapsulation techniques: The techniques, which are used to produce the semipermeable microcapsule membranes, are classified as phase inversion, polyelectroyte coacervation, and interfacial precipitation (Hunkeler, 1997). Phase inversion involves the induction of phase separation in a previously homogeneous polymer solution by a temperature change or by exposing the solution to a nonsolvent component either in a bath (wet process) or in a saturated atmosphere (dry process). Polymer precipitation time, polymer-diluent compatibility, and diluent concentration all influence phase separation and membrane porosity. Examples of polymer materials are polyacrylate and poly(hydroxyethylmethacrylate-co-methyl methacrylate) (Feng and Sefton, 2002).

In the polyelectrolyte coacervation process, a hydrogel membrane is formed by the complexation of oppositely charged polymers to yield an interpenetrating network (Strand et al., 2004). Mass transport characteristics can be modulated by osmotic conditions, diluents, and the molecular-weight distribution of the polyionic species (Chaikof, 1999). To reduce membrane permeability and to improve biocompatibility as well as mechanical properties, one or more additional coating layers with oppositely charged polymer can be added.

The interfacial precipitation technique involves the coating of a hydrogel bead with a semipermeable membrane (Lacik, 2004). A lot of research has been focused on the alginate-poly-L-lysine (PLL)-polyethyleneimine (PEI) microcapsule, originally developed by Lim and co-workers in the early 1980s (Lim and Sun, 1980; Lim and Moss, 1981). Immunoisolation of the islets of Langerhans (bioartificial pancreas) has been studied extensively and evaluated thoroughly. Examples of applications of cell immobilization in microcapsules are listed in Table 10.5.

10.2.3.2 Cell immobilization using membranes

The main advantage of membrane bioreactors is that they provide simultaneous bioconversion and product separation, which is especially attractive for the production of high-value biological molecules (Obradovic et al., 2004). As compared to conventional reactor types, the design of membrane reactors is relatively more complex and more expensive (due mainly to the high cost of the membrane material). However, since the attainment of highly concentrated products could eliminate the need for some steps of costly product purification, the utilization of these reactors can be favorable. For low-value biological products, conventional bioreactors are usually more appropriate. An exception is the case where co-factors need to be co-entrapped to perform co-enzyme-dependent bioconversion reactions.

Membrane reactors can be configured as flat sheet or hollow fiber modules. Hollow fiber modules provide a higher surface to volume ratio without the need for membrane support. However, the geometry of flat sheet modules is simpler, providing an accurate regulation of the distances between the membranes. Additionally, these modules can be easily disassembled, providing an easy access to module compartments and options for membrane cleaning and replacement.

Membranes can be used for three types of cell immobilization as illustrated in panels A, B and C of Figure 10.4, i.e. immobilization within the membrane, immobilization on the membrane where a biofilm is formed, immobilization in a cell compartment, which is separated by a membrane.

Immobilization within the membrane: Cell immobilization in a membrane can give very high cell densities. The permselective membrane permits the transport of nutrients from the bulk medium to the cells and the removal of products, while the release of cells into the bulk liquid is prevented. Various set-ups can be used, and depending on the membrane type even convective transport through the membrane is possible. Preformed asymmetric polymetric membranes are usually used. A hollow fiber module consists of a bundle of porous hollow fiber membranes potted at both ends in a cylindical module. They were originally developed for separation processes.

Immobilization on the membrane: Membranes can act as support for biofilm development with direct oxygen transfer through the membrane wall in one direction and nutrient diffusion from the bulk liquid phase into the biofilm in the other direction. This type of immobilization has successfully been employed in aerobic cultivation of mammalian and microbial cells.

Immobilization in a cell compartment: In this case, a high suspended cell concentration is obtained by preventing the cells to escape from the bioreactor. Membrane filters, which can be positioned internally (e.g. a spin filter for the continuous cultivation of mammalian cells) or externally, are used to keep the cells in the

Table 10. 5 Examples of cell immobilization in microcapsules

Material	Cell type	Application	Reference
Bacterial cells			
Alginate-PLL[e]-alginate	*Escherichia coli*	Urea and ammonia removal	Prakash and Chang, 1995
Alginate-PLL[e]-alginate	*Erwinia herbicola*	Tyrosine production	Lloyd-George and Chang, 1993.
Alginate-PLL[e]-alginate	Various bacteria	Therapeutic delivery of live bacteria	Chang, 2005; Prakash and Jones, 2005
Alginate-alginate	*Lactococcus lactis*	Bacteriocin production	Scannell et al., 2000
Cellulose acetate phthalate	*Lactobacillus acidophilus* + *Bifidobacterium lactis*	Probiotics	Favaro-Trindade and Grosso, 2002
Fungal cells			
1,6-hexanediamine-poly-(allylamone)-dodecane-dioyl dichloride	*Saccharomyces cerevisiae*	Bioconversion in organic solvents	Green et al., 1996
CS-PDMDAAC[f]	*Yarrowia lipolytica*	Citrate production	Förster et al., 1994
Animal cells			
Alginate/agarose-PLL-alginate	BHK fibroblast, C_2C_{12} myoblast	Viability assessment	Orive et al., 2003

Alginate-PLL-alginate	Engineered mouse myoblast C_2C_{12}sFvIL-2	Tumor suppression	Cirone et al., 2002; 2003; 2004; 2005
Alginate-PLL-alginate	Engineered C_2C_{12} myoblast to secrete erythropoietine	In vivo erythropoietine delivery	Orive et al., 2005
Alginate-PLL-alginate	Engineered 293-EBNA JN3 myeloma	Endostatin production Hepatocyte growth factor production	Rokstad et al., 2002
Alginate-PLL-alginate	Islets of Langerhans	Diabetes treatment	de Vos et al., 2003
Alginate-PLL-alginate	Murine fibroblast	High throughput GMP encapsulation using JetCutter technology	Schwinger et al., 2002
Alginate/agarose/cellulose sulfate/pectin-PLL-alginate	GDNF[a] secreting 3T3 fibroblast	Treatment central nervous system diseases	Ponce et al., 2005
Alginate-poly-L-ornithine-alginate	HEK 293, HCT 116 and Hep G2 cell spheroids	Cell implantation	Leung et al., 2005
Alginate-CS[b]-pDADMAC[c]	Engineered CHO	Erytropoietine production	Weber et al., 2004
Collagen-HEMA-MMA-MAA-MeOCPMA[d]	Rat hepatocytes	Bioartificial liver	Quek et al., 2004

[a] GDNF = glial cell line-derived neurotrophic factor; [b] CS = cellulose sulphate; [c] pDADMAC = poly-diallyl-dimethyl-ammoniumchloride; [d] HEMA-MMA-MAA-MeOCPMA = hydroxyethylmethacrylate-methacrylate-mathacylic acid-4-(4-methoxycinnamoyl)phenyl methacrylate; [e] PLL = poly-L-lysine; [f] PDMDAAC = poly(dimethyldiallylammonium chloride)

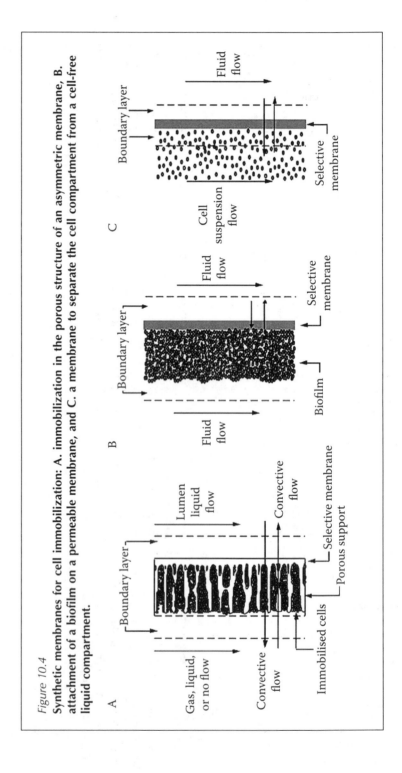

Figure 10.4

Synthetic membranes for cell immobilization: A. immobilization in the porous structure of an asymmetric membrane, B. attachment of a biofilm on a permeable membrane, and C. a membrane to separate the cell compartment from a cell-free liquid compartment.

reactor. Alternatives for an external membrane filter are a centrifuge or a settling tank (Table 10.6).

Very high cell densities (and high productivities) have been reported for membrane-based cell recycle systems. These systems have also the advantage of being homegeneous. Although none of the intrinsic benefits of cell immobilization is obtained (e.g. shear protection), most other advantages such as high productivity, avoiding washout and simpler product recovery are retained. Some of the drawbacks of immobilization, and especially the substrate transport limitation or product inhibition can be avoided or reduced. Oxygen transport can be a major problem at high cell densities and limits the attainable cell densities.

10.2.4 Self-aggregation of cells

Cells that naturally aggregate, clump, form pellets or flocculate can also be considered as immobilized (Figure 10.3). Many industrially important products are produced during secondary metabolism by fungal pellets, e.g., antibiotics using *Penicillium* species (Elander, 2003). Microbial aggregates can be encountered in wine-making and brewing, where yeast cells flocculate at the end of fermentation. The culturing of algae, plant cells, and animal cells can also result in aggregation phenomena. Simple serum-free medium appeared to be adequate to support the growth of anchorage-dependent animal aggregates of several commercially important cell types such as African Green Monkey Cells (Vero), Baby Hamster Kidney Cells (BHK), and Chinese Hamster Ovary Cells (CHO) (Litwin, 1992). During callus culture — in the absence of shear fields — aggregates may reach several centimetres across. Consequently, plant cell aggregates are very susceptible to the hydrodynamic conditions in the bioreactor.

The cell-wall region is influenced by biological and environmental factors directly and indirectly through metabolism. Biological factors that affect microbial aggregation are cell wall, extracellular secretions, genetics, growth rate, nutrition, and physiological age. Environmental factors can be subdivided into physical (hydrodynamic properties, interfacial phenomena, ionic properties, and temperature), chemical (presence of chelating agents, C/N ratio, enzymes, ferrocyanide, nitrogenous substances, oils, sugars and trace metals), and biological factors (inoculum size, presence of other organisms or strains). Artificial flocculating agents or cross-linkers may be added to enhance the aggregation process for cells that do not naturally flocculate. Polyelectrolytes, coupling agents by covalent bond formation, or inert powders can be used as linking agents.

An example of self-aggregation is the flocculation of yeast cells. Many fungi contain a family of cell wall glycoproteins (called "adhesines") that confer unique adhesion properties.

Table 10.6 Examples of cell immobilization using a membrane as a barrier

Material	Cell type	Application	Reference
Immobilization within the membrane			
Polypropylene	Bovine aortic endothelial cells	Artificial lung	Takagi et al., 2003
Polysulphone	*Saccharomyces cerevisiae*	Ethanol production	Inloes et al., 1985.
Polyvinyl chloride-polyvinylidine chloride-polyacrylonitrile	*Escherichia coli*	Production of β-lactamase	Inloes et al., 1983
Silicon carbide	*Saccharomyces cerevisiae*	Beer production	Andries et al., 2000
Immobilization on the membrane			
Polypropylene	*Aspergillus niger*	Citric acid production	Chung and Chang, 1988
Polylysine-coated polysulfone	H1 fibroblast	Aerobic cell growth	Tharakan and Chau, 1986
Immobilization in a cell compartment (cell recycle)			
Ceramic	*Bacillus stearothermophilus*	Lactic acid production	Danner et al., 2002.
Polydimethylsiloxane layer on polysolfone	*Saccharomyces cerevisiae*	Ethanol production	O'Brien and Craig, 1996
Polyethersulfone	*Leuconostoc mesenteroides*	D-mannitol production	von Weymarn et al., 2002
Polysulfone	*Halobacterium halobium*	Bacteriorhodopsin production	Lee et al., 1998
	Lactobacillus rhamnosus	Lactic acid production	Kwon et al., 2001

These molecules are required for the interactions of fungal cells with each other (flocculation and filamentation) (Viyas et al., 2003), inert surfaces such as agar and plastic (Reynolds and Fink, 2001), and mammalian tissues (Li and Palecek, 2003); they are also crucial for the formation of fungal biofilms (Green et al., 2004). The adhesin protein family responsible for its flocculation in *Saccharomyces cerevisiae* is encoded by *FLO1, FLO5, FLO9,* and *FLO10.* These proteins are called flocculins (Caro et al., 1997) because these proteins promote cell-cell adhesion to form multicellular clumps that sediment out of solution. Table 10.7 lists some examples of cell immobilization by self-aggregation.

10.3   Design of immobilized cell reactors

10.3.1   *Mass transport phenomena in immobilized cell systems*

The analysis of the influence of mass transfer on the reactor performance in immobilized-cell reactors can be quite useful since the performance of these reactors are often limited by the rate of transport of reactants to and products (external mass transfer limitation), and by the rate of transport inside the immobilized cell system (internal mass transfer limitation). External mass transfer limitations can be reduced or eliminated by a proper choice or design of the reactor and immobilized cell system. Internal mass transfer limitations are often more difficult to eliminate. An estimation of the significance of mass transport limitations is a prerequisite to optimise the performance of an immobilized-cell bioreactor.

10.3.1.1   *Diffusion coefficient*

Mass transport by molecular diffusion is defined by Fick's law, i.e., the rate of transfer of the diffusing substance through a unit area is proportional to the concentration gradient measured normal to the section:

$$J = -D\frac{\partial C}{\partial x} \tag{10.1}$$

where J is the mass transfer rate per unit area of a section, C the concentration of diffusing substance (amount per total volume of the system), x the space coordinate, and D the diffusion coefficient. It is general practice to use an effective diffusion coefficient (D_e), which can be readily used in the expression for the Thiele modulus and for the determination of the efficiency factor of a porous biocatalyst. When D is replaced by D_e in Eq.

Table 10.7   Examples of cell immobilization by self-aggregation of cells

Micro-organism	Application	Bioreactor type	Reference
Bacteria			
Zymomonas mobilis	Ethanol production	Fluidised-bed reactor	Scott, 1983
Fungi			
Aspergillus awamori	Enzyme production	Stirred tank reactor	Hellendoorn et al., 1998
Aspergillus oryzae	α-Amylase production	Stirred tank reactor	Carlsen, 1994
Penicillium chrysogenum	Penicillin production	Stirred tank reactor	Nielsen and Carlsen, 1996
Saccharomyces cerevisiae	Beer production	Cylindro-conical tank (batch)	Verstrepen et al., 2003
Trichoderma reesei	Cellulolytic enzymes	Stirred tank reactor	Lejeune and Baron, 1995
Mammalian cells			
BHK cells	Recombinant protein	Stirred tank reactor	Moreira and Alves, 1995
Neural stem cells	High-cell-density expansion	Stirred tank reactor	Kallos and Behie, 1999
Vero cells	Recombinant protein	Stirred tank reactor	Litwin, 1992
293 cells	Recombinant protein	Stirred tank reactor	Goetghebeur and Hu, 1991

(10.1), the corresponding concentration (C_L) is expressed as the amount of solute per unit volume of the liquid void phase. Concentration C may be correlated with C_L by using the void fraction (ε), which is the accessible fraction of a porous particle to the diffusion solute as $C = \varepsilon C_L$. Hence, the relationship between the effective diffusion coefficient and the diffusion coefficient is

$$D_e = \varepsilon D \qquad\qquad (10.2)$$

10.3.1.2   *Diffusion in immobilized cell systems*

The effective diffusion coefficient through a porous support material (matrix) is lower than the corresponding diffusion coefficient in the aqueous phase (D_a) due to the exclusion and obstruction effect. By the presence of the support, a fraction of the total volume $(1 - \varepsilon)$ is excluded for the diffusing solute. The impermeable support material obstructs the movement of the solute and results in a longer diffusional path length, which can be represented by a tortuosity factor (τ), which equals the square of the tortuosity (Epstein, 1989). The influence of both effects on the effective diffusion coefficient can be represented by

$$D_e = \frac{\varepsilon}{\tau} D_a \tag{10.3}$$

This equation holds as long as there is no specific interaction of the diffusion species with the porous carrier. In the case of gel matrices, predictions using the polymer volume fractions are recommended by the following equation, since neither ε nor τ can be measured for a gel in a simple way:

$$D = \frac{(1-\phi_p)^2}{(1+\phi_p)^2} D_a \tag{10.4}$$

where ϕ_p is the polymer volume fraction. For low-molecular-weight solutes in cell free gels, an approximate measure of ε can be given as

$$\varepsilon = 1 - \phi_p \tag{10.5}$$

D_e can also be expressed as a function ϕ_p by combining Eqs. (10.2), (10.4), and (10.5) to give

$$D_e = \frac{(1-\phi_p)^3}{(1+\phi_p)^2} D_a \tag{10.6}$$

10.3.1.3 External mass transfer

In the case of permeable spheres, the effect of the external mass transfer resistance on the overall uptake and/or release rate by the beads may be quantitatively evaluated by calculating the time constant for the external film (τ_e), and to compare it to the time constant for diffusion in the sphere (τ_i). The internal time constant can be calculated by the following equation:

$$\tau_i = \frac{R^2}{15D_e} \tag{10.7}$$

where R is the radius of the bead. The Biot number (Bi) for beads is defined by the following equation as the ratio of the characteristic film transport rate to the characteristic intraparticle diffusion rate:

$$Bi = \frac{k_s R}{D_e} \tag{10.8}$$

An estimation of the external mass transfer coefficient (k_s) is required to calculate τ_e, Bi, or the film thickness. The value of k_s can be calculated by a procedure recommended by Harriot and co-workers in the mid-1970s (Sherwood et al., 1975). Merchant and co-workers (1987) determined Bi for a rotating sphere. Using the empirical correlation of Noordsij and Rotte (1967), k_s could be estimated using the following equation:

$$Sh = \sqrt{4 + 1.21\left(Re_p Sc\right)^{0.67}} \qquad\qquad (10.9)$$

where Sh is the Sherwood number, Re_r the rotational number of Reynold, and Sc the Schmidt number. In the case of diffusion through a membrane or thin disc, Bi can also be calculated. For free-moving particles k_s can be determined using the following correlations (van't Riet and Tramper, 1991):

$$Sh = \sqrt{4 + 1.21\left(Re_p Sc\right)^{0.67}} \qquad \text{for } Re_p Sc > 10^4 \qquad (10.10)$$

$$Sh = 2 + 0.6 Re_p^{0.5} Sc^{0.33} \qquad\qquad \text{for } Re_p < 10^3 \qquad (10.11)$$

where Re_p is the (particle) Reynolds number, which can be estimated using the following correlations:

$$Re_p = \frac{Gr}{18} \qquad\qquad \text{for } Gr < 36 \qquad (10.12)$$

$$Re_p = 0.153 Gr^{0.71} \qquad\qquad \text{for } 36 < Gr < 8 \; 10^4 \qquad (10.13)$$

$$Re_p = 1.74 Gr^{0.5} \qquad\qquad \text{for } 8 \; 10^4 < Gr < 3 \; 10^9 \qquad (10.14)$$

where Gr is the Grashof number. Another correlation used to estimate k_s for gel beads in agitated reactors is (Kikuchi et al., 1988)

$$Sh = 2 + 0.52\left(e_s^{1/3} d_p^{4/3} / v\right)^{0.59} Sc^{1/3} \qquad\qquad (10.15)$$

where d_p is the average diameter of the particle, v is the kinematic viscosity, e_s is the energy dissipation given as $e_s = N_p n_i^3 D_i^5 / V$ for a stirred tank (where N_p is the power number, n_i the impeller speed, D_i the impeller diameter, and V the volume of the reactor). The ranges of validity for this correlation are

$$10 < \left(e_s^{1/3} d_p^{4/3} / v\right) < 1500 \quad \text{and} \quad 120 < Sc < 1450 \qquad (10.16)$$

Also, a correlation, which was originally developed for fluidised particles (Calderbank and Moo-Young, 1961), has been recommended for agitated dispersions of small, low-density solids (Øyaas et al., 1995):

$$k_s = \frac{2 D_{e0}}{d_p} + 0.31\left(Sc\right)^{-2/3} \left(\frac{\Delta \rho v g}{\rho_l}\right)^{1/3} \qquad\qquad (10.17)$$

where $\Delta\rho$ is the particle/liquid density difference and ρ_l the density of the bulk liquid. For spherical particles in a packed bed, k_s depend on the liquid velocity around the particles. For the range

$10 < Re_p < 10^4$, the Sherwood number has been correlated by the following equation (Moo-Young and Blanch, 1981):

$$Sh = 0.95 Re_p^{0.5} Sc^{0.33} \tag{10.18}$$

An estimation of k_s can be calculated if the stirred chambers have the shape of flat cylinders using the following correlation (Sherwood et al., 1975):

$$k_s = 0.62 D_a^{2/3} v^{-1/6} \omega^{1/2} \tag{10.19}$$

where v is the kinematic viscosity and ω the rotational speed of the stirrer (in rad/s). Other correlations can be adapted from heat transfer correlations (Axelsson and Westrin, 1991).

The external mass transfer limitation can be experimentally investigated by observing the concentration-time profile at different mixing regimes in the bioreactor (e.g., rotation speeds of the stirrer) (Sun et al., 1989).

10.3.2 *Reaction and diffusion in immobilized cell systems*

In immobilized cell systems, cellular reactions can take place in the presence of significant concentration gradients. These reactions are also called heterogeneous reactions. Reactions can only take place when the substrate molecules are transported to the reaction place, i.e., mass transport phenomena can have a profound effect on the overall conversion rate. The concentration in each internal position has to be known to determine the local rates. In most cases, these internal concentrations cannot be measured but can be estimated using a reaction-diffusion model.

10.3.2.1 *Reaction-diffusion models*

The major issues involved in modeling immobilized cell reactors are very similar to those in heterogeneous chemical reactors. This analogy has encouraged a rapid development in the model building for immobilized cell systems, even if the level of understanding of biocatalysts is lower than for chemical catalysis. The ability to predict the behavior of immobilized cell systems is required for the understanding, the design and optimization of an appropriate bioreactor. It is necessary to consider both the bioreactor performance and the microbial kinetics. Description of the bioreactor performance involves modeling of mass transfer effects and the flow pattern in both gas and liquid phases while microbial kinetics deals both with the kinetics on the individual cell level and on the level on the whole cell population. Single-cell kinetics can be described either with an unstructured model (no intracellular components considered) or with a structured model (intracellular components considered). The population model may be either unstructured (all cells in the whole population assumed to be identical, i.e., only one morphological form) or morphologically structured (with an infinite number of morphological forms the term segregated population model is often used). The models describing immobilized cell behavior are usually of the unstructured type.

Initially, models that described immobilized cell kinetics were based on the steady-state models for immobilized enzymes. Steady-state models can give valuable information for design purposes but fail to describe transient phenomena (like the start-up dynamics and response to changing conditions in the reactor) encountered in growing immobilized cell systems. Therefore, dynamic models have been developed to simulate the transient behavior of growing immobilized cells.

In general, gel-immobilized cell systems are considered as effective continua. However, it has been observed that when gel beads are inoculated with a low cell concentration, each growing cell will be the origin of a microcolony and growth results in the formation of expanding microcolonies (e.g., Willaert and Baron, 1993; Picioreanu et al., 1998). A rigorous modeling approach of this microcolonies system requires consideration of the microstructure of the immobilized cell system: diffusion in the gel phase, and reaction and diffusion in the microcolony ("two-phase" system).

10.3.2.1.1 Intrinsic kinetics

Intrinsic kinetics describes the growth and product formation rates of cells in the immobilized (or free) state as a function of the local concentrations. A typically simple unstructured model of microbial kinetics for growth on a single substrate can be described by the following three equations:

Biomass growth: $\qquad \mu = f(C_S)$ $\qquad\qquad\qquad$ (10.20)

Substrate consumption: $\quad q_s = \dfrac{1}{Y_{X/S}}\mu + m_s$ $\qquad\qquad$ (10.21)

Product formation: $\qquad q_p = \dfrac{1}{Y_{X/P}}\mu + m_p$ $\qquad\qquad$ (10.22)

where μ is the specific growth rate of the cells (gDW/gDW/h); C_S is the substrate concentration; q_s is the specific substrate utilisation rate (g substrate used/gDW/h); q_p is the specific product formation rate (g product formed/gDW/h); $Y_{X/S}$ (gDW/g substrate) and $Y_{X/P}$ (gDW/g product) are the yield coefficients; m_s (g substrate/gDW/h) and m_p (g product/gDW/h) are the specific maintenance rates for substrate and product, respectively. In some cases, these maintenance coefficients may be omitted or combined with a "cell death coefficient" (e.g., Doran, 1995). The specific growth rate is a function of the substrate concentration and is usually of the Monod kinetics form. The model can also be extended to include growth inhibition by the product (and biomass), and $f(C_S)$ becomes some function of substrate and product (and biomass). The Monod equation is bound by zero-order (at high substrate concentrations relative to the Monod constant K_s) and first-order (at vanishingly small substrate concentrations) kinetics. The solutions of reaction-diffusion problems with these two simple rate equations are valuable in that they can be applied as lower or upper bounds to the general problem without requiring detailed knowledge of the rate expressions and thus considerably facilitating the calculations.

In the interpretation of kinetic data for immobilized cells, it is important to assess the significance of mass transfer limitations. If negligible mass transfer limitation is present, the externally observed kinetics is the intrinsic cell kinetics. Any external or internal mass transfer limitation will lead to externally observed lower conversion rates. Mass transfer limitations may appear either in the external film around the support matrix or within the gel matrix, or in both.

A variety of claims have been made regarding changes in the intrinsic growth rate of immobilized cells, primarily regarding cells adhering to a surface. It has been asserted that the growth rate for immobilized cells is much higher than that for free-living cells. For gel-immobilized cell systems, it has been observed that the metabolic rates of gel-immobilized cells depend only on the local solution concentrations and are identical to those for free cells if diffusional limitations are absent although some reports showed a decreased growth rate upon entrapment. Some researchers found no significant difference between the maximum specific growth rates for immobilized yeast and bacteria, and those for free cells (Table 10.8). On the contrary, a significant decrease has been noted by other researchers for the same micro-organisms. This change of metabolic activity upon immobilization may be due to diffusional limitations or a change in cellular physiology.

10.3.2.1.2 Modeling

By the entrapment of cells in gel matrices, an additional barrier to mass transfer relative to free cells is introduced. This tends to lower the overall reaction rate, as well as creating a specific micro-environment around the cells. Immobilized cells can grow in the gel matrix and the mass transfer limitations on substrate delivery and product removal lead to time-dependent spatial variations in growth rates and biomass densities, which may be accompanied by alterations in cellular physiology and biocatalytic activity. Since the local effective diffusion coefficient

Table 10.8 Comparison of the specific growth rates for gel immobilized (μ_I) and free (μ_f) cells (Willaert et al., 2004).

Micro-organism	Gel-system	μ_I (h⁻¹)	μ_f (h⁻¹)	Reference
Bacteria				
Escherichia coli	Carrageenan (2%)	2.04[a]	2.08	Marin-Iniesta et al., 1988
		1.69[b]	1.63	
E. coli B/pTG201	Carrageenan (2%)	0.24[c]	0.30	Hooijmans et al., 1990a
			0.18[d]	
E. coli B/pTG201	Carrageenan (2%)	0.18	0.36	Huang et al., 1990
Fungi				
Candida guilliermondii	Ba-alginate	0.021	0.029	Dias et al., 2001
Saccharomyces cerevisiae	Ca-alginate (2%)	0.30[e]	0.31	Willaert and Baron, 1993
		0.27[f]		
S. cerevisiae	Gelatin (25-30%)	0.28	0.51	Doran and Bailey, 1986
S. cerevisiae	Ca-alginate (1.5%)	0.115	0.126	Agrawal and Jain, 1986
	Carrageenan (2.5%)	0.100		
S. cerevisiae	Ca-alginate (2%)	0.25	0.41	Galazzo and Bailey, 1990
S. cerevisiae	Ca-alginate (2%)	0.46	0.50	Vives et al., 1993
Thiosphaera pantotropha	Agarose (5%)	0.45[g]	0.45	Hooijmans et al., 1990b
		0.58[h]		

[a] supply of 21% oxygen; [b] supply of 100% oxygen; [c] growth in gel slabs; [d] growth in gel beads; [e] single immobilized cells; [f] cells in a microcolony; [g] growth in stirred tank reactor; [h] growth in Kluyver flask.

depends on the local biomass density, this nonhomogeneous growth will influence the local diffusive rates. The existence of chemical environmental gradients in immobilized cell systems has been verified experimentally with various microprobe techniques (Willaert et al., 2004).

Dynamic modeling: A general dynamic model, which describes the growth of the immobilized cells and the resulting time dependent spatial variation of substrate and product in the system, can be constructed by writing the mass balances over the immobilization matrix (it is usually assumed that diffusion in the system is governed by Fick's law, the cells are initially distributed homogeneously over the carrier and there is no deformation of the matrix due to cell growth or gas production):

$$\frac{\partial}{\partial t}(\varepsilon\beta C_i) = z^{-n}\frac{\partial}{\partial z}\left(z^n D_{e,i}\frac{\partial C_i}{\partial z}\right) \pm \varepsilon\beta r_i \qquad (10.23)$$

where ε is the ratio of the volume of the pores of the matrix to the total volume; β is the ratio of the volume of the pores minus the volume of the cells to the volume of the pores in the matrix; C_i is the substrate ($i = S$) or product ($i = P$) concentration (expressed per volume available for substrate; n is a shape factor of value 0 for planar, 1 for cylindrical or 2 for spherical geometry; and $D_{e,i}$ is the effective diffusion coefficient for species i. The substrate consumption rate (r_s) and the product formation rate (r_p) are linked to the growth rate (r_x) by the following equations:

$$\frac{\partial}{\partial t}(\varepsilon C_X) = \varepsilon r_x = \varepsilon \mu C_X \tag{10.24}$$

$$r_s = \frac{1}{Y_{X/S}} r_x \quad \text{and} \quad r_p = \frac{1}{Y_{X/P}} r_x \tag{10.25}$$

where C_X is the biomass concentration expressed per volume available for the cells. Since ε is constant with time, the left-hand side of Eq. (23) can be written as

$$\frac{\partial}{\partial t}(\varepsilon \beta C_i) = \varepsilon \beta \frac{\partial C_i}{\partial t} + \varepsilon C_i \frac{\partial \beta}{\partial t} \tag{10.26}$$

If a dry weight cell density ρ_c is defined as the ratio of the cell dry mass per cell volume, β can be expressed as function of C_X:

$$\beta = 1 - \frac{C_X}{\rho_c} \tag{10.27}$$

Using the relationship $Y_{X/S} = dC_X/dC_S$, the second term on the right side of Eq. (10.26) is negligible when C_S is much smaller than $\beta \rho_c / Y_{X/S}$, and under those conditions substitution into Eq. (10.23) gives

$$\varepsilon \beta \frac{\partial C_i}{\partial t} = z^{-n} \frac{\partial}{\partial z}\left(z^n D_{e,i} \frac{\partial C_i}{\partial z} \right) \pm \varepsilon \beta r_i \tag{10.28}$$

These equations can be integrated to yield the substrate and biomass profiles as a function of time (usually together with the reactor model) using the correct initial and boundary conditions. These equations are valid for a wide range of immobilized cell systems.

Pseudo-steady-state modeling: Under certain conditions, the full dynamic modeling to describe transient behavior can be simplified to "pseudo-steady-state" modeling. Therefore, the biomass growth and the substrate consumption rate and/or product formation rate are treated separately. This approach is valid as long as the time scale for growth is much larger than the time scale for consumption and product formation. Hence, a pseudo-steady-state substrate/product distribution is assumed at each instant. As a result, the system of partial differential equations is reduced to a system of ordinary differential equations, which facilitates the numerical solution.

Steady-state modeling: If the cell mass does not vary rapidly, or is fairly uniform, the concentration profiles in a gel matrix with entrapped cells can be simulated using a steady-state model at any point in time. These models can give valuable information for design purposes or can be combined with experimental *in situ* measurements (e.g., microelectrodes). In this case, mathematical calculations can be very simple, and straightforward analytical solutions can be obtained for simple reaction kinetics (e.g., zero- or first-order kinetics).

Effectiveness factor: The effectiveness factor (η) can be calculated to obtain a numerical measure of the influence of mass transfer on the reaction rate. The effectiveness factor is defined as

$$\eta = \frac{\text{observed reaction rate}}{\substack{\text{rate which would be obtained} \\ \text{without mass transfer resistance}}} \tag{10.29}$$

Effectiveness factor calculations can be based on steady-state or dynamic reaction-diffusion models with the assumption of a homogeneous distribution of cells over the carrier (e.g., Willaert and Baron, 1994). The effectiveness factor for substrate consumption can mathematically be expressed as the volume averaged reaction rate relative to the rate at bulk phase concentration:

$$\eta = (n+1) \frac{\left(D_e \frac{dC_S'}{dz'} \right)_{z'=1}}{r_s(1)} \tag{10.30}$$

or

$$\eta = (n+1) \frac{\int_0^1 z'^{n'} r_s(C_S') dz'}{r_s(1)} \tag{10.31}$$

where n is a shape factor (0 for planar, 1 for cylindrical, or 2 for spherical geometry) and z' is the dimensionless position coordinate; r_s is the substrate reaction rate, which is a function of the dimensionless substrate concentration (C_S') (and also of the position in the case of transient effectiveness factor).

10.3.3 *Bioreactor design*

A classification of immobilized cell reactors with their advantages and disadvantages is given in Table 10.9 (for further details, see Chapter 13). Three categories can be distinguished: i.e., bioreactors filled with

- mixed, suspended particles,
- fixed particles or large surfaces, or
- moving surfaces.

Most bioreactors (Figure 10.5) contain three phases: solid (the carrier or cell aggregate), bulk liquid, and gas (air/oxygen or gas feed, gaseous products).

Criteria for the selection of an appropriate bioreactor for immobilized cells and reactor types satisfying certain criteria are summarized in Table 10.10. The cell aggregate can only be fully active if the external supply or removal rates match the internal transport, utilization, and production rates. The high cell densities in the reactors put higher demands on nutrient supply and transport rates.

In aerobic fermentations, oxygen is often the limiting substrate since the liquid film around the gas bubbles is a major resistance to oxygen transfer. Gas-liquid mass transport is characterized by the liquid-phase mass transfer coefficient k_L (usually expressed as $k_L a$, with a the area of the bubble). Additionally, intraparticle resistance of oxygen in the cell aggregate can also become significant depending on the Thiele modulus. Consequently, the Thiele modulus and $k_L a$ are important parameters for the design and scale-up of immobilized cell bioreactors. The parameters, which are grouped in the Thiele modulus, are the size of the aggregate, cell kinetics (kinetic parameters), and the diffusion coefficient. Therefore, the particle size or biofilm thickness is a major design variable. As substrate concentration is often not a free parameter, only aggregate size and biomass loading are engineering parameters. Unless the substrate concentration is very low (e.g., degradation of pollutants), diffusion limitation for substrate only occurs for particles of more than several millimeters in diameter. In contrast, the penetration depth of oxygen in particles for aerobic processes is only 50 to 100 μm.

10.4 Physiology of immobilized microbial cells

The micro-environment of microbial cells is altered upon immobilization depending on the method of immobilization. The changed chemical composition and/or the physical interaction between the matrix material and the cell can have a profound effect on the physiology of the cells. Some examples of the effect of immobilization upon the cell's physiology are discussed for bacteria and fungi. An overview of observed effects, discussed below, upon immobilization of bacterial and fungal cells is summarized in Tables 10.11 and 10.12, respectively.

Table 10.9   Classification of immobilized cell reactors
(Chang and Moo-Young, 1988; Baron et al., 1996).

Reactor type	Advantages	Disadvantages
Mixed, suspended particles:		
Stirred tank reactor	Flexible	High power consumption
	Variable mixing intensity	Shear damage to sensitive matrices
	Suitable for high viscosity	High cost
Gas/air lift reactor	No moving parts	Low local mixing intensity
Bubble column reactor	Simple	Only for low viscosities
	High solids fraction possible	Excessive foaming possible
	High gas transfer	
	Good heat transfer	
Fluidised bed reactor	No moving parts in reactor	Difficult matching of feed and
	Simple, low cost	fluidisation rates
	Very high solids contents	Requirements on particle density
	Good heat transfer	(dense support)
	Variable mixing characteristics for	Good local mixing intensity
	liquid and solid	Only for low viscosity
Fixed particles or large surfaces:		
Packed bed reactor	Simple, low cost	Plugging by solids at low flow rates
Monolith reactor	Plug flow characteristics possible	High pressure drop
	Large surface to volume ratio	Aeration only externally possible
		Channelling and maldistribution
		Gas build-up and formation of
		'dry' spots
		Low external mass transfer rate at
		low flow rates
Trickle bed reactor	Simple, very low cost	Plugging by excessive growth
	Plug flow approached	(cleaning possible)
	Large surface to volume ratio	Only large supports possible
	High oxygen transfer rate	Only for low viscosities
	Suitable for gas cleaning (biofilter)	
Solid (surface) culture	Simple	High cost (often manual)
	Flexible	Only batch
	Low humidity (mould fermentation)	Difficult control over operation
	Low contamination risk	Usually limited to solid substrates
	(by bacteria or yeasts)	
Membrane reactor	Very high cell densities	Sterilization problems
	Very high productivities	Microbial damage (membrane
	Perfusion operation possible	perforation)
	Simultaneous product separation	Low capacities only
	possible	High cost

Table 10.9 (continued) Classification of immobilized cell reactors (Chang and Moo-Young, 1988; Baron et al., 1996).

Reactor type	Advantages	Disadvantages
	Separate feed of gas and liquid possible	
	Low shear	
Moving surfaces:		
Rotating surfaces (disc, cylinder or packing)	Low shear on biofilm	Power consumption
	Batch or continuous	Maintenance
	Excellent aeration	
	High productivity	
	Suitable for high viscosity	

Table 10.10 Criteria for the selection of the reactor type (modified from Baron et al., 1996).

Criterion		Stirred tank	Air lift Bubble column	Fluidised bed	Packed bed	Trickle bed	Membrane reactor	Rotating biological contactor
Matrix	strong	x	x	x	x	x		
	weak		x	x				
Biocatalyst	biofilm				x	x		x
	particles	x	x	x	x			
	membrane						x	
Sterisability	good	x	x	x			x	
	poor				x	x		x
High capacity		x	x	x	x	x		
High feed viscosity		x						x
Solids in feed		x	x	x				x
Flexibility required		x	x					
Equipment cost	high	x		x			x	x
	low		x		x	x		
High oxygen (gas) requirement		x	x		x	x		
High cell growth rate		x	x	x				
High gas production		x	x	x		x		x
Cell free effluent								
Low shear rate	low		x		x	x	x	

Figure 10.5
Immobilized cell bioreactors: A. Stirred tank reactor; B. bubble column reactor;
C. gas(air) lift reactor; D. fluidized-bed reactor (left), tapered fluidised reactor (right);
E. packed bead reactor with optional recycle loop; F. rotating drum reactor;
G. tricle bed reactor; and H. membrane cell-recycle reactor.

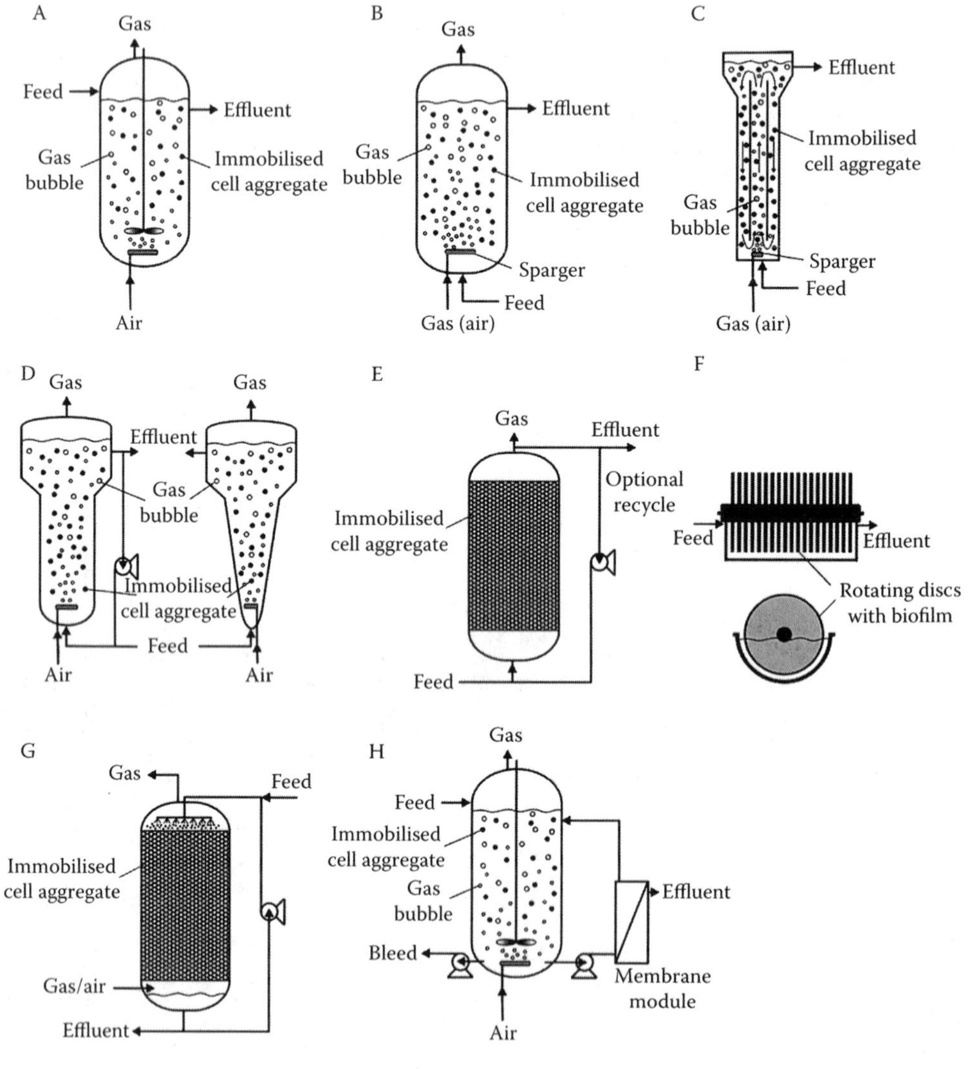

10.4.1 *Bacterial cells*

Plasmid stability: Recombinant plasmid stability in host cells can be increased upon immobilization (Table 10.10). The increased plasmid stability may have resulted from the mechanical properties of the immobilization system that allows only a limited number of cell divisions to occur.

Protective micro-environment: Immobilization can confer protection to cells exposed to toxic or inhibitory substrates or environments (Table 10.10). Gel immobilized *Trichosporium* sp. and *Pseudomonas putida* showed higher rates of phenol degradation and phenol tolerance. Alginate entrapped *E. coli* cells grown entrapped in calcium alginate showed low lipid-to-protein ratios even without phenol in the growth medium. Immobilization of cells also markedly changed the protein pattern of the outer membrane. Ca-alginate immobilized cultures of *Streptococcus* cells were protected from bacteriophage attack due to the exclusion of phage particles from the gel matrix. These cultures were also functionally proteinase-deficient when immobilized and grown in milk, which resulted in a lower acid production due in part to the inability of the immobilized cells to hydrolyse milk proteins, and to diffusional limitations of substrate into the beads. A different protein profile was found after submitting agar entrapped *E. coli* to a cold shock (Perrot et al., 2001). It was suggested that such induction of specific molecular mechanisms in immobilized bacteria might explain the high resistance of sessile-like organisms to stresses. The degradation rate of alkyl benzene sulphonate by Ca-alginate entrapped *Pseudomonas aeruginosa* was considerably increased compared to free cells; and could be further increased using low-intensity ultrasonic irradiation (Lijun et al., 2005), since ultrasound can improve the osmosis of the cell membrane, cell growth, and enzyme activities. Recently, a new ultrasound-based cell immobilization technique was described that allows manipulation and positioning of cells/particles within various gel matrices before polymerization (Gherardini et al., 2005). Proteomic analysis of agar-entrapped *P. aeruginosa* showed that the immobilized bacteria were physiologically different from free cells (Vilain *et al.*, 2004).

Effect of mass transport limitation: Mass transport limitations can have an influence on the morphology of immobilized *Lactobacillus helveticus* due to the long response time of entrapped cells to random pH changes. The citrate metabolism (Cachon and Divies, 1993) and lactate production (Klinkenberg et al., 2001) of *Lactococcus lactis* were altered due to the concentration and pH gradients in the gel beads. The germination time of *Bacillus subtilis* cells was significantly longer than for free cells; after a time lag due to encapsulation, the growth of the cells was uninhibited and no differences between entrapped and free cells were found (Brito et al., 1990).

Enhanced productivity of enzymes and other products: It has been observed that the production of several enzymes is increased upon immobilization (Table 10.11). In one case (α-amylase by alginate entrapped *B. amyloliquefaciens*), enzyme production

Table 10.11 Effects of bacterial cell immobilization on the physiology of micro-organisms

Immobilization effect	Micro-organism	Immobilization system	Reference
Increased plasmid stability			
	Escherichia coli	Ca-alginate	Georgiou et al., 1985; Vieth, 1989; Dincbas et al., 1993
		Agarose	Birnbaum et al., 1988
		κ-Carrageenan	Barbotin et al., 1990; Ryan and Parulekar, 1991
		Cotton cloth	Joshi and Yamazaki, 1987
		Hollow fiber	Inloes et al., 1983
		Polyacrylamide/hydrazide	Kanayama et al., 1988
		Polyvinyl alcohol	Ariga et al., 1991
		Silicone beads	Oriel, 1988
	Lactococcus lactis	κ-Carrageenan/locust bean gum	D'Angio et al., 1994
	Myxococcus xanthus	κ-Carrageenan	Jaoua et al., 1986
Protective micro-environment			
Increased phenol degradation	Pseudomonas putida	Ca-alginate, polyacryl-amide-hydrazide	Bettmann and Rehm, 1984
Increased antibiotics tolerance	Trichosporium sp.	Ca-alginate	Santos et al., 2001
Increased phenol tolerance	Pseudomonas aeruginosa	Ca-alginate	Coquet et al., 1999
	Escherichia coli	Ca-alginate	Keweloh et al., 1989
	Stapylococcus aureus		
Increased benzene degradation	Pseudomonas putida and Pseudomonas fluorescens	Cotton terry cloth	Shim and Yang, 1999
Increased degradation of linear alkyl benzene sulphonate	Pseudomonas aeruginosa	Ca-alginate	Lijun et al., 2005
Increased stress tolerance	Bifidobacterium longum and Lactococcus lactis	κ-Carrageenan/locust bean gum	Doleyres et al., 2004
Metabolic differences	Marinobacter sp.	Porous glass	Bonin et al., 2001
Degradation of DMP[a] at higher concentration	Bacillus sp.	Ca-alginate, polyurethane foam	Niazi and Karegoudar, 2001

Effect	Organism	Support material	Reference
Protection from bacteriophage attack	Streptococcus lactis	Ca-alginate	Steenson et al., 1987
	Streptococcus cremoris		
Different protein profile after cold shock	Escherichia coli	Agar	Perrot et al., 2001
Increased survival rate at low pH	Bifidobacterium longum	Gellan-xanthan	Sun and Griffiths, 2000
		Ca-alginate	Lee and Heo, 2000
Influence of mass transport limitation			
Decreased pH dependent morphology change	Lactobacillus helveticus	κ-Carrageenan/locust bean gum	Norton et al., 1993b
Changed citrate metabolism and lactate production	Lactococcus lactis	Ca-alginate	Cachon and Divies, 1993
			Klinkenberg et al., 2001
Loss of SDS[b] metabolisation	Pseudomonas fluorescens	Polyacrylamide	White and Thomas, 1990
Changed productivity of enzymes and other products			
Increased α-amylase production	Bacillus sp.	κ-Carrageenan	Shinmyo et al., 1982
		κ-Carrageenan	Chevalier and de la Noüe, 1987
	Bacillus subtilis	Polyacrylamide	Kokubu et al., 1978
Reduced α-amylase production	B. amyloliquefaciens	Ca-alginate	Argirakos et al., 1992
Increased dextransucrase production	Leuconostoc mesenteroides	Ca-alginate	El-Sayed et al., 1990a,b
Increased β-galactosidase Production	Escherichia coli	Ca-alginate	Zhang et al., 1989
Increased proteinase production	Humicola lutea	Polyhydroxyethyl-methacrylate	Aleksieva et al., 1991
Increased protease production	Myxococcus xanthus	Ca-alginate	Fortin and Vuillemard, 1990
Increased alginate production	Pseudomonas aeruginosa	κ-Carrageenan	Brito et al., 1990
Increased gellan gum production	Pseudomonas elodea	κ-Carrageenan	Brito et al., 1990
Changed morphology			
	Escherichia coli	Agar	Shapiro, 1987
	Lactobacillus helveticus	κ-Carrageenan/locust bean gum	Norton et al., 1993b
	Pseudomonas putida	Cotton terry cloth	Shapiro, 1985
	Pseudomonas putida	Cotton terry cloth	Shim and Yang, 1999
	P. fluorescens	Cotton terry cloth	Shim and Yang, 1999

[a] DMP = dimethylphtalate; [b] SDS = sodium dodecyl sulphate

was decreased (Argirakos et al., 1992). Protease production by alginate entrapped *Myxococcus xanthus* cells was increased due to a reduced inhibition by peptone and gelatin as result of mass transfer limitation in the gel (Fortin and Vuillemard, 1990).

10.4.2 *Fungal cells*

Plasmid stability: As in the case of recombinant bacteria, improved plasmid stability in immobilized *Saccharomyces cerevisiae* has also been demonstrated.

Protective micro-environment: Immobilization can create a protective micro-environment for the cells (Table 10.11). In the case of co-immobilization of *S. cerevisiae* cells with vegetable oils, cells could be protected against inhibitory substances due to a better solubility of the inhibitory compounds in the oil phase. An increased tolerance for ethanol by immobilized brewer's yeast (Norton et al., 1995) and a decreased inhibition of ethanol productivity in entrapped *Kluyveromyces marxianus* at high osmolality (Dale et al., 1994) have been observed. Immobilized *S. cerevisiae* contained significantly higher percentages of saturated fatty acids due to altered osmotic conditions in the micro-environment of the cells; other examples are shown in Table 10.11.

Influence of mass transport limitation: A high invertase activity, which was exhibited by immobilized cells, was due to a maintained expression of the *SUC2* gene and a reduced susceptibility of the enzyme to endogenous proteolytic attack (de Alteriis et al., 1999). These results have been interpreted in terms of diffusional limitations and changes in the pattern of invertase glycosylation due to growth of yeast in an immobilized state.

Enhanced productivity: The increase in ethanol productivity by gel-entrapped and co-entrapped cells of *Saccharomyces cerevisiae* was attributed to stimulation in cell permeability by Si^{4+} while higher ethanol production by κ-carrageenan-entrapped *S. bayanus* was attributed to a favorable media supplement in the aqueous phase of the matrix (Brito et al., 1990). On the other hand, however, the reduced yield of ethanol in Ca-alginate-entrapped *S. cerevisiae* was due to lower substrate concentrations toward the center of the bead due to mass transport limitations (Gilson and Thomas, 1995).

Enhanced enzyme stability: A hydroxylase from entrapped *Mortierella isabellina* was found to retain its activity over a longer period compared with free mycelia. The effect of immobilization on the physiology of yeast cells and fungi is given in Table 10.12.

10.5 Beer production using immobilized cell technology: A case study

10 5.1 Flavor maturation of green beer

One of the objectives of the maturation (or secondary fermentation) of green beer is the removal of unwanted aroma compounds. Especially the removal of the vicinal diketones diacetyl and 2,3-pentanedione is important since these compounds have very low flavor thresholds. Diacetyl is quantitatively more important than 2,3-pentanedione and is therefore used as a marker compound. It has a taste threshold around 0.10–0.15 mg/ml in lager beer, approximately 10 times lower than that of pentanedione. These compounds impart a "buttery," "butterscotch" aroma to the beer. During the primary fermentation, these flavor-active compounds are produced as by-products of the synthesis pathway of isoleucine, leucine, and valine (ILV pathway) and are linked to the amino acid metabolism and the synthesis of higher alcohols. The excreted α-acetohydroxy acids are overflow products of the ILV pathway and are nonenzymatically converted to the corresponding vicinal diketones (Figure 10.6). This nonenzymatic oxidative decarboxylation step is the rate-limiting step and proceeds faster at a high temperature and lower pH. The produced amount of α-acetolactate is very dependent on the used strain. The production also increases with increasing yeast growth. For classical lager fermentation, 0.6 ppm α-acetolactate is typically formed. At high aeration, this value can raise to 0.9 ppm and

GREEN BEER
During the beer fermentation, wort is fermented to beer. Wort is the carbohydrate extract from grinded barley malt. The fermentation proceeds in two stages. The first stage is called the primary fermentation and the second one is the secondary fermentation or maturation. The obtained beer after the primary fermentation is called "green" beer.

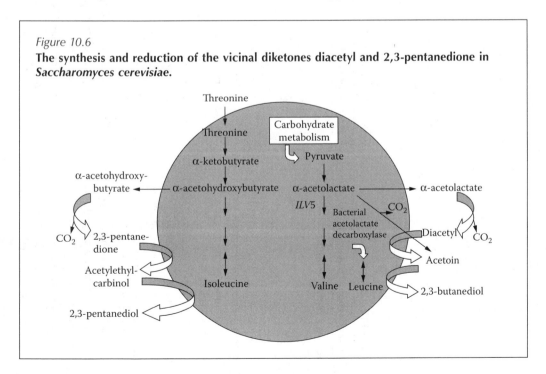

Figure 10.6

The synthesis and reduction of the vicinal diketones diacetyl and 2,3-pentanedione in *Saccharomyces cerevisiae*.

Table 10.12 Effects of fungal cell immobilization on the physiology of micro-organisms

Immobilization effect	Micro-organism	Immobilization system	Reference
Increased plasmid stability			
	Saccharomyces cerevisiae	Ca-alginate bead	Sode et al., 1988; Roca et al., 1996
		Cotton cloth sheet	Zhang et al., 1997
		Gelatin beads	Walls and Gainer, 1989
		Porous glass beads	Shu and Yang, 1996
Protective micro-environment			
Altered osmotic conditions in micro-environment	*Saccharomyces cerevisiae*	Ca-alginate	Hilge-Rotmann and Rehm, 1991
Decreased ethanol tolerance	*Pichia stipitis*	κ-Carrageenan	Chamy et al., 1994
Increased ethanol tolerance	*Saccharomyces cerevisiae*	κ-Carrageenan	Norton et al., 1995
Increased organic solvent tolerance	*Saccharomyces cerevisiae*	Polyhydroxylated silane	Desimone et al., 2003
Decreased inhibition of cell growth and metabolism by better CO_2 removal	*Saccharomyces cerevisiae*	Stainless steel fiber cloth	Shen et al., 2004
Increased tolerance for nitriles and amides	*Candida guilliermondii*	Ba-alginate	Dias et al., 2001
More efficient treatment of dairy effluents	*Candida pseudotropicalis*	Ca-alginate	Marwaha et al., 1990
More sensitive for sucrose	*Aspergillus niger*	Ca-alginate	Honecker et al., 1989
Protection against inhibitory substances	*Saccharomyces cerevisiae*	Ca-alginate	Ohta et al., 1994; Tanaka et al., 1994
Retention of high metabolic activity during long-term fermentation	*Saccharomyces cerevisiae*	Ca-alginate	Melzoch et al., 1994

Influence of mass transport limitation

Process	Organism	Carrier	Reference
Less susceptible of enzyme to endogenous proteolysis	Saccharomyces cerevisiae	Gelatin	de Alteriis et al., 1999
Lower specific productivity	Kluyveromyces lactis	Ca-alginate	de Alteriis et al., 2004

Changed productivity

Process	Organism	Carrier	Reference
Increased ethanol production	Saccharomyces formosensis	Ca-alginate and co-entrapped sand	Fang et al., 1983
	Saccharomyces bayanus	κ-Carrageenan	Brito et al., 1990
Increased laccase production	Agaricus sp.	Polyurethane foam, straw, textile strips	Kaluskar et al., 1999
Increased penicillin production	Penicillium chrysogenum	κ-Carrageenan	Mussenden et al., 1993
Shifted phosphate concentration optimum for alkaloid production	Claviceps purpurea	Ca-alginate	Lohmeyer et al., 1990
Changed secondary metabolite production	Fusarium moniliforme	Alginate, carrageenan, polyurethane	Nava Saucedo et al., 1989a; 1990

Changed morphology/composition

Process	Organism	Carrier	Reference
	Aspergillus niger	Ca-alginate	Honecker et al., 1989
	Claviceps fusiformis	Ca-alginate	Kren et al., 1987
	Gibberella fujikuroi	Ca-alginate	Nava Saucedo et al., 1989a, b
	Saccharomyces cerevisiae	Ca-alginate, agar, gelatine	Svoboda and Ourednicek, 1990
	Saccharomyces cerevisiae	Ca-alginate	Simon et al., 1990

Enhanced enzyme stability

Process	Organism	Carrier	Reference
Hydroxylase	Mortierella isabellina	Ca-alginate	Kutney et al., 1985
		Polyurethane foam	Kutney et al., 1988

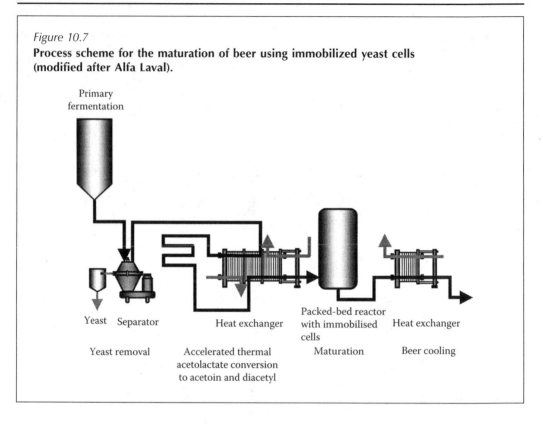

Figure 10.7

Process scheme for the maturation of beer using immobilized yeast cells (modified after Alfa Laval).

Primary fermentation

Yeast   Separator

Yeast removal

Heat exchanger

Accelerated thermal acetolactate conversion to acetoin and diacetyl

Packed-bed reactor with immobilised cells

Maturation

Heat exchanger

Beer cooling

in cylindro-conical fermentation tanks even to 1.2–1.5 ppm. Yeast cells possess the necessary enzymes (reductases) to reduce diacetyl to acetoin and further to 2,3-butanediol, and 2,3-pentanedione to 2,3-pentanediol. These reduced compounds have much higher taste thresholds and have no impact on the beer flavor. The reduction reactions are yeast-strain-dependent and occurs during the course of maturation after fermentation. Sufficient yeast cells in suspension are necessary to obtain an efficient reduction. Yeast strains, which flocculate early during the main fermentation, need a long maturation time to reduce the vicinal diketones.

The traditional maturation process is characterized by a near-zero temperature, low pH, and low yeast concentration, resulting in a very long maturation period of 3 to 4 weeks. Nowadays, the maturation phase is considerably reduced to around one week since several strategies are used to accelerate the vicinal diketone removal. Using immobilized cell technology, this maturation period can be further reduced to a time range of a few hours. Examples of ICT maturation processes are illustrated in Table 10.13. An example of a process scheme is illustrated in Figure 10.7.

It should also be mentioned that accelerated beer maturation can also be performed by increasing the maturation temperature, which can be realized by integrating the secondary fermentation in the primary fermentation. In this way, the production of beer can be accomplished in less than 2 weeks, in one cylindroconical vessel.

10.5.2 *Production of alcohol-free or low-alcohol beer*

The technology to produce alcohol-free or low-alcohol beer is based on the suppression of alcohol formation by arrested (restricted) free-cell batch fermentation. The resulting beers are, however, characterized by an undesirable wort aroma, since the wort aldehydes have only been reduced to a limited degree. An alternative method of producing these beers is based on the removal of ethanol from stronger beers by using membrane, destillation or vacuum evaporation processes. These methods have the disadvantage that the production cost is increased.

Controlled ethanol production for low-alcohol and alcohol-free beers have been successfully achieved by partial fermentation using immobilized yeast (Table 10.14). The reduction of the wort aldehydes can be quickly achieved by a short-contact with the immobilized yeast cells. The process is performed at a low temperature to avoid undesirable cell growth and ethanol production. A disadvantage of this short contact process is the production of only a small amount of desirable esters. Nuclear mutants of *S. cerevisiae,* which are defective in the synthesis of tricarboxylic acid cycle enzymes, i.e. fumarase or 2-oxoglutarate dehydrogenase, can be used for the production of non-alcohol beer since they produce minimal amounts of ethanol, but much lactic acid. These mutants have been used in a continuous immobilized-cell process (Navrátil et al., 2000).

10.5.3 *Continuous main fermentation*

Traditional beer fermentation technology uses freely suspended yeast cells to ferment wort in a non-stirred batch reactor. These fermentations are very time consuming. The traditional primary fermentation for lager beer takes approximately 7 days with a subsequent secondary fermentation (maturation) of several weeks. However, the resulting beer has a well-balanced flavor profile, which is very well accepted and appreciated by the consumer. Nowadays, large breweries use an accelerated fermentation scheme, which is based on using a higher fermentation temperature and specific yeast strains, thus facilitating the speedy production of beer; approximately 15 days.

Narziss and Hellich (1971) developed one of the first well-described ICT processes for beer production. Yeast cells were immobilized in kieselguhr (which is widely used in the brewing industry as a filter aid) and a kieselguhr filter was employed as bioreactor (called the "bio-brew bioreactor"). This process was characterized by a very low residence time of 2.5 h, but required

Table 10.13 ICT for beer maturation

Immobilization Method	Immobilization Matrix	Reactor Type	Scale	Reference
Surface attachment	DEAE-cellulose beads	Packed-bed	Industrial ($1\ 10^6$ hl/year)	Pajunen, 1995
Entrapment	Porous glass beads	Packed-bed	Industrial ($4\ 10^5$ hl/year)	Dillenhöfer, 1996; Mensour et al., 1997; Back et al., 1998
Entrapment	Ca-alginate beads	Fluidised-bed	Laboratory	Shindo et al., 1994

Table 10.14 ICT for the production of alcohol-free or low-alcohol beer

Immobilization Method	Immobilization Matrix	Reactor Type	Scale	Reference
Surface attachment	DEAE-cellulose	Packed-bed	Laboratory	Colin et al., 1991; Van Dieren, 1995
Surface attachment	DEAE-cellulose	Packed-bed	Industrial	Pittner et al., 1993; Mensour et al., 1997
Entrapment	Porous glass	Fluidised-bed	Pilot	Aivasidis et al., 1991; Breitenbücher et al., 1995
Entrapment	Silicon carbide rods	Cartridge loop reactor	Pilot	Van De Winkel et al., 1995
Entrapment	Ca-alginate	Gas lift	Lab	Nedovic et al., 1996, 1997a, 1997b
Entrapment	Ca-pectate	Packed-bed	Lab	Kaclíková et al., 1992; Mockovciaková et al., 1993; Navrátil et al., 2000

Table 10.15 ICT for the main beer fermentation

Immobilization Method	Immobilization Matrix	Reactor Type	Reference
Entrapment	Ca-alginate beads	Stirred tank[a]	Inoue, 1995; Yamauchi et al., 1994;
	Ceramic beads	+ 2 packed-bed[b]	Yamauchi et al., 1995
	Ca-alginate microbeads	Gas lift	Nedovic et al., 2005
Entrapment	Silicon carbide rods	Cartridge loop reactor[b] + Stirred tank[a]	Andries et al., 2000
Entrapment	Porous organic or glas beads	Packed-bed	Hamdy, 1989
Entrapment	κ-Carrageenan beads	Gas lift	Mensour et al. 1995, 1996, 1997; Neufeld et al., 1997
Entrapment	Chitosan beads	Fluidised-bed	Shindo, 1994.
Entrapment	Ca-pectate beads	Gas lift	Smogrovicová et al., 1997
Entrapment	Polyvinyl alcohol beads	Gas lift	Smogrovicová et al., 2001; Nedovic et al., 2005
Adsorption	Woodchips	Packed-bed	Kronlöf and Virkajärvi, 1999
Adsorption	DEAE-cellulose	Packed-bed	Andersen et al., 1999
Adorption	Gluten pellets	Fluidised-bed	Bardi et al., 1997
Adsorption	Spent grains	Gas lift	Brányik et al., 2002; 2004

[a] Free cells; [b] Immobilized cells

the addition of viable yeast and a 7-day maturation period to reduce the high concentration of vicinal diketones in the green beer. Although this result looked very good, the bio-brew bio-reactor gave overall no satisfying result as it contained a high amount of α-acetolactate; this was later optimized (Dembowski et al., 1993). An aerobic reactor was installed in front of the bio-brew reactor, the beer flow though the filter was optimized, and a cooling plate was installed in the filter reactor to control the temperature, thus increasing cell viability and improving the organolyptic qualities of beer. However, the concentration of the low-molecular-weight nitrogenous substances in the beer remained too high.

Baker and Kirsop (1973) were the first to use heat treatment of green beer to considerably accelerate the chemical conversion of α-acetolactate to diacetyl. They designed a two-step continuous process. The first reactor was a packed-bed reactor, containing also kieselguhr as immobilization matrix, to perform the primary fermentation. The green beer was heated using a heating coil to accelerate the α-acetolactate conversion. It was next cooled before it entered a smaller packed-bed reactor to perform the secondary fermentation. Problems associated by this process were a gradual blocking of the packed bed and a changed beer flavor.

The design and optimization of an ICT process for primary and secondary fermentations remain a challenging task, although encouraging results have been obtained recently albeit on lab and pilot scales only (Table 10.15). The reasons why these ICT processes have not yet been adopted in the brewing industry include complexity of operations compared to batch processes, flavor problems (due to a lack of understanding and controllability of the changed metabolism), yeast viability, and carrier price. Especially the altered metabolism and the knowledge to tune the metabolism to the desired flavor await further investigation.

Summary

- Selections of immobilization matrix and method as well as bioreactor type and design are usually interrelated.
- For optimization, knowledge of fluid dynamics and external/internal mass transfer is a good starting point.
- The use of mathematical models to quantitatively describe and analyze the behavior of immobilized cells is an important component; steady-state, pseudo-steady state, as well as dynamic models are discussed.
- The physiology of living cells upon immobilization can be changed since they are present in a different micro-environment compared to free-living cells.
- Different physiological effects due to the changed chemical composition and/or the physical interaction between the matrix material and immobilized bacterial and fungal cells are discussed.
- The production of beer using immobilized cell technology has been described by way of a case study.
- Cell immobilization has been implemented for the production of alcohol-free lager, enhancing of green beer, and continuous main stream fermentation.
- The use of cell immobilization technology has been successful at the industrial level in the production of biological reagents from animal cell cultures and tissue engineering. On the other hand, the full potential of cell immobilization for the production of biopharmaceuticals using microbial cells has yet to be fully realized and exploited. A few reasons include
  - The biopharmaceutical (fermentation) industry is very conservative and reluctant to replace the existing robust systems using free cells with this new technology;
  - An immobilized cell system is more complex and means a higher risk.
  - Validation, regulatory, and health and safety issues are both costly and time-consuming.
- Successful industrial/commercial fermentation processes/products based on immobilized living cells have recently been introduced. Examples include
  - The production of alcohol-free beer,
  - Stabilized probiotic products,

- Protective bacterial cultures for meat products,
- Production of bio-ethanol.
- For the production of low-added-value fermentation products, the added cost of immobilization techniques and in some cases higher complexity *versus* free cell culture will not be justified, unless the volumetric productivity is strongly increased.
- Future research will increase our understanding of these "complex" systems and will lower the barrier for exploitation on an industrial/commercial scale.

Acknowledgments

Financial support from the European Space Agency (Prodex program), the Flanders Interuniversity Institute for Biotechnology (VIB), and the Research Council of the Vrije Universiteit Brussels is acknowledged.

References

Agrawal, D., V. K. Jain (1986). Kinetics of repeated batch production of ethanol by immobilized growing yeast cells. *Biotechnol. Lett.* 8:67–70.

Aivasidis, A., C. Wandrey, H. G., Eils, M. Katzke (1991). Continuous fermentation of alcohol-free beer with immobilized yeast cells in fluidized bed reactors. Proceedings European Brewery Convention Congress, pp. 569–576.

Aleksieva, P., E. Petricheva, E. Konstantinov, C. Robeva, S. Mutafov (1991). Acid proteinases production by *Humicola lutea* cells immobilized in polyhydroxyethylmethacrylate gel. *Acta Biotechnol.* 11:255.

Andersen, K., J. Bergin, B. Ranta, T. Viljava (1999). New process for the continuous fermentation of beer. Proceedings of the European Brewery Convention Congress, pp. 771–778.

Andries, M., P. C. Van Beveren, O. Goffin, P. Rajotte, C. A. Masshelein (2000). Results on semi-industrial continuous top fermentation with the Meura-Delta immobilized yeast fermenter. *Brauwelt Int. II*: 134–136.

Argirakos, G., K. Thayanithy, D. A. John Wase (1992). Effect of immobilization on the production of α-amylase by an industrial strain of *Bacillus*. *J. Chem. Technol. Biotechnol.* 53:33–38.

Ariga, O., Y. Ando, Y. Fujishita, T. Watari, Y. Sano (1991). Production of thermophilic α-amylase using immobilized transformed *Escherichia coli* by addition of glycine. *J. Fermnt. Bioeng.* 71:397–402.

Axelsson, A., B. Westrin (1991). Application of the diffusion cell for the measurement of diffusion in gels. *Chem. Eng. Sci.* 46:913–915.

Back, W., M. Krottenthaler, T. Braun (1998). Investigations into continuous beer maturation. *Brauwelt Int. III*: 222–226.

Baker, D. A., B. H. Kirsop, (1973) Rapid beer production and conditioning using a plug fermentor. *J. Inst. Brew.* 79: 487–494.

Barbotin, J. N., S. Sayadi, M. Nasri, F. Berry, D. Thomas (1990). Improvement of plasmid stability by immobilization of recombinant micro-organisms. *Ann. N.Y. Acad. Sci.* 589: 41–53.

Bardi, E., A. A. Koutinas, M. Kanellaki (1997). Room and low temperature brewing with yeast immobilized on gluten pellets. *Process Biochem.* 32:691–696.

Baron, G. V., R. G. Willaert (2004). Cell immobilization iinn pre-formed porous matrices. In: Nedovic V., Willaert R. (eds.), *Fundamentals of Cell Immobilization Biotechnology*. Kluwer Academic Publishers, The Netherlands; pp. 229–244.

Baron, G. V., R. G, Willaert, L. De Backer (1996). Immobilized cell reactors. In: Willaert R.G., Baron G.V., De Backer L. (eds.), *Immobilized Living Cell Systems: Modeling and Experimental Methods*. John Wiley & Sons, Chichester; pp. 67–95.

Bekers, M., J. Laukevics, A. Karsakevich, E. Ventina, E. Kaminska, D. Upite, I. Vina, R. Linde, R. Scherbaka (2001). Levan-ethanol biosynthesis using *Zymomonas mobilis* cells immobilized by attachment and entrapment. *Process Biochem.* 36:979–986.

Bergmaier, D., C. P. Champagne, C. Lacroix (2005). Growth and exopolysaccharide production during free and immobilized cell chemostat culture of *Lactobacillus rhamnosus* RW-9595M. *J. Appl. Microbiol.* 98:272–284.

Bettmann, H., H. J. Rehm (1984). Degradation of phenol by polymer entrapped micro-organisms. *Appl. Microbiol. Biotechnol.* 20:285–290.

Birnbaum, S., L. Bulow, K. Hardy, K. Mosbach (1988). Production and release of human proinsulin by recombinant *Escherichia coli* immobilized in agarose microbeads. *Enzyme Microb. Technol.* 10:601–605.

Bonin, P., J. F. Rontani, L. Bordenave (2001). Metabolic differences between attached and free-living marine bacteria: Inadequacy of liquid cultures for describing in situ bacterial activity. *FEMS Microbiol. Lett.* 194:111–119.

Brányik, T., A. A. Vicente, J. M. M. Cruz, J. A. Texeira (2004). Continuous primary fermentation of beer with yeast immobilized on spent grains – the effect of operational conditions. *J. Am. Soc. Brew. Chem.* 62:29–34.

Brányik, T., A. A. Vicente, R. Oliveira, J. Teixeira (2004). Physicochemical surface properties of brewing yeast influencing their immobilization onto spent grains in a continuous reactor. *Biotechnol. Bioeng.* 88:84–93.

Breitenbücher, K., M. Mistler (1995). Fluidized-bed fermenters for the continuous production of non-alcoholic beer with open-pore sintered glass carriers. EBC Monograph XXIV, EBC Symposium on immobilized yeast applications in the brewing industry, pp. 77–89.

Brito, L. C., A. M. Vieira, J. G. Leitão, I. Sá-Correia, J. M. Novais, J. M. S. Cabral (1990). Effect of the aqueous soluble components of the immobilization matrix on ethanol and microbial exopolysaccharides production. In: de Bont J.A.M., Visser J., Mattiasson B., Tramper J. (eds.), *Physiology of Immobilized Cells*. Elsevier, Amsterdam, The Netherlands; pp. 399–404.

Brown, C.D., A. S. Hoffman (2002). Modification of natural polymers: chitosan. In: Atala A., Lanza, R.P. (ed.), *Methods of Tissue Engineering*. Academic Press, San Diego; pp. 565–574.

Buitelaar, R.M., A. C. Hulst, J. Tramper (1990), Cell immobilization in thermogels: activity retention after gelling in various organic solvents. In: de Bont J.A.M., Visser J., Mattiasson B., Tramper J. (eds.), *Physiology of Immobilized Cells*. Elsevier, Amsterdam; pp. 205–208.

Cachon, R., C. Divies (1993). Localization of *Lactococcus lactis* ssp. *lactis* bv. *diacetylactis* in alginate gel beads affects biomass density and synthesis of several enzymes involved in lactose and citrate metabolism. *Biotechnol. Techn.* 7:453–456.

Calderbank, P. H., M. G. Moo-Young (1961) The continuous phase heat and mass-transfer properties of dispersions. *Chem. Eng. Sci.* 16:39–54.

Camelin, I., C. Lacroix, C. Paquin, H. Prévost, R. Cachon, C. Divies (1993). Effect of chelatants on gellan gel rheological properties and setting temperature for immobilization of living bifidobacteria. *Biotechnol. Prog.* 9:291–297.

Carenza, M., F, M. Veronese (1994). Entrapment of biomolecules into hydrogels obtained by radiation-induced polymerization. *J. Control. Rel.* 29:187–193.

Carlsen, M. (1994). α-Amylase production by *Aspergillus oryzae*. Ph.D. dissertation, The Technical University of Denmark.

Caro, L. H., H. Tettelin, J. H. Vossen, A. F. Ram, H. van den Ende, F. M. Klis (1997) In silicio identification of glycosyl-phosphatidylinositol-anchored plasma-membrane and cell wall proteins of *Saccharomyces cerevisiae*. *Yeast* 13:1477–1489.

Chaikof, E. L. (1999). Engineering and material considerations in islet cell transplantation. *Ann. Rev. Biomed. Eng.* 1:103–127.

Champagne, C. P., N. J. Gardner, L. Soulignac, J. P. Innocent (2000). The production of freeze-dried immobilized cultures of . *Streptococcus thermophilus* and their acidification properties in milk. *J. Appl. Microbiol.* 88:124–131.

Chamy, R., M. J. Nunez, J. M. Lema (1994). Product inhibition of fermentation of xylose to ethanol by free and immobilized *Pichia stipitis*. Enzyme Microb. Technol. 16: 622-626.

Chang, T. M. S. (2005). Therapeutic applications of polymeric artificial cells. *Nature Rev.* 4:221–235.

Chang, H. N., M. Moo-Young (1988). In: Moo-Young, M. (ed.), *Bioreactor Immobilized Enzymes and Cells*. Elsevier Applied Science, London, pp. 33.

Chang, T. M. S. (1964). Semipermeable microcapsules. *Science* 146:908-909.

Chen, K. C., Y. F. Lin (1994) Immobilization of microorganisms with phosphorylated polyvinyl alcohol (PVA) gel. *Enzyme Microb. Technol.* 16: 79.

Chen, J. P., J. B. Wang (1997). Wax ester synthesis by lipase-catalyzed esterification with fungal cells immobilized on cellulose biomass support particles. Enzyme Microb. Technol. 20:615–622.

Chevalier, P., J. de la Noüe (1987). Enhancement of α-amylase production by immobilized *B. subtilis* in an airlift fermenter. *Enzyme Microb. Technol.* 9:53–56.

Chung, B. H., H. N. Chang (1988). Aerobic fungal cell immobilization in a dual hollow-fiber bioreactor: continuous production of citric acid. *Biotechnol. Bioeng.* 32:205–212.

Cirone, P., J. M. Bourgeois, R. C. Austin, P. L. Chang. (2002) A novel approach to tumor suppression with microencapsulated recombinant cells. *Human Gene Therapy* 13:1157–1166.

Cirone, P., J. M. Bourgeois, P. L. Chang (2003). Antiangiogenic cancer therapy with microencapsulated cells. *Human Gene Therapy* 14:1065–1077.

Cirone P., J. M. Bourgeois, F. Shen, P. L. Chang P.L. (2004). Combined immunotherapy an antiangiogenic therapy of cancer with microencapsulated cells. *Human Gene Therapy* 15:945–959.

Cirone P., F. Shen, P. L. Chang (2005). A multiprong approach to cancer gene therapy by co-encapsulated cells. *Cancer Gene Therapy* 12:369–380.

Collin, S., M. Montesinos, E. Meersman, W. Swinkels, B. Van Dieren, H. Lomni (1991). Yeast dehydrogenase activities in relation to carbonyl compounds removal from wort and beer. *Proceedings of the European Brewery Convention Congress*, pp. 409–416.

Coquet, L., G. A. Junter, T. Jouenne (1999). Resistance of artificial biofilms of *Pseudomonas aeruginosa* to imipenem and tobramycin. *J. Antimicrob. Chemother.* 42:755–760.

Dale, M. C., A. Eagger, M. R. Okos (1994). Osmotic inhibition of free and immobilized *K. marxianus* anaerobic growth and ethanol productivity in whey permeate concentrate. *Process Biochem.* 29:535–544.

D'Angio, C., C. Béal, C. Y. Boquien, G. Corrieu (1994). Influence of dilution rate and cell immobilization on plasmid stability during continuous cultures of recombinant strains of *Lactococcus lactis* subsp. *lactis. J. Biotechnol.* 34:87–95.

Danner, H., L. Madzingaidzo, C. Thomasser, M. Neureier, R. Braun (2002). Thermophilic production of lactic acid using integrated membrane bioreactor systems coupled with monopolar electrodialysis. *Appl. Microbiol. Biotechnol.* 59:160–169.

de Alteriis, E., P. M. Alepuz, F. Estruch, P. Parascandola (1999). Clues to the origin of high external invertase activity in immobilized growing yeast: prolonged *SUC2* transcription and less susceptibility of the enzyme to endogenous proteolysis. *Can. J. Microbiol.* 45:413–417.

de Alteriis, E., G. Silvestro, M. Poletto, V. Romano, D. Capitanio, C. Compago, P. Parascandola (2004). *Kluyveromyces lactis* cells entrapped in Ca-alginate beads for the continuous production of a heterogous flucoamylase. *J. Biotechnol.* 109:83–92.

Dembowski, K., L. Narziss, H. Miedaner (1993). Technologisch optimierte Bierherstellung im Festbettfermentor bei sehr kurzer Produktionszeit. *Proceedings 24th European Brewery Convention Congress*, Oslo; pp. 299–306.

Dempsey, M. J. (1990). Ethanol production by *Zymomonas mobilis* in a fluidized bed fermenter. In: de Bont J.A.M., Visser J., Mattiasson B., Tramper J. (eds.), *Physiology of Immobilized Cells.* Elsevier, Amsterdam; pp. 137–148.

Demuyakor, B., Y. Ohta (1992). Promotive action of ceramics on yeast ethanol production, and its relationship to pH, glycerol and ethanol dehydrogenase activity. *Appl. Microbiol. Biotechnol.* 36:717–721.

Desimone, M. F., J. Degrossi, M. D'Aquino, L. E. Diaz (2003) Sol-gel immobilization of *Saccharomyces cerevisiae* enhances viability in organic media. *Biotechnol. Lett.* 25:671–674.

de Vos, P., C. G. van Hoogmoed, J. van Zanten, J. H. Strubbe, H. J. Busscher (2003). Long-term biocompatibility, chemistry, and function of microencapsulated pancreatic islets. *Biomaterials* 24:305–312.

Dias, J. C., R. P. Rezende, V. R. Linardi (2001). Bioconversion of nitriles by *Candida guilliermondii* CCT 7207 cells immobilized in barium alginate. *Appl. Microbiol. Biotechnol.* 56:757–761.

Dillenhöfer, W., D. Rönn (1996). Secondary fermentation of beer with immobilized yeast. *Brauwelt Int.* 14:344–346.

Di Martino, A., M. Sittinger, M. V. Risbud (2005). Chitosan: A versatile biopolymer for orthopedic tissue-engineering. *Biomaterials* 26:5983–5990.

Dincbas, V., A. Hortascu, A. Camurdan (1993). Plasmid stability in immobilized mixed cultures of recombinant *Escherichia coli*. *Biotechnol. Prog.* 9:218–220.

Doleyres, Y., I. Fliss, C. Lacroix (2004). Increased stress tolerance of *Bifidobacterium longum* and *Lactococcus lactis* produced during continuous mixed-strain immobilized-cell fermentation. *J. Appl. Microbiol.* 97:527–539.

Doran P.M. (1995). *Bioprocess Engineering Principles*. Academic Press, London.

Doran, P., J. E. Bailey (1986) Effects of immobilization on growth, fermentation properties, and macromolecular composition of *Saccharomyces cerevisiae* attached to gelatin. *Biotechnol. Bioeng.* 28:73–87.

Eberl, L., S. Molin, M. Givskov (1999). Surface motility of *S. liquefaciens* MG1. *J. Bacteriol.* 181:1703–1712.

Elander, R. P. (2003). Industrial production of beta-lactam antibiotics. *Appl. Microbiol. Biotechnol.* 61:385–392.

El-Sayed, A. M. M., W. M. Mahmoud, R. W. Coughlin (1990a). Production of dextransucrase by *Leuconostoc mesenteroides* immobilized in calcium-alginate beads: I. Batch and fed-batch fermentations. *Biotechnol. Bioeng.* 36:338–345.

El-Sayed, A. M. M., W. M Mahmoud, R. W. Coughlin (1990b). Production of dextransucrase by *Leuconostoc mesenteroides* immobilized in calcium-alginate beads: II. Semicontinuous fed-batch fermentations. *Biotechnol. Bioeng.* 36:346–353.

Epstein, N. (1989). On tortuosity and the tortuosity factor in flow and diffusion through porous media. *Chem. Eng. Sci.* 44:777–779.

Fang, B. S., H. Y. Fang, C. S. Wu, C. T. Pan (1983). High productivity ethanol production by immobilized yeast cells. *Biotechnol. Bioeng. Symp.* No. 13:457–464.

Favaro-Trindade, C. S., C. R. Grosso (2002) Microencapsulation of *L. acidophilus* and *B. lactis* and evaluation of their survival at pH values of the stomach and in bile. *J. Microencapsul.* 19:485–494.

Feng, M., M. V. Sefton (2002). Microencapsulation methods: polyacryaltes. In: Atala A., Lanza, R.P. (eds.), *Methods of Tissue Engineering*. Academic Press, San Diego; pp. 825–839.

Förster, M., J. Mansfield, J. Schellenberger, H. Dautzenberg (1994). Immobilization of citrate-producing *Yarrowia lipolytica* cells in polyelectrolyte complex capsules. *Enzyme Microb. Technol.* 16:777–784.

Fortin, C., J. C. Vuillemard (1990). Elucidation of the mechanism involved in the regulation of protease production by immobilized *Myxococcus xanthus* cells. *Biotechnol. Lett.* 12:913.

Freed, L. E., J. C. Marquis, G. Vunjak-Novakovic, J. Emmanual, R. Langer (1994). Composite of cell-polymer cartilage implants. *Biotechnol. Bioeng.* 43:605–614.

Galazzo, J. L., J. E. Bailey (1990). Growing *Saccharomyces cerevisiae* in calcium-alginate beads induces cell alterations which accelerate glucose conversion to ethanol. *Biotechnol. Bioeng.* 36:417–426.

Gbewonyo, K., J. Meier, D. I. C. Wang. (1987) Immobilization of mycelial cells on celite. *Methods. Enzymol.* 135:318–333.

Gemeiner, P., L. Texova-Benkova, F. Svec, O. Norrlöw (1994). Natural and synthetic carriers suitable for immobilization of viable cells, active organelles, and molecules. In: Veliky I.A., McLean R.J.C. (eds.), *Immobilized Biosystems: Theory and Practical Applications*. Blackie Academic & Professional, Glasgow; pp. 1–128.

Georgiou, G., J. J. Chalmers, M. L. Shuler, D. B. Wilson (1985) Continuous immobilized recombinant protein production from *E. coli* capable of selective protein excretion: a feasibility study. *Biotechnol. Prog.* 1:75–79.

Gerin, P. A., M. Asther, P. G. Rouxhet (1997). Peroxidase production by the filamentous fungus *Phanerochaete chrysosporium* in relation to immobilization in "filtering" carriers. *Enzyme Microb. Technol.* 20:294–300.

Gherardini, L., C. M. Cousins, J. J. Hawkes, J. Spengler, S. Radel, H. Lawler, B. Devcic-Kuhar, M. Groschl, W. T. Coakley, A. J. McLoughlin (2005). A new immobilization method to arrange particles in a gel matrix by ultrasound standing waves. *Ultrasound Med. Biol.* 31:261–272.

Gilson, C., A. Thomas (1995). Ethanol production by alginate immobilized yeast in a fluidised bed bioreactor. *J. Chem. Technol. Biotechnol.* 62:38–45.

Glassner, D. A., E. A. Grulke, P. J. Oriel (1989). Characterization of immobilized biocatalyst system for production of thermostable amylase. *Biotechnol. Prog.* 5:31–39.

Goetghebeur, S., W. S. Hu (1991) Cultivation of anchorage-dependent animal cells in microphere-induced aggregate culture. *Appl. Microb. Biotechnol.* 34:735–741.

Goosen, M. F. A. (ed.) (1993). Fundamentals of Animal Cell Encapsulation and Immobilization. CRC Press, Boca Raton, Florida.

Green, C. B., G. Cheng, J. Chandra, P. Mukherjee, M. A. Ghannoum, L. L. Hoyer (2004). T-PCR detection of *Candida albicans ALS* gene expression in the reconstituted human epithelium (RHE) model of oral candidiasis and in model biofilms. *Microbiol.* 150:267–275.

Green, K. D., I. S. Gill, J. A. Khan, E. N. Vulfson (1996) Micro-encapsulation of yeast cells and their use as a biocatalyst in organic solvents. *Biotechnol. Bioeng.* 49:535–543.

Gutowska, A., B. Jeong, M. Jasionowski (2001). Injectable gels for tissue engineering. *Anat. Rec.* 263:342–349.

Hamdy, M. K. (1989). Method for rapidly fermenting alcoholic beverages. US patent 4,929–452.

Hellendoorn, L., H. Mulder H., J. C. van den Heuvel, S. P. P. Ottengraf (1998). Intrinsic kinetic parameters of the pellet forming fungus *Aspergillus awamori*. *Biotechnol. Bioeng.* 58:478–485.

Hilge-Rotmann, B., H. J. Rehm (1991). Relationship between fermentation capability and fatty acid composition of free and immobilized *Saccharomyces cerevisiae*. *Appl. Microbiol. Biotechnol.* 34:502–508.

Ho, M. H., D. M. Wang, H. J. Hsieh, H. C. Liu, T. Y. Hsien, J. Y., Lai, L. T. Hou (2005). Preparation and characterization of RGD-immobilized chitosan scaffolds. *Biomaterials* 26:3197–3206.

Hoffman, A. S. (2004). Applications of "smart polymers" as biomaterials. In: Ratner B.D., Hoffman A.S., Schoen F.J., Lemons J.E. (eds.), *Biomaterials Science – An Introduction to Materials in Medicine*. Elsevier Academic Press, London; pp. 107–115.

Honecker, S., B. Bisping, Z. Yang, H. J. Rehm (1989). Influence of sucrose concentration and phosphate limitation on citric acid production by immobilized cells of *Aspergillus niger*. *Appl. Microbiol. Biotechnol.* 31:17–24.

Hooijmans, C. M., C. A. Briasco, J. Huang, B. G. M. Geraats, J. N., Barbotin, D. Thomas, K.Ch.A. M. Luyben (1990a). Measurement of oxygen concentration gradients in gel-immobilized recombinant *Escherichia coli*. *Appl. Microbiol. Biotechnol.* 33:611–618.

Hooijmans, C. M., S. G. M. Geraats, E. W. J. van Neil, L. A., Robertson, J. J. Heijnen, K. Ch. A. M. Luyben (1990b). Determination of the growth and coupled nitrification/denitrification by immobilized *Thiospaera pantotropha* using measurement and modelling of oxygen profiles. *Biotechnol. Bioeng.* 36:931–939.

Huang, J., C. M. Hooijmans, C. A., Briasco, S. G. M. Geraats, K. Ch. A. M., Luyben, D. Thomas, J. N. Barbotin (1990). Effect of free-cell growth parameters on the oxygen concentration in gel-immobilized recombinant *Escherichia coli*. *Appl. Microbiol. Biotechnol.* 33:619–632.

Hunkeler, D. (1997). Polymers for bioartificial organs. *Trends Polymer Sci.* 5:286–293.

Inloes, D. S., A. S. Michaels, C. R. Robertson, A. Matin (1985). Ethanol production by nitrogen-deficient yeast cells immobilized in a hollow-fiber membrane bioreactor. *Appl. Microbiol. Biotechnol.* 23:85–91.

Inloes, D. S., W. J. Smith, D. P. Taylor, S. N. Cohen, A. S. Michaelis, C. R. Robertson (1983). Hollow fiber membrane bioreactors using immobilized *E. coli* for protein synthesis. *Biotechnol. Bioeng.* 25:2653–2681.

Inoue, T. (1995). Development of a two-stage immobilized yeast fermentation system for continuous beer brewing. In: *Proceedings of the European Brewery Convention Congress*; pp. 25–36.

Iqbal, M., A. Saeed (2005). Novel method for cell immobilization and its application for the production of organic acids. *Lett. Appl. Microbiol.* 40:178–182.

Jaoua, S., A. M. Breton, G. Younes, J. F. Guespin-Michel (1986). Structural instability and stabilization of IncP-1 plasmids integrated into the chromosome of *Myxococcus xanthus*. *J. Biotechnol.* 4:313–323.

Jianlong, W. (2000). Production of citric acid by immobilized *Aspergillus niger* using a rotating biological contactor (RBC). *Bioresource Technol.* 75:245–247.

Joshi, S., H. Yamazaki, (1987). Stability of pBR322 in *Escherichia coli* immobilized on cotton cloth during use as resident inoculum. Biotechnol. Lett. 9:825–830.

Kaclíková, E., T. M. Lachovicz, Y. Gbelská, J. Subík (1992). Fumaric acid overproduction in yeast mutants deficient in fumarase. *FEMS Microbiol. Lett.* 91:101–106.

Kallos, M. S., L. A. Behie (1999). Inoculation and growth conditions for high-cell-density expansion of mammalian neural stem cells in suspension bioractors. *Biotechnol. Bioeng.* 20:473–483.

Kaluskar, V. M., B. P. Kapadnis, C. Jaspers, M. J. Penninckx (1999). Production of laccacase by immobilized cells of *Agaricus* sp.: induction effect of xylan and lignin derivatives. *Appl. Biochem. Biotechnol.* 76:161–170.

Kanayama, H., K. Sode, I. Karube (1988). Continuous hydrogen evolution by immobilized recombinant *E. coli* using a bioreactor. *Biotechnol. Bioeng.* 32:396–399.

Keweloh, H., H. J. Heipieper, H. J. Rehm (1989). Protection of bacteria against cytotoxicity of phenol by immobilization in calcium alginate. *Appl. Microbiol. Biotechnol.* 31:383–389.

Kikuchi, K. I., T. Sugarawa, H. Ohashi (1988) Correlation of liquid-side mass transfer coefficient based on the new concept of specific power group. *Chem. Eng. Sci* 43:2533–2540.

Klausen, M., A. Heydorn, P. Ragas, L. Lambertsen, A. Aaes-Jørgensen, S. Molin, T. Tolker-Nielsen (2003). Biofilm formation by *Psudomonas aeruginosa* wild type, flagella and type IV pili mutants. *Mol. Microbiol.* 48:1511–1524.

Klinkenberg, G., K. Q. Lystad, D. W. Levine, N. Dyrset (2001). pH-controlled cell release and biomass distribution of alginate-immobilized *Lactococcus lactis* subsp. *lactis. J. Appl. Microbiol.* 91:705–714.

Kokubu, T., I. Karube, S. Suzuki (1978). α-Amylase production by immobilized whole cells of *Bacillus subtilis*. *Eur. J. Appl. Microbiol. Biotechnol.* 5:233–240.

Koutinas, A. A., M. Kanellaki, A. Lykourghiotis, M. A. Typas, C., Drainas (1988). Ethanol production by *Zymomonas mobilis* entrapped in alumina pellets. *Appl. Microbiol. Biotechnol.* 28:235–239.

Kren, V., J. Ludvík, O. Kofronová, J. Kozová, Z. Rehácek (1987). Physiological activity of immobilized cells of *Claviceps fusiformis* during long-term semicontinuous cultivation. *Appl. Microbiol. Biotechnol.* 26:219–226.

Kronlöf, J., I. Virkajärvi (1999). Primary fermentation with immobilized yeast. *Proceedings European Brewery Convention Congress;* pp. 761–770.

Kutney, J. P., J. D. Berset, G. M. Hewitt, M. Singh (1988) Biotransformation of dehydroabietic, abietic, and isopimaric acids by *Mortierella isabellina* immobilized in polyurethane foam. *Appl. Environ. Microbiol.* 54:1015–1022.

Kutney, J. P., L. S L. Choi, G. M. Hewitt, P. J. Salisbury, M. Singh (1985). Biotransformation of dehydroabietic acid with resting cell suspensions and calcium alginate-immobilized cells of *Mortierella isabellina*. *Appl. Environ. Microbiol.* 49:96–100.

Lacik, I. (2004) Polyelectrolyte complexes for microcapsule formation. In: Nedovic V., Willaert R. (eds.), *Fundamentals of Cell Immobilization Biotechnology*. Kluwer Academic Publishers, The Netherlands; pp. 103-120.

Lee, K. Y., T. R. Heo (2000) Survival of Bifidobacterium longum immobilized in calcium alginate beads in simulated gastric juices and bile salt solution. *Appl. Environ. Microbiol.* 66: 869–873.

Lee, S. Y., H. N. Chang, Y. S. Um, S. H. Hong (1998). Bacteriorhodopsin production by cell recycle culture of *Halobacterium halobium*. *Biotechnol. Lett.* 20:763–765.

Lejeune, R., G. V. Baron (1995). Effect of aggitation and enzyme production of *Trichoderma reesei* in batch fermentation. *Appl. Microbiol. Biotechnol.* 43:249–258.

Le-Tien, C., M. Millette, M. A. Mateescu, M. Lacroix (2004). Modified alginate and chitosan for lactic acid bacteria immobilization. *Biotechnol Appl Biochem.* 39:189–198.

Leung, A., Y. Ramaswamy, P., Munro, G. Lawrie, L. Nielsen, M. Trau (2005). Emulsion strategies in the microencapsulation of cells: pathways to thin coherent membranes. *Biotechnol. Bioeng.*, in press.

Li, F., S. P. Palecek (2003). *EAP1*, a *Candida albicans* gene involved in binding human epithelial cells. *Eukaryot. Cell* 2: 1266–1273.

Lijun, X., W. Bochu, L. Zhimin, D. Chuanren, W. Qinghong, L. Liu (2005). Linear alkyl benzene sulphonate (LAS) degradation by immobilized *Pseudomonas aeruginosa* under low intensity ultrasound. *Colloids Surf. B. Biointerfaces* 40:25–29.

Lilly, M. D., T. Keshavarz, ,C. Bucke, A. T. Bull, G. Holt (1990). Pilot-scale studies of immobilized *Penicillium chrysogenum*. In: de Bont J.A.M., Visser J., Mattiasson B., Tramper J. (eds.), *Physiology of Immobilized Cells*. Elsevier, Amsterdam; pp. 369–376.

Lim, F., R. D. Moss (1981). Microencapsulation of living cells and tissues. *J. Pharm. Sci* 70:351–354.

Lim, F., A. M. Sun (1980). Microencapsulated islets as a bioartificial endocrine pancreas. *Science* 210:908–910.

Litwin, J. (1992). The growth of Vero cells in suspension as cell aggregates in serum-free medium. *Cytotechnol.* 10:169–174.

Lloyd-George, I., T. M. S. Chang (1993). Free and microencapsulated Erwinia herbicola for the production of tyrosine. *Biomat. Art. Cells Immob. Biotech.* 21:323–333.

Lohmeyer, M., W. Dierkes, H. J. Rehm (1990). Alkaloid production by high-performance strains of *Claviceps purpurea*. In: de Bont J.A.M., Visser J., Mattiasson B., Tramper J. (eds.), *Physiology of Immobilized Cells*. Elsevier, Amsterdam, The Netherlands; pp. 503–512.

Loo, C. Y., D. A. Corliss, N. Ganeshkumar (2000). *S. gordonii* biofilm formation: identification of genes that code for biofilm phenotypes. *J. Bacteriol.* 182:1374–1382.

Lusta, K. A., N. G. Starostina, B. A. Fikhte (1990). Immobilization of micro-organisms: Cytophysiological aspects. In: de Bont J.A.M., Visser J., Mattiasson B., Tramper J. (eds.), *Physiology of Iimmobilized Cells*. Elsevier, Amsterdam; pp. 557–563.

Lysaght, M. J., D. Rein (2004). Immunoisolation. In: Ratner B. D., Hoffman A.S., Schoen F.J., Lemons J.E. (eds.), *Biomaterials Science – An Introduction to Materials in Medicine*. Elsevier Academic Press, London; pp. 728–734.

Ma, T., Y. Li, S.T. Yang, D. A. Kniss (2000). Effects of pore size in 3-D fibrous matrix on human trophoblast tissue development. *Biotechnol. Bioeng.* 70:606–618.

Marin-Iniesta, F., M. Nasri, P. Dhulster, J. N. Barbotin, D. Thomas (1988). Influence of oxygen supply on the stability of recombinant plasmid pTG201 in immobilized *E. coli* cells. *Appl. Biotechnol. Biotechnol.* 28:455–462.

Martins, R. F., F. M. Plieva, A. Santos, R. Hatti-Kaul (2003). Integrated immobilized cell reactor-adsorption system for β-cyclodextrin production: a model study using PVA-cryogel entrapped *Bacillus agaradhaerens* cells. *Biotechnol. Lett.* 25: 1537–1543.

Marwaha, S. S., J. F. Kennedy, P. K., Khanna, H. K. Tewari, A. Redhu (1990). Comparative investigations on the physiological parameters of free and immobilized yeast cells for effective treatment of dairy effluents. In: de Bont J.A.M., Visser J., Mattiasson B., Tramper J. (eds.), *Physiology of Immobilized Cells*. Elsevier, Amsterdam, The Netherlands; pp. 265–273.

Mavituna, F. (2004) Pre-formed carriers for cell immobilization. In: Nedovic V., Willaert R. (eds.), *Fundamentals of Cell Immobilization Biotechnology*. Kluwer Academic Publishers, The Netherlands; pp. 121–139.

Melzoch, K., M. Rychtera, V. Habova (1994). Effect of immobilization upon the properties and behavior of *Saccharomyces cerevisiae* cells. *J. Biotechnol.* 32:59–65.

Mensour, N., A. Margaritis, C. L. Briens, H. Pilkington, I. Russell (1995). Gas lift systems for immobilized cell systems. In: *EBC Monograph XXIV*, EBC Symposium on immobilized yeast applications in the brewing industry, pp. 125–133.

Mensour, N., A. Margaritis, C. L. Briens, H. Pilkington, I. Russell (1996). Applications of immobilized yeast cells in the brewing industry. In: Wijffels R.H., Buitelaar R.M., Bucke C., Tramper J. (eds.), *Immobilized Cells: Basics and Applications*, Amsterdam, Elsevier, pp. 661–671.

Mensour, N., A. Margaritis, C. L. Briens, H. Pilkington, L. Russell (1997). New developments in the brewing industry using immobilized yeast cell bioreactor systems. *J. Inst. Brew.* 103:363–370.

Merchant, F. J. A., A. Margaritis, J. B. Wallace (1987) A novel technique for measuring solute diffusivities in entrapment matrices used in immobilization. *Biotechnol. Bioeng.* 30: 936–945.

Millette, M., W. Smoragiewicz, M. Lacroix (2004), Antimicrobial potential of immobilized *Lactococcus lactis* subsp. *lactis* ATCC 11454 against selected bacteria. *J. Food Prot.* 67:1184–1186.

Mockovciaková, D., V. Janitorová, E. Kaclíková, M. Zagulski, J. Subík (1993). The *ogd1* and *kgd1* mutants lacking 2-oxoglutarate dehydrogenase activity in yeast are allelic and can be differentiated by the cloned amber suppressor. *Curr. Genet.* 24:377–381.

Moo-Young, M., H. W. Blanch (1981). Design of biochemical reactors: mass transfer criteria for simple and complex systems. *Adv. Biochem. Eng.* 19:1–69.

Moreira, J. L., P. M. Alves (1995). Hydrodynamic effects on BHK cells grown as suspended natural aggregates. *Biotechnol. Bioeng.* 46:351–360.

Mussenden, P., T. Keshavarz, G. Saunders, C. Bucke (1993). Physiological studies related to the immobilization of *Penicillium chrysogenum* and penicillin production. *Enzyme Microb. Technol.* 15:2–7.

Najafpour, G., H. Younesi, Ku Syahidah, K. Ismail (2004). Ethanol fermentation in an immobilized cell reactor using *Saccharomyces cerevisiae*. *Bioresour. Technol.* 92:251–260.

Nakaoka, R., T. Tsuchiya, A. Nakamura (2003). Neural differentiation of midbrain cells on various protein-immobilized polyethylene films. *J. Biomed. Mater. Res A.* 64:439–446.

Narziss, L., P. Hellich (1971) Ein Beitrag zur wesentlichen Beschleunigung der Gärung und Reifung des Bieres. Brauwelt 111: 1491–1500.

Nava Saucedo, J. E., J. N. Barbotin, D. Thomas (1989a). Physiological and morphological modifications in immobilized *Gibberella fujikuroi* mycelia. *Appl. Environ. Microbiol.* 55:2377–2384.

Nava Saucedo, J. E., J. N. Barbotin, M. Velut, D. Thomas (1989b). Ultrastructure examination of *Gibberella fujikuroi* mycelia: Effect of immobilization in calcium alginate beads. *Can. J. Microbiol.* 35:1118–1131.

Nava Saucedo, J. E., J. N. Barbotin, D. Thomas (1990) Fusarium moniliforme: A paradigm for physiological studies of immobilized filamentous microorganisms. In: de Bont J.A.M., Visser J., Mattiasson B., Tramper J. (eds.), *Physiology of Immobilized Cells*. Elsevier, Amsterdam; pp. 577–582.

Navrátil, M., P. Gemeiner, E. Sturdík, Z. Dömény, D. Smogrovicová, Z. Antalova (2000). Fermented beverages produced by yeast cells entrapped in ionotropic hydrogels of polysaccharide nature. *Minerva Biotec.* 12:337–344.

Nedovic, V. A., I. Leskosek-Cukalovic, G. Vunjak-Novakovic (1996). Short-time fermentation of beer in an immobilized yeast air-lift bioreactor. In: Hervey J. (ed.), *Proc. 24th Conv. Inst. Brew.*, Singapore, Winetitles, Adelaide, p. 245.

Nedovic, V. A., I. Leskosek-Cukalovic, V. Milosevic, G. Vunjak-Novakovic (1997a). Flavor formation during beer fermentation with immobilized *Saccharomyces cerevisiae* in a gas-lift bioreactor. In: Godia F., Poncelet D (eds.), *Proc. Int. Workshop Bioencapsulation VI*, Barcelona, Spain; pp. 1–4.

Nedovic, V. A., R. Pesic, I. Leskosek-Cukalovic, D. Laketic, G. Vunjak-Novakovic (1997b). Analysis of liquid axial dispersion in an internal loop gas-lift bioreactor for beer fermentation with immobilized yeast cells. In: Olazar M., San José M.J. (eds.), *Proc. II Eur. Conference on Fluidization*, The University of the Basque Country Press Service, Bilbao; pp. 627–635.

Neufeld, R. J., D. J. C. M. Poncelet, S. D. J. M. Norton (1997). Immobilized-cell carrageenan bead production and a brewing process utilizing carrageenan bead immobilized yeast cells. US patent 6217916.

Niazi, J. H., T. B. Karegoudar (2001). Degradation of dimethylphtalate by cells of *Bacillus* sp. immobilized in calcium alginate and polyurethane foam. *J. Environ. Sci. Health Part A Tox. Hazard Subst. Environ. Eng.* 36:1135–1144.

Nielsen, J., M. Carlsen (1996). Fungal pellets. In: Willaert R.G., Baron G.V., De Backer L. (eds.), *Immobilized Living Cell Systems: Modelling and Experimental Methods*. John Wiley & Sons, Chichester; pp. 273–293.

Nilsson, K., P. Brodelius, K. Mosbach (1987). Entrapment of microbial and plant cells in beaded polymers. *Methods Enzymol.* 135:222–230.

Noordsij, P., Rotte J.W. (1967) Mass transfer coefficients to a rotating and a vibrating sphere. *Chem. Eng. Sci.* 22: 1475–1481.

Norton, S., C. Lacroix, J. C. Vuillemard (1993b) Effect of pH on the morphology of *Lactobacillus helveticus* in free-cell batch and immobilized-cell continuous fermentation. *Food Biotechnol.* 7: 235–251.

Norton, S., K. Watson, T.. D'Amore (1995). Ethanol tolerance of immobilized brewers' yeast cells. *Appl. Microbiol. Biotechnol.* 43:18–24.

Obradovic, B., V. A. Nedovic, B. Bugarski, R. G. Willaert, G. Vunjak-Novakovic (2004). Immobised cell reactors. In: Nedovic V., Willaert R. (eds.), *Fundamentals of Cell Immobilization Biotechnology*. Kluwer Academic Publishers, The Netherlands; pp. 411–436.

O'Brien, D. J., J. C. Craig Jr. (1996). Ethanol production in a continuous fermentation/membrane pervaporation system. *Appl. Microbiol. Biotechnol.* 44:699–704.

Ogbonna, J. C., S. Tomiyama, H. Tanaka (1996). Development of a method for immobilization of non-flocculating cells in loffa (*Luffa cylindrica*) sponge. *Process Biochem.* 31: 37–744.

Oh, Y. S., J. Maeng, S. J. Kim (2000). Use of micro-organism-immobilized polyurethane foams to absorb and degrade oil on water surface. *Appl. Microbiol. Biotechnol.* 54:418–423.

Ohta, T., J. C. Ogbonna, H. Tanaka, M. Yajima (1994). Development of a fermentation method using immobilized cells under unsterile conditions. 2. Ethanol and L-lactic acid production without heat and filter sterilization. *Appl. Microbiol. Biotechnol.* 42:246–250.

Oriel, P. (1988). Immobilization of recombinant *Escherichia coli* in silicone polymer beads. *Enzyme Microb. Technol.* 10:518–523.

Orive, G., M. De Castro, S. Ponce R. M. Hernandez, A. R. Gascon, J. L. Pedraz (2005). Long-term expression of erythropoietin from myoblasts immobilized in biocompatible and neovascularized microcapsules. Mol. Therapy, in press.

Orive, G., R. M. Hernandez, A. R. Gascon, J. L. Pedraz (2003). Survival of different cell lines in alginate-agarose microcapsules. Eur. J. Pharmac. Sci. 18:23–30.

Orive, G., R. M. Hernandez, A. R. Gascon, R. Calafiore, T. M. S. Chang, P. De Vos, G. Hortelano, D. Hunkeler, I. Lacík, A. M. J., Shapiro, J. L., Pedraz (2004). History, challenges and perspectives of cell microencapsulation. *Trends Biotechnol.* 22:87–92.

Øyaas, J., I. Storrø, M. Lysberg, H. Svendsen, D. W. Levine (1995). Determination of effective diffusion coefficients and distribution constants in polysaccharide gels with non-steady-state measurements. *Biotechnol. Bioeng.* 47:501–507.

Pajunen, E. (1995) Immobilized yeast lager beer maturation: DEAE-cellulose at Synebrychoff. *EBC Monograph XXIV*, EBC Symposium on immobilized yeast applications in the brewing industry; pp. 24–40.

Perrot, F., M. Hebraud, R., Charlionet, G. A. Junter, T. Jouenne (2001). Cell immobilization induces changes in the protein response of *Escherichia coli* K-12 to a cold shock. *Electrophoresis* 22:2110–2119.

Picioreanu, C., M. C. M. van Loosdrecht, J. J. Heijnen (1998). A new combined differential-discrete cellular automaton approach for biofilm modeling: Application for growth in gel beads. *Biotechnol. Bioeng.* 57:718–731.

Pittner, H., W. Back, W. Swinkels, E. Meersman, B. Van Dieren, H. Lomni (1993). Continuous production of acidified wort for alcohol-free-beer with immobilized lactic acid bacteria. *Proceedings of the European Brewery Convention Congress*; pp. 323–329.

Ponce, S., G. Orive, A. R. Gascon, R. M., Hernandez, J. L. Pedraz (2005). Microcapsules prepared with different biomaterials to immobilize GDNF secreting 3T3 fibroblasts. *Int. J. Pharma.* 293:1–10.

Prakash, S., T. M. S. Chang (1995). Preparation and in vitro analysis of microencapsulated genetically engineered *E. coli* DH5 cells for urea and ammonia removal. *Biotechnol. Bioeng.* 46:621–626.

Prakash, S., M. L. Jones (2005). Artificial cell therapy: new strategies for the therapeutic delivery of live bacteria. *J. Biomed. Biotechnol.* 1:44–56.

Quek, C. H., J. Li, T. Sun, M. L. H. Chan, H. Q. Mao, L. M. Gan, K. M. Leong, H. Yu (2004). Photo-crosslinkable microcapsules formed by polyelectrolyte copolymer and modified collagen for rat hepatocyte encapsulation. *Biomaterials* 25:351–354.

Reynolds, T. B., G. R. Fink (2001). Baker's yeast, a model for fungal biofilm formation. *Science* 291:878–881.

Riddle, K. W., D. Mooney (2004) Biomaterials for cell immobilization: a look at carrier design. In: Nedovic V., Willaert R. (eds.), *Fundamentals of Cell Immobilization Biotechnology*. Kluwer Academic Publishers, The Netherlands; pp. 15–32.

Roca, E., N. Meinander, B. Hahn-Hägerdal (1996). Xylitol production by immobilized recombinant *Saccharomyces cerevisiae* in a continuous packed-bed bioreactor. *Biotechnol. Bioeng.* 51:317–326.

Rokstad, A. M., S. Holtan, B. Strand, B. Steinkjer, L. Ryan, B. Kulseng, G. Skjåk-Bræk, T. Espevik (2002). Microencapsultaion of cells producing therapeutic proteins: Optimizing cell growth and secretion. *Cell Transplant.* 11: 313–324.

Ryan, W., S. J. Parulekar (1991). Immoblization of *Escherichia coli* JM103(pUC8) in κ-carrageenan coupled with recombinant protein release by in situ cell membrane permeabilization. *Biotechnol. Prog.* 34:99–100.

Sanderson, G. R., V. L. Bell, D. A. Ortega (1989) A comparision of gellan gum, agar, κ-carrageenan and algin. *Cereal Foods* World 34:991–998.

Santos, V. L., N. M. Heilbuth, V. R. Linardi (2001). Degradation of phenol by *Trichosporon* sp. LE3 cells immobilized in alginate. *J. Basic Microbiol.* 41:171–178.

Santos, J. C., I. R. Pinto, W. Carvalho, I. M. Mancilha, M. G. Felipe, S. S. Silva (2005). Sugarcane bagasse as raw material and immobilization support for xylitol production. *Appl. Biochem. Biotechnol.* 121-124:673–683.

Sauer, K., A. K. Camper, G. D. Ehrlich, J. W. Costerton, D. G. Davies (2002). *Pseudomonas aeruginosa* displays multiple phenotypes during development as a biofilm. *J. Bacteriol.* 184:1140–1154.

Scannell, A. G. M., C. Hill, R. P. Ross, S., Marx, W. Hartmeier, E. K. Arendt (2000). Continuous production of lacticin 3147 and nisin using cells immobilized in calcium alginate. *J. Appl. Microbiol.* 89:573–579.

Schwinger, C., S. Koch, U. Jahnz, P. Wittlich, N. G. Rainov, J. Kressler (2002). Hightroughput encapsulation of murine fibroblasts in alginate using the JetCutter technology. *J. Microencapsil.* 19: 273–280.

Scott, C .D. (1983). Fluidized-bed bioreactors using flocculation strain of Zymomonas mobilis for ethanol production. *Ann. N.Y. Acad. Sci.* 413:448–456.

Shahbazi, A., M. R. Mims, V. Shirley, S. A. Ibrahim, A. Morris (2005). Lactic acid production from cheese whey by immobilized bacteria. Appl. Biochem. Biotechnol. 122:529–540.

Shapiro, J. A. (1985). Scanning electron microscope study of *Pseudomonas putida* colonies. *J. Bacteriol.* 164: 1171–1181.

Shapiro, J. A. (1987). Organisation of developing *Escherichia coli* colonies viewed by scanning electron microscopy. *J. Bacteriol.* 169:142–156.

Sharma, D. K., H. S. Saini, M. Singh, S. S. Chimni, B. S. Chadha (2004). Biological treatment of textile dye Acid violet-17 by bacterial consortium in an up-flow immobilized cell bioreactor. *Lett. Appl. Microbiol.* 38:345–350.

Shen, H. Y., S. De Schrijver, N. Moonjai, K. J. Verstrepen, F. R. Delvaux (2004). Effects of CO_2 on the formation of flavor volatiles during fermentation with immobilized brewer's yeast. *Appl. Microbiol. Biotechnol.* 64:636–643.

Sherwood, T. K. A., R. L. Pigford, C. R. Wilke (1975). *Mass Transfer*, McGraw-Hill Book Company, London, UK.

Shim, H., S. T. Yang. (1999). Biodegradation of benzene, toluene, ethylbenzene, and o-xylene by a coculture of *Pseudomonas putida* and *Pseudomonas fluorescens* immobilized in a fibrous-bed bioreactor. *J. Biotechnol.* 67:99–112.

Shindo, M. (1994) Method of producing liquor. European patent 0669393.

Shindo, S., H. Sahara, S. Koshino (1994). Suppression of α-acetolactate formation in brewing with immobilized yeast. *J. Inst. Brew.* 100:69–72.

Shinmyo, A., H. Kimura, H. Okada (1982). Physiology of α-amylase production by immobilized *Bacillus amyloliquefaciens*. *Eur. J. Appl. Microbiol. Biotechnol.* 14:7–12.

Shu, C. H., S. T. Yang (1996). Effect of particle loading on GM-SF production by *Saccharomyces cerevisiae* in a three-phase fluidized bed bioreactor. *Biotechnol. Bioeng.* 51:229–236.

Shuler, M. L., T. Cho, T. Wickham, O. Ogonah, M. Kool, D. A. Hammer, R. R. Granados, H. A. Wood (1990). Bioreactor development for the production of viral pesticides or heterogous proteins in insect cell cultures. *Ann. N.Y. Acad. Sci.* 589:399–422.

Simon, J. P., T. Benoot, J. P. Defroyennes, B. Deckers, D. Dekegel, J. Vandegans (1990). Physiology and morphology of Ca-alginate entrapped *Saccharomyces cerevisiae*. In: de Bont J.A.M., Visser J., Mattiasson B., Tramper, J. (eds.), *Physiology of Immobilized Cells*. Elsevier, Amsterdam, The Netherlands; pp. 583–590.

Sirkar, K., W. Kang (2004). Whole cell immobilization in chopped hollow fibers. In: Nedovic V., Willaert R. (eds.), *Fundamentals of Cell Immobilization Biotechnology*. Kluwer Academic Publishers, The Netherlands; pp. 245–256.

Smogrovicová, D., Z. Dömény, P. Gemeiner, A. Maloviková, E. Sturdík (1997). Reactors for continuous primary beer fermentation using immobilized yeast. *Biotechnol. Techn.* 11:261–264.

Smogrovicová, D., Z. Dömény, M. Navrátil, P. Dvorák (2001). Continuous beer fermentation using polyvinyl alcohol entrapped yeast. *Proceedings European Brewery Convention Congress*; 50, pp. 1–9.

Sode, K., T. Morita, A. Peterhan, F. Meussdoerfer, K. Mosbach, I. Karube (1988). Continuous production of α-peptide using recombinant yeast cells. A model for continuous production of foreign peptide by recombinant yeast. *J. Biotechnol.* 8:113–122.

Steenson, L. R., T. R. Klaenhammer, H. E. Swaisgood (1987). Calcium alginate-immobilized cultures of lactic streptococci are protected from bacteriophages. *J. Dairy Sci.* 70:1121–1127.

Stoodley, P., K. Sauer, D. G. Davies, J. W. Costerton (2002). Biofilms as complex differentiated communities. *Ann. Rev. Microbiol.* 56:187–209.

Strand, B. L., O. Skjåk-Bræk, O. Gåserød (2004), Microencapsule formulation an formation. In: Nedovic V., Willaert R. (eds.), *Fundamentals of Cell Immobilization Biotechnology*. Kluwer Academic Publishers, The Netherlands; pp. 165–183.

Sun, Y., S. Furusaki, A. Yamauchi, K. Ichimura (1989). Diffusivity of oxygen into carriers entrapping whole cells. *Biotechnol. Bioeng.* 34:55–58.

Sun, W., M. W. Griffiths (2000). Survival of bifidobacteria in yogurt and simulated gastric juices following immobilization in gellan-xanthan beads. Int. J. Food 61:17–25.

Sungur, S., U. Akbulut (1994). Immobilization of β-galactosidase onto gelatin by glutaraldehyde and chromium(III) acetate. *J. Chem. Technol. Biotechnol.* 59:303–306.

Suwannakham, S., S. T. Yang (2005). Enhanced propionic acid fermentation by *Propionibacterium acidipropionici* mutant obtained by adaptation in a fibrous-bed bioreactor. *Biotechnol. Bioeng.* In press.

Svoboda, A., P. Ourednicek (1990). Yeast protoplasts immobilized in alginate: cell wall regeneration and reversion to cells. *Curr. Microbiol.* 20:335–338.

Takagi, M., K. Shiwaku, T. Inoue, Y. Shirakawa, Y. Sawa, H. Matsuda, T. Yoshida (2003). Hydrodynamically stable adhesion of endothelial cells onto a propylene hollow fiber membrane by modification with adhesive protein. *J. Artif. Organ.* 6:222–226.

Tanaka, H., T. Ohta, S. Harada. J. C. Ogbonna, M. Yajima (1994) Development of a fermentation method using immobilized cells in unsterile conditions. 1. Protection of immobilized cells against anti-microbial substances. *Appl. Microbiol. Biotechnol.* 41:544–550.

Tharakan, J. P., P. C. Chau (1986). A radial hollow fiber bioreactor for the large-scale culture of mammalian cells. *Biotechnol. Bioeng.* 18:329–342.

Tsai, W. B., M. C. Wang (2005). Effects of an avidin-biotin binding system on chondrocyte adhesion and growth om biodegradable polymers. *Macromol. Biosc.* 5:214–221.

Urek, R. O., N. K. Pazarlioglu (2004). A novel carrier for *Phanerochaete chrysosporium* immobilization. *Artif. Cells Blood Substit. Immobil. Biotechnol.* 32:563–574.

van der Sluis, C., C. J. P. Stoffelen, S. J. Castelein, G. H. M. Engbers, E. G. ter Schure, J. Tramper, R. H. Wijffels (2001). Immobilized salt-tolerant yeasts: application of a new polyethylene-oxide support in a continuous stirred-tank reactor for flavor production. *J. Biotechnol.* 88:129–139.

Van De Winkel, L. (1995). Design and optimization of a multipurpose immobilized yeast bioreactor system for brewery fermentations. *Cerevisia* 20(1):77–80.

Van Dieren, D. (1995) Yeast metabolism and the production of alcohol-free beer. *EBC Monograph XXIV*, EBC Symposium on immobilized yeast applications in the brewing industry; pp. 66–76.

van 't Riet, K., J. Tramper (1991). *Basic Bioreactor Design.* Marcel Dekker, New York, USA.

Varfolomeyev, S D., E. I. Rainina, V. I. Lozinsky, S. V. Kalyuzhny, A. P. Sinitsyn, T. A. Makhlis, G. P. Bachurina, I. G. Bokova, O. A. Sklyankina, E. B. Agafonov (1990). Application of polyvinyl alcohol cryogels for immobiliztion of mesophilic and thermophilic micro-organisms. In: de Bont J.A.M., Visser J., Mattiasson B., Tramper J. (eds.), *Physiology of Immobilized Cells*. Elsevier, Amsterdam; pp. 325–330.

Verstrepen, K. J., G. Derdelinckx, H. Verachtert, F. R., Delvaux (2003). Yeast flocculation: What brewers should know. Appl. *Microbiol. Biotechnol.* 61:197–205.

Vieth, W. R. (1989). Inducible recombinant cell cultures and bioreactors. In: Fiechter A., Okada H., Tanner A. (eds.), *Bioproducts and Bioprocesses*. Springer-Verlag, Berlin; pp. 51–70.

Vilain, S., P. Cosette, M. Hubert, C. Lange, G. A. Junter, T. Jouenne (2004). Proteomic analysis of agar gel-entrapped *Pseudomonas aeruginosa*. *Proteomics* 4: 1996–2004.

Visted, T., T. Furmanek, P. Sakariassen, W. B. Foegler, K. Sim, H. Westphal, R. Bjerkvig, M. Lund-Johansen (2003). Prospects for delivery of recombinant angiostatin by cell-encapsulation therapy. *Human Gene Therapy* 14:1429–1440.

Vives, C., C. Casas, F. Godia, C. Sola (1993). Determination of the intrinsic fermentation kinetics of *Saccharomyces cerevisiae* cells immobilized in Ca-alginate beads and observations on their growth. *Appl. Microbiol. Biotechnol.* 38:467–472.

Viyas, V. K., S. Kuchin, C. D. Berkely, M. Cralson (2003). Snf1 kinases with different β-subunit isoforms play distinct roles in regulating haploid invasive growth. *Mol. Biol. Cell* 23:1341–1348.

von Weymarn, N., K. Kiviharju, M. Leisola (2002). High-level production of D-manitol with membrane cell-recycle bioreactor. *J. Ind. Microbiol. Biotechnol.* 29:44–49.

Walls, E. L., J. L. Gainer (1989). Retention of plasmid bearing cells by immobilization. *Biotechnol. Bioeng.* 34:717–724.

Wang, L., R. M. Shelton, P. R. Cooper, M. Lawson, J. T. Triffitt, J. E. Barralet (2003). Evaluation of sodium alginate for bone marrow cell tissue engineering. *Biomaterials* 24:3475–3481.

Weber, W., M. Rinderknecht, M. Daoud-El Baba, F. N. de Glutz, D. Aubel, M. Fussenegger (2004). CellMAC: A novel technology for encapsulation of mammalian cells in cellulose sulphate/pDADMAC capsules assembled on a transient alginate/Ca^{2+} scaffold. *J. Biotechnol.* 114:315–326.

White, G. F., O. R. T. Thomas (1990). Immobilization of the surfactant-degrading bacterium *Pseudomonas* C12B in polyacrylamide gel beads: I. Effect of immobilization on the primary and ultimate biodegradation of SDS, and redistribution of bacteria within beads during use. *Enzyme Microb. Technol.* 12:697–705.

Willaert, R., G. Baron (1993). Growth kinetics of gel-immobilized yeast cells studied by on-line microscopy. *Appl. Microbiol. Biotechnol.* 39:347–352.

Willaert, R. G., G. V. Baron (1994). Effectiveness factor calculations for immobilized growing cell systems. *Biotechnol. Techn.* 8:695–700.

Willaert, R. G., G. V. Baron (1996). Gel entrapment and microencapsulation: methods, applications and engineering principles. *Rev. Chem. Eng.* 12:1–205.

Willaert, R., V. Nedovic, G. V. Baron (2004). Physiology of immobilized cells. In: Nedovic V., Willaert R. (eds.), *Fundamentals of Cell Immobilization* Biotechnology. Kluwer Academic Publishers, The Netherlands; pp. 469–492.

Wittlich, P., E. Capan, M. Schlieker, K. D. Vorlop, U. Jahnz (2004). Entrapment in Lentikats: Encpsulation of various biocatalysts – bateria, fungi, yeast or enzymes into polyvinyl alcohol based hydrogel particles. In: Nedovic V., Willaert R. (eds.), *Fundamentals of Cell Immobilization Biotechnology.* Kluwer Academic Publishers, The Netherlands; pp. 53–63.

Wu, K. Y. A., K. D. Wisecarver (1992). Cell immobilization using PVA crosslinked with boric acid. *Biotechnol. Bioeng.* 39:447–449.

Yamada, K. M., B. Geiger (1997). Molecular interactions in cell adhesion complexes. *Curr. Opin. Cell Biol.* 9:76–85.

Yamauchi, Y., T. Kashihara (1995). Kirin immobilized system. *EBC Monograph XXIV*, EBC Symposium on immobilized yeast applications in the brewing industry; pp. 99–117.

Yamauchi, Y., T. Okamoto, H. Murayama, A. Nagara, T. Kashihara, K. Nakanishi (1994). Beer brewing using an immobilized yeast bioreactor design of an immobilized yeast bioreactor for rapid beer brewing system. *J. Ferm. Bioeng.* 78:443–449.

Yang, S. T., J. Luo, C. Chen (2004). A fibrous-bed bioreactor for continuous production of monoclonal antibody by hybridoma. *Adv. Biochem. Eng. Biotechnol.* 87:61–96.

Zhang, X., S. Bury, D. Dibiasio, J. Miller (1989). Effects of immobilization on growth, substrate consumption, β-galactosidase induction, and byproduct formation in *Escherichia coli. J. Ind. Microbiol.* 4:239.

Zhang, Z., J. Scharer, M. Moo-Young (1997). Protein production using recombinant yeast in an immobilized-cell-film airlift bioreactor. *Biotechnol. Bioeng.* 55:241–251.

Zheng, S., Z. X. Xiao, Y. L. Pan, M. Y. Han, Q. Dong (2003). Continuous release of interleukin 12 from microencapsulated engineered cells for colon cancer therapy. *World J. Gastroenterol.* 9:951–955.

11

Biosensors in Bioprocess Monitoring and Control: Current Trends and Future Prospects

Chris E. French and Marco F. Cardosi

11.1   Introduction

Fermentation processes are complex systems, with many interacting variables. Continuous monitoring of parameters such as pH, temperature, and dissolved oxygen levels, with direct feedback to control systems, is standard even in the smallest bioreactors and the enabling technologies are well established, as described in Chapters 12, 13, and 14 in this volume. Monitoring of other parameters, such as the levels of substrates and products, is also extremely useful but presents greater difficulties. Products and metabolites can be monitored off-line by analyzing samples using techniques such as HPLC, GC, or ELISA, but this presents inevitable delays. In this chapter, we will examine the potential benefits offered by biosensors for process monitoring.

11.2   Biosensors in process monitoring

The term "biosensor" is used in slightly different ways by different authors. The common theme is that a biosensor couples a biological element, which provides specific recognition of the target analyte, to a transducer, which generates an electrical signal. The electrical signal can be directed to a digital readout, or can be logged or used as an input to a control system. The biological recognition element may be an enzyme, an antibody or similar binding molecule, or a living cell. The transducer may detect a redox reaction, light emission, altered levels of an ion, a luminescent or fluorescent signal, or a change in mass associated with a surface. Whatever the nature of the initial response, it is

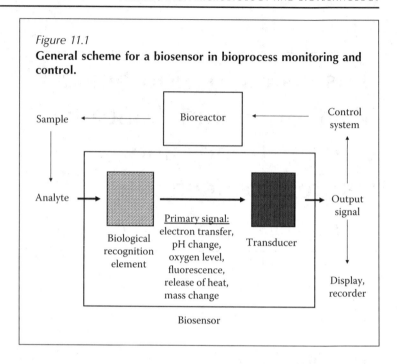

Figure 11.1

General scheme for a biosensor in bioprocess monitoring and control.

converted into an electrical signal, which can be used in process monitoring and control (Figure 11.1). The aim of a biosensor is to provide a rapid, highly specific, quantitative response, with a minimal or zero requirement for sample preprocessing.

Why are biosensors potentially superior to analysis by removal of samples for manual or automatic analysis by techniques such as HPLC or ELISA? The simple answer is specificity and response time. In the case of HPLC, for example, a relatively nonspecific detector, such as an ultraviolet absorption or refractive index instrument, is used, and specificity is supplied by separating compounds on a column; compounds are essentially differentiated based on retention time. Thus, there is an intrinsic and unavoidable lag in obtaining results. By contrast, a biosensor uses a specific biological detector and is expected to supply a steady-state response within about 60 seconds or so. Techniques such as ELISA also use a biological detection element, but require considerable sample processing over a number of steps, again leading to delays. The ideal of a biosensor (not always achieved!) is to apply such a biological detection element in a configuration that gives a rapid automatic output without such sample processing.

The biological element of the biosensor is critical in that it provides the specificity of response. This relies on the evolved ability of biological molecules (especially proteins) to recognize their interaction partners with extremely high affinity. The main types of recognition element used are enzymes and antibodies. In some cases nucleic acids or whole living cells, usually bacteria, can be used.

The use of a biological sensing element is both the strength and the weakness of biosensors. Biological sensing elements are relatively labile and are generally not stable to high temperatures or other sterilizing agents, and sterilizing them is likely to be difficult or impossible. They are therefore generally unsuitable for in-dwelling probes inside the bioreactor, unless separated from the fermentation broth by a cell-impermeable membrane. They also generally become denatured to some degree over time under normal reaction conditions, so their performance must be monitored, and they must be replaced on a regular basis. Expected lifetime of the biological element is an important parameter to bear in mind when considering the use of a biosensor.

Biosensors can be applied to many fields. The major application, historically and currently, is the quantitation of glucose in body fluids for the control of diabetes, using biosensors based on glucose oxidase (described in more detail in the case study in Section 11.13). Biosensors are also applied to the detection of many other analytes in clinical samples, as well as to detection of pollutants in the environment, monitoring of various components in the food industry, and monitoring of metabolites during fermentation processes, the topic of this chapter. This review is aimed at the nonspecialist and will give a basic outline of the principles of operation of some well-established types of biosensor which are potentially suitable for process monitoring and control. For a recent review of developments in process control technology, including biosensors, see Schugerl (2001). A number of other reviews describe recent advances and novel configurations in biosensors (see, for example, Nakamura and Karube, 2003; Lowe, 1999).

Biosensors can be applied to process monitoring in a variety of ways. As mentioned above, lability of the detection element makes sterilization problematic, so direct insertion of the sensor into the bioreactor after the fashion of an oxygen electrode is not generally feasible. Biosensors can be used for *in situ* measurements if coupled with a suitable device for separation from the cells, such as a tangential flow microfiltration unit. However, to judge from the recent literature, perhaps the most popular way of using biosensors in bioprocess monitoring is **flow injection analysis (FIA)**. This refers to automatic abstraction of samples from the test system and their injection into a flow of a carrier liquid, which is constantly passed through the detection element. For a more detailed description, see Schugerl (2001). Some types of biosensors, such as some amperometric enzyme electrodes, are potentially suited to continuous monitoring; others, especially those based on antibodies, may require regeneration of the detection surface by removal of bound ligand, or even replacement of the detection surface, between measurements. Again, some types of biosensors, particularly those based on some redox enzymes, may require the addition of small soluble co-factors or co-substrates along with the sample, and are therefore not suitable for *in situ* use. Sensors that do not require such additions are referred to as "reagentless."

11.3 Overview of transduction methods

The transduction method to be used depends on the type of signal generated by the biological recognition element. The main types of transducer are electrical (potentiometric, amperometric, conductimetric, and capacitive); optical (based on absorbance, fluorescence, luminescence, or surface plasmon resonance); thermal (calorimetric); and mechanical (piezoelectric). The most commonly used techniques in biosensors suitable for process monitoring are

- **Amperometric.** In an amperometric transducer, an electrode is held poised at a constant voltage relative to a **reference electrode** (usually Ag/AgCl) by the use of a circuit known as a **potentiostat**. Redox-active molecules can be oxidized or reduced at the electrode surface, generating a current, which is measured. The potential at which the electrode is held is chosen so as to allow oxidation or reduction of the target analyte whilst minimizing the oxidation or reduction of other substances that may be present. Interference by redox-active molecules such as ascorbic acid, which may be present in fermentation broths, must always be considered. Amperometric transduction is ideal where an enzyme reaction generates or consumes a redox-active substance such as O_2, H_2O_2, NAD(P)H, or PQQ. For recent reviews describing electrochemical biosensors, see May (1999), Zhang et al. (2000), Chaubey and Malhotra (2001), and Mehrvar and Abdi (2004).

- **Potentiometric.** In a potentiometric transducer, an **ion-selective electrode** is used, and the build-up of the appropriate ion causes a change in potential (voltage) with very little current flow. The standard pH electrode is an example of such an ion-selective electrode. A variety of other ion or gas-selective electrodes are available, and these can easily be used as the basis for a biosensor if an enzyme is available that acts on the target analyte to generate or consume the appropriate ion or gas. Potentiometric transduction may be suitable when the reaction generates or consumes H^+, NH_4^+ (NH_3), or CO_3^{2-} (CO_2). Potentiometric biosensors are described along with amperometric biosensors in the reviews listed above. The **ENFET** (Enzyme Field Effect Transistor) can be considered as a special case of potentiometric transduction, since it detects similar signals. An ENFET is a robust solid-state device consisting of an **ISFET** (ion-selective field effect transistor) with an immobilized enzyme layer over the gate electrode.

For recent reviews of ENFET technology, see Miao et al. (2003), Schoning et al. (2002), and Sandifer and Voycheck (1999).

- **Thermal.** In this case, an extremely sensitive temperature-sensing element (**thermistor**) is used to detect the increase in temperature due to heat released by an enzyme-catalyzed reaction. This technique is potentially more widely applicable than amperometric or potentiometric transduction, in that it can potentially be used with any enzyme-catalyzed reaction, though obviously it is likely to be most sensitive when the reaction is highly exothermic. For a recent review of thermal biosensors, see Ramanathan and Danielsson (2001).

- **Optical.** Optical transduction in optical fiber biosensors is based on the transmission of light along an optical fiber to a detector. Light may be generated by fluorescent or luminescent reactions associated with enzymes immobilized at the tip of the fiber optic bundle. Compared to electrical transduction, optical transduction has the advantage of not being affected by electrically noisy environments, but the disadvantage that it may be affected by turbidity. Optical transduction is suitable where an enzyme consumes or generates a fluorescent cofactor such as NAD(P)H, or where a reaction product can be used to generate light in a chemiluminescent reaction, as in the peroxidase-catalyzed reaction of H_2O_2 with luminol. Another application is the bioluminescent detection of ATP based on the firefly luciferase reaction. Optical fibers can also be applied to antibody-based detection using fluorescently labeled competing ligands or second antibodies. For recent reviews of optical biosensors, see Marazuela and Moreno-Bondi (2002), Ulber et al. (2003), Monk and Walt (2004), Choi (2004), and Gauglitz (2005).

- **Surface Plasmon Resonance (SPR).** SPR is an altogether different class of optical technique and is applied to the detection of small changes in mass due to ligand binding in a film of immobilized antibodies (or other binding molecules). The technique is described in more detail below. Essentially, a ligand (the analyte) binds to a film of immobilized antibody, and the change in mass of the film due to ligand binding is detected. SPR is most suitable for direct detection of relatively large analytes such as proteins; it may be useful, for example, in monitoring the production of recombinant proteins in a bioreactor. Small molecules can also be detected using indirect (competitive) assays. For recent reviews of SPR biosensors, see Homola (2003) and Rich and Myszka (2000). SPR is also discussed in the reviews of optical biosensors listed above.

- **Piezoelectric.** Like SPR, piezoelectric transduction is applied to the detection of small changes in mass due to ligand binding in a film of immobilized antibodies. In this case, the film is immobilized on the surface of a piezo-electric crystal such as quartz, and detection is based on changes in the mechanical properties of the surface film as its mass increases, measured by changes in its vibration properties when an electrical potential is applied. The most commonly used device for biosensors is the Quartz Crystal Microbalance (QCM). For recent reviews of piezo-electric biosensors, see Su et al. (2000) and O'Sullivan and Guilbault (1999).

11.4   Catalytic biosensors: Enzymes as biological sensing elements

Enzymes form the basis of the most widely used biosensors. The physiological role of an enzyme is to recognize its target substrate and to catalyze some reaction, generating a new product. Enzymes thus provide a good basis for biosensors, in that they both recognize a target compound and cause (or, more correctly, greatly enhance the rate of) some reaction that can generate a signal suitable for transduction.

When compared to binding molecules such as antibodies, enzymes offer a second layer of recognition specificity; a non-target compound with roughly a similar shape to the target analyte may be able to bind to the enzyme's active site, but unless it can also undergo the same reaction, it will not be detected, so will not give a false positive. The downside of this is that enzyme reactions are subject to **inhibition**, by both **competitive** and **noncompetitive inhibitors**. A competitive inhibitor is a compound similar in shape to the target analyte (the normal enzyme substrate), which may bind to the active site and reduce the quantity of active sites available for target analyte binding, thereby decreasing the signal; a noncompetitive inhibitor is one that binds elsewhere on the enzyme, decreasing the reaction rate by other means. The possible presence of inhibitors in the sample must always be considered when using an enzyme in a biosensor or any other type of enzyme-based assay, and some sample pretreatment may be necessary to remove any inhibitors which may be present.

Any biosensor device must be calibrated to establish its **sensitivity** (the lowest concentration of analyte for which it gives a detectable response) and the **dynamic range** of the response (the range of substrate concentrations at which the biosensor will give a response from which the substrate concentration can be estimated). Figure 11.2 shows a simplified enzyme reaction scheme, as well as standard equations, based on certain simplifying assumptions, which are used to describe enzyme activity. One important parameter is K_m, the **Michaelis constant**. This is

Figure 11.2
Enzyme kinetics. (a) Scheme for single-substrate enzyme reactions. The enzyme and substrate interact to form an enzyme-substrate complex, which reacts to form an enzyme-product complex, from which the product is released. The enzyme-substrate complex and enzyme-product complex are not indistinguishable by steady-state kinetics. (b) The Michaelis-Menten equation for simple enzyme-catalysed reactions. V = rate of reaction; Vmax = maximum rate of reaction at saturating substrate concentrations; [S] = substrate concentration; Km = Michaelis constant, the substrate concentration at which V is half of Vmax. This equation makes several assumptions, one of which is that the enzyme concentration is much smaller than the substrate concentration. (c) Graph showing the shape of the rate dependence on substrate concentration.

(a)
$$\text{Enzyme + Substrate} \underset{k_{-1}}{\overset{k_1}{\rightleftharpoons}} \text{Enzyme:substrate Complex} \overset{k_2}{\longrightarrow} \text{Enzyme + Product}$$

(b) $\quad V \text{ (rate)} = \dfrac{V_{max} \cdot [S]}{K_m + [S]}$

(c)

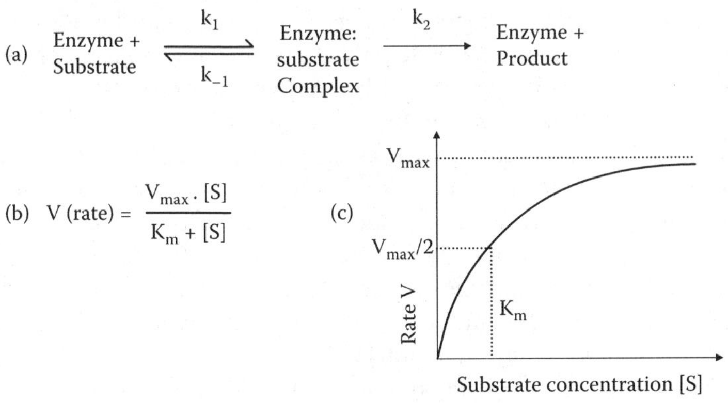

a measure of the affinity of the substrate for the enzyme. In the limiting case where substrate binding and unbinding is much more rapid than reaction, it is equal to K_s, the dissociation constant; in more realistic cases, it also takes account of the rate of reaction. In practical terms, K_m is the substrate concentration at which the rate of reaction, and therefore the signal, is half the maximal rate under the conditions used. At substrate concentrations below K_m, reaction rate (and signal) is approximately proportional to substrate concentration, provided that the substrate concentration is much higher than the enzyme concentration; at substrate concentrations much higher than K_m (in practice, perhaps three times K_m or more), the reaction rate, and signal, are no longer dependent on substrate concentration. If the expected substrate concentration is much higher than K_m, a dilution step may be necessary. This simplified analysis assumes that diffusion of substrates or products is not rate-limiting. In a biosensor context, where the enzyme and perhaps cofactors are immobilized, kinetics may be controlled by the rate of diffusion of substrate to the sensor surface, which is proportional to the difference in substrate concentration between the bulk solution and the local medium at the sensor surface.

The use of an enzyme as a recognition element in a biosensor depends on its ability to generate a signal that can easily be detected and converted into an electrical response. In this regard, not all enzymes are equally convenient. For example, isomerases, which simply alter the configuration of substrate molecules, are unlikely to generate a useful signal. Generally, the most useful class of enzymes for biosensors are **oxidoreductases**, which oxidize or reduce substrates; by their very nature they are involved in the movement of electrons, and lend themselves to amperometric detection methods. **Hydrolytic** and **lyase** enzymes split substrate molecules into smaller parts; they may be useful if they generate or consume a substance that can be detected potentiometrically. Thermal detection is based on the release of energy during an enzyme-catalyzed reaction, as detected by an extremely sensitive thermistor. In principle, this type of device can be used for almost any type of enzyme-catalyzed reaction, though clearly it is likely to work better for highly exothermic reactions.

In some cases, enzyme-based systems require additional small soluble molecules, such as co-factors or co-substrates, to be added for enzyme activity. This necessity for the addition of reagents departs from the ideal concept of the biosensor, in which the sensor is simply exposed to the analyte solution and generates a rapid specific response. Systems that do not require the addition of reagents are called **"reagent-free"** or **"reagentless"** systems. Many ingenious concepts have been applied to convert reagent-requiring systems to reagentless systems. Some of these are discussed below.

A principal disadvantage of enzymes is that enzyme-based biosensors can only be developed for analytes for which a suitable enzyme can be found. While studies of enzyme structure-function relationships are proceeding at a great rate, the current state of the art does not yet allow us to rationally modify or design enzymes to bind to a particular substrate and catalyze a particular reaction. Sometimes a similar effect can be achieved by **"directed evolution,"** in which large numbers of random mutants are screened for a desired change in substrate specificity (see, for example, Arnold and Volkov, 1999), but this remains a lengthy and uncertain procedure.

11.5   Affinity biosensors: Antibodies as biological detection elements

Compared to enzymes, antibodies and similar binding molecules have the advantage of much greater versatility. In principle, a biosensor can be generated to quantify any molecule for which an antibody or other binding molecule can be generated.

Figure 11.3
Affinity-based assays. (a) Direct assay. Binding of the analyte ligand (A) to the immobilised capture molecule is detected directly through some change in the properties of the immobilised film due to ligand binding. This is suited to Surface Plasmon Resonance and piezoelectric transduction. (b) Sandwich assay. A second, labelled affinity molecule (eg antibody) binds to the captured analyte ligand. Detection is based on the quantity of label (L) at the capture surface. The label (L) may be a fluorophore or an enzyme; detection may be optical or electrical. (c) Competitive assay. There are a number of variants of this. In the version depicted here, the analyte ligand and a labelled competing ligand (CL) compete for binding sites on the film of immobilised capture molecules. Increased quantities of analyte in the sample mean that less label is detected at the capture surface.

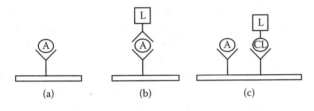

(a)   (b)   (c)

The disadvantage is that, unlike enzymic reactions, a binding event does not intrinsically generate a strong specific signal, so that binding events are basically harder to detect than reaction events. In the context of biosensors, binding events can be detected directly, as in Surface Plasmon Resonance or piezoelectric systems, or indirectly, using competitive assays with labeled competing ligands, or sandwich assays with labeled second antibodies. Biosensors based on antibodies are sometimes called **immunosensors**. For recent reviews of immunosensor technology, see Bange et al. (2005), Medyantseva et al. (2001) and Luppa et al. (2001).

Indirect detection of antibody-binding events is based on the use of labeled molecules that are detected optically or electrically. The two major types of label-based assay are sandwich assays and competitive assays (Figure 11.3). The label may be a fluorescent molecule, in which case detection is by fluorescence, or an enzyme, in which case a variety of transduction methods is possible, including chemiluminescence (using peroxidase as the label and adding luminol as substrate) and amperometric detection. In **sandwich assays**, one antibody (the capture antibody) is bound to the surface, and a second, labeled antibody, which binds to another epitope on the same antigen, is present in solution. Ligand (analyte) binds to the immobilized antibody, and the labeled antibody then binds to this. Detection is based on the concentration of label bound at the film surface. Sandwich assays only work where the analyte ligand is large enough to bind two antibody molecules simultaneously. **Competitive assays** can be used for smaller analytes. A variety of configurations are possible, but all rely on competition between the analyte and a labeled ligand for binding sites. In one common version, the capture antibody is immobilized, and a mixture of analyte ligand and a known concentration of competing labeled ligand are applied to the system. These compete for binding sites. As the amount of analyte ligand increases, the amount of labeled ligand that becomes bound decreases. Detection may be based either on the amount of label that ends up bound at the surface (in which case a larger signal corresponds to a smaller amount of analyte ligand present in the assay) or on the amount of label that remains unbound (in which case a larger signal corresponds to a higher concentration of analyte). In an alternative configuration, the competing ligand is immobilized, and a labeled antibody is present in solution. The presence of the analyte (ligand) in solution reduces the amount of labeled antibody that binds at the surface.

Direct detection methods are sometimes referred to as **label-free** since they detect the binding event directly rather than via a labeled additive. Label-free systems are intrinsically preferable for on-line monitoring, since they do not require addition of a reagent to the system. However, label-dependent systems can also be used for process monitoring, if the necessary reagents are provided as needed.

Direct detection of binding by SPR or piezoelectric methods has the advantage of being intrinsically reagentless, but is more demanding than detecting a label, since the mass change will usually only be a small fractional change compared to the mass of antibody bound to the surface. Thus, these techniques are perhaps best suited to quantitation of relatively large analytes such as proteins. A major problem with such techniques is **nonspecific binding**. In contrast to label-based methods, which detect only binding at the proper binding site, mass-based techniques will detect any binding to the surface whatsoever. Fermentation broths are often complex and contain many large molecules, such as proteins, which are capable of binding nonspecifically to surfaces or to other proteins. Such potential interference must always be considered.

Binding of a ligand to a binding molecule, assuming a homogeneous preparation, is described by the single parameter K_d, the **dissociation constant**. This assumes that sufficient time is allowed for the system to reach equilibrium. In unsteady-state systems, the important parameters are the **association rate k_1**, and the **dissociation rate k_{-1}**; these are related to K_d by the simple formula $K_d = k_{-1} / k_1$. This also assumes that diffusion of the ligand to the binding molecule is not rate-limiting. If the interactions between the ligand and binding molecule are relatively weak, then continuous monitoring of the ligand concentration may be possible by following the instantaneous level of bound ligand (Ohlson et al., 2000). Where binding interactions are stronger, it is necessary to take a series of measurements, replacing or regenerating the detection film between measurements.

Binding molecules other than antibodies may also be used in affinity sensors in precisely the same way. Where the target analyte is a nucleic acid, DNA can be used as the capture molecule. For other types of analyte, an alternative to antibodies is the use of **aptamers**. These are nucleic acids generated by an evolutionary process specifically for the ability to bind some target molecule (for an example, see Liss et al., 2002). Another possibility is polymers generated by **molecular imprinting**. In this case a polymer is generated by polymerization of a monomer in the presence of the target analyte, or a suitable analogue. On removal of this molecule, the polymer retains cavities of a suitable size and conformation to bind the target analyte specifically (for an example, see Feng et al., 2004). It is also possible to use specific binding proteins other than antibodies, such as receptor proteins (Carmon et al., 2004) or small molecules with a specific affinity for the target analyte (Yang and Chen, 2002). These examples are discussed in more detail below in Section 11.12.

11.6   Immobilization of the biological recognition element

For practical use in a biosensor, the biological recognition element must generally be immobilized on a surface. This might be an electrode surface for amperometric or potentiometric transduction, an optical fiber for optical transduction, or a gold film on a crystal surface for surface plasmon resonance or piezo-electric transduction. Immobilization is essential for retention of the recognition element within the system, and may also have a strong effect on the lifetime of the recognition element.

In the great majority of cases applicable to process control, the recognition element is a protein, either an enzyme or an antibody. Immobilization of proteins may be achieved in several different ways.

- **Adsorption.** In this case the enzyme or antibody is simply exposed to the surface and attaches by adsorption. The advantage of this method is simplicity; the disadvantage is that attachment is relatively weak and reversible, leading to loss of the protein from the system over time. Adsorption may be useful for preliminary experiments, but may not be ideal for production systems.
- **Retention within a membrane-bounded compartment.** In this case the electrode is surrounded by a dialysis membrane, and the enzyme solution is contained between this

and another dialysis membrane. This method is also easily applied and has the advantage that the protein is essentially in solution, so its activity should not be affected. However, the increased stability often seen with more decisive immobilization methods would not be expected.

- **Covalent attachment.** The surface is coated with an activated coating that can react with functional groups on the protein to form covalent bonds. This provides irreversible attachment, at the risk that a certain fraction of protein molecules may be inactivated or become unavailable due to steric restrictions, because of the orientation in which they have bound.

- **Protein cross-linking.** The protein is applied to the surface and then covalently cross-linked to form a macromolecular film, using a bifunctional reagent such as glutaraldehyde. Sometimes a nonfunctional protein such as albumin is included to provide extra protein for binding. This may be combined with other methods of immobilization, for enhanced permanence.

- **Entrapment in a polymer film.** This appears to be the most popular method used in recent reports. A solution of protein plus monomer is applied to the surface, and the monomer is polymerised *in situ*, entrapping the protein. A variant of this, widely used for cell immobilization, is the use of **alginate**, a polysaccharide derived from algae or bacteria. Sodium alginate is soluble, but when calcium ions are added, an insoluble calcium alginate gel is formed. **Polyacrylamide** is another popular polymer, formed by chemical polymerization of a mixture of acrylamide and bis-acrylamide, which forms cross links between chains. Recent biosensor reports have mainly favored the use of **electrically conductive polymers**, such as polypyrrole and polyaniline, which may be easily polymerized on an electrode surface by application of an appropriate potential. For recent reviews regarding the use of electrically conductive polymers in biosensors, see Cosnier (1999), Gerard et al. (2002), Vidal et al. (2003), and Cosnier (2003). Nonconducting polymers may also be generated by electropolymerization at an electrode surface, with the potential advantage that very thin films are generated since the film growth is self-limiting (Miao et al., 2004).

- **Attachment to a gold film via thiol-gold interactions.** Where transduction is by surface plasmon resonance or piezoelectric methods, the relevant surface is usually a gold film on the surface of a crystal. A noble metal film can also be added to an electrode surface for amperometric transduction. Proteins may be attached to gold films via the interaction of thiols with gold atoms. In some cases, surface-exposed thiols on the protein may serve this purpose. In other cases, one can use an "adapter," consisting of a molecule with a thiol at one end and a covalent or high-affinity protein-binding active group at the other. Usually a **self-assembled monolayer (SAM)** of a substituted alkanethiol is used. For a recent review, see Chaki and Vijay-amohanan (2002).

11.7 Amperometric biosensors based on redox enzymes

As described above, amperometric biosensors measure the current generated between a test electrode and a reference electrode when the potential between them is maintained at a constant level by means of a potentiostat. The reference electrode most commonly used is Ag/AgCl. The test electrode is usually platinum or carbon, with a coating of a suitable immobilized enzyme. In order to generate a detectable current, the enzyme must catalyze a redox reaction specific to the target analyte. Electrochemical biosensors have been recently reviewed by Mehrvar and Abdi (2004) and in somewhat less detail by May (1999). Zhang et al. (2000) provide a description of materials and techniques used in the construction of electrochemical biosensors.

The enzymes most commonly used in such biosensors are **oxidases**. "Oxidase" is a general term for an enzyme that reacts with a substrate and with molecular oxygen (O_2) to generate an oxidized substrate and hydrogen peroxide (H_2O_2). Oxidases generally contain a tightly bound flavin (flavin adenine dinucleotide, FAD) as a prosthetic group. In a typical reaction cycle, the substrate XH_2 binds to the enzyme active site and interacts with the flavin, FAD, to generate the oxidized product X and the reduced flavin, $FADH_2$ (Figure 11.4). The oxidized product then dissociates. Molecular oxygen (O_2) then binds to the active site, and reacts with the reduced flavin, regenerating FAD and being reduced to H_2O_2. Oxidases are highly suitable for biosensors, since they are simple single component enzymes that do not require soluble co-factors, and they both consume and generate redox-active substances suitable for amperometric transduction (H_2O_2 may also be detected optically by chemiluminescence, as described below). Oxidase-based biosensors are also historically important in the development of the biosensor industry, since the largest market for biosensors was, and is, in the quantitation of glucose in body fluids for the control of diabetes (discussed in more detail in the case study at the end of this chapter). The earliest enzyme-based biosensors, which established the industry in its present form, were enzymic glucose electrodes, initially developed by Leland Clark and commercially pioneered by Yellow Springs Instruments, using the enzyme glucose oxidase, derived from fungi of the genus *Aspergillus*. In addition to its other virtues, this enzyme is secreted,

Figure 11.4

Enzyme-catalysed redox reactions used in biosensors. (a) Oxidases. The reduced substrate, XH_2, reduces the bound flavin of the oxidase, FAD, to $FADH_2$. Oxygen then interacts with the reduced flavin, re-oxidising it to FAD and being reduced to hydrogen peroxide, H_2O_2. The flavin remains bound to the enzyme throughout the catalytic cycle. (b) Dehydrogenases. The reduced substrate XH_2 and an oxidised cofactor, such as $NAD(P)^+$ or PQQ, form a tertiary complex, and reducing equivalents are transferred to the cofactor, generating the reduced form, such as $NAD(P)H$ or $PQQH_2$, which then dissociates from the enzyme. In a living cell, the reduced cofactor would be re-oxidised by a separate enzyme.

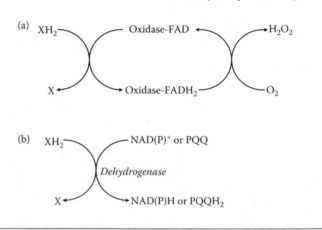

therefore relatively easy to purify from fermentation broths, and when immobilized, is fairly stable, with a working lifetime of several weeks. While glucose oxidase kick-started the industry, many other oxidases suitable for monitoring relevant molecules have been described in the literature, and a number are commercially available, including glutamate oxidase, alcohol oxidase, and lactate oxidase.

The oxidase reaction cycle lends itself to many variations on amperometric detection (Figure 11.5). The earliest oxidase-based biosensors simply used Clark oxygen electrodes with a layer of enzyme immobilized at the surface. The Clark oxygen electrode is simply a platinum electrode held at around -600 mV (relative to an Ag/AgCl counterelectrode); oxygen is reduced to water at the electrode surface, leading to a detectable current proportional to the oxygen concentration in the bulk fluid. When a layer of glucose oxidase is immobilized at the electrode surface, the glucose oxidase reaction reduces the oxygen concentration, leading to a reduced signal from the oxygen electrode. The electrode is held at a potential such that the reaction is diffusion-controlled, so that the reduction in signal is related to the glucose concentration. This transduction method

Figure 11.5

Use of redox enzymes in amperometric biosensors. (a) Oxidase, with detection by an oxygen electrode. Presence of the substrate (analyte) leads to consumption of oxygen, and a reduced signal from the oxygen electrode. This method was used in the earliest amperometric glucose sensors, but is no longer widely used. (b) Oxidase, with detection of hydrogen peroxide. Hydrogen peroxide is oxidised at the electrode. In an alternative version, hydrogen peroxide can be reduced to water at an electrode held at about -50 mV relative to Ag/AgCl; this reduces interference from redox-active substances in the sample.

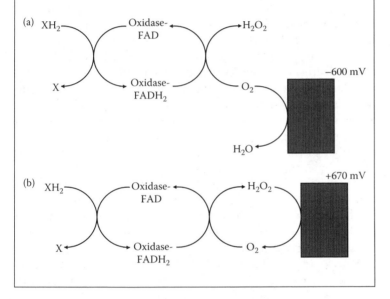

is no longer widely used. Later generations of oxidase-based biosensors detected hydrogen peroxide produced in the reaction. In general, detection of a reaction product always allows for higher sensitivity than monitoring decreased concentration of a substrate, since at low reaction rates, one is detecting a small positive signal against a low background, rather than a small negative signal against a high background. Hydrogen peroxide itself is a redox-active molecule, and may be detected at an electrode surface by oxidation to O_2 (at a potential of around +670 mV). However, the high oxidizing potential of the electrode can lead to interference from other redox-active molecules. In the context of glucose sensing in body fluids, the major problems are ascorbic and uric acids, but redox-active molecules may also be present in fermentation broths. Alternatively, the oxidase can be co-immobilized with horseradish peroxidase, a heme-containing enzyme which reduces hydrogen peroxide to water. The enzyme can then be reduced, either directly or via a suitable mediator, by an electrode at around 0 to -50 mV (see, for example, Dock et al., 2001; Elekes et al., 1995).

All of these systems require oxygen as a reactant. Oxygen is poorly soluble in water, and its concentration is difficult to control. More recent generations of oxidase-based biosensor have removed oxygen from the equation altogether, transferring electrons directly from $FADH_2$ to the electrode surface. This may be accomplished in several ways. If the enzyme is immobilized sufficiently close to the electrode surface, simple proximity may allow direct electron transfer, but this is generally not efficient, since the FAD is located some distance from the surface of the protein. More generally, a redox-active molecule known as a **mediator** is used to transfer electrons between the FAD and the electrode. Ferrocenes were the first class of mediators to be used in this way, and are still widely used. A soluble mediator can carry electrons between the enzyme and electrode, but this is not altogether satisfactory, since the mediator is not immobilized and may escape from the system. In this case, it would be necessary to supply the mediator to the system to replace losses. A better solution is to immobilize the mediator as well as the enzyme, giving a reagentless system. Most recent reports describe systems where the oxidase and mediator are co-immobilized by entrapment within a film of an electrically conductive polymer, which conducts electrons from the $FADH_2$ via the mediator to the electrode. The use of mediators in electrochemical biosensors has been reviewed by Chaubey and Malhotra (2001). For more detail of mediated amperometric biosensors, see the case study in Section 11.13.

Since oxidases all work in the same way and generate the same signal, but recognize different substrates, arrays of oxidases may be used for simultaneous quantitation of multiple analytes. For example, Moser et al. (2002) describe a microflow sensor using glucose oxidase, lactate oxidase, glutamate oxidase, and glutamate oxidase plus glutaminase, for the simultaneous determination of glucose, lactate, glutamate, and glutamine, suitable for monitoring a mammalian cell culture process.

The use of mediators in amperometric biosensors also allows the use of other classes of redox enzymes (oxidoreductases), such as **dehydrogenases**. Dehydrogenases are enzymes that catalyze reaction of a substrate XH_2 with a (usually soluble) cofactor C as follows: $XH_2 + C \rightarrow X + CH_2$ (Figure 11.4). The reaction may be freely reversible or essentially irreversible, depending on the relative reducing potentials of the substrate and co-factor. Among the most common co-factors are the pyridine nucleotides NAD (nicotinamide adenine dinucleotide) and NADP (nicotinamide adenine dinucleotide phosphate). The oxidized forms of these cofactors are written as NAD^+ and $NADP^+$, or collectively $NAD(P)^+$; the reduced forms as NADH and NADPH, or collectively, NAD(P)H. Pyridine nucleotide cofactors themselves do not readily react at electrodes, but inclusion of a suitable mediator in the system allows transfer of electrons from NAD(P)H to an electrode surface. Alternatively, the dehydrogenase can be co-immobilized with a **diaphorase**, a general term for a (usually) flavin-containing enzyme, in which the flavin is reduced by NAD(P)H and then re-oxidized by oxygen or some other electron acceptor. The diaphorase flavin can thus facilitate transfer of electrons between NAD(P)H and a mediator (see, for example, Takamizawa et al., 2000). The NAD(P) itself can also be immobilized, by covalent attachment to a larger molecule; for example, Mak et al. (2003) reported the use of NAD covalently linked to polyethylene glycol (PEG) in a formate biosensor based on formate dehydrogenase. If the co-factor is not immobilized, then it must be supplied to the system to replace losses.

A number of recent reports describe biosensors based on NAD(P)-dependent dehydrogenases. For example, Nicelescu et al. (2003) described a glycerol biosensor suitable for the analysis of glycerol levels in wine, based on NAD-dependent glycerol dehydrogenase, co-immobilized with a mediator (phenazine methosulphate, PMS) or embedded in a redox-active polymer film. Santos et al. (2003) described a biosensor for ethanol based on NAD-dependent alcohol dehydrogenase, with a working lifetime exceeding three months. Zhao et al (2002) described the generation and use of films of Prussian blue as immobilized mediators in amperometric biosensors using NAD-dependent formate dehydrogenase immobilized in a polypyrrole film. Maestre et al (2001) reported a coupled sensor for sucrose determination using three enzymes: sucrose phosphorylase, phosphoglucomutase, and glucose-6-phosphate dehydrogenase (G6PDH). This sensor was used in a flow injection analysis system. Saidman et al. (2000) described a sorbitol biosensor based on sorbitol dehydrogenase, and Takamizawa et al. (2000) reported a sensor for xylitol.

Other dehydrogenases use pyrroloquinoline quinone (PQQ) as a co-factor rather than NAD(P). PQQ-dependent alcohol dehydrogenases and glucose dehydrogenases (derived from various bacteria) have been used in biosensors. For example, Malinauskas et al. (2004) reported the construction of a mediator-less glucose biosensor based on PQQ-dependent glucose dehydrogenase immobilized within a redox-active polymer on a carbon electrode. Razumiene et al. (2001) described the construction of enzyme electrodes based on PQQ-dependent alcohol dehydrogenase and glucose dehydrogenase immobilized on a carbon electrode, with 4-ferrocenylphenol as a mediator, and Tkac et al. (2001) described a fructose sensor based on PQQ-dependent fructose dehydrogenase immobilized in a cellulose acetate film. Okuda et al. (2002) reported that co-immobilization of cytochrome c or cytochrome b562 with the PQQ-dependent glucose dehydrogenase greatly enhanced the signal.

A number of manufacturers sell complete instruments suitable for monitoring glucose and other molecules by amperometric

LIVERPOOL JOHN MOORES UNIVERSITY
LEARNING SERVICES

methods. For example, Trace Analytics (www.trace.de) supplies the ProcessTRACE system, designed specifically for monitoring glucose levels aseptically in laboratory and industrial cultures. The company that first introduced a commercial amperometric glucose biosensor in 1975, Yellow Springs Instruments (www.ysi.com), is still a market leader in the technology. For example, the YSI 2700 SELECT instrument is designed for process analysis in the food and bioprocess industries and can provide off-line or on-line measurements of glucose, galactose, lactose, sucrose, lactate, glutamine, glutamate, and ethanol. An attachment is also available for automated aseptic sampling and control of a feed pump based on results. Typical working lifetimes for the biological components are quoted as ranging from 5 days for alcohol oxidase to 21 days for glucose oxidase. For those who wish to build their own electrochemical biosensor, Zhang et al (2000) give an outline of the materials and techniques required.

11.8   Potentiometric biosensors and enzyme field effect transistor (ENFET)

Hydrolytic enzymes and lyases cleave a substrate bond, usually releasing a small molecule. In the former case, the cleavage is hydrolytic; in the latter case, a carbon-carbon double bond is generated. Such cleavage reactions do not intrinsically generate a strong signal suitable for transduction. However, in some cases, one of the products generated may be detectable by a second enzyme, usually an oxidoreductase, or may be detected potentiometrically. Potentiometric detection is based on the generation and detection of a voltage at an electrode surface. The best-known potentiometric sensor is the common laboratory pH electrode. Ion-selective electrodes are also available, consisting basically of pH electrodes surrounded by gas-permeable membranes that allow passage of, for example, ammonia (NH_3) or carbon dioxide (CO_2). The majority of potentiometric biosensors discussed in the recent literature are for the detection of urea in clinical contexts, or pesticides in environmental contexts, but some devices suitable for bioprocess monitoring have also been described.

The simplest case is where the reaction alters the local pH sufficiently that the reaction can be detected using a pH electrode; however, the buffering capacity of the fermentation medium must be considered. For example, Park et al. (2004) described a potentiometric biosensor using penicillinase (beta-lactamase) to quantify penicillin G and derivatives. The enzyme was immobilized on a cellulose nitrate membrane on a pH electrode; in a 2-mM phosphate buffer, pH 7.2, as little as 1 micromolar penicillin could be detected. Glucose can also be detected potentiometrically using glucose oxidase immobilized on a pH electrode, since the glucose oxidase reaction generates gluconic acid, which decreases the local pH. For example, Tinkilic et al. (2002) described the construction of miniaturized glucose and urea biosensors based on glucose oxidase and urease immobilized on pH and ammonium electrodes respectively, obtaining a linear response between 0.1 and 50 mM glucose.

In other cases, an ammonium or carbonate-responsive electrode can be used to detect a reaction product. For example, Kim and Kim (2003) describe a potentiometric biosensor based on a carbonate electrode for the quantization of isocitrate in foodstuffs. Similarly, ammonium-responsive electrodes have been used to make potentiometric biosensors for lysine, using lysine oxidase (Garcia-Villar et al., 2003).

The ENFET (Enzyme Field Effect Transistor) can be considered as a special case of potentiometric transduction. An ENFET is derived from a standard FET (field effect transistor). This is a solid-state device, in which the potential at a gate electrode determines the conductivity, and thus the current drawn, between source and drain connections held at different potentials. The drain current is measured and gives an indication of the potential at the gate electrode. An ISFET (Ion Selective Field Effect Transistor) is a version of this device in which the gate electrode is replaced by an arrangement similar to a pH electrode (that is, an ion-selective membrane and electrolyte solution with a reference electrode, usually Ag/AgCl). Changes in pH alter the potential at the gate electrode, changing the conductivity of the transistor and causing a measurable change in the drain current. Thus, an ISFET can be used as a robust, solid-state pH sensing device. Other alterations in the amount of charge near the gate electrode can also potentially be detected. An ENFET is an ISFET with a layer of enzyme immobilized at the gate electrode surface. Enzyme reactions altering the local pH can be detected, as described above for standard potentiometric biosensors. For example, Poghossian et al (2001) reported a penicillin-sensitive ENFET (dubbed a PenFET) based on immobilized beta-lactamase. Glucose can also be detected by an ENFET with glucose oxidase (for example, Luo et al., 2004). ENFET technology has recently been reviewed by Miao et al. (2003), Schoning et al. (2002), and Sandifer and Voycheck (1999).

Interestingly, it has also been reported than ENFETs can be used to detect dehydrogenase-catalyzed redox reactions. In this case, an NAD-dependent dehydrogenase is immobilized at the gate surface along with PQQ, which acts as a mediator. The dehydrogenase reaction reduces NAD^+ to NADH, which reduces PQQ to $PQQH_2$. $PQQH_2$ can be reoxidized by O_2. Increased NADH levels change the steady-state ratio of $PQQH_2$ to PQQ at the gate, altering the local pH and the conductivity of the FET. Devices of this nature have been used to quantify lactate and ethanol using lactate dehydrogenase and alcohol dehydrogenase (Pogoroleva et al., 2003).

11.9   Thermal biosensors

Thermal biosensors are based on the use of a thermistor to detect the tiny amounts of heat released by an enzyme-catalyzed reaction. Almost all enzyme catalyzed reactions release small amounts of heat detectable by a suitably sensitive thermal sensor such as a thermistor. Usually, the enzyme is immobilized on resin beads packed in a column attached to the thermistor. A control column lacking enzyme can be used to control for background temperature changes. The combination of thermistor and immobilized enzyme is sometimes referred to as an **enzyme thermistor**. For a recent review, see Ramanathan and Danielsson (2001). The use of thermal biosensors in process monitoring was reviewed by Ramanathan et al. (1999).

In principle, thermal detection of enzyme reactions is possible with any class of enzyme, provided that the heat released by the reaction can be measured with sufficient sensitivity. For example, Rank et al. (1992) reported the use of a thermal biosensor for monitoring penicillin V production in a Novo-Nordisk production-scale fermentor (160 cubic meters). Energy release due to penicillin hydrolysis was measured in a column packed with immobilized beta-lactamase or penicillin acylase. To correct for background temperature changes, a control column lacking active enzyme was used. Values were reported to correlate well with those obtained by HPLC. More recently, Lawung et al. (2001) described flow-injection analysis for the quantization of both penicillins and cephalosporins using two different β-lactamases immobilized on an affinity column. In another example, Navratil et al. (2001) used an on-line thermal biosensor with immobilized glycerokinase and galactose oxidase to monitor the bioconversion of glycerol to dihydroxyacetone by cells of *Gluconobacter oxydans* in an airlift fermentor. Finally, thermal biosensors provide yet another option for the quantitation of glucose with glucose oxidase. Ramanathan et al. (2001) described the use of immobilized glucose oxidase in an enzyme thermistor. To increase the amount of heat released per molecule of glucose oxidized, a second enzyme, catalase, was co-immobilized with the glucose oxidase to catalyze degradation of the hydrogen peroxide formed in the oxidase reaction.

11.10   Optical biosensors based on redox enzymes

Optical biosensors are based on the detection of colorimetric, fluorescent, or luminescent signals at the tip of a bundle of optical fibers. The fibers relay the light signal to a detector, which generates an appropriate electrical signal. A sensor based on optical fibers is sometimes referred to as an **optode**. Enzymes may be immobilized in a film on the tip of the fiber bundle. For some enzymes, a choice between optical and electrical transduction is possible. Optical transduction has the advantage that it is not affected by electrically noisy environments, and an optical fiber can be inserted, for example, directly into a bioreactor. However, optical transduction systems may be affected by turbidity. Optical biosensors have recently been reviewed by Marazuela and Moreno-Bondi (2002), Ulber et al. (2003), Monk and Walt (2004), Choi (2004), and Gauglitz (2005).

While oxidases are ideally suited to amperometric detection, as described above, they are also suitable for use in optical fiber biosensors. Hydrogen peroxide may be detected with great sensitivity by chemiluminescence, based on its peroxidase-catalyzed reaction with luminol, enhanced by p-iodophenol. A variety of optical biosensors based on this reaction has been reported. For example, Blankenstein et al. (1994) reported simultaneous on-line analysis of multiple analytes including glucose, lactate, and glutamate in animal cell culture using a multichannel oxidase-based flow-injection analysis system. Another way of optically transducing oxidase reactions involves the ability of oxygen to quench the fluorescence of certain metal complexes. In the presence of the oxidase substrate, the local oxygen concentration at the optical fiber surface is reduced, leading to increased fluorescence.

The activity of NAD(P)-dependent dehydrogenases may also be detected by optical fiber biosensors. The signal in this case is based on the fact that the reduced form of the co-factor, NAD(P)H, absorbs light at around 340 nm, fluorescing at around 450 nm. Thus, the presence of a dehydrogenase substrate leads to an increase in the NAD(P)H concentration and increased fluorescence. Enzymes that cause a change in the local pH, such as beta-lactamase, can also be transduced optically, using indicator dyes, which alter their fluorescence characteristics according to the local pH. This provides an alternative to potentiometric transduction. For further details, consult the reviews cited above.

11.11   Indirect affinity sensors: Optical and electrical biosensors based on antibodies

Detection of binding events can be accomplished by displacement of a labeled competing ligand, as in indirect immunoassays, or by binding of a second, labeled antibody, as in a sandwich-type immunoassay. This can be detected optically. The label may be a fluorescent molecule, in which case optical transduction is possible, or an enzyme, which can be detected either optically, via a chemiluminescent reaction, or electrically.

One configuration reported frequently in the literature is the use of antibodies labeled with horseradish peroxidase (HRP). Such antibodies are routinely used in laboratories for colorimetric or fluorimetric detection protocols based on the reaction of HRP with luminol or chromogenic substrates such as 4-amino-phenazone. When used in a biosensor, horseradish peroxidase can be detected optically, by chemiluminescence. For example, Jain et al. (2004) reported a generic competitive immunosensor using a flow analysis configuration, where analyte and a horseradish peroxidase-labeled analogue compete for antibody binding sites, with the amount of peroxidase being detected by enhanced chemiluminescence following the addition of luminol and p-iodophenol (which intensifies the chemiluminescent reaction). Alternatively, as noted above (Section 8.7), HRP in the presence of a suitable mediator, or in some cases even without a mediator, can transfer electrons between hydrogen peroxide (H_2O_2) and an electrode poised at around 0 to –50 mV with respect to an Ag/AgCl reference electrode. Thus, following its capture at an electrode surface, the HRP-labeled antibody or analyte-competitor can be quantified amperometrically by the addition of H_2O_2 and a suitable mediator. For example, Lopez et al. (1998) reported an amperometric immunosensor for the detection of atrazine using a competitive configuration, where atrazine in the sample competed with peroxidase-labeled atrazine for binding sites on an antibody layer immobilized on an electrode. In an extension to this approach, Darain et al. (2003) reported an amperometric immunosensor device in which the electrode surface was modified with horseradish peroxidase and antibody, and the analyte was detected following competition with glucose oxidase-labeled analyte; hydrogen peroxide produced by the glucose oxidase reaction was reduced by the electrode at -350 mV via the peroxidase. Keay and McNeil (1998) reported a similar system for detection of atrazine.

An interesting variation on this was reported by Ikebukuro et al. (2005), using aptamers (artificially evolved ligand-binding nucleic acids) rather than antibodies, in a sandwich-type assay. The capture aptamer was immobilized on a gold electrode surface, and the detection aptamer was labeled with

PQQ-dependent glucose dehydrogenase (described above). The amount of enzyme captured near the electrode surface was detected amperometrically following the addition of glucose.

Potentiometric biosensors can also be used to detect suitably labeled antibodies. For example, Campanella *et al.* (1999) compared the use of antibodies labeled with glucose oxidase (detected amperometrically) and urease (detected potentiometrically) for quantitation of human immunoglobulins. Such a system could potentially be adapted for detection of protein products in a biological process.

11.12   Direct affinity detection using surface plasmon resonance and piezoelectric biosensors

In direct affinity detection systems, the binding of ligand (analyte) to immobilized capture molecule (for example, antibody or aptamer) is detected directly by monitoring the amount of mass associated with the capture surface. Attachment of large ligands to the capture molecule increases the mass of the film to a detectable degree. The main methods used for such systems are surface plasmon resonance (SPR) and piezoelectric transduction, mainly using the quartz crystal microbalance (QCM). Such systems have the advantage of being intrinsically label-free and reagentless, but are prone to interference from nonspecific binding of nonanalyte molecules to the detection surface.

Surface plasmon resonance (SPR) is an effect in which a beam of laser light is shone through a transparent medium coated with a thin film of gold, with a fluid layer on the other side of the gold film. Above a certain angle of incidence, total reflectance from the surface will occur, because of the difference in refractive index between the transparent medium and the fluid film. The reflected radiation intensity can be easily measured. During the process of reflectance, an evanescent wave, that is essentially an electrical field, is generated on the fluid side of the interface. At one particular narrow band of angles of incidence in this range, the phenomenon of surface plasmon resonance occurs, as photons are absorbed and converted to surface plasmons, leading to a sharp reduction in the intensity of the reflected light. The angle of incidence at which this effect peaks is strongly affected by the mass of material, such as antibodies plus any bound ligands, which is bound at the surface. Thus, as more ligand binds to a layer of immobilized antibody in such a system, the angle at which resonance occurs changes fractionally, and this can be detected with very high sensitivity. The measurement is given in **resonance units (RU)**, where 1 RU corresponds to a change of 0.0001 degree in the angle of incidence at which the reflected radiation intensity minimum occurs. For recent reviews of biosensors based on surface plasmon resonance, see Homola (2003) and Rich and Myszka (2000).

As discussed above, the sensitivity of SPR detection depends on the fractional change in mass caused by binding of the ligand to the detection surface. Thus, SPR is better suited to the detection of large molecules, such as proteins, than to small molecules. SPR may be suited to monitoring the production of recombinant proteins. Ivansson et al. (2002) described the use of an SPR instrument for monitoring intracellular production of recombinant human superoxide dismutase in *Escherichia coli*. Initially, samples of bacteria were withdrawn from the reactor and disrupted by sonicated prior to analysis. A later paper (Tkac et al., 2004) reported disruption of cells by exposure to a non-ionic surfactant prior to SPR analysis, a technique more suited to automated sample processing. Hsieh et al. (1998) reported off-line monitoring of the production of *Clostridium perfringens* beta toxin by applying fermentation broth directly to an SPR sensor chip. McCormick et al. (2004) developed an SPR assay for recombinant factor VIII, a blood-clotting factor, and suggested that their assay might be suitable for monitoring the recombinant production process.

SPR can also be applied to the detection of small molecules using sandwich or competition assays similar to those described above. Sandwich assays are essentially the same as those described above except that the second antibody need not be labeled; the increase in surface mass due to the mass of the second antibody is sufficient for detection. For competitive assays, instead of an enzyme or fluorescent label, a large protein is used, to provide a sufficient mass change for detection. It has also been reported that recent SPR instruments are sufficiently sensitive for direct detection of the binding of small molecules to immobilized proteins (Rich and Myszka, 2000).

Some reports have also described integration of multiple sensors of different types in a single chip. For example, Suzuki et al. (1999) reported the construction of an SPR immunosensor chip for human IgG, also bearing integrated amperometric sensors for glucose and lactate; this was used for simultaneous on-line monitoring of glucose, lactate, and IgG in a cell culture bioreactor.

The first and best-known manufacturer of SPR instrumentation is **Biacore** (www.biacore.com), formerly a division of Pharmacia Biotechnology. The first commercial SPR biosensor system was introduced by Biacore in 1990, and the technology has now matured to the point where it is suitable for routine monitoring purposes. Biacore supplies a range of SPR instruments, as well as sensor chips. The standard sensor chip (CM5) is essentially a glass slide coated with a 50-nm layer of gold, covered with a film of carboxymethylated dextran suitable for covalent attachment of antibodies or other binding molecules. Other types of sensor chips are also available (see the Biacore Web site or catalogue for further details). Another supplier of

SPR instrumentation is Texas Instruments, which has recently introduced a small, relatively cheap SPR device known as **Spreeta**. For a discussion of the properties of the Spreeta system, see Chinowsky et al. (2003). Other commercial suppliers of SPR instrumentation are listed by Homola (2003) and Rich and Myszka (2000).

Acoustic wave (piezoelectric) systems can also be used to measure mass changes due to ligand binding in a film of immobilized antibody. These devices are generally based on detection of changes in the nature of vibration of a piezoelectric material such as quartz when an electrical potential is applied. In this case, the antibody film is immobilized on the surface of a piezoelectric crystal. An applied voltage causes vibration of the crystal, and various characteristics of this vibration can be detected and used to infer the quantity of bound ligand. A wide variety of configurations is available. The most widely discussed type for biosensor use is the **Bulk Acoustic Wave (BAW) Thickness Shear Mode** device, also known as a **Quartz Crystal Microbalance (QCM)**. Piezoelectric biosensors have been recently reviewed by Su et al. (2000) and O'Sullivan and Guilbault (1999). Proteins are typically immobilized on the surface via thiol-gold interactions, as with SPR (Chaki and Vijayamohanan, 2002). The most obvious bioprocess monitoring application for QCM, as with SPR, is in the direct detection of relatively large molecules, such as recombinant proteins. To cite a single example, Saha et al. (2002) reported the use of a QCM-based sandwich assay for the quantitation of insulin.

Some interesting recent reports have discussed the use of ligands other than antibodies as capture molecules in piezoelectric biosensors. As discussed above, piezoelectric immunosensors, like SPR immunosensors, are only expected to show a strong response for relatively large ligands, where binding of the ligand causes a significant mass shift at the detection surface. However, in some cases even small ligands may provoke a large response. For example, Carmon et al. (2004) reported an unexpectedly strong signal from the binding of glucose to an *Escherichia coli* glucose receptor protein immobilized on a QCM surface. They postulated that this was due to a conformational change on glucose binding leading to a change in the film structure from a viscous to a more rigid state. Thus, it appears that piezoelectric sensors may sometimes be suitable for direct detection of small molecules, with an appropriate choice of capture molecule, where ligand binding is associated with a conformational change. Another interesting point in this report is that the glucose receptor protein was modified by site-directed mutagenesis to incorporate a surface-exposed thiol-bearing cysteine residue, allowing simple immobilization by direct binding of the thiol to the gold surface.

Another option for detection of small molecules is the use of a molecularly imprinted polymer as the capture molecule, rather than an antibody. This is a polymer that was polymerized in the presence of the analyte, so that removal of the analyte leaves vacant spaces in the polymer of the correct size and conformation to bind the analyte specifically. For example, Feng et al. (2004) reported the use of an electropolymerized molecularly imprinted polymer for the detection of sorbitol. Yet another possibility is an aptamer, a nucleic acid specifically evolved to bind the target ligand. For example, Liss et al. (2002) compared the use of an aptamer and an antibody for the detection of human immunoglobulin E, and reported that the aptamer showed similar sensitivity and specificity to the antibody, along with improved dynamic range and a greater tolerance for the regeneration procedure. Finally, the capture molecule can be a small molecule with a specific affinity for the target analyte. For example, Yang and Chen (2002) reported the use of immobilized polymyxin B on a QCM for the quantitation of endotoxin (toxic lipopolysaccharide shed from the cell walls of Gram negative bacteria, a great concern in the manufacture of injectable biological products).

Interestingly, piezoelectric sensors can also be used as a transduction system to detect oxidase activity. In this case the oxidase generates hydrogen peroxide, which is used by a peroxidase to oxidize and polymerize a small molecule such as diaminobenzidine or o-dianisidine, forming a polymer that is deposited on the detection surface, leading to an increase in mass, which is detected as described above. This has been described for the detection of cholesterol using cholesterol oxidase (Martin et al., 2003) and glucose using glucose oxidase (Reddy et al., 1998).

11.13   Amperometric glucose biosensors for blood glucose monitoring: A case study

Diabetes mellitus (hereafter referred to as diabetes) is one of the most common chronic diseases in both Western and developing countries. It has a prevalence of approximately 8% in much of Europe and the United States and exacts a high cost in terms of morbidity and mortality (Hadden and Harris, 1987). This metabolic disorder is characterized by hyperglycemia and disturbances of carbohydrate, protein, and fat metabolism, secondary to an absolute or relative lack of the hormone insulin (Alberti and Zimmet, 1998). Vital for life, insulin is a hormone produced by the pancreas that helps the glucose to enter the cells where it is used as fuel by the body. The classification and diagnostic criteria used for the diagnosis of diabetes and disorders of glucose homeostasis has been the subject of much debate (Alberti and Zimmet, 1998; Gavin et al., 1997). Currently, there are five major clinical categories of disordered glucose homeostasis, which are summarized below.

- Type 1 diabetes, also known as insulin-dependent diabetes
- Type 2 diabetes, also known as noninsulin-dependent diabetes
- Impaired glucose tolerance/impaired fasting glucose
- Gestational diabetes
- Other rare forms, which include maturity-onset diabetes of the young (MODY), pancreatic diseases

Type 1 diabetes develops if the body is unable to produce any insulin. This type of diabetes usually appears before the age of 40. It is treated by insulin injections, and diet and regular exercise is recommended. Type 1 diabetes is characterized by beta cell destruction, thereby resulting in an absolute deficiency of insulin. It has an autoimmune basis, and islet cell antibodies and glutamic acid decarboxylase antibodies are present in up to 80% of patients (Rehman, 2001). Research suggests that in genetically susceptible individuals, unknown environmental agents such as viruses may initiate the process, which is possibly mediated via cytokines (Zimmet, 1995). Type 1 diabetes is classically a disease of the young and is generally of rapid onset, but it can occur at any age. Diabetes should therefore be suspected in a patient of any age who presents with sudden weight loss and raised blood glucose.

By contrast, type 2 diabetes is characteristically a disease of the middle aged or elderly and usually begins insidiously. Insulin resistance and a family history are the hallmarks. More than 90% of all identified cases of diabetes are classified as type 2 (Nathan et al., 1997). Type 2 diabetes is frequently undiagnosed with perhaps as many as half the individuals who have this type of diabetes being unaware of their condition (de Courten and Zimmet, 1997). Most type 2 diabetes is preceded by an asymptomatic period of impaired glucose tolerance, which is characterized by a response to an oral glucose challenge that is not normal, but is not diagnostic of diabetes. Type 2 diabetes develops when the body can still make some insulin, but not enough, or when the insulin that is produced does not work properly (known as insulin resistance). It is treated by diet and exercise alone or by diet, exercise, and tablets or by diet, exercise, and insulin injections. Impaired glucose tolerance and impaired fasting glucose refer to a metabolic state between normal glucose homeostasis and diabetes (Gavin et al., 1997). It has been suggested that these disorders should be considered as a stage in the natural history of disordered carbohydrate metabolism, indicating risk of future diabetes and/or cardiovascular risk (Alberti, 1996), rather than discrete clinical entities.

Impaired glucose tolerance is a common problem affecting approximately 17% of the U.K. population aged 40 to 65 years (Brown et al., 1991). The importance of this condition is debatable. Controversy has arisen partly because of the high variability in the response to the oral glucose tolerance test. Up to 50% of persons initially classified as having impaired glucose tolerance are reclassified on repeat testing as having normal glucose tolerance. Nonetheless, impaired glucose tolerance is a strong risk factor for the development of type 2 diabetes (Meigs et al., 2000). It also identifies a group of individuals at higher risk of developing coronary heart disease (CHD) (Fuller et al., 1983).

More recently, impaired fasting glucose (fasting plasma glucose 6.1 to 6.9 mmol/l) has been introduced as a new category of impaired glucose regulation (Gavin et al., 1997). The exact significance of impaired fasting glucose is yet to be determined, but it would appear to be a stronger predictor of the risk of progression to diabetes than impaired glucose tolerance (Shaw et al., 2000). The importance of impaired fasting glucose as a risk factor for cardiovascular disease is, however, less clear (Tominga et al., 1999; Lim et al., 2000).

Gestational diabetes is a form of carbohydrate intolerance first identified during pregnancy. Therefore, this definition may also include glucose intolerance present prior to pregnancy but unrecognized. It is generally agreed that gestational diabetes is a prediabetic state and associated with an increased cardiovascular risk.

Diabetes can result from a number of rare genetic, endocrine, and infectious processes that affect the exocrine pancreas. MODY is associated with abnormal beta cell function, inherited in an autosomal dominant fashion. It is genetically and clinically heterogeneous. Unlike type 2 diabetes, the underlying defects do not produce insulin resistance, and individuals with MODY are seldom obese.

The incidence of diabetes continues to rise throughout the world. Indeed, by 2010 it has been estimated that the diabetic population will increase to 221 million (3 million in the UK) from 110 million in 1994 (Orchard, 1998; Zimmet et al., 2001). The majority of the new cases will be those with type 2 diabetes and most of these will be in China, the Indian subcontinent, and Africa. It is estimated that from 65 million cases of type 2 diabetes in Asia and Oceania in 1995, the number will double to 135 million by 2010 (Zimmet, 2003). Type 2 diabetes has a significant subclinical phase that may last several years, and it is estimated that there are approximately one million people in the UK currently undiagnosed with the disease.

The main symptoms of untreated diabetes are increased thirst, increased frequency of urination, extreme tiredness, weight loss, genital itching or regular episodes of thrush, and blurred vision.

11.13.1 Home blood glucose monitoring: The glucose meter

The main aim of treatment of diabetes is to achieve blood glucose and blood pressure levels as near to normal as possible. This, together with a healthy lifestyle, will help to improve well-being and protect against long-term damage to the eyes, kidneys, nerves, heart, and major arteries.

An important tool in the homeostatic control of diabetes is Home Blood Glucose Monitoring (HBGM). It can help to maintain day–to-day control, detect hypoglycaemia, assess control during any illness, and help to provide information that can be used in the prevention of long-term complications. Blood glucose monitoring gives a direct measure of the glucose concentration at the time of the test and can detect hypoglycemia as well as hyperglycemia. The Diabetes Control and Complications Trial (DCCT) (Diabetes Control and Complications Trial Research Group, 1993), a 10-year nationwide study of 1,441 diabetics, conclusively demonstrated that improved control of blood sugar delayed or prevented many of the aforementioned complications at least 50% better than with poorly controlled subjects. Subsequent studies have corroborated this conclusion. This good control is enabled by frequent, consistent, and accurate self-testing of blood glucose to optimize therapy. Blood glucose meters that utilize an enzyme electrode (biosensor) as the glucose sensing element are particularly suitable medical devices for HBGM. The advantages offered by biosensors in HBGM arise for the following reasons. Blood is a complex fluid and glucose levels vary widely over time in a single patient; many factors besides glucose vary in blood from healthy patients (hematocrit, oxygen levels, and metabolic by-products); therefore, great specificity is a prime requirement. In addition, patients with diabetes may have a wide range of other medical problems, creating even greater variation in their blood. Finally, biosensors can be used directly in the blood without requiring major modifications to the biological sample (increased temperature or pressure, dramatic pH changes, addition of highly reactive chemicals, etc.). Commercial examples of biosensors used for HBGM are shown in Figure 11.6.

There are several essential elements in a medical device designed for patient self-monitoring. Because these systems are medical devices, used to make medical decisions or avoid potentially life-threatening incidents every day, they must be of very high quality, and the information displayed must be accurate. The sensors must be easily manipulated by sight-impaired users, and the system must be very user-friendly to encourage more frequent testing for better control. In the hospital or doctor's office there are additional quality requirements, and the possibility of multiple sample types (e.g., capillary, arterial, venous, and neonatal blood). Many commercially available systems,

Figure 11.6

Examples of commercially available glucose biosensors for self-monitoring by diabetic patients. (In some examples the "top tape" has been removed to expose the underlying electrode configuration) Example 1 and example 2 are the OneTouch Ultra and the Accu-Check Advantage strip, which are discussed in more detail in the body of the text. (Photograph kindly provided by Dr Maria Teodorczyk, LifeScan, Milpitas, USA.)

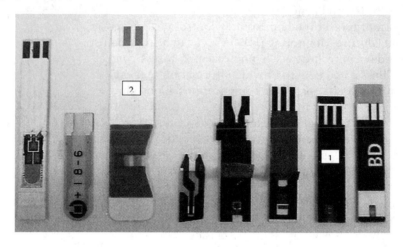

meter plus sensing electrode (see Table 11.1) meet these needs by providing accurate and dependable glucose measurement using a tiny drop of blood, for many blood sample types over a wide range of hematocrit. (Hematocrit is the ratio of red blood cells to plasma in the blood sample).

11.13.2 *Enzymes used in glucose biosensors*

Traditionally, two glucose oxidizing enzymes have been used in the manufacture of glucose biosensors for HBGM. Although both enzymes oxidize the pyranose form of β-D-glucose (at the C_1 position) to the corresponding lactone, they use different cofactors to carry out the redox process. One is the flavo-protein glucose oxidase (GOx), which has flavin adenine dinucleotide (FAD) as co-factor, and the other is the quinoprotein-dependent glucose dehydrogenase PQQ-GDH). Although both enzymes catalyze the oxidation of glucose, their biological properties differ, so choice of enzyme is normally determined by the intended use of the sensor under specific clinical conditions.

Glucose oxidase is a very specific enzyme (99.9% specific for β-D glucose) that is commercially available in highly purified form and is therefore the enzyme of choice in most devices. It does, however, suffer from the problem of cross reactivity with oxygen, which means that under situations of high oxygen partial pressure, the reading obtained from the meter can be lower than expected. Recognition of this fact is of clinical

Table 11.1 Examples of commercially available HBGM test kits. (Data adapted from Diabetes UK Web site.)

Company	Meter	Test strip required	Blood range in mmol/l	Sample volume in μl
Abbott Diabetes Care	Precision QID	MediSense G2 Sensor electrodes	1.1–33.3	3.5
Therasense	FreeStyle Classic	FreeStyle	1.1–27.8	0.3
	FreeStyle Mini	FreeStyle	1.1–27.8	0.3
Bayer Diagnostics	Ascensia Contour	Ascensia Microfill	0.6–33.3	0.6
	Ascensia Breeze	Ascensia Autodisc (10-test disc)	0.6–33.3	2.0–3.0
DiagnoSys Medical	Prestige Qx Smart System	Prestige Smart System	1.3–33.3	4
Hypoguard	Supreme Plus	Hypoguard Supreme	2.0–25.0	7
LifeScan	OneTouch Ultra	OneTouch Ultra	1.1–33.3	1
	OneTouch UltraSmart	OneTouch Ultra	1.1–33.3	1
Menarini Diagnostics	GlucoMen Glyco	GlucoMen Sensors	1.1–33.3	2
	GlucoMen PC	GlucoMen Sensors	1.1–33.3	2
Roche Diagnostics	Accu-Chek Compact	Accu-Chek Compact	0.55–33.3	1.5
	Accu-Chek Advantage	Accu-Chek	0.55–33.3	4
	Accu-Chek Active	Accu-Chek Active	0.55–33.3	2

http://www.diabetes.org.uk/products/index.html

Table 11.2 Examples of commercially available glucose test strips and
the associated chemical reactions. (Data adapted from *Crit Care Med*
(2001) Vol. 29, No. 5.)

Advantage H, Comfort Curve

    glucose + GD/PQQ → gluconic acid + GD/PQQH$_2$ (Eq. 1)

    GD/PQQH$_2$ + ferricyanide → GD/PQQ + ferrocyanide   (Eq. 2)

    ferrocyanide → ferricyanide + e$^-$ (Eq. 3)

Precision PCx, Precision QID

    glucose + GO/FAD → Gluconic acid + GO/FADH$_2$ (Eq. 4)

    GO/FADH$_2$ + ferricinium → GO/FAD + ferrocene   (Eq. 5)

    ferrocene → Ferricinium + e$^-$ (Eq. 6)

    GO/FADH$_2$ + O$_2$ GO/FAD + H$_2$O$_2$ (Eq. 7)

SureStep

    glucose + GO/FAD gluconic acid + GO/FADH$_2$ (Eq. 8)

    GO/FADH$_2$ + O$_2$ → GD/FAD + H$_2$O$_2$ (Eq. 9)

    H$_2$O$_2$ + MBTH-R + HRP → MBTH-R$^+$ (Eq. 10)

    MBTH-R$^+$ + ANS + ½ O$_2$ → blue-green dye   (Eq. 11)

Glucometer Elite

    glucose + GO/FAD → gluconic acid + GO/FADH$_2$ (Eq. 12)

    GO/FADH$_2$ + ferricyanide → GO/FAD + ferrocyanide (Eq. 13)

    ferrocyanide → ferricyanide + e$^-$ (Eq. 14)

    GO/FADH$_2$ + O$_2$ → GO/FAD + H$_2$O$_2$ (Eq. 15)

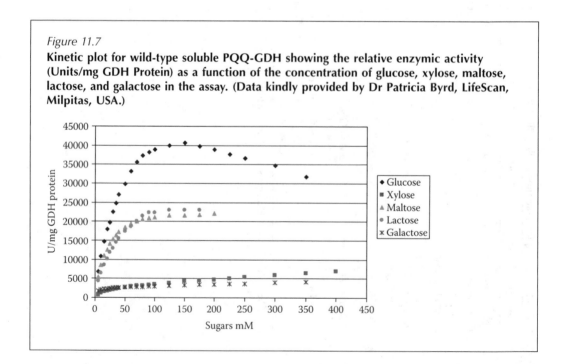

Figure 11.7

**Kinetic plot for wild-type soluble PQQ-GDH showing the relative enzymic activity
(Units/mg GDH Protein) as a function of the concentration of glucose, xylose, maltose,
lactose, and galactose in the assay. (Data kindly provided by Dr Patricia Byrd, LifeScan,
Milpitas, USA.)**

Figure 11.8
Summary of the kinetics of sugar oxidation by wild-type soluble PQQ-GDH.

	K_m (mM)	V_{max} (U/mg, %)	V_{max}/K_m (%)
Glucose	25.0	4610 (100)	184 (100)
3-O-m-glucose	28.7	3596 (78)	123 (67)
Allose	35.5	2997 (65)	84.6 (46)
Maltose	26.0	2305 (50)	88.3 (48)
Lactose	18.9	1982 (43)	105 (57)
Galactose	5.3	277 (6)	52.3 (28)

importance because high blood oxygen tensions are frequently observed in the critically ill and in patients receiving oxygen therapy or undergoing surgery. Hence, changes in $_pO_2$ levels can falsely lower glucose meter measurements and may mislead medical decision-making. Mechanisms by which oxygen affects GOx-based amperometric test strips are suggested by the reactions in Table 11.2. Oxygen, a known natural electron acceptor, competes with electron mediators, such as ferricinium used with the Precision PCx and Precision QID (Table 11.2, Eq. 11.5) and ferricyanide used with the Glucometer Elite (Table 11.2, Eq. 11.13), for the reoxidation of GOx/FADH$_2$. At high $_pO_2$, oxygen outcompetes the electron mediator for the re-oxidation of GOx/FADH$_2$, which results in reduced chemical reaction between the electron mediator and GOx/FADH$_2$, leading to fewer electrons produced. As a result, low glucose measurements are obtained at high $_pO_2$ levels.

On the other hand, PQQ-GDH does not react with oxygen and so is suitable for use as the biocatalyst in the clinical scenarios described above. However, the wild type enzyme is not as specific as GOx, as illustrated by the kinetic data in Figures 7 and 8. Consequently, patients that have high levels of, for example, maltose in the blood (which could result as a side effect of peritoneal dialysis) or have an inbred genetic disorder resulting in impaired carbohydrate metabolism, would obtain an inaccurate high reading when testing with glucose electrodes incorporating this enzyme. Because of the oxygen insensitivity of PQQ-GDH, there is much commercial interest in producing a mutant form of the enzyme that retains its nonreactivity to oxygen but shows improved specificity with respect to β-D glucose.

11.13.3 *Mediated electrochemistry*

Although enzymes offer many advantages as the biocatalysts in glucose sensors, they rarely exchange electrons directly with electrodes. The same is true for many other biological molecules, such as glucose. An electrochemical measurement often requires a substance to facilitate (or mediate) this transfer; such reagents are termed *mediators*. A well-known mediator is potassium ferricyanide.

The reaction process is

1. Glucose first reacts with the enzyme (GOx or PQQ-GDH). Glucose is oxidized to gluconic acid and the enzyme is temporarily reduced by two electrons transferred from glucose to the enzyme.
2. The reduced enzyme next reacts with the mediator, transferring a single electron to each of two mediator ions. The enzyme is returned to its original state, and the two oxidized mediator molecules are reduced.
3. Ferricyanide and ferrocyanide (the oxidized and reduced forms of the mediator) are capable of rapidly transferring electrons with an electrode. The electrons may thus be transferred between glucose and the electrode *via* enzyme and mediator.

Now we have a chemical mechanism for transferring electrons from glucose to the electrode. The biosensor reagent is actually more complex, containing a number of other active ingredients (e.g., stabilizers, processing aids, etc.), but for simplicity we shall assume it is simple a deposited layer or enzyme and mediator.

11.13.4 *The electrochemical measurement*

An extremely useful method is *amperometry*. This technique sets the electrode potential (a potential difference represents a difference) in the *average electron energy* at two points at a level where every molecule or ion reaching the electrode surface undergoes an electron transfer reaction. The current (rate of electron transfer) is thus limited by how rapidly the reactants arrive, a *diffusion-controlled* current, since diffusion is the primary transport mechanism. Because reactant is being converted (consumed) at the electrode surface, its average concentration will be decreasing in the vicinity of the electrode, so current should *decrease* with time. Under certain boundary conditions this behavior is described by the *Cottrell equation*:

$$i_t = \frac{nFAC_0 D^{\frac{1}{2}}}{(\pi t)^{\frac{1}{2}}}$$

where

i_t = current, in *amps*

n = number of electrons transferred in the reaction (for ferrocyanide, $n = 1$)

F = Faraday constant (the quantity of charge carried by 1 mole of electrons = 96,485 *Coulombs/mol*)

A = electrode area *(cm²)*

D = diffusion coefficient (a measure of how rapidly reactant is transported; for ferrocyanide the diffusion coefficient is ~ 7×10^{-6} *cm²/sec* at room temperature but varies with the medium)

C_0 = concentration of reactant *(mol/cm³)*, the model assumes uniform concentration before the potential difference is applied

t = time *(seconds)*

11.13.4.1 *The measurement protocol*

The measurement protocol may be summarized as follows.

1. An enzyme (GOx or PQQ-GDH) rapidly transfers electrons from glucose to a mediator such as ferricyanide, which then transfers them to an electrode.

2. An electrode potential is imposed on the system. In any circuit, current must be identical at all points. If we put two electrodes in a circuit, electron flow rate *out* of one electrode (*working electrode*) has to equal the electron flow rate *into* the other (*counter electrode*). Since ferrocyanide is oxidized at the working electrode, an equal quantity of material must be reduced at the counter. Since ferricyanide is normally present in large quantities, this species normally provides the counter reaction. This is called *bi-amperometry*. In bi-amperometry the counter and working electrodes are normally made of the same material. For example, in the Accu-Check Advantage systems the electrodes are made of palladium metal. In the OneTouch Ultra system, the electrodes are made of carbon.

3. The exact measurement sequence that is used for detecting glucose can vary from manufacturer to manufacturer. For example, in the Accu-Check Advantage system, inserting the glucose electrode automatically switches on the meter. When the sample is detected, an incubation period follows (under conditions of zero current flow) wherein glucose reacts with the enzyme (PQQ-GDH) to generate ferrocyanide. At the end of the incubation period, the meter applies a potential difference and the current is measured. Current data are analyzed, the result is recorded, and then it is displayed to the user. The OneTouch Ultra system, on the other hand, does not have an incubation period.

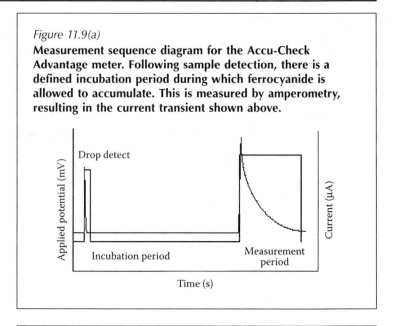

Figure 11.9(a)

Measurement sequence diagram for the Accu-Check Advantage meter. Following sample detection, there is a defined incubation period during which ferrocyanide is allowed to accumulate. This is measured by amperometry, resulting in the current transient shown above.

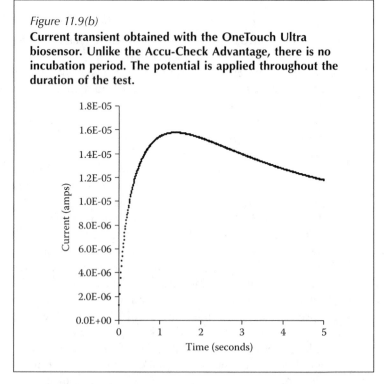

Figure 11.9(b)

Current transient obtained with the OneTouch Ultra biosensor. Unlike the Accu-Check Advantage, there is no incubation period. The potential is applied throughout the duration of the test.

After sample detection, the current flows throughout the 5-second sample test time. At the end of this period the current is analyzed and the level of glucose in the sample calculated. Unlike the aforementioned system, there is no separate incubation period. These two protocols are summarized in Figure 11.9.

Summary

- Biosensors offer a method for automated monitoring of metabolites and reaction products during fermentation and can be used in control systems. The analyte of interest is recognized specifically by a biological molecule such as an enzyme or antibody, and an electrical signal is generated by a transducer. A very wide range of transduction methods have been reported in the literature, though relatively few seem to have been commercially adopted for process control as yet.

- For small molecules such as sugars, alcohols, and hydroxyacids for which oxidase or dehydrogenase enzymes are available, the most obvious choice is an amperometric enzyme electrode. This is a well-established technology. Indeed, amperometric biosensors for the detection of common substrates and metabolites such as glucose and other sugars, ethanol, lactate, and glutamate are commercially available in systems specifically designed for process monitoring and control.

- For small molecules where no redox enzyme is available, but which can be acted upon by an enzyme (such as hydrolase) yielding or consuming H^+, NH_4^+ (NH_3), or HCO_3^- (CO_2), a potentiometric enzyme electrode or ENFET would be a reasonable choice.

- For small molecules for which no enzyme is available, but against which an antibody could be raised, a competitive or sandwich immunoassay, using either optical or electrical transduction, might be suitable. For larger molecules, such as recombinant proteins, a direct affinity sensor based on SPR or piezoelectric transduction might offer a suitable detection system.

- Undoubtedly, HBGM has represented the most successful commercialization of biosensor technology to date. The specificity offered by enzymes (the biological recognition element) has been paramount in the successful implementation of the technology in whole blood without the need for any sample pretreatment. Because of advances in electronics, safe and easy-to-use meters can be produced for use by the diabetic in the home after little formal training. However, because the strips are designed to be used only once problems due to drift, memory effects, etc. are inconsequential to this type of sensor application.

References

Alberti, K. G. (1996). The clinical implications of impaired glucose tolerance. *Diabetic Medicine* 13, 927–937.

Alberti, K. G., P. Z. Zimmet (1998). New diagnostic criteria and classification of diabetes – again? *Diabetic Medicine* 15, 535–536.

Arnold, F. H., A. A. Volkov (1999). Directed evolution of biocatalysts. *Current Opinion in Chemical Biology*, 3, 54–59.

Bange, A., H. B. Halsall, W. R. Heineman (2005). Microfluidic immunosensor systems. *Biosensors and Bioelectronics,* **20**, 2488–2503.

Blankenstein, G., U. Spohn, F. Preuschoff, J. Thommes, M. R. Kula (1994). Multi-channel flow injection analysis biosensor system for on-line monitoring of glucose, lactate, glutamine, glutamate and ammonia in animal cell culture. *Biotechnology and Applied Biochemistry,* **20**, 291–307.

Brown, D. C., C. D. Byrne, P. M. S. Clark, B. D. Cox, N. E. Day, C. N. Hales, J. R. Shackleton, T. W. M. Wang, D. R. R. Williams (1991). Height and glucose tolerance in adult subjects. *Diabetologia* **34**, 531–533.

Campanella, L., R. Attioli, C. Colapiccioni, M. Tomassetti (1999). New amperometric and potentiometric immunosensors for anti-human immunoglobulin G determination. *Sensors and Actuators B – Chemical,* **55**, 23–32.

Carmon, K. S., R. E. Baltus, L. A. Luck (2004). A piezoelectric quartz crystal biosensor: The use of two single cysteine mutants of the periplasmic *Escherichia coli* glucose/galactose receptor as target proteins for the detection of glucose. *Biochemistry,* **43**, 14249–14256.

Chaki, N. K., K. Vijayamohanan (2002). Self-assembled monolayers as a tunable platform for biosensor applications. *Biosensors and Bioelectronics,* **17**, 1–12.

Chaubey, A., B. D. Malhotra (2001). Mediated biosensors. *Biosensors and Bioelectronics,* **17**, 441–456.

Chinowsky, T. M., J. G. Quinn, D. U. Bartholomew, R. Kaiser, J. L. Elkind (2003). Performance of the Spreeta 2000 integrated surface plasmon resonance affinity sensor. *Sensors and Actuators B – Chemical,* **91**, 226–274.

Choi, M. M. F. (2004). Progress in enzyme-based biosensors using optical transducers. *Microchimica Acta,* **148**, 107–132.

Cosnier, S. (1999). Biomolecule immobilization on electrode surfaces by entrapment or attachment to electrochemically polymerized films: A review. *Biosensors and Bioelectronics,* **14**, 443–456.

Cosnier, S. (2003). Biosensors based on electropolymerized films: new trends. *Analytical and Bioanalytical Chemistry,* **377**, 507–520.

de Courten, M., P. Zimmet (1997). Screening for non-insulin-dependent diabetes mellitus: Where to draw the line? *Diabetic Medicine* **14**, 95–98.

Darain, F., S. U. Park, Y. B. Shim (2003). Disposable amperometric immunosensor system for rabbit IgG using a conducting polymer modified screen-printed electrode. *Biosensors and Bioelectronics,* **18**, 773–780.

Diabetes Control and Complications Trial Research Group (1993). The effect of intensive treatment of diabetes on the development and progression of long-term complications in insulin-dependent diabetes. *New England Journal of Medicine* **329**, 977–993.

Dock, E., A. Lindgren, T. Ruzgas, L. Gorton (2001). Effect of interfering substances on current response of recombinant peroxidase and glucose oxidase – recombinant peroxidase modified graphite electrodes. *Analyst,* **126**, 1929–1935.

Elekes, O., D. Moscone, K. Venema, J. Korf (1995). Bienzyme reactor for electrochemical detection of low concentrations of uric acid and glucose. *Clinica Chimica Acta,* **239**, 153–165.

Feng, L., Y. Liu, Y. Tan, J. Hu. (2004). Biosensor for the determination of sorbitol based on molecularly imprinted electrosynthesized polymers. *Biosensors and Bioelectronics,* **19**, 1513–1519.

Fuller, J. H., M. J. Shipley, G. Rose, R. J. Jarret, H. Keen (1983). Mortality from coronary heart disease and stroke in relation to the degree of glycaemia: The Whitehall Study. *BMJ* **287**, 867–870.

Garcia-Villar, N., J. Saurina, S. Hernandez-Cassou (2003). Flow injection differential potentiomtric determination of lysine by using a lysine biosensor. *Analytica Chimica Acta,* **477**, 315–324.

Gauglitz, G. (2005). Direct optical sensors: Principles and selected applications. *Analytical and Bioanalytical Chemistry,* **381**, 141–155.

Gavin, J. R., K. G. M. M. Alberti, M. B. Davidson, R. A. DeFronzo, A. Drash, S. G. Gabbe, S. Genuth, M. I. Harris, R. Kahn, H. Keen, W. C. Knowler, H. Lebovitz, N. K. Maclaren, J. P. Palmer, P. Raskin, R. A. Rizza, M. P. Stern (1997). Report of the expert committee on the diagnosis and classification of diabetes mellitus. *Diabetes Care* **20**, 1183–1197.

Gerard, M., A. Chaubey, B. D. Malhotra (2002). Application of conducting polymers to biosensors. *Biosensors and Bioelectronics,* **17**, 345–359.

Hadden, W. C., M. I. Harris (1987). Prevalence of diagnosed diabetes, undiagnosed diabetes, and impaired glucose tolerance in adults 20-74 years of age. *Vital & Health Statistics – Series 11: Data from the National Health Survey* **237**, 1–55.

Homola, J. (2003) Present and future of surface plasmon resonance biosensors. *Analytical and Bioanalytical Chemistry,* **377**, 528–539.

Hsieh, H. V., B. Stewart, P. Hauer, P. Haaland, R. Campbell (1998). Measurement of *Clostridium perfringens* beta toxin production by surface plasmon resonance immunoassay. *Vaccine,* **16**, 997–1003.

Ikebukuro, K., C. Kiyohara, K. Sode (2005). Novel electro-chemical system for protein using the aptamers in sandwich manner. *Biosensors and Bioelectronics*, **20**, 2168–2172.

Ivansson, D., K. Bayer, C. F. Mandenius (2002). Quantitation of intracellular recombinant human superoxide dismutase using surface plasmon resonance. *Analytica Chimica Acta*, **456**, 193–200.

Jain, S. R., E. Borowska, R. Davidsson, M. Tudorache, E. Ponten, J. Emneus (2004). A chemiluminescence flow immunosensor based on a porous monolithic methacrylate and polyethylene composite disc modified with Protein G. *Biosensors and Bioelectronics*, **19**, 795–803.

Keay, R. W., C. J. McNeil (1998). Separation-free electro-chemical immunosensor for rapid determination of atrazine. *Biosensors and Bioelectronics*, **13**, 963–970.

Kim, M., M. J. Kim (2003). Isocitrate analysis using a potentiometric biosensor with immobilized enzyme in a FIA system. *Food Research International*, **36**, 223–230.

Lawung, R., B. Danielsson, V. Prachayasittikul, L. Bulow (2001). Calorimetric analysis of cephalosporins using an immobilized TEM-1 beta-lactamase on Ni^{2+} chelating sepharose fastflow. *Analytical Biochemistry*, **296**, 57–62.

Lim, S. C., E. S. Tai, B. Y. Tan, S. K. Chew, C. E. Tan (2000). Cardiovascular risk profile in individuals with borderline glycaemia. *Diabetes Care* **23**, 278–282.

Liss, M., B. Petersen, H. Wolf, E. Prohaska (2002). An aptamer-based quartz crystal biosensor. *Analytical Chemistry*, **74**, 4488–4495.

Lopez, M. A., F. Ortega, E. Dominguez, I. Katakis (1998). Electrochemical immunosensor for the detection of atrazine. *Journal of Molecular Recognition*, **11**, 178–181.

Lowe, C. R. (1999). Chemoselective biosensors. *Current Opinion in Chemical Biology*, **3**, 106–111.

Luo, X. L., J. J. Xu, W. Zhao, H.Y. Chen (2004). Glucose biosensor based on ENFET doped with SiO_2 nanoparticles. *Sensors and Actuators B – Chemical*, **97**, 249–255.

Luppa, P. B., L. J. Sokoll, D. W. Chan (2001). Immunosensors – principles and applications to clinical chemistry. *Clinica Chimica Acta*, **314**, 1–26.

Maestre, E., I. Katakis, E. Dominguez (2001). Amperometric flow-injection determination of sucrose with a mediated tri-enzyme electrode based on sucrose phosphorylase and electrocatalytic oxidation of NADH. *Biosensors and Bioelectronics*, **16**, 61–68.

Mak, K. K. W., U. Wollenberger, F. W. Scheller, R. Renneberg (2003). An amperometric bi-enzyme sensor for determination of formate using cofactor regeneration. *Biosensors and Bioelectronics*, **18**, 1095–1100.

Malinauskas, A., J. Kuzmarskyte, R. Meskys, A. Ramanavicius (2004). Bioelectrochemical sensor based on PQQ-dependent glucose dehydrogenase. *Sensors and Actuators B – Chemical*, **100**, 387–394.

Marazuela, M. D., M. C. Moreno-Bondi (2002). Fiber-optic biosensors – an overview. *Analytical and Bioanalytical Chemistry*, **372**, 664–682.

May, S. W. (1999) Applications of oxidoreductases. *Current Opinion in Biotechnology*, **10**, 370–375.

Martin, S. P., D. J. Lamb, J. M. Lynch, S. M. Reddy (2003). Enzyme-based determination of cholesterol using the quartz crystal acoustic wave sensor. *Analytica Chimica Acta*, **487**, 91–100.

McCormick, A. N., M. E. Leach, G. Savidge, A. Alhaq (2004). Validation of a quantitative SPR assay for recombinant FVIII. *Clinical and Laboratory Haematology*, **26**, 57–64.

Medyantseva, E. P., E. V. Khaldeeva, G. K. Budnikov (2001). Immunosensors in biology and medicine: Analytical capabilities, problems and prospects. *Journal of Analytical Chemistry*, **56**, 886–900.

Mehrvar, M., M. Abdi (2004). Recent developments, characteristics and potential applications of electrochemical biosensors. *Analytical Sciences*, **20**, 1113–1126.

Meigs, J. B., M. A. Mittleman, D. M. Nathan, G. H. Tofler, D. E. Singer P. M. Murphy-Sheehy, I. Lipinsky, R. D'Angostino, P. W. F. Wilson (2000). Hyperinsulinaemia, hyperglycaemia and impaired glucose homeostasis – the Framingham offspring study. *JAMA* **283**: 221–228.

Miao, Y. Q., J. G. Guan, J. R.Chen (2003). Ion sensitive field effect based biosensors. *Biotechnology Advances*, **21**, 527–534.

Miao, Y. Q., J. R. Chen, X. H. Wu (2004). Using electropolymerised non-conducting polymers to develop enzyme amperometric biosensors. *Trends in Biotechnology*, **22**, 228–231.

Monk, D. J., D. R. Walt (2004). Optical fiber-based biosensors. *Analytical and Bioanalytical Chemistry*, **379**, 931–945.

Moser, I., G. Jobst, G. A. Urban (2002). Biosensor arrays for simultaneous measurement of glucose, lactate, glutamate, and glutamine. Biosensors and Bioelectronics, **17**, 297–302.

Nakamura, H., I. Karube (2003). Current research activity in biosensors. *Analytical and Bioanalytical Chemistry*, **377**, 446–468.

Nathan, D. M., J. Meigs, D. E. Singer (1997). The epidemiology of cardiovascular disease in type 2 diabetes mellitus: how sweet it is ... or is it? *The Lancet* **350**: SI4–9.

Navratil, M., J. Tkac, J. Svitel, B. Danielsson, E. Sturdik (2001). Monitoring of the bioconversion of glycerol to dihydroxy-acetone with immobilized *Gluconobacter oxydans* cell using thermometric flow injection analysis. *Process Biochemistry*, **36**, 1045–1052.

Nicelescu, M., S. Sigina, E. Csoregi (2003). Glycerol dehydrogenase based amperometric biosensor for monitoring of glycerol in alcoholic beverages. *Analytical Letters*, **36**, 1721–1737.

O'Sullivan, C. K., G. G. Guilbault (1999). Commercial quartz crystal microbalances – theory and applications. *Biosensors and Bioelectronics*, **14**, 663–670.

Ohlson, S., C. Jungar, M. Strandh, C. F. Mandenius (2000). Continuous weak affinity immunosensing. *Trends in Biotechnology*, **18**, 49–52.

Okuda, J., J. Wakai, K. Sode (2002). The application of cytochromes as the interface molecule to facilitate the electron transfer for PQQ glucose dehydrogenase employing mediator type glucose sensor. *Analytical Letters*, **35**, 1465–1478.

Orchard, T. (1998) Diabetes: A time for excitement and concern. Hopeful signs exist that the ravages of diabetes can be tamed. *British Medical Journal* **317**, 691–692.

Park, I. S., D. K. Kim, N. Kim (2004). Characterization and food application of a potentiometric biosensor measuring beta-lactam antibiotics. *Journal of Microbiology and Biotechnology*, **14**, 698–706.

Poghossian, A., M. J. Schoning, P. Schroth, A. Simonis, H. Luth (2001). An ISFET-based penicillin sensor with high sensitivity, low detection limit and long lifetime. *Sensors and Actuators B – Chemical*, **76**, 519–526.

Pogoroleva, S. P., M. Zayats, A. B. Kharitnov, E. Katz, I. Willner (2003). Analysis of $NAD(P)^+$ cofactors by redox-functionalized ISFET devices. *Sensors and Actuators B – Chemical*, **89**, 40–47.

Ramanathan, K., M. Rank, J. Svitel, A. Dzgoev, B. Danielsson (1999). The development and applications of thermal biosensors for bioprocess monitoring. *Trends in Biotechnology*, **17**, 499–505.

Ramanathan, K., B. Danielsson (2001). Principles and applications of thermal biosensors. *Biosensors and Bioelectronics*, **16**, 417–423.

Ramanathan, K., B. R. Jonsson, B. Danielsson (2001). Sol-gel based thermal biosensor for glucose. *Analytica Chimica Acta*, **427**, 1–10.

Rank, M., B. Danielsson, J. Gram (1992). Implementation of a thermal biosensor in a process environment: On-line monitoring of penicillin V in production scale fermentations. *Biosensors and Bioelectronics*, **7**, 631–635.

Razumiene, J., V. Gureviciene, V. Laurinavicius, J. V. Grazu-
levicius, (2001). Amperometric detection of glucose and
ethanol in beverages using flow cell and immobilized on
screen-printed carbon electrode PQQ-dependent glucose
or alcohol dehydrogenases. *Sensors and Actuators B –
Chemical*, **78**, 243–248.

Reddy, S. M., J. P. Jones, T. J. Lewis, P. M. Vadgama (1998).
Development of an oxidase based glucose sensor using
thickness-shear-mode quartz crystals. *Analytica Chimica
Acta*, **363**, 203–213.

Rehman, H. (2001). Diabetes mellitus in the young. *Journal of
the Royal Society of Medicine* **94**, 65–67.

Rich, R. L., D. G. Myszka (2000). Advances in surface plasmon
resonance biosensor analysis. *Current Opinion in Biotechnology*, **11**, 54–61.

Saha, S., M. Raje, C. R. Suri (2002). Sandwich microgravimet-
ric immunoassay: sensitive and specific detection of low
molecular weight analytes using piezoelectric quartz crystal. *Biotechnology Letters*, **24**, 711–716.

Saidman, S. B., M. J. Lobo-Castanon, A. J. Miranda-Ordieres,
P. Tunon-Blanco (2000). Amperometric detection of D-
sorbitol with NAD^+-D-sorbitol dehydrogenase modified
carbon paste electrode. *Analytica Chimica Acta*, **424**,
45–50.

Sandifer, J. R., J. J. Voycheck (1999). A review of biosensor and
industrial applications of pH-ISFETs and an evaluation
of Honeywell's "Dura-FET." *Mikrochimica Acta*, **131**,
91–98.

Santos, A. S., R. S. Freire, L. T. Kubota (2003). Highly stable
amperometric biosensor for ethanol based on Meldola's
blue adsorbed on silica gel modified with niobium oxide.
Journal of Electroanalytical Chemistry, **547**, 135–142.

Schoning, M. J., A. Poghossian (2002). Recent advances in
biologically sensitive field-effect transistors. *Analyst*, **127**,
1137–1151.

Schugerl, K. (2001). Progress in monitoring, modelling and control of bioprocesses during the last 20 years. *Journal of
Biotechnology*, **85**, 149–173.

Shaw, J .E., P. Z. Zimmet, D. McCarty, M. de Courten (2000).
Type 2 diabetes worldwide according to the new classification and criteria. *Diabetes Care* **23**: B5–B10.

Su, X.-D., F. T. Chew, S. F. Y. Li (2000). Design and application
of piezoelectric quartz crystal-based immunoassay. *Analytical Sciences*, **16**, 107–114.

Suzuki, M., Y. Nakashima, Y. Mori (1999). SPR immunosensor integrated two miniature enzyme sensors. *Sensors and
Actuators B – Chemical*, **54**, 176–181.

Takamizawa, K., S. Uchida, M. Hatsu, T. Suzuki, K. Kawai (2000). Development of a xylitol biosensor composed of xylitol dehydrogenase and diaphorase. *Canadian Journal of Microbiology*, **46**, 350–357.

Tominaga, M., H. Eguchi, H. Manaka, K. Igarashi, T. Kato, A. Sekikawa (1999). Impaired glucose tolerance is a risk factor for cardiovascular disease but not impaired fasting glucose. *Diabetes Care* **22**: 920–924.

Tinkilic, N., O. Cubuk, I. Isildaki (2002) Glucose and urea biosensors based on all solid-state $PVC-NH_2$ membrane electrodes. *Analytica Chimica Acta*, **452**, 29–34.

Tkac, J., I. Vostiar, E. Sturdik, P. Gemeiner, V. Mastihuba, J. Annus (2001). Fructose biosensor based on D-fructose dehydrogenase immobilized on a ferrocene-embedded cellulose acetate membrane. *Analytica Chimica Acta*, **439**, 39–46.

Tkac, J., I. Vostiar, C. F. Mandenius (2004). Evaluation of disruption methods for the release of intracellular recombinant protein from *Escherichia coli* for analytical purposes. *Biotechnology and Applied Biochemistry*, **40**, 83–88.

Ulber, R., J.G. Frerichs, and S. Beutel (2003). Optical sensor systems for bioprocess monitoring. *Analytical and Bioanalytical Chemistry*, **376**, 342–348.

Vidal, J.C., E. Garcia-Ruiz, J.R. Castillo (2003). Recent advances in electropolymerized conducting polymers in amperometric biosensors. *Microchimica Acta*, **143**, 93–111.

Yang, M. X., J. R. Chen (2002). Self assembled monolayer based quartz crystal biosensors for the detection of endotoxin. *Analytical Letters*, **35**, 1775–1784.

Zhang, S., G. Wright, Y. Yang (2000). Materials and techniques for electrochemical biosensor design and construction. *Biosensors and Bioelectronics*, **15**, 273–282.

Zhao, H., Y. Yuan, S. Adeljou, G. G. Wallace (2002). Study on the formation of the Prussian Blue films on the polypyrrole surface as a potential mediator system for biosensing applications. *Analytica Chimica Acta*, **472**, 113–121.

Zimmet, P. Z. (1995). The pathogenesis and prevention of diabetes in adults. Genes, autoimmunity, and demography. *Diabetes Care* **18**: 1050–1064.

Zimmet, P., K. G. Alberti,, J. Shaw (2001). Global and societal implications of the diabetes epidemic. *Nature* **414**, 782–787.

Zimmet, P. (2003) Diabetes and obesity worldwide: Epidemics in full flight. *International Diabetes Institute*, Australia. (www. medforum.nl/reviews/diabetes_and_obisity.htm – accessed September 2003).

12

Fermentors: Design, Operation, and Applications

A.R. Allman

12.1 Batch culture fermentation

The purpose of this chapter is to give an introduction to the methodology of batch fermentation using a small, autoclavable, bench-top fermentor system. Specifically, this chapter should provide an understanding of the following:

- The engineering and biological concepts of fermentors;
- The principles behind the instrumentation used with a modern fermentor;
- The assembly and preparation of a fermentor vessel and ancillary equipment for autoclaving;
- How samples are taken from the fermentor vessel during operation;
- How simple continuous culture can be accomplished.

12.2 The main components of a fermentor and their uses

The main subdivisions of a bench-scale fermentor are as follows:

- Base components including drive motor, heaters, pumps, gas control, etc.;
- Vessel and accessories;
- Peripheral equipment such as reagent bottles;
- Instrumentation and sensors.
- These components combine to perform the following functions:
  - provide operation free from contamination;
  - maintain a specific temperature;
  - provide adequate mixing and aeration (see Box 12.1);
  - control the pH of the culture;

MODERN FERMENTOR
A system consisting of a few pieces of equipment that provide controlled environmental conditions for the growth of microbes (and/or production of specific metabolites) in liquid culture whilst preventing entry and growth of contaminating microbes from the outside environment.

CONTINUOUS CULTURE
A method of allowing culture to be grown in a fermentor at a specific growth rate. The growth rate is determined by the flow of medium through the vessel (dilution rate).

– allow monitoring and/or control of dissolved oxygen;
– allow feeding of nutrient solutions and reagents;
– provide access points for inoculation and sampling;
– use fittings and geometry relevant to scale-up;
– minimize liquid loss from the vessel;
– facilitate the growth of a wide range of organisms.

The example used to illustrate these points is the smallest, simplest system that displays all these components and attributes, i.e., a small, autoclavable, bench-top fermentor. Mention is made of larger, *in situ* sterilizable units and special features that can be added to achieve certain objectives such as containment work. Also, a range of basic applications are illustrated, e.g., mammalian cell culture as a starting point for anyone contemplating the use of small-scale fermentors for these purposes. The specific information provided should be treated with a great deal of circumspection as the parameter values required for any particular microbes or cell line could be completely different.

12.3   Component parts of a typical vessel

The vessel can be constructed either as a single-walled cylinder of borosilicate glass or as a glass-jacketed system, which typically has a round bottom. The top plate is made from "316"L stainless steel and is compressed onto the vessel flange by nuts or a quick-release clamping system. A seal separates the vessel glass from the top plate. Port fittings of various sizes are provided for insertion of probes, inlet pipes, exit gas cooler, cold fingers, sample pipes, etc. These work by compressing the sides of the probe/pipe against an O-ring seal. A special inoculation port will have a membrane seal held in place with a collar. Culture can be withdrawn into a sampling device or a reservoir bottle via a sample pipe situated in the bulk of the fermentor fluid. A gas sparger is also fixed into the top plate and this terminates in a special assembly that ensures that incoming air is dispersed efficiently within the culture by the flat-bladed "Rushton-type" impellors fixed to the drive shaft. A drive motor provides stirring power to the drive shaft and is usually fitted directly to the drive hub on the vessel top plate. An exit gas cooler works as a condenser to remove as much moisture as possible from the gas leaving the fermentor to prevent excessive liquid losses during the fermentation and wetting of the exit air filter.

PT-100 TEMPERATURE SENSOR
A platinum resistance electrode used to give an accurate indication of vessel temperature by relating changes in electrical resistance of the sensor to temperature.

A narrow platinum resistance (Pt-100) temperature sensor completes the list of minimum essential fittings. Temperature control is either by direct heating using a heater pad or by circulating warm water around the vessel jacket. If direct heating is used, a cold finger is used to cool the vessel contents (more than one could be used, if required).

Box 12.1 **Mixing and aeration**

MIXING is important for:

- Heat transfer
- Homogeneous environment
- Keeping solids in suspension

Rheology describes how a culture behaves when subjected to shear forces. There are two main types of mixing behavior:

1. Newtonian — the viscosity of the culture is independent of shear force
2. Non-Newtonian — the viscosity varies with shear forces — these are usually fungal broths that can alter in apparent viscosity during a fermentation. Several different types of non-Newtonian behavior have been described, e.g., Bingham plastic and Casson body. The increasing viscosity can effect bubble retention and therefore, gas transfer.

AERATION

The key parameter is Oxygen Transfer Rate (OTR), often expressed as a related parameter, Kla as hr^{-1}. Kla is a combination of the film resistance between the gas and liquid phases in an aerated fermenter and the surface area in contact between the gas and the liquid. Essentially, the more & smaller gas bubbles you have, the better the gas will be dissolved and the better the KLa value achieved for a given vessel configuration. For maximum time for gas exchange with sparging, a tall vessel is better (hence the typical aspect ratio of 3:1). With the use of oxygen supplementation, a shorter vessel can be used. Similarly, a low aspect ratio 2:1 (or even 1:1) was used for cell culture vessel for gas transfer via the head-space but now sparging at a low gas flow rate has been almost universally adopted so a taller vessel is an advantage.

The sensors that fit into the vessel do so either by direct coupling with a thread on the body of the electrode, as is the case in the gel-filled type of pH electrode, or by a special fitting on the vessel top plate, as in the case of 19-mm diameter dissolved oxygen electrodes (polarographic type). Another system involves the use of a simple compression fitting that holds the body of the electrode, as with the foam probe. In this case, the height is variable and the tip of the probe is above the culture fluid. Figure 12.1 illustrates all the main components of a fermentor.

12.4 Peripheral parts and accessories

12.4.1 Reagent pumps

Pumps are normally part of the instrumentation system for pH and antifoam and control. Peristaltic pumps are used and the flow rate is usually fixed with a timed "shot and delay" feed system of control. Flow rates will depend on the bore of the tubing used and it is a good idea to move the tubing every few days so that one piece does not become worn out by the pump rollers. The peristaltic tubing links the reservoir bottles with the vessel multiway inlet. The tubing is clamped shut during autoclaving, already connected to the vessel, and opened for active addition of reagent. Alternatively, the bottles can be connected following autoclaving by using an aseptic coupling to join the tubing between the pump and the multiway inlet. A dip tube goes to the bottom of the bottle. A filter in the top of the reservoir bottle is kept open during autoclaving to prevent a

Figure 12.1
Major components of a fermentor.

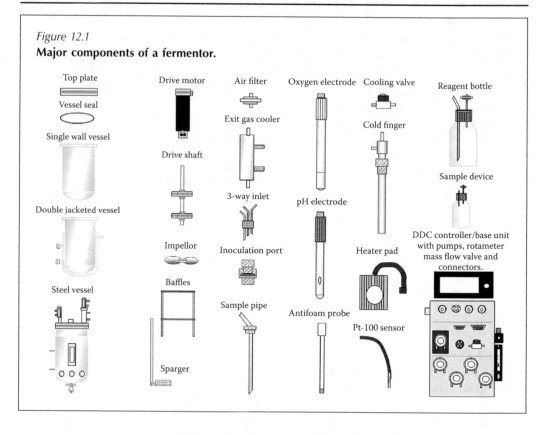

build-up of pressure. For extremely high accuracy of addition, the reservoir bottles can be placed on analytical balances used to determine how much reagent has been pumped in a given time.

12.4.2 Medium feed pumps and reservoir bottles

Medium feed pumps are often variable speed to give the maximum possible range of feed rates. The speed of operation of the pump can be set manually or the whole system put under computer control. The reservoir bottles are usually larger, e.g., 5–20 L, but are prepared in the same way as normal reagent bottles. These bottles may have to be changed several times during a long, continuous culture experiment, so it is usual to fit an aseptic coupling in the tubing. A harvest pump is often used to remove culture fluid from the fermentor vessel into a storage reservoir bottle.

ROTAMETER
A variable-area flow meter that indicates the rate of gas flow into a fermentor. A manual valve is adjusted until an indicator ball rises up a tube of increasing width until the required flow rate value is reached on a calibrated scale marked on the glass wall of the tube. The bottom of the ball should rest on the calibration line.

12.4.3 Rotameter/gas supply

Gas input, usually air, can be provided by use of a laboratory air supply or from a separate pump (which must be oil-free). A variable-area flow meter or rotameter is used to control the air flow rate into the fermentor vessel. A pressure regulator valve before the rotameter ensures safer operation. A sterile filter (usually 0.22 μm) is used as a bridge between the tubing from the rotameter and that connected to the air sparger of the fermentor. A second filter on the exit gas cooler (often 0.45 μm) stops microbes from being released into the laboratory air as the gas leaves the fermentor under a slight positive pressure.

12.4.4 Sampling device

This allows culture fluid to be removed aseptically during the fermentation at intervals decided by the user. The frequency of sampling and the size of each sample are determined empirically according to the needs of the experiment.

12.5 Alternative vessel designs

There are alternatives to using a conventional, "stirred tank" fermentor. Usually, alternative vessel designs are tried when the standard vessel configuration does not allow adequate growth of the organism (e.g., animal cells are often disrupted by shear forces in fermentors with turbine impellors) or the scale-up criteria require a different design of bioreactor (e.g., production of large quantities of single-cell protein is cheaper on a large scale if air lift fermentors are used to eliminate energy costs associated with a drive system). A number of these special designs are available at the bench/pilot scale of operation to allow small-scale research into the suitability of a particular method.

12.5.1 Air lift

This vessel design eliminates the need for a stirrer system. A tall, thin vessel is the best shape with an aspect ratio of around 10:1 (height to base diameter). Sometimes a "conical" section is used in the top part of the vessel to give the widest possible area for gas exchange. Sensors can be mounted in a steel base section, from a collar at the base of the conical section or from the vessel top plate.

The culture fluid is both mixed and aerated by a stream of air, which enters near the base of the vessel. A hollow pipe or draft tube in the center of the vessel provides a "riser" for the air (which is full of bubbles) to move upwards to the top of the vessel. If a very large vessel is used, the hydrostatic head of the fluid provides a pressurizing effect to the lowest region of the culture where the air enters and so increases the dissolved oxygen concentration. The draft tube is usually double-walled to allow heating and cooling using a thermocirculator system.

ASPECT RATIO
The ratio of the height of a fermentation vessel to its diameter. Typically, vessels for microbial work have an aspect ratio of 2.5Ð3:1, while vessels for animal cell culture tend to have an aspect ratio closer to 1:1.

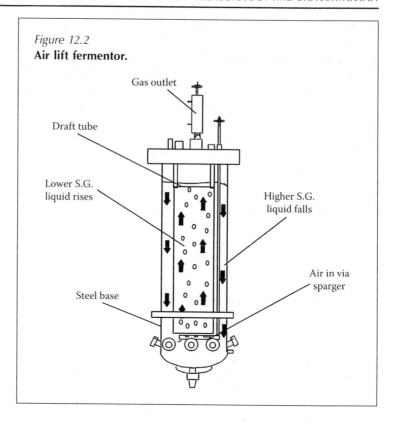

Figure 12.2
Air lift fermentor.

When the aerated culture fluid reaches the top of the draft tube, it "spills over" and begins to fall towards the bottom of the vessel via the space between the outer wall of the draft tube and the inner wall of the vessel. A large head space above the top of the draft tube allows for easy gas transfer from the liquid to the gas phase, which causes the specific gravity of the liquid to increase and so it descends to the bottom of the vessel. The descending liquid returns to the base of the vessel where it is re-gassed and begins to rise again (see Figure 12.2).

A common use of air lift fermentors is the growth of shear-sensitive cells such as plant and animal cultures. Also, the design has been used for the production of large amounts of biomass as single-cell protein.

12.5.2   Fluidized bed, immobilized and solid-state systems

This important area of bioreactor design and application involving bioreactor design will only be mentioned in outline here, as a complete chapter is devoted to this subject (*see Williart, R. Cell Immobilisation and Its Application in Biotechnology: Current Trends and Future Prospects*).

The microbes/cells are trapped in a physical medium (e.g., alginate beads) and held in the vessel by a mesh. Medium is recirculated via a pump; this can be easily adapted to give a continuous/semicontinuous flow to allow the trapped cells to effect chemical changes on constituents of the medium without being washed out along with spent medium. This system is well suited to growth of animal cells on the smaller scale and has large-scale application in effluent/decontamination treatment plants.

Animal cells benefit from being immobilized on polymer beads in a stirred tank reactor system as it provides a surface for attachment of anchorage-dependent cells and affords protection to the cells against damage by bubbles and shear forces. Bacterial and fungi are also grown in immobilized systems, using entrapment and also growth on solid surfaces (membrane bioreactors).

Solid-state fermentation is increasing in importance in several areas of application. Here the substrate for growth is a solid, such as soil (or semisolid, e.g., fats). The substrate is rotated or mechanically mixed and temperature control achieved by both air and water circulation. As fitting conventional instrumentation is difficult when the whole vessel rotates, exit gas analysis is commonly used to follow the progress of a process (see Section 12.10.2). Applications include the bio-remediation of soils to remove toxins or waste products, improving the efficiency of composting for agriculture and production of enzymes in solid fermentation, where yields in some cases can be an order of magnitude higher than those achieved in liquid fermentation.

12.5.3 Hollow fiber

This is a similar idea but the cells are now embedded in fibers contained in a cartridge that is bathed in circulating culture medium. This is often used for mammalian cell culture, where anchorage-dependent cell lines can be perfused with oxygenated medium. An extension of this method is to use cartridges containing two different bundles of fibers as a separation "membrane" between, e.g., a pair of fermentor vessels and allows dissolved gases and metabolites to be exchanged without cells crossing the barrier. This arrangement could allow, for example, toxicity or interactive studies. A closely related application is micro-filtration, where cells are passed through an external cartridge with a pore size of e.g. 0.1 microns and small quantities of medium eluted. This removal can be continuous for subsequent on-line biochemical analysis or larger amounts can be taken regularly for cell-free harvest of product, e.g., a protein.

12.5.4 *In situ* sterilizable fermentors

The use of autoclavable vessels of greater than 5–10 l working volume quickly becomes impractical. For safety and insurance considerations, vessels above this size are usually made of stainless steel (316L) and are designed to be sterilized *in situ* using house steam or an electrical steam generator (which may be built into the base unit). The heating for the vessel is normally provided via a double jacket, which can either be the full length of the vessel or cover just the bottom third. The bottom section contains large (25 mm) port fittings for electrodes and usually has some kind of steam-sterilizable sampling device/harvest valve. The mechanical seal and drive shaft usually enter from the bottom on this size of vessel. As the vessel body is steel, a sight window and a light have to be fitted in order to see the culture. The vessel top plate has port fittings that use a membrane seal and port closure. Unlike the bench units, anything that dips into the culture liquid or is used as an addition port must be autoclaved separately and pushed through a membrane using aseptic techniques after the vessel has been autoclaved. In some fermentors, this even includes the air inlet filter and connecting pipe to the air sparger. The steam for sterilization is either supplied from an in-house source and used to heat the vessel jacket or can be raised electrically from a separate steam generator. The medium in the vessel is heated to 121°C and often supplies the steam for sterilization of the exit gas filter (see Figure 12.3).

12.5.5 Containment

An *in-situ* sterilizable fermentor may have to be altered in certain ways if the organism to be cultured is pathogenic or genetically modified. The alterations are designed to contain any release of microbes into the environment by using features such as a double mechanical seal, magnetic coupling, additional air filters, extra foam control systems, and special sampling devices. More elaborate precautions include steam traces to all vessel fittings and discharge of any released liquid directly to a tank of disinfectant (see Figure 12.3).

There are several different categories of containment. Requirements differ at each stage and relate to building and working protocols as much as the design of the fermentor.

Applications for this sort of technology are medical research and vaccine manufacture.

ANALOGUE
INSTRUMENTS
Measurement or control
modules that do not
contain processors. All
actions are "hard-wired"
as electronic components,
and any adjustment
is made using
potentiometers.

12.6 Different types of instrumentation

Fermentor instrumentation can range from simple analogue control modules arranged as a "stack," to powerful, embedded microprocessors that operate as on-board computers to directly operate the heaters, pumps, valves, and other control actuators of a fermentor. The analogue controllers have now largely given way to microprocessor based systems and so will not be described.

Figure 12.3
Major components of an in situ sterilizable fermentor vessel.

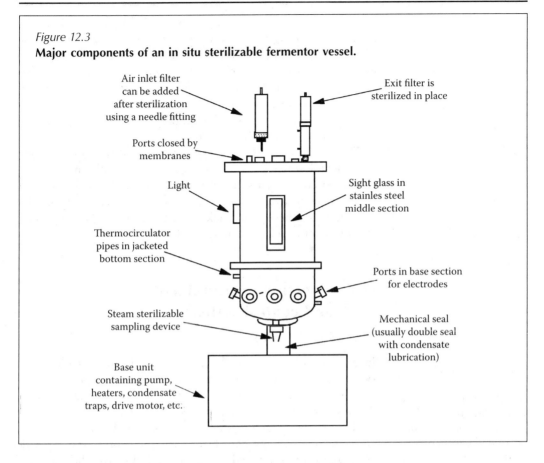

12.6.1   Digital controllers — embedded microprocessor

The measurement sensors link directly to the single control module and several parameters are displayed immediately on a single screen. Control is by direct action on heaters, valves, etc. (Direct Digital Control or DDC). The microprocessor is permanently embedded in the instrumentation and may even be a single chip. Operation is usually via a simple menu system.

12.6.2   Digital controllers — process controllers

A complete process controller (usually from a production control environment) is added to a housing containing all the signal processing and control actuators for the fermentors. It exerts control in the same way as an embedded controller but is essentially a "plug-in" component. The controller is usually programmed using simple commands to input set-point values, etc.

12.6.3   Digital controllers — direct computer control

In this case, there is no external instrumentation or processor between the actuators and the computer. A printed circuit board

DIGITAL
CONTROLLER
A digital controller uses a processor to store information about control output characteristics as mathematical algorithms. Consequently, changing the characteristics of such a controller is achieved by reprogramming the processor.

with operational amplifiers for the probe input signals is the only electronic part that may be present in the fermentor base unit. A special Input/Output (I/O) card is needed for the computer, which allows the input values to be accessed by the measurement and control software and sends signals out to operate relays, e.g., to turn a heater on or off. The processing power and speed of modern PCs make them more than capable of replacing separate PLCs or embedded controllers.

The advantage with this system is that the computer display and control software are integrated totally with the fermentor. Fewer components mean that these systems are usually less expensive. However, the fermentor cannot be used without the computer and there is no backup for the control systems should the computer develop a fault and "hang."

12.7　Common measurement and control systems

12.7.1　Speed control

TACHOMETER
An electronic device usually integrated into a drive motor to provide feedback about rotational speed in the form of an analogue signal.

Speed control relies on the feedback from a tachometer located within the drive motor determining the power delivered by the speed controller to maintain the speed set-point value set by the user. Actual speed in rpm is displayed, as determined by a tachometer signal. A power meter is sometimes included to indicate how hard the motor has to work to maintain the set speed and thereby, indirectly, the viscosity or "density" of the culture fluid. A DC, low-voltage (24–50 V) motor is often used for safety reasons. Speed range is typically from 50 to 1500 rpm for bacterial systems and 10 to 300 rpm for cell culture units.

Where speed is used to control the level of dissolved oxygen, an external signal from the oxygen controller can have an effect on stirrer speed. In this case, an absolute maximum and minimum value for speed can be set on the speed control module to limit the effects of the oxygen controller (which could set either too low a speed and impair mixing or too high a speed and cause excessive foaming) (see Figure 12.4).

12.7.2　Temperature control

COLD FINGER
A closed pipe or coil that passes through the fermentor top plate and allows cooling water to circulate, to act as a heat exchanger with the culture.

A thermocirculation system around a vessel jacket has been chosen as an example here because it is the most complex of all the methods of temperature control. For simple direct systems such as a heater pad, it is simply a matter of fitting the heater, setting the desired temperature, and switching on. Cooling is normally via a cold finger and flow of cooling water is controlled via the action of a solenoid valve. The Pt-100 sensor provides the

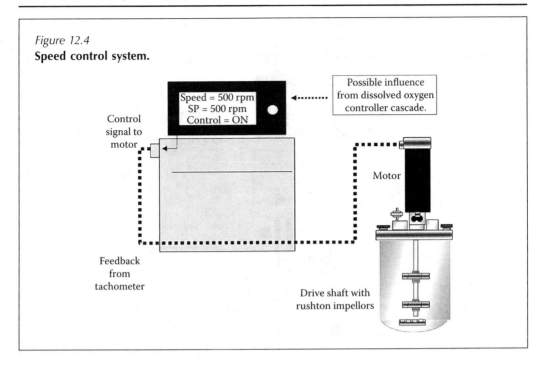

Figure 12.4
Speed control system.

feedback signal that causes the controller to take one of the following actions:

- Heat at full power as the actual temperature is some way below set-point;
- Pulse the heater power as the actual temperature is close to set-point;
- Turn on the cooling valve as the actual temperature is above set-point.

There is usually some indication to show which action the controller is taking at any given moment. A circulation pump and pipe work are added to the system for water circulation and any heating is indirect, i.e., on the water circulating in the vessel jacket. In this case, the connections to cold tap water must be made (securely, using jubilee clips or cable ties) and a drain pipe provided from the overflow point to a sink with a clear fall to the drain, i.e., the sink must be the lowest point for the whole length of this pipe. The water should be delivered from the mains at a minimum pressure of 1.5–2 bar and a flow rate of greater than 5 l min^{-1}. The water hardness should be no more than 50 ppm suspended solids to protect the heating elements from "furring." The vessel must be connected to the circulation loop (normally by rapid coupling connectors and flexible pressure tubing). Water is first supplied by opening a manual valve until the jacket is filled. The heating and cooling is controlled in exactly the same way as a directly heated system, but only the water in the

Figure 12.5
Temperature control system using water circulation.

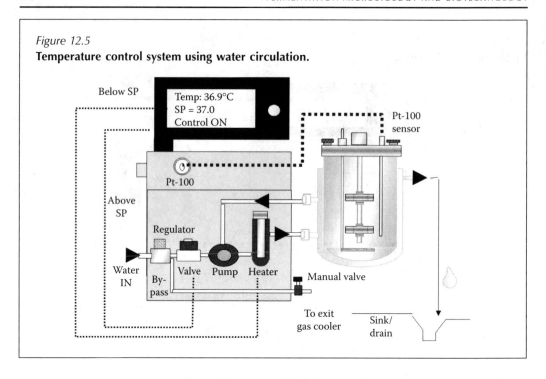

jacket is affected. The jacket provides a large surface area in contact with the vessel wall for heat exchange. Good temperature control can be achieved from approximately 5–8°C above the ambient temperature or above the temperature of the cooling water. Counter-cooling with water ensures stable temperature control when operating near ambient temperatures. Measured range is typically from 0 to 60°C (exceptionally up to 90°C) (see Figure 12.5).

12.7.3 Control of gas supply

The supply of gas (normally air) to the fermentor vessel is provided from a compressed air supply (oil-free) that may provide air for more than one vessel, depending upon its size and output. A pressure regulation valve ensures that air reaches the vessel at a maximum of 0.5–0.75 bar.

The rotameter controls the actual flow rate of air through the fermentor. This should not exceed 1.5 vessel volumes per minute or droplets of water may become entrained in the stream of gas leaving the fermentor and so wet the exit gas filter, causing it to block. A valve at the bottom of the rotameter is turned and the indicator ball in the rotameter tube rises or falls in proportion to the valve position. A scale on the tube gives flow rates in ml min^{-1} or l h^{-1}.

The air passes through the inlet air filter (Figure 12.6), which prevents any microbes from entering the vessel via this path. The end of the sparger is typically a ring with small holes through

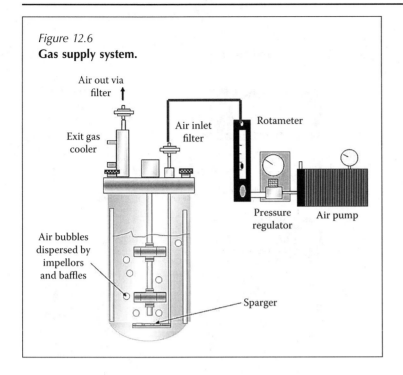

Figure 12.6
Gas supply system.

which the air is forced. The bubbles are immediately broken up and dispersed by the impellors on the drive shaft and the baffles, which can be fitted near the wall of the vessel. The use of several impellors ensures all regions of the vessel receive good aeration. A "head space" of around 20–30% is normally left between the culture level and the vessel top plate. Sometimes, gas can also be introduced into this region via a short pipe in the fermentor top plate, e.g., CO_2. For certain types of fermentation, e.g., mammalian cell culture, a gas mixing station can be used to premix several gases before they are introduced into the fermentor.

12.7.4 Control of pH

The hydrogen ion concentration (pH) is controlled by the addition of either acid or alkali as the conditions change with growth. The controller uses a pH electrode (either glass or gel type) to sense these pH changes and provide a feedback signal. The pH measurement system is identical to a bench pH meter and is calibrated in the same way before the electrode is autoclaved. Temperature compensation during operation may have to be set manually or may be automatic. The calibration normally requires that the temperature is manually set to equal the temperature of the buffer solutions. Autoclavable electrodes have a limited life (20–50 sterilizations typically). The pumps supplying the acid and alkali are normally built into the control module or the base part of the fermentor. Measured range is typically pH 2–12.

GAS MIXING STATION
A device used for animal cell culture that allows a mixture of air, oxygen, nitrogen, and carbon dioxide gases to be blended into any desired combination before they are introduced into a fermentor. This allows great flexibility in how dissolved oxygen concentration and pH are controlled within the culture.

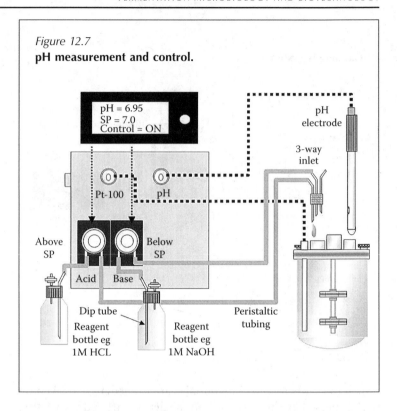

Figure 12.7
pH measurement and control.

The reagent bottles are connected to the fermentor via transfer lines of silicone tubing. The bore of tubing selected will determine the volume of acid or base added by each activation of the relevant pump by the controller. Selecting the concentration of the acid or alkali will determine how much effect each dose has on the vessel contents. Normally, solutions of between 0.5 and 2 M of acid and base are used as a starting point. Using an ammonium salt for alkali addition can provide an extra nitrogen source for the growing culture.

A set-point value is set on the controller and/or an upper and lower limit to provide a "dead band" range in which the controller is not active. This band should normally be around 0.5 pH units each side of the desired value to prevent overdosing of acid and alkali sequentially. A proportional band adjustment may be present to widen or tighten the range of pH value over which the controller acts (see Figure 12.7).

12.7.5   Control of dissolved oxygen

Dissolved oxygen is one of the most important and yet one of the most difficult parameters to control properly during a fermentation. The electrodes used to measure dissolved oxygen are of two distinct types:

DEAD BAND
An area around a set-point value which can be set where no control action will take place even if the actual value deviates from the set-point. This is used especially for a parameter such as pH where small changes are often not critical and attempts to control them would lead to excessive controller action.

1. Galvanic electrodes, which are generally simple, cheap, slow and have a limited life.
2. Polarographic electrodes, which are generally more complex, accurate, rapid, and robust but expensive.

The polarographic electrode is used more commonly and will be the example provided. The key point with this type of electrode is that it requires a voltage to polarize the anode and cathode of the detecting cell. This polarization can take between 2 and 6 hours to complete. During this time, the electrode must be connected to its relevant module, which in turn must be switched on.

The membrane of the polarographic electrode needs periodic replacement, and a special cartridge kit is available from the manufacturer to make this a simple task.

The electrode can be zeroed after autoclaving by passing nitrogen gas through the vessel for some minutes and then adjusting the zero point.

The 100% value is a relative setting made after autoclaving and polarization of the electrode. The air flow and stirrer speed are turned to the maximum needed for the fermentation and the vessel left for some minutes. The display is now adjusted until it reads 100%.

Control of dissolved oxygen can be purely by speed control, by control of a proportional air valve, or by a combination of both. Speed adjustment may interfere with good mixing or increase the degree of foaming. Adjusting the air flow rate may also affect levels of foam. The most accurate form of flow control is to use a thermal mass flow control valve that measures and controls air flow based on the cooling effect the gas exerts when passed over a heated element.

An additional option for **high-density cultures**, e.g., *Ecolab* or *Pichia pastoris*, is the use of a solenoid valve to supplement the air flow with pure oxygen in pulses (see Figure 12.8). For larger fermentors, pressure control is also used to increase the amount of oxygen that can dissolve in the medium. Several control strategies are often used sequentially during fermentation as a cascade. This can include the option to control feeding to match the available dissolved oxygen if a limit has been reached.

Range is typically 0–120%.

12.7.6 Antifoam control

The control of foam is based on its detection by a conductance probe in the vessel head space, which leads to the controller delivering a dose of liquid antifoam reagent via a peristaltic pump. A delay timer ensures the antifoam reagent has adequate time to reduce the foam level before another shot of antifoam

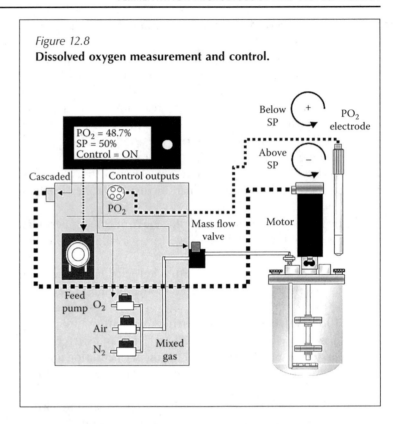

Figure 12.8
Dissolved oxygen measurement and control.

is added. The sensitivity of the probe to foam can be adjusted by the user to suit the conditions prevailing in the fermentor. A sheath of inert material around the probe prevents splashes of foam thrown up from the culture from giving "false positives" (see Figure 12.9). Normally, the metal top plate is used to provide the electrical circuit for the probe to operate so a flying lead is provided that fits into a socket somewhere on the top plate.

Antifoam reagents can be mineral oils, vegetable oils, or certain alcohols. Commercial preparations are available for use with cultures that ultimately provide materials for pharmaceutical manufacture. The key thing with using oils is that they can form a skin on the surface of the culture and interfere with gas transfer at the liquid/air interface. If foam is allowed to build up unchecked, it can eventually reach the exit gas filter, thus blocking it and providing a path for contamination.

12.8 Additional sensors

Several additional process values can be measured and controlled within a fermentor. The most common are listed below:

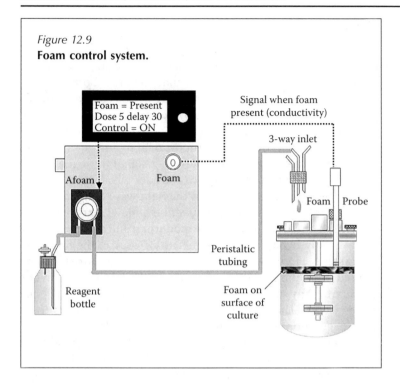

Figure 12.9
Foam control system.

12.8.1 Redox

This refers to the reduction/oxidation (redox) potential of a system and can be applied to biological cultures as well as chemical reactions. Redox values are given in millivolts and usually refer to the oxidation or reduction potential of a given chemical relative to a standard hydrogen half-cell (which is defined as having a redox potential of 0.00 V). A half-cell, which is more reducing than the hydrogen standard, has a negative value and a stronger oxidizing agent will have a positive value. Reducing agents donate an electron to oxidizing agents. Depending on conditions (the pH can affect redox potential), this reaction is reversible.

The complexity of a typical fermentation broth (in terms of the sheer number of separate reactions going on at any one moment) means that the actual measured value is only of limited importance. However, for anaerobic cultures, the redox electrode provides a very sensitive "watchdog" for the presence of oxygen. Tiny amounts of oxygen will have a relatively large effect on the redox value. A redox value below –200 mV is a good indicator of anaerobic conditions.

The redox electrode is very similar to a pH electrode and the conditions for handling and care are almost identical. An electrical zero can usually be set and shorting out the electrode connections should give a reading close to 0 mV.

12.8.2 Air flow

This has been mentioned previously in connection with dissolved oxygen control. Most fermentors now offer the possibility to use a thermal mass flow control valve that combines both the measurement element and the control valve in one compact unit. In this case, measurement of flow is by detecting the change in temperature of a heating element before and after gas passes over it. The temperature difference is proportional to flow rate, and this can be processed to give a measurement signal for logging and/or control. The practical points to consider when using this system are to ensure that the valve is the right way round (an arrow on the metal section shows the direction of flow; however, this is sometimes obscured by the valve mounting) and that the control valve usually bypasses the rotameter. Of course, the flow control systems can measure other gas flows, not just air. They can be used in combination under computer control to give a precise gas mix for any total flow rate using one mass control valve per gas and post-blending the mixture. However, this tends to be an expensive option. A simpler system uses solenoid valves to introduce each gas into a premixing chamber at manually set flow rates and then controls the final gas flow into the fermentor with a single mass flow control valve.

12.8.3 Weight

This parameter is useful for continuous culture or fed-batch work where the rate of addition of feed must be known and controlled very accurately. It is possible to mount a whole fermentor on a balance and "tare" out its weight. However, a more precise approach is to use a system with a small load cell mounted in such a way that only the vessel and its contents are measured. The deformation of the load cell provides an output that can be transduced into an electrical signal. Control is normally by setting a maximum and/or minimum weight such that a feed pump operates whenever the value is less than the maximum. At the maximum value, a second pump operates to remove culture until the minimum weight is reached.

A variant of this system can be used to measure very accurately the amount of reagent, e.g., alkali, added to a fermentor. In this case, the reagent bottle is placed on an analytical balance that has a computer output (normally RS232) and the decrease in weight used as a measure of the amount of reagent added.

LOAD CELL
A method for measuring changes in weight in a fermentor vessel using deformation of a crystal as an indicator of load. Load cells are used where it would be impractical to use a conventional balance.

12.8.4 Pressure

The use of pressure as a control parameter is more common in larger, *in situ* sterilizable fermentors, for the obvious reasons that the vessel is made of steel and is already adapted to working at pressures of several bars. Also, it is typically in the larger vessel that maintaining the dissolved oxygen concentration at a high level presents a problem (an increased pressure increases the amount of oxygen that can be dissolved in the fermentation broth). However, even a small glass vessel can often operate at some overpressure (e.g., up to 0.7–1 bar), but it is vital that this is first confirmed by the manufacturer.

The measurement and control are the same whatever the size and type of vessel, except that the size of the control valve would alter. Measurement is normally provided by a piezo-electric sensor, which can be mounted into a port closure and fitted to the vessel top plate. As with a load cell, it is the deformation of the crystal by the internal vessel pressure which gives a proportional electrical signal, typically in the range of 0–2 bar. The control element is a proportional valve that restricts the flow of gas out of the fermentor and thereby creates a back-pressure. A mechanical overpressure valve or burst-disc is an essential safety requirement if overpressure is to be used.

12.8.5   On-line measurement of biomass

A direct measurement of the number of organisms in a culture is clearly desirable for the control of feed rate, oxygenation, and general process optimization. Normally, measurements of total cell numbers, wet weight, dry weight, and viable cell counts would be made by taking samples of the culture at set times and performing the relevant laboratory analysis. The results would be entered some minutes, hours, or even days later, when they would be of little value for real-time control. Fluorescence microscopy has also been used as a measure of viable cell density either with a probe or externally through the vessel glass. Probe-based systems that are reasonably accurate and can be used with bench-scale vessels will be described by way of examples below.

12.8.5.1   Optical density (OD)/turbidity systems

This relies on either the increasing absorbance of light over time by the culture (OD) or the scattering of light by particles in the culture (i.e., microbes). In either case, the sensor is directly mounted in the vessel and connected to the electronic, usually via a fiber optic cable. A two-point calibration procedure is used and results can be expressed in a variety of units. A range up to 200 g l-1 dries weight or higher is measurable with some systems. These methods have the advantage of being cost-effective and work well for cultures in the logarithmic phase of growth, when viability is almost 100%. However, both live and dead cells (plus any other particulate material) are measured by these techniques.

PROPORTIONAL VALVE
A valve that can be adjusted electrically or pneumatically from 0 to 100% open or closed. The action of the valve is in proportion to the degree of change required by the controller to, for example, maintain a certain level of dissolved oxygen by adjusting the air flow rate.

12.8.5.2 Capacitance/conductance-based biomass monitor

This uses a totally novel measurement system to give viable cell numbers for all types of organisms (bacteria, yeast, filamentous fungi, and animal cells). A probe in the vessel uses a radio-frequency electrical field to measure the natural capacitance of living cells with an intact plasma membrane. This build-up of charge is proportional to the cell concentration and specific to different types of cell. The signal is processed and can be expressed as dry weight or concentration in cells ml–1. An anti-fouling system is used to maintain the ability of the probe to give accurate readings over long time periods.

This type of instrument makes possible the control of feed pumps by biomass concentration and could form the basis for an automated transfer system for delivering a seed culture to a larger vessel after an inoculum of sufficient concentration has been produced in a smaller fermentor.

12.9 Simple continuous culture

There are a number of different ways of setting up a continuous culture. The method of choice has the virtue of being able to be added to any fermentor vessel without modification or the addition of extra controllers beyond the acquisition of a feed pump and a harvest pump. The feed pump is variable speed but the harvest pump can be fixed speed as long as its speed is greater than that of the feed pump. The harvest pipe, being sited at the surface of the culture, has the disadvantage of taking off only the top of the culture (which may not be representative of the bulk culture) but the advantage of helping to prevent build-up of foam. If foam is a problem, then antifoam reagent can be added to the medium feed at a background level, e.g., 1 part in 20,000. Samples can be taken by putting a "Y" piece in the take-off tubing and placing a sampling device in one arm of the Y (see Figure 12.10).

12.10 Additional accessories and peripherals

12.10.1 Feed pumps

A separate feed pump is necessary for fed-batch or continuous applications. The type of pump used will depend on the size of the vessel, the required flow rates, the nature of the additive (e.g., viscous, corrosive, etc.), and the speed at which the addition should be made. Often, a feed pump is included as part of the basic fermentor or provision is made for its inclusion as an optional accessory.

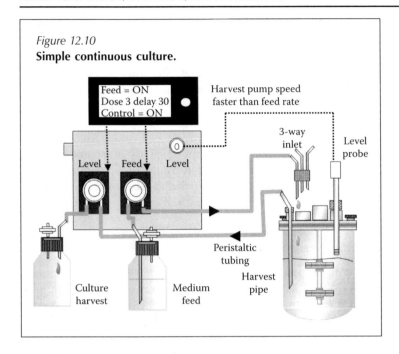

Figure 12.10
Simple continuous culture.

Feed = ON
Dose 3 delay 30
Control = ON

Harvest pump speed
faster than feed rate

3-way
inlet

Level
probe

Level   Feed   Level

Peristaltic
tubing

Harvest
pipe

Culture
harvest

Medium
feed

12.10.1.1 Peristaltic pump

This is the most common type of addition pump used in fermentor applications and is often found in a fixed-speed version for addition of reagents such as acid and alkali. The pump uses rollers to occlude sections of silicone tubing sequentially. This action creates a peristaltic flow of liquid through the tubing (analogous to the movement of food through the body by waves of peristalsis). The more rollers a peristaltic pump has, the smoother the flow. The pumps used for medium feed are typically variable speed units that can accept signals from an external source such as a computer or process controller to allow adjustment of feed rates based on, for example, respiratory quotient (RQ) or optical density (OD). The key advantages of peristaltic pumps are as follows:

- The liquid being pumped is not contaminated by any part of the pump.
- The pump is not contaminated by the liquid being pumped.
- A wide range of flow rates is possible using tubing of different bores.
- The accuracy of delivery is good over a wide range of flow rates.
- The pump can run safely even when the tubing is empty.

The disadvantages are

RESPIRATORY
QUOTIENT (RQ)
RQ is a mathematically
derived value related
to the use of oxygen
by a microbial culture
as compared with the
evolution of carbon
dioxide. This value
can be used to adjust
feed rates of sugars to
manipulate microbial
physiology.

- The flow is pulsed and cannot "push" against high back-pressure.
- The tubing wears over time and flattens, reducing accuracy and increasing the risk of bursting.
- Silicone tubing is not suitable for all liquids and alternative tubing material may be too rigid to allow correct peristaltic action.

The problem with tubing wear can be partly overcome by leaving extra length in the tubing when connecting it to the fermentor vessel and moving a new length into the pump head every few days to stop wear in just one small section.

12.10.1.2 Syringe pump

This type of pump works on a completely different principle and is useful for fed-batch applications rather than continuous medium feed. In this case, a syringe (of various sizes from µl to 100 ml, typically) is filled with the liquid to be added and secured onto the syringe pump. This is done in such a way that the syringe plunger is linked to a movable piston and the syringe barrel is held firmly to the main body of the pump. A flow rate is set (in mm min^{-1} or directly in ml min^{-1} or ml h^{-1} if the pump can be calibrated). The piston is then moved at the desired speed by the action of toothed gears linked to a stepper motor engaging with teeth on the underside of the piston. When the syringe is empty, an alarm condition is indicated and the pump automatically stops or, on very sophisticated pumps, a signal can be sent that allows automatic refilling from a reservoir. This type of pump is typically used in medical applications for the precise delivery of drugs. Most now allow for computer control.

The advantages of syringe pumps are

- Very precise and accurate delivery; nonpulsile flow.
- Very low flow rates possible.
- Can work against back-pressure.
- Can pump very viscous fluids, e.g.. oils.

The disadvantages are

- Must be replaced or refilled regularly.
- Range of flow rates not as wide as with peristaltic pumps.
- Usually more expensive.

12.10.1.3 Diaphragm/positive displacement pump

This type of pump is a hybrid of the two previous types in that it uses a flexible membrane and soft tubing to displace the liquid and control its flow but also employs a piston and pump heads of fixed volume. The flow rate is controlled by the frequency of the pump stroke and the size of the pump chamber (which can usually be exchanged).

Advantages of this type of pump include

- Accurate delivery.
- More choice in type of tubing for corrosive liquids, for example.

Disadvantages are

- Less flexibility on flow rate ranges for each pump head.
- Stainless steel chamber and diaphragm may not be suitable for some liquids.
- Pulsed flow even more pronounced than with a peristaltic pump.

12.10.2 Exit gas analysis

Measurement of the amounts of different gases leaving a fermentor vessel can provide valuable information about the metabolic processes taking place under the conditions in which the culture is growing. For example, the ratio of oxygen and carbon dioxide entering and leaving the vessel, or respiratory quotient (RQ), can be used to determine whether a yeast is producing biomass or alcohol. The RQ can be calculated by most fermentation software packages and can then be used as part of a control algorithm so that the flow rate of a feed pump can be altered accordingly (see Box 12.2). The entry gas does not need to be analyzed providing it is air, because the amounts of oxygen and carbon dioxide will be those of the atmosphere. The flow rate of the air into the vessel will need to be accurately measured, normally by using a thermal mass flow controller. The exit gas may need to be conditioned (e.g., moisture removed) before going into the analyzer, depending on the type of instrument used.

12.10.2.1 Infra-red carbon dioxide analyzer

A hot wire is used to generate a source of infra-red radiation that passes through the gas coming from the fermentor into the sample chamber. An infra-red detector measures the amount of radiation reaching it after some is absorbed by the gas. An optical filter is used to make sure the detector only responds to the gas of interest. A "chopper" or rotating shutter is used to allow the detector to see a reference source at regular intervals. This type of detector allows continuous measurement with good sensitivity, accuracy, and selectivity. Output signals are normally provided either as analogue signals (e.g., 0–10 V) or a string of ASCII characters for printing or transfer to computer software.

12.10.2.2 Paramagnetic oxygen analyzer

Oxygen has a particular physical property that this type of analyzer can utilize, i.e., that it is more susceptible to a magnetic field than other gases. This property is measured using a finely balanced test apparatus suspended in the test chamber of the analyzer. A dumb-bell of gas-filled spheres is linked to a support mechanism suspended in a magnetic field created by permanent magnets. If oxygen is present in the test gas, its attraction to the magnetic field will cause the dumb-bell to be displaced. The movement is detected using a mirror in the center of the balance system which displaces a beam of light shone via the mirror onto a photocell. The signal from the photocell will be proportional to the amount of oxygen present in the sample. A low flow rate is needed for this system to work properly, so inlet gas is typically pumped into the detector cell with most being discarded through bypass pipework. Once again, suitable analogue and computer outputs are normally provided.

Both infra-red detectors and paramagnetic oxygen analyzers need to undergo a calibration procedure before use. They are often combined in one instrument.

12.10.2.3 Mass spectrometer

This represents a step upwards in versatility, capability and also cost! A wide range of gases can be analyzed (both in a gas stream and as dissolved gases in the vessel), e.g., oxygen, carbon dioxide, argon, nitrogen, ammonia, hydrogen, methanol, ethanol, and several other organic volatiles.

The analysis is rapid, thus allowing a multi-inlet system to be used that "shares" the analyzer between separate vessels within a bank of fermentors. Mass spectrometers can be magnetic sector, multi-collector, or quadrupole. Each has its own particular advantages, but the magnetic sector can be used to illustrate the principles upon which they all work. The physical principle involved is the ionization of gas molecules by an electron source, usually a hot filament, followed by their separation in a magnetic field (the quadrupole analyzer uses a combination of RF and DC electrical fields to do this job) according to their mass/charge ratio. This separation takes place in a virtual vacuum to minimize collisions prior to sorting. The magnetic field is tuned so that only the ions of interest will be focused onto the detector system, a Faraday Cup. This metal plate generates a tiny electric current whenever a gas ion strikes it. This signal can be amplified and displayed. If you were to retune the magnetic field very quickly (milliseconds), different ions can be detected and measured in a sequential way that is so fast it appears to give an almost simultaneous measurement.

Box 12.2 **Calculation of RQ**

When data are obtained from an exit gas analysis, it is generally required that the parameter RQ is calculated in order to gain an insight into the metabolism of the organism under investigation. This is normally done by programming a definition into a computer software package for data-logging and control. Here is a generalized outline of the steps needed:

RQ = CPR (Carbon dioxide production rate)/ OUR (Oxygen uptake rate)

CPR = [Flow CO_2 out] × [Conc'n CO_2 out] – [Flow CO_2 in] × [Conc'n CO_2 in] gmol/O_2/h

OUR = [Flow O_2 in] × [Conc'n O_2 in] – [Flow O_2 out] × [Conc'n O_2 out] gmol/O_2/h

Concentrations of gases in inlet gas (AIR) O_2 = 20.95% CO_2 = 0.03%

The working volume of the fermentor (WV) and the gas flow rate (FR) must be known to make the calculation:

VUO_2 (Volumetric uptake O_2) = FR × (0.2095 – (Exit O_2/100)) /WV

$VPCO_2$ (Volumetric production CO_2) = FR × (Exit CO_2/100) – 0.0003))/ WV

Typical values for Exit O_2 could reach approximately 16% and for Exit CO_2 around 5%.

Values of RQ for yeast biomass fermentation, for instance, would need to be kept around 1 for optimal biomass production.

A worked example of the above definitions:

WV = 1 l, FR = 1 l min–1 (1 vessel volume of gas per minute, a typical flow rate).

Exit O_2 = 16% and Exit CO_2 = 5%

VUO_2 = 1/1 × (0.2095 – 0.16) = 0.0494

$VPCO_2$ = 1/1 × (0.05 – 0.0003) = 0.497

RQ = 0.497/0.494 = 1.006

To achieve control, a sequence along the lines of the following would have to be set up:

if RQ < 0.95, then feedrate = feedrate × 1.05

if RQ > 1.05, then feedrate = feedrate × 0.95

12.10.3 Substrate sensors

This is a convenient title for grouping a number of different types of sensor that provide information about the concentration of substrates in a fermenter vessel (see Chapter 11 for more details). They usually depend on chemical or enzymic reactions for the sensing element and can either take a small sample directly from the fermenter or detect volatile components being released into the exit gas. Two of the most common examples are given below:

12.10.3.1 Glucose measurement and control

The more sophisticated versions incorporate a small dialysis unit in a sterilizable probe to take minute quantities of liquid from the medium for analysis. An enzymic reaction linked to a transducer provides a measurable current that can be amplified to give an output signal. This can provide a measurement of glucose from less than one gram/liter of glucose to over 100 g/l.

12.10.3.2 Methanol measurement and control

In this case, a chemical sensor is used to measure the concentration of methanol vapor in the exit gas stream (the sensors are usually variants of the type used in "breathalyzer" systems). The concentration of methanol in the vapor can be directly related to the concentration in the culture medium. A signal transducer and amplifier system provides the output measured value. The type of sensor is typically used for the control of the growth and protein expression by the yeast Pichia pastoris.

In each case, the measured value can be linked to a pump control system (either directly in the one box or by external control software) to allow the accurate dosing of substrate feed against selected criteria.

12.11   Fermentor preparation and use

12.11.1 Disassembly of the vessel

The fermentation is shut down from the control unit and transfer lines plus cable connections removed. Following a fermentation, the vessel should be re-autoclaved, ensuring that inlets and outlets are properly prepared. The culture should be disposed of as laid down by departmental safety procedures The clip/clamps/bolts that retain the vessel top plate are undone until the whole assembly is free. The top plate can now be lifted upwards away from the glass section of the vessel, taking care that the air sparger, drive shaft/impellors, and Pt-100 temperature probe completely clear the vessel. The flanged top section of the glass vessel and seal can now be seen.

12.11.2 Cleaning

The pH and dissolved oxygen electrodes should be removed and stored in suitable reagents according to the manufacturer's instructions. Periodic cleaning and regeneration of the electrodes are also covered by these instructions. Doing this maintenance is very cost-effective. The vessel should be rinsed several times in distilled water to remove any loose culture residues. Cleaning of growths of culture on the vessel walls may require disassembly and light brushing of the glass. At this point, an examination of any chips or cracks in the vessel glass can be carried out and a replacement made if necessary. Vessels must be stored clean and dry. In use, any spillages of reagents or medium should be wiped up immediately with a damp cloth and not be allowed to dry out. Contact between the top plate and liquids with a high chloride ion content (e.g., common salt solutions, hydrochloric acid) should be avoided to prevent corrosion. The pump heads and covers must be thoroughly cleaned if a tube breaks and reagent leaks out. Peristaltic tubing should be sterilized using water in the line and not strong acid or base.

12.11.3 Preparations for autoclaving

The vessel seal should be removed and lightly greased with suitable silicone grease. On replacing the seal, it must be correctly located so that there is no chance of any part lifting or kinking. At this point, the vessel can be filled with medium to a maximum of 70–80% full (if active aeration is to be used, this space is vital for gas exchange). The minimum medium volume is the amount needed to cover the electrodes adequately. The vessel top plate can now be replaced and any clamping ring or bolts tightened firmly. The ports for electrodes have O-ring seals which should be lightly greased with a silicone preparation before autoclaving. Electrodes normally push directly into the port fitting and the collar is tightened down to compress the O-ring seal. All other fittings such as pipes are fitted in the same way. Ports not in use have "stoppers" fitted, and their O-ring seals should have a light coating of silicone grease applied. The pH electrode should be calibrated in appropriate buffers, i.e., pH 7, then either pH 4 or pH 9 (or just 4 and 9 for a "high reference/low reference" calibration of a DDC controller channel). The pH and dissolved oxygen electrodes should be fitted, taking care not to damage them by careless insertion into the port. Both must be tightly capped to prevent moisture getting into the electrical contacts. For the dissolved oxygen electrode, a cap may have to be improvised from aluminum foil. The Pt-100 temperature sensor must be fitted and capped. If used, the foam probe is fitted so that it is above the liquid level in the head space (Figure 12.11).

Reagent bottles are prepared in a similar way to the fermentor vessel. A cap or head plate (including a seal) is fitted with a short tube and longer dip tube. A disposable filter is then connected to the short tube with silicone tubing. The shorter pipe must not dip into the liquid and nothing must block the free passage of air through the filter. The long pipe dips into the liquid as far as possible, usually with a plastic/silicone tubing extension. This pipe should now have a length of silicone tubing fitted to it which is long enough to reach the peristaltic pump heads of the fermentor instrumentation for which it is intended. The tubing is clamped so that no liquid can escape during autoclaving. A similar procedure is used for sampling and/or harvest bottles except that two short pipes are used so neither dips into the collected culture. Some preparation is needed at the autoclaving stage for the aseptic coupling of silicone tubing ready for operation. For correcting larger bottles, see Box 12.3.

Figure 12.11
Vessel and accessories prepared for autoclaving.

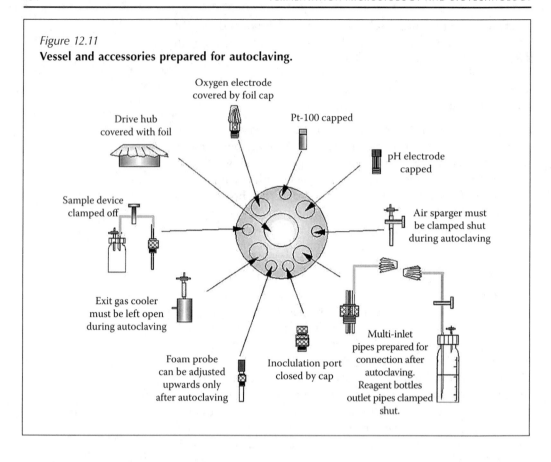

Oxygen electrode
covered by foil cap

Drive hub
covered with foil

Pt-100 capped

pH electrode
capped

Sample device
clamped off

Air sparger must
be clamped shut
during autoclaving

Exit gas cooler
must be left open
during autoclaving

Foam probe
can be adjusted
upwards only
after autoclaving

Inoclulation port
closed by cap

Multi-inlet
pipes prepared for
connection after
autoclaving.
Reagent bottles
outlet pipes clamped
shut.

Box 12.3 **Preparation of tubing for aseptic connection**

Any silicone tubing to be joined needs the two ends to be prepared as follows:

1. The end of the tubing that remains open is simply covered in aluminum foil held in place with autoclave tape. For added security, the open end can be sealed with a short piece of glass rod flattened at one end to make a "bung."

2. The other side has a short length of stainless steel pipe pushed into the silicone tubing. The exposed length of pipe is covered with foil which is taped closed.

3. Silicone tubing is connected to any of the inlet pipes intended for use for reagent addition. The tubing must stretch to the peristaltic pump heads when the vessel is in place.

4. The tubing is closed with clamps or any other closure system which will withstand autoclaving (the arterial clamps used in medicine are often employed where a connection must be opened and closed many times, e.g., to a sampling device). The tubing which goes to the sampling device from the top plate must be clamped shut during autoclaving.

The exit gas cooler should be fitted to one of the wider ports. A short length of silicone tubing should be attached to the top of the air outlet and a small 0.22- or 0.45-µm filter fitted. The air outlet line must be kept open during autoclaving. A short length of silicone tubing must be fitted to the air sparger inlet pipe with a 0.2-µm disposable filter mounted on top. The tubing between

the sparger pipe and the filter must be clamped shut during autoclaving. If a port is to be used for inoculation or piercing with a
needle, a silicone membrane must be fitted into the empty port
and a clamping collar/cap used to hold it in place.

12.11.4 Autoclaving

The vessel and any reagent/sampling bottles already connected
by silicone tubing are assembled together on a steel tray or in
an autoclave basket. A final check should be made that at least
one route is available for air to enter and leave the vessel(s) and
that all lines dipping into liquid are clamped closed. If the vessel
has top drive and a mechanical seal, the seal must be lubricated
(normally with glycerine).

   If the medium cannot be autoclaved, a suitable volume of
distilled water should be used, e.g., 10–20 ml to keep the electrodes in a moist environment. If a larger volume of, e.g., Phosphate Buffered Saline (PBS) is used, this must be removed via a
sample line before the actual medium and inoculum are aseptically transferred. A quantity of liquid is certain to be lost during autoclaving (approximately 10%) so either the medium is
overdiluted to compensate for this or sterile, distilled water is
added afterwards to restore the volume. Some form of indicator
such as autoclave tape should be included to provide a warning if the correct sterilization procedure has not been carried
out. Autoclaving at 121°C for up to one hour is normally considered adequate for vessel sterilization to minimize damage to
the constituent chemicals of the medium. However, some work
may require temperatures of 134°C for several hours to ensure
sterility. If in doubt, your safety committee should be consulted.
Also, the autoclave used must have good pressure equalization
during the cooling-down phase of operation to prevent medium
being boiled off. The vessel and any accessory bottles must be
allowed to cool completely before handling.

12.11.5 Set-up following autoclaving

First, the silicone tubing prepared prior to autoclaving must
be connected to the fermentor vessel (if the bottles they are
connected to were autoclaved separately). Figure 12.12 shows
how this is done.

   The air sparger is connected to the rotameter by a piece of
silicone tubing from the top of the filter to the air outlet of the
rotameter. The air sparger line is unclipped between the metal
pipe and the air filter. The exit gas cooler is connected to the
water supply, either directly or via the fermentor base unit. The
tubing for water in, out, and drain is connected to the vessel
jacket for a water system, and the water turned on so that the
vessel jacket is filled. Alternatively, any pads or heater cartridges

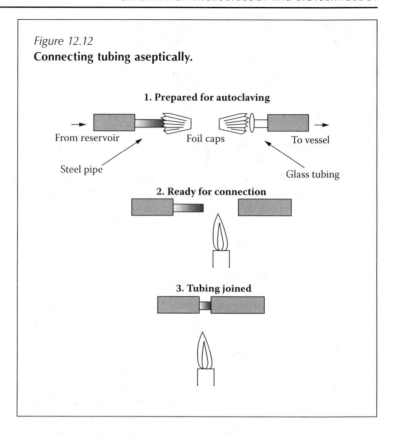

Figure 12.12
Connecting tubing aseptically.

1. Prepared for autoclaving

From reservoir   Foil caps   To vessel

Steel pipe   Glass tubing

2. Ready for connection

3. Tubing joined

are connected to the base unit or temperature control module; the cold finger is connected to the water supply.

The tubing from the reagent bottles is connected to the multi-way inlet (if necessary), and the silicone tubing from the reagent bottles is located in the relevant peristaltic pump. Any aseptic connections must be made first if the reagent bottles were auto-claved separately from the vessel. The clamps are removed so liquid can flow freely. A manual switch is often fitted, which allows the pumps to be primed with liquid prior to use. The drive motor is located onto the top plate (if appropriate), ensuring a good connection is made to the drive shaft. The Pt-100 tempera-ture sensor is connected to the control module and the pH elec-trode by removing the shorting cap and screwing in the cable. The dissolved oxygen electrode is connected to the appropriate cable (this requires some care, but the connector should lock firmly when it is correctly positioned). Connections to the foam probe are made, usually one wire into the electrode and one on the vessel top plate to make a circuit. The mains electricity to the instrumentation modules is switched on, allowing some hours (minimum of 2, preferably more than 6) for the dissolved oxy-gen electrode to polarize properly. This delay is not needed for galvanic electrodes. Setting the temperature control at this stage will ensure the fermentor is ready to inoculate after calibration of the dissolved oxygen electrode. After polarization the dissolved oxygen electrode is calibrated for the zero point using nitrogen gas and then the air supply turned on. The maximum stirrer speed to be used and the maximum air flow required on the rotameter are set. After leaving for about 15 min, the 100% level is set. The fermentor is now ready to inoculate.

12.11.6 Inoculation of a fermentor vessel

This section assumes that all the set-up procedures listed above have been carried out. The simplest way to inoculate is to have a dedicated port fitted with a membrane that is capped off before autoclaving. The inoculum (which should normally be no more than 5–10% of the total culture volume) is aseptically transferred to a sterile, disposable syringe of a suitable size. The port fitting is removed and held vertically to prevent contamination of the bottom end. The syringe needle is quickly pushed through the membrane and the inoculum transferred into the vessel. The vessel may be actively aerated during this procedure to minimize the risk of a contaminant getting into the vessel (safety considerations permitting). The syringe needle is quickly withdrawn and the silicone membrane reseals. The port fitting is now replaced. For added security, a little 70% ethanol can be placed on the membrane surface before piercing and ignited to provide a thermal barrier. Alternatively, an "aseptic connection" can be made to an inlet pipe. If the line connected has a "Y" coupling in it, then the same aseptic connection could be used to introduce medium. This technique is useful if a dedicated inoculation port cannot be provided. Figure 12.13 shows the alternative methods.

12.11.7 Sampling from a fermentor vessel

All sampling starts with a sample pipe, which should dip into the bulk of the culture liquid. The simplest system is to have a length of silicone tubing connected to the sample pipe (and clamped off until a sample is to be taken), which is plugged and covered at the open end. After autoclaving, the tubing is quickly unwrapped, the plug removed, and the open end dipped in a container of 70% ethanol. When a sample is to be taken, a sterile syringe is quickly coupled to the tubing and the sample withdrawn into the syringe. Aeration is stopped during this process to prevent surging of culture into the syringe. Any culture in the transfer line is discarded into a kill jar and the tube replaced in the ethanol. Clearly, this method can be prone to contamination and is not suitable for organisms with any risk to operators. A better approach is to use a sampling device. At its simplest, this is a bottle with two metal needles/pipes permanently fixed through both the metal and rubber seals of its cap. One pipe has

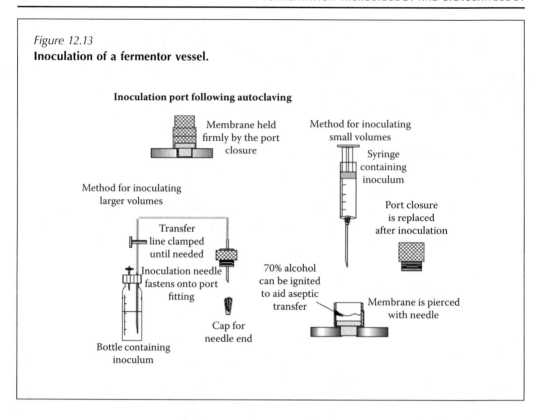

Figure 12.13

Inoculation of a fermentor vessel.

a 0.22-µm air filter connected, and the second is linked to the vessel by silicone tubing (clamped off until a sample is needed) to the sample pipe. A syringe is fitted to the air filter after autoclaving. The sample device is usually attached to the vessel top plate so that the glass bottle hangs down vertically beneath the cap and a supply of bottles of the same size is autoclaved ready for use. When a sample is to be taken, the clamp on the sample pipe tubing is released and the syringe pulled back to create a partial vacuum in the sample bottle; consequently, the culture will flow into the sample bottle. When the required volume of sample has been taken (never more than 75% of the total bottle volume), the syringe is pushed in to clear the line and the sample tubing is clamped shut again. The sample bottle with the culture is quickly removed and capped with a sterile cap from a new glass bottle, which replaces it under the sampling device. Figure 12.14 illustrates the method.

It is often preferable simply to discard the first few milliliters of the next sample by using it as a "wash" for the transfer line, putting a fresh bottle under the sampling device to actually take the sample.

Alternative types of sampling devices are now being used, especially for cell cultures that use aseptic connectors for syringes as used in medical applications.

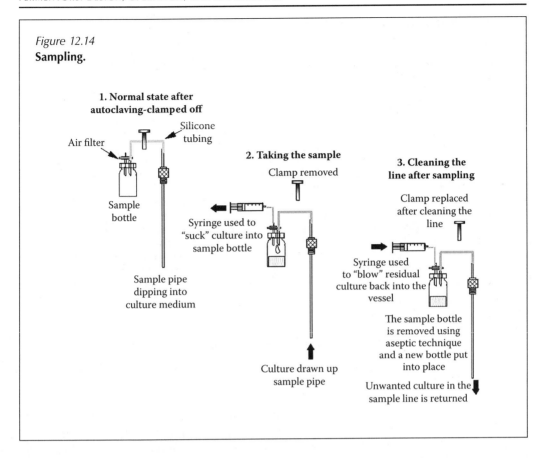

Figure 12.14
Sampling.

1. Normal state after autoclaving-clamped off

Air filter

Silicone tubing

Sample bottle

Sample pipe dipping into culture medium

2. Taking the sample

Clamp removed

Syringe used to "suck" culture into sample bottle

Culture drawn up sample pipe

3. Cleaning the line after sampling

Clamp replaced after cleaning the line

Syringe used to "blow" residual culture back into the vessel

The sample bottle is removed using aseptic technique and a new bottle put into place

Unwanted culture in the sample line is returned

12.11.8   Routine maintenance of fermentor components

12.11.8.1   *pH electrodes*

pH electrodes are a semi-consumable item. Typically, an electrode will need replacing after 20–50 autoclave cycles. This broad range reflects the different results sometimes found between different brands and types of electrode and the harshness of the environment in which they are used, e.g., high temperatures, frequency of autoclaving, etc. Aging of electrodes can show up as a sluggish response time, and a smaller slope on calibration, especially in the alkaline region. The aging process is sped up by high temperatures, so an autoclavable electrode will invariably have a shorter working life if repeatedly heated to sterilization temperatures. However, during its working life, the pH electrode may need attention to ensure optimal performance. In this respect, the gel-filled pH electrodes are normally less demanding and are physically more robust than the glass, liquid-filled variety. They also clearly show the aging process by the transparent gel becoming progressively darker brown (see Box 12.4).

Box 12.4   **Routine Maintenance of pH Electrodes**

1. Always store an electrode which is not in use in liquid — preferably 3 M KCl solution, although pH 7 buffer or even fermentation medium/PBS will do (in descending order of preference).
2. An electrode that has completely dried out will need regenerating — a procedure that normally involves a brief immersion in a hydrofluoric acid solution (typically supplied by the manufacturer) followed by prolonged soaking in storage solution. This procedure may also regenerate "aged" electrodes for a time. (Note that glass vessels cannot be used for this task due to the effects of the hydrofluoric acid.)
3. A build-up of protein material on the electrode membrane can be dealt with by soaking the electrode in a pepsin/hydrochloric acid solution for about an hour.
4. If a discoloration (blackening) of the diaphragm has been caused by sulphides in the medium, soaking for several hours in an HCl/thiourea solution will recover its shiny surface.
5. Never wipe the tip of the electrode with a paper towel, as this causes scratching.
6. Glass/liquid electrodes should have the reference electrolyte topped up periodically.
7. To test if the pH meter/cable are functioning correctly, a pH simulator can be used. If such a device is not at hand, a paper clip can be used to short the center pin of the electrode plug and the outside metal ring of the cap. The pH reading should be very close to 7.0.

12.11.8.2 *Dissolved oxygen electrodes*

Galvanic electrode: This is a simple, self-powered system that relies on a lead anode and a silver cathode that are shorted together. Oxygen is reduced at the silver cathode, and electrical current flows in proportion to its partial pressure. The response time is slower than a polarographic electrode, but the initial cost is far lower. The electrode is conditioned for the first time by the autoclaving process. The probe is relatively robust except if the lead anode is exposed to air due to a drop in the level of electrolyte in the probe. Therefore, all air bubbles must be removed when electrolyte is added and probes must always be stored upright with an adequate level of electrolyte and with the electrode shorted out (cap on). Periodically, the electrode membrane will need replacing. This involves soaking a section of new silicone rubber tubing in chloroform to swell it (safety hazard!). While this is happening, the old tubing and membrane is cut away. A new membrane is stretched tightly across the electrode tip and the silicone tubing placed over it to hold it in place without any wrinkling. The end away from the tip is trimmed off and secured with a little Araldite glue and the electrode tapped gently to remove air bubbles. Galvanic electrodes are rarely used these days.

Polarographic electrode: These electrodes work on a different principle (the Clark Cell). The electrode consists of a silver anode/reference electrode with a platinum cathode encased in a glass rod. The end of the glass rod is shaped to accept a replaceable membrane cartridge. A temperature sensor is included for automatic compensation. The whole assembly is usually encased in stainless steel housing. A chloride solution is used as an electrolyte. A polarization voltage must be applied to the electrode, which can then reduce the oxygen passing through the membrane and generate a current. For this reason, the electrode must be autoclaved shorted out and be connected to a source of polarizing voltage for an absolute minimum of 2 h before use. The membrane must be replaced periodically by unscrewing the bottom section of the electrode and gently removing the old cartridge. A new cartridge is fitted, having been filled with fresh electrolyte, and pushed onto the inner glass rod. If necessary, the small O-ring in the bottom section of the housing can be renewed before replacing the cartridge. New electrodes are supplied dry and the cartridge must be half-filled with the supplied electrolyte before first use to prevent damage.

A functional check on the electrode can be performed by setting the reading of a polarized electrode in water saturated with air to 100%. The air is replaced with nitrogen gas and the value should fall to less than 1.5% after about 5 min. If the electrode is now removed into the air, the value should return to 98%+ within 90 s. An inability of the electrode to zero properly is often a sign of malfunction.

12.11.8.3 *Changing a mechanical seal*

Different arrangements of mechanical seal are used for different makes of fermentor. The type described is a good example as it shows many of the features used with the mechanical seals employed on both bench-scale and *in situ* sterilizable fermentors. Seals should be replaced regularly (at least every 6–12 months, depending on use) (see Box 12.5).

12.12 Major types of organisms used in fermentation

A wide range of organisms can be used in fermentation for the production of various chemicals and antibiotics.

12.12.1 Bacteria/yeast/fungi

Either a natural isolate of the culture is made or an uncontaminated "master" culture is revived from a freeze-dried state or from liquid nitrogen storage. Bacteria, yeasts, and filamentous fungi are the usual starting materials. The starting cultures are grown in Petri dishes on agar containing the required nutrients and incubated at an optimum temperature, humidity, and gaseous atmosphere. Fungal cultures are often started as spores in liquid medium. If necessary, shake-flask cultures are prepared in an incubator shaker to produce larger volumes of culture

Box 12.5   **The procedure for replacing a typical single mechanical seal is as follows:**

1. Unscrew the top part of the drive shaft and remove. Measure the distance between the bottom of the retaining ring and the inner surface of the top plate. This allows the new seal to be correctly tensioned by setting the same gap at the end of the procedure.

2. Remove any retaining ring by loosening the small screw(s) holding it in place. Disassemble the retaining ring, hard seal, soft seal and bottom O-ring in order. Examine the faces of the static and rotating seals. Scoring of the surface indicates the seal has worn out. Never try to replace the old seal if it looks undamaged — it cannot be reused.

3. Insert a new O-ring into the "hole" in the vessel plate which leads to the drive shaft bearings. This may be a tight fit but it must seat properly to prevent leaks. Replace the static seal by pushing it into the O-ring to obtain a snug fit. The protective rubber cover over the top face must be removed at this point. If you accidentally touch the exposed seal face, it can be wiped clean using a little acetone on a lint-free cloth.

4. Fit the rotating seal onto the drive shaft. Ensure the inner O-ring and metal washer are firmly pushed down into place as the tensioning spring will rest on these. Fit the new tensioning spring into place then push onto this firmly with the retaining ring until the spring is tightly compressed. Measure the gap with a ruler/calipers, and only tighten the screw(s) on the retaining ring when the correct distance has been achieved.

5. Replace the drive shaft and gently turn the shaft by hand a small distance to ensure it moves freely. Lubricate with glycerine. A "whistling" sound can usually be heard if a mechanical seal is running dry.

for analysis or for use as an inoculum for a fermentor. Flasks are typically filled to 20% to provide a large area for gas exchange and are 25 ml to 5 l in total capacity. An inoculum is prepared either from the shake-flask culture or by suspending colonies from growth on solid media in a nutrient/buffer solution. This is aseptically transferred to a fermentor vessel. The inoculum is typically 5–10% of the volume of medium in the fermentor and is often diluted in order to introduce a known number of cells/ml in this volume. A laboratory scale fermentor can be used to provide larger volumes (typically 1–5 l of culture). Growth is normally as batch fermentation with the following parameters measured and controlled as necessary:

Temperature typically 20–40°C (4–90°C at the extremes)
Speed typically 150–1500 rpm (10–2000 rpm at the extremes)
pH typically 4–8 (2–12 at the extremes)
Dissolved oxygen typically 40–80% (0–100% at the extremes)

In some cases, a continuous culture system may be used to draw off culture of a known cell concentration for long periods. A concentration of a growth-limiting metabolite (chemostat) or an indirect estimate of cell numbers (e.g., turbidity, a turbidostat) is used to keep the culture density constant for many days/weeks. Samples are taken from the fermentor for counting, culturing on solid medium or biochemical assay for a particular metabolite, sugar utilization, etc. On completion of the culture, the whole culture can be centrifuged, e.g., $4000 \times g$ for 15 min to separate whole cells from the culture supernatant fluid. Alternatively, a membrane or hollow-fiber system can be used for this initial separation.

The separated cells can be washed in buffer, re-suspended, and can then be disrupted using a variety of methods, e.g., ultrasonic disruption, freeze-thaw, shaking with beads, etc. Supernatant fluids can be further separated and analyzed using separation columns, fraction collectors and spectrophotometery. Gel electrophoresis, ELISA assay, etc., are also used. Applications are almost too numerous to mention, but obvious categories include brewing (yeasts), antibiotic production (fungi), enzyme production (bacteria), and food manufacture such as yogurt (bacteria). Environmental uses include removal of toxic wastes, biological control of crop pests, and soil regeneration.

12.12.2 Plant cells

The whole plant is grown by conventional means. A sterile knife is used to cut a small section of root, stem, or leaf that is transferred onto solid medium to produce a ball of cells (a thallus). The thallus is separated with a sterile blade and the cells grown in shake-flask culture in an incubator shaker at 20–30°C and 100–200 rpm. The cells are transferred to a fermentor under the following conditions: e.g.,

- stirrer speed ≈ 100–150 rpm;
- temperature 20–30°C (cooling needed if lights used);
- pH not generally controlled as it tends to cycle naturally;
- dissolved oxygen rarely controlled as it is not generally a limiting factor;
- foaming can be a problem unless controlled.

Cells may be cycled between the shaker and fermentor several times to build up shear resistance. The cell growth is normally too slow for effective continuous culture. For material inside the plant cell a "Potter" type homogenizer can be used for gentle disruption. Centrifugation at approx. 1000 rpm for whole cells will separate them from the culture supernatant. Higher speeds will be needed to deposit cell fragments. The cellular pellets are discarded or extracted depending on where the required component is located. Plant cell cultures are often used for production of specific biochemicals such as flavinoids and alkaloids used in foods and chemical production. Culture of plant cells for agrochemical uses is another obvious application. Air lift systems are also a popular choice for plant cell culture.

POTTER HOMOGENIZER
A device for breaking open cells, e.g., plant cells, by using the rotation of a tightly fitting piston inside a glass cylinder.

12.12.3 Mammalian cell culture

There are many different cell lines of human, mammalian, avian, and even insect origin. For simplicity, a single type,

HYBRIDOMA CELL
LINE
A cell derived from
the artificial fusion
of a normal, e.g., an
antibody-producing cell
line from the spleen,
with a transformed
(immortal) cell line, e.g.,
a myeloma. The resulting
hybrid is immortal and
can produce a specific
antibody if the correct
spleen cell was selected
after challenge of an
animal with a specific
antigen.

hybridomas, will be used as an example here. Hybridoma cell lines are prepared using cells from mouse spleen which secrete desired antibodies crossed with a continuous cell line grown *in vitro*. The cells are placed in suspension culture in an incubator shaker at 37°C and 40–100 rpm speed. The cells are transferred to a specially adapted fermentor. Large, broad-bladed impellors are used to minimize shear forces, and typical stirrer speeds are 20–100 rpm. A gas mixing station can provide an air/carbon dioxide mixture, and CO_2 gas can be used for pH control via a solenoid valve as replacement for a peristaltic pump. Oxygen and nitrogen can be introduced for dissolved oxygen control. For many cell lines, the control of foam is vital as its presence can contribute to cell damage. A very low flow rate for gas supply often helps in this respect, e.g., 0.05 VVM.

The cells can sometimes grow freely in suspension but more often are anchored to beads or a hollow-fiber system. A special mesh can be connected to the drive shaft to provide a "spin-filter," or to the sample pipe for taking samples of culture fluid relatively cell-free. Some alternative approaches to growth of mammalian cells are as follows:

- Air lift: There is no drive motor and it is the pumping of liquid around the draught tube that provides for mixing and gas exchange.
- Hollow-fiber cartridges: They remain outside the fermentor vessel but culture fluids can circulate between the fibers and the vessel. All mixing and gas exchange is done in the vessel, and the enriched medium passes to the cells anchored to the hollow fibers via a pump.
- A gassing basket inside a stirred fermentor: Where cells need good levels of dissolved oxygen but are very sensitive to shear, a large amount of thin-walled silicone tubing supported on a frame can be placed in the vessel to provide a large area for gas exchange. Air is bubbled through the pores in the silicone tubing to give good gas transfer rates without forming large bubbles.
- Applications of mammalian cell culture include monoclonal antibody production, culture of viruses for vaccine manufacture, and large quantities of specific cells for medical research.

SPIN FILTER
This device is normally
attached to the drive
shaft of an animal cell
fermentor and allows
culture liquid to be
removed while leaving
the cells inside the
fermentor. It allows
for slow-growing cells
to produce antibodies,
for example, over a
long period by regular
harvesting of culture
supernatant and
replenishment by a
controlled medium feed.

12.12.4 Algae

This category covers a diverse range of micro-organisms that differ vastly in their growth conditions and metabolic requirements. Some are halophiles (salt-loving), barophiles (survive under high pressure), thermophiles (grow at high temperatures), and psychrophiles (grow at low temperatures). Most need special culture conditions, and so it is best to seek out published literature with specific details. Applications for the use of algae include food additives and chemical production by biotransformation. Large-scale systems for photosynthetic algae are now commercially available that rely on long banks of tubes for circulation of the culture past banks of lights.

12.13 Subfermentor systems
— a new approach

A noticeable trend is the volume of culture required from each vessel to become smaller. At the same time, there is an increasing need for sophisticated experimental protocols and rapid statistical analysis from a large number of cultures. Typical applications requiring this approach are the profiling of feed strategies, medium optimization, toxicity testing, and rapid identification/selection of microbes with desirable characteristics. The conventional systems used for production have both advantages and disadvantages for this type of work:

1. The Shake Flask.
   Advantages:
   - Large numbers can be used in the same environmental conditions — statistical validity!
   - Set-up is simple — one person can look after many experiments using familiar equipment.
   - Small culture volumes can be used, e.g., <1000 ml.
   - Low cost.
   Disadvantages:
   - Sampling not easy.
   - No substrate or reagent feeds.
   - No measurement probes in vessel, e.g., pH.
   - No control of important parameters, e.g., pH.
   - Growth may be suboptimal, e.g., oxygen-limited.
   - Recording of experimental data is limited.
2. The Bench-Scale Fermentor
   Advantages:
   - Sampling is easy.
   - Sophisticated measurement and control of each process parameter separately, including feeds.
   - Links to data-logging and control systems.
   Disadvantages:
   - Set-up is time-consuming — limits experiments.
   - Only a single (or few) experiment(s) at one time — lack of statistical validity.
   - Cost.
   - Working volumes typically one liter or more.

For screening and rapid development applications, a system is needed that provides the measurement, control, feeding, sampling capability, and flexibility of the bench-scale fermentor with the ease of use, multiple operation, and low working volumes of the shake flask culture.

In fact, several approaches have been developed. These include shake flasks with ports for electrodes and sampling, multiple bubble columns, and instrumented spinner flasks for cell culture. Many of these simple alternatives began with large numbers of replicates in one unit but seem to have converged on systems with 4 to 8 vessels.

Applications involving cell cultures have introduced another new approach — the use of disposable vessels or bags which can also be instrumented, sampled and provide with feed/perfused. The bag systems can even be produced to volumes of several hundred liters for production.

12.13.1 Parallel small fermentor systems

This alternative is the closest to the conventional fermentor. The vessel has a similar construction, configuration, and fittings to a conventional bench-scale unit, but they are significantly smaller (typical working volumes of 300–1000 ml) and housed in a base unit capable of accepting 4 to 8 vessels (possibly in groups of 2 to 4).

Multiple peristaltic pumps provide reagent and medium feeds independently to each vessel. Separate heaters/cooling actuators and stirrer systems provide individual control of temperature and stirrer speed. A single DDC control unit provides for different process parameters, profiles, and links to external supervisory software.

These systems are usually easier to set up than a conventional fermentor, provide measurement/control of all the major process parameters, and can be adapted for the growth of virtually all types of microbes and cells.

12.13.2 Simplified fermenter systems

A further alternative is to make fermentation volumes and capabilities available without the complexity of standard systems. A major improvement in quality of information and productivity can be achieved by simply adding those key elements listed in the comparison with shake flasks above, e.g., volumes >1L, pH control, feeding and sampling.

The trend for protein production from genetically modified organisms has created a need for molecular biologists to generate quantities of material that are beginning to go beyond the capacity of shake flasks. What is needed is a cost-effective way to supply, e.g., 1–5 liters of culture at a higher cell density than can be produced in shake flasks. This requires a minimum of adequate mixing/aeration, temperature control, one-way pH control (addition of base) a feed line and a sampling system. Mammalian and insect cell cultures would also need dissolved oxygen control by gas supplementation/mixing and slow-speed stirring.

Key areas where the conventional fermenter could be improved for the new or occasional user would include:

- Improved handling of the vessel and peripheral bottles for autoclaving and set-up
- Simplified instrumentation with standard parameters and limited options
- Costs more in line with an incubator shaker than a fermenter
- Software for data logging and analysis integral to the system
- Templates and prepared methodologies to make the transition as simple as possible

12.14 Solutions to common problems in fermentation

As the number of different types of fermentors and applications are legion, only the most general advice can be given. However, certain common themes do occur, and this information may be of help in finding solutions:

12.14.1 General hardware problems

Working through a checklist of the simple things often solves fundamental problems where things appear not to work at all:

- Is the parameter switched on?
- Is the sensor for the parameter connected?
- Was calibration successful?
- How old is the sensor?
- If you can exchange sensors or swap to another unit, what happens -- i.e. does the problem move with the sensor?
- Is control switched on?
- Is something acting on the parameter externally, e.g., a cascade?
- Are services switched on, e.g., air and water?
- Are reagents available, connected, and unclamped, e.g., acid, base?
- Is control erratic or slowly drifting? – check PID values.
- Are all cables connected and firmly in place?

12.14.2 Contamination problems

- Is the seed uncontaminated? (check by plating out on rich media)
- Is the autoclave functioning properly? (check with spore strips)
- Is the autoclaving time adequate? (plastic bottles may need longer). Use of a relocatable temperature sensor can help.
- If the vessel is double-jacketed, is water kept in the jacket during autoclaving? (This may be necessary for good heat transfer.)
- Filling tubing with water before autoclaving may help.
- Check all O-rings for signs of wear, and replace as necessary.
- Check the vessel top plate and seal fit together properly.
- Check the mechanical seal (if fitted). Does it retain lubricant?

12.14.3 Poor growth of the microbe

- Is the quantity of inoculum (in cells/ml) adequate for the vessel volume?
- Is the inoculum fresh and in logarithmic growth?
- Is the fermentor already at the correct temperature, etc.?
- Is the strain sensitive or not suited to growth in fermentors?
- Are all supplements and additions in the medium correct?
- Some cultures need a short time with very low oxygen to allow a small build-up of carbon dioxide to aid growth.
- Is the carbon source at an adequate concentration for exponential growth? If necessary, a fed-batch regime can be implemented.
- Is the quantity of dissolved oxygen adequate for exponential growth? Adjust stirrer speed and gas flow rate accordingly.
- Are toxins, e.g., acid/lactate, building up and preventing growth?
- Are starting conditions optimal, e.g., lower stirrer speed and gas flow in the lag phase than in the log phase (important for some recombinant *E. coli*) strains?

12.1.4.4 Poor productivity, e.g., metabolites or proteins

- Does production need a change in operating conditions and/or the exhaustion of the primary carbon source?
- Does production depend on increasing biomass? In this case, a fed batch regime may improve yields by prolonging the culture.

- Is the strain capable of delivering the production levels required?

12.1.4.5 Other problems during fermentation

- If the culture volume goes down significantly over time, check the exit gas cooler water supply is on and the flow rate is enough to be effective. Ensure gas flow rate is no higher than 1.5 VVM.
- If foaming is a problem, then reducing gas flow rate, adding antifoam into the medium at the beginning of the fermentation, and using a liquid antifoam additive will all help.
- Should the exit gas filter keep blowing off, then reducing the gas flow rate, increasing the size (either pore size and/or physical dimensions), or changing to a depth-type filter are all possible.
- If peristaltic pump flow rates vary, ensure the calibration of the tubing is not made until it has been used for at least 20 minutes to allow stretching, check for flattening of the tubing in the pump head, and check for any kinks or sharp bends in the tubing/clamps not fully released.

Summary

- Fermentors are composed of a number of different components that can be grouped by their functions, i.e., temperature control, speed control, continuous culture accessories, etc. A wide range of peripheral devices can enhance the basic facilities of the fermentor,,e.g., feed pumps, exit gas analyzers, etc.
- Fermentor instrumentation can be of several different types, depending on the age, cost, and sophistication of the fermentor. Microprocessor-based systems offer advantages in terms of the flexibility of control and the ability to store operational protocols.

- Many different types of vessels exist such as air lift, fluidized bed, hollow fiber, and specially modified stirred tank reactors for containment or *in situ* sterilization.
- Practical steps for autoclaving, use, inoculation, and sampling ensure safe and reliable operation for any make or type of fermentor.
- Routine maintenance of electrodes, glass, etc., is repaid in longevity of components and safe operation.
- The range of organisms that can be adapted for growth in fermentors is wide and includes microbes and plant and animal cells.

References and suggested reading

Collins, C. H., A. J. Beale (eds.) (1992). *Safety in Indus-trial Microbiology and Biotechnology.* London: Butterworth-Heinemann.

Freshney, R. I. (1994). *Culture of Animal Cells: A Manual of Basic Techniques.* New York: A.R. Liss.

Stafford, A., G. Warren (eds.) (1991). *Plant Cells and Tissue Culture.* Buckingham: Open University Press.

Winkler, M. A. (ed.) (1990). *Chemical Engineering Problems in Biotechnology.* London: SCI/Elsevier.

Wiseman, A. (ed.) (1983). *Principles of Biotechnology.* Surrey: Surrey University Press.

13

Control of Fermentations: An Industrial Perspective

Craig J.L. Gershater

13.1 Requirement for control

Control is generally defined as the power to direct or influence. In the case of fermentation control, the requirement to control a given biotechnological process is generally dictated by the need to bring about a desired outcome such as maximizing output of product formation. Micro-organisms used in industrial processes have been isolated by virtue of their ability to overproduce certain commercially significant attributes. These attributes can range from the ability to produce carbon dioxide for leavening activity in the case of *Saccharomyces cerevisiae* (baker's yeast) to the production of useful microbial metabolites such as alcohol from yeast or antibiotics from filamentous bacteria.

Whatever the reason for commercial exploitation, the micro-organism will have been obtained directly or indirectly from an ecosystem very remote from that of the fermentation laboratory and process plant. The metabolic attribute to be exploited will have evolved as a result of environmental factors unknown to the fermentation scientist and therefore process optimization must seek to replicate those environmental factors responsible for expression of the desired attribute in the totally "artificial" environment of the industrial bioreactor or fermentor.

13.1.1 *Microbial growth*

The principal fermentation control requirement is population growth (see Chapter 3 for more details) of the micro-organism of interest. In its natural habitat, the microbe will respond to environmental stimuli such as excess nutrients by synthesizing enzymes and biomass capable of exploiting the resource as effectively as possible. In the fermentor system, the microbe will be inoculated into the fermentation medium "feast," and will thus attempt to colonize this environment through rapid growth.

The role of the fermentation control strategy is to provide, by control of environmental effectors such as temperature, aeration, pH, and dissolved oxygen, the optimum conditions for growth and "colonization." The fermentation scientist uses an environment that is capable of being controlled to a limited degree, i.e., the fermentor, to develop a control strategy that will modify inputs to the fermentor system in order to achieve the desired outputs. This system of modifying inputs to obtain a desired output from a control system is often described as a control loop.

13.1.2 *Nature of control*

A fermentation development program seeks to establish what control set-points are needed for the control loops in the control system. It is easy to forget that the control system can be human! There are many examples of accurate human control loop systems including the motor car. In the motor car the driver observes the speed of the vehicle (the output from the system) by looking at the speedometer (sensor). If the car is traveling too fast, the control system (the driver) reacts by reducing pressure on the accelerator, thus reducing the flow of fuel to the engine (the input to the system). The car slows and the driver observes whether the speed matches the desired outcome (speed limit); if it does, the adjustment to the accelerator will be modified again, and so on. Control of a fermentation system can similarly be by manual intervention of control valves adjusted as a result of changes observed on gauges; however, this chapter will concentrate on the elements of automatic (computer) control.

13.1.3 *Control loop strategy*

The basic element of a control system is the control loop. Figure 9.1 summarizes the various components that make up the control loop. At the centre of the control loop is the system requiring control, the fermentor. The system can be affected by a number of influences, and in this case it is the temperature of the system that is to be controlled. A very simple control loop is shown where only cooling may be applied to the fermentor principally to remove metabolic heat during incubation. The temperature of the fermentor is measured using a thermometer and either a human operator or a control system can monitor this measurement and make adjustments if the desired temperature (the set-point) and the actual temperature are not equal (error in the system). If the temperature of the fermentor is above the set-point, the flow of cooling water is turned on (input to the system) and once the temperature equals the set-point, the control valve is switched off. In reality, dynamic fermentor systems are never fully stable and constant monitoring and control are required,

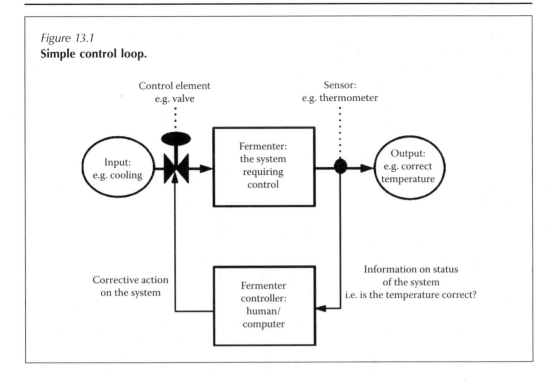

Figure 13.1
Simple control loop.

Control element
e.g. valve

Sensor:
e.g. thermometer

Input:
e.g. cooling

Fermenter:
the system
requiring
control

Output:
e.g. correct
temperature

Corrective action
on the system

Fermenter
controller:
human/
computer

Information on status
of the system
i.e. is the temperature correct?

hence the advantage of automated systems, which maintain a constant vigil on the process and adjust inputs to the system as and when required automatically.

13.2 Sensors

Development of fermentation processes thus far has been accomplished primarily in large-scale reactors (> 10 liters) with the aid of relatively few sensors. The problem with monitoring the fermentation system is maintaining the sterile integrity of the fermentor hence any sensor has to be capable of sterilization either *in situ* during steam sterilization or remotely and then aseptically introduced to the fermentor.

13.2.1 *Historical perspective*

In the 1940s the sensors were largely based on manual sampling and off-line analysis, control was improved through the 1950s and 1960s by the use of limited electrical signals controlling pneumatic outputs (valves, etc.). In the 1970s new sensors capable of being sterilized were introduced including pH and dissolved oxygen probes. In addition, more accurate methods of measuring flow rates (both liquid and gas) became available and other engineering parameters such as motor power to the agitator and enhanced nutrient feed addition systems were developed.

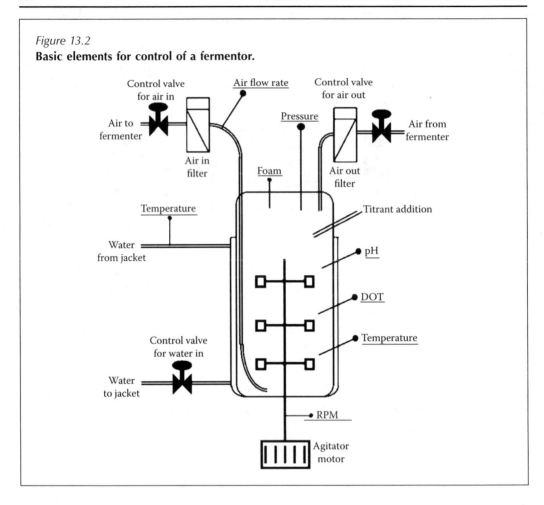

Figure 13.2
Basic elements for control of a fermentor.

As well as improvements to sensors and other measuring devices mini-computers were introduced to provide simple control and data logging.

Recently, a wide range of new sensors (see Chapter 11 for more details) have emerged for on-line analysis of fermentation parameters. Recent techniques that have become available include biomass probes, on-line liquid chromatography systems, near infra-red spectroscopy, and so on. For the most part, the reliability and relevance of some of the measurements together with prohibitive cost tend to preclude these sensors from everyday fermentations.

13.2.2 *Typical fermentation sensors*

The control elements (sensors) that should be considered routine for most (aerobic) fermentation systems are

1. Temperature: measured using a platinum resistance thermometer (PRT probe), where the increase in temperature

is proportional to the increase in electrical resistance in the probe.

- Temperature will be controlled by the addition of cooling water to a jacket or cooling finger of a fermentor; heat will be added by direct heating of the vessel or its contents (electrical heating mantle or "hot finger") or by the injection of hot water or steam to the circulating water in a jacket or heat exchanger.

2. Air flow rate: measured using a standard pressure drop device such as variable area flow meters or more often a mass flow sensor.

- Air flow will generally be controlled using a proportional (0 to 100% open) valve upstream of the sterile inlet filter on a fermentor.
- Air flow is frequently quoted as VVM; volume of gas per volume of liquid per minute, fermentor design generally permits up to 2 VVM.

3. Vessel pressure: measured using diaphragm protected Bourdon gauges or strain gauge pressure transducers.

- Pressure in the vessel is induced during *in situ* steam sterilization and during normal incubation with the introduction of air. Control of vessel pressure is by regulation of the vent gas from a fermentor.
- Pressure is generally a negatively acting loop in that a fully driven output valve (100% open) results in minimal pressure in the fermentor.
- The units of pressure are generally bar gauge, i.e., pressure within the reactor above atmospheric.

4. Vessel agitation rate: measured using proximity detectors to detect the speed of the shaft.

- Impellors or mixers are controlled by standard motor controllers and the units are revolutions per minute.
- Agitation power is sometimes measured using current transformers measuring the electrical power consumption of the motor; the units of power are Watts.
- pH is measured using steam sterilizable combined glass electrodes.
- pH is controlled by the use of buffers or by the addition of acids or base titrants.

5. Dissolved oxygen: measured using polargraphic type probes, here galvanic voltages on a membrane-covered oxygen reducing cathode induce a current (amperometric) proportional to the amount of oxygen diffusing through the membrane.

- Dissolved oxygen is controlled by altering the status of those control loops affecting dissolved oxygen generation. An increase in dissolved oxygen is induced in fermentation medium by
  1. increasing the air flow rate (volumetric increase in oxygen),

2. increasing agitator speed (smaller air bubbles increasing the surface area available for diffusion),
3. increasing over pressure in the fermentor (generally increasing the residence time for air bubbles).

- Dissolved oxygen is measured as the partial pressure of oxygen at the electrode surface and hence is quoted in percentage saturation (often referred to as dissolved oxygen tension, DOT). To obtain a mass, a fully saturated solution of oxygen in water is approximately equivalent to 1.2 mM/liter.

6. Foam: detected by observation, or conductance or capacitance probes completing an electric circuit when foam is contacted.

- Foam is controlled by reducing the cause of foaming, i.e. high aeration and/or agitation rates or by the addition of antifoaming agents.

Figure 13.2 shows the main control elements of a fermentor, there are many different configurations that may be specified but the one shown would be capable of providing data and control options for an aerobic fermentation.

13.2.3 *Control action*

- Temperature control would be achieved by the regulated supply of temperature controlled water to the jacket of the fermentor. In the case illustrated there are two thermometers in the system, one indicating the temperature of the medium/broth, the other indicating the temperature of the return water flow. In this configuration the two signals may be compared and control options could include linking their function to provide a regulated flow of temperature controlled water to the jacket via the control valve shown.

- Airflow control would be achieved by regulating the linear gas flow rate to the system by adjusting the airflow control valve shown in response to signals coming from the air flow sensor.

- Pressure control is achieved by regulating the flow of off gas from the fermentor. It is probably beneficial to have the airflow and pressure control functions independent of each other. For the most part these control loops will act to the limit of their engineering configuration and if a particular set-point, say for airflow, cannot be achieved because of excess back-pressure, then this may be noted as a process constraint and the operating protocol adjusted accordingly.

- Agitator speed is controlled by direct feedback of the revolutions per minute using a suitable tachometer.

- pH is generally controlled by the addition of acid or base titrant in response to changes in pH value during the fermentation. One important aspect of the pH loop is the calibration of the probe before and after sterilization. These are achieved before sterilization by the adjustment of slope and intercept values on the pH controller to be used during the fermentation. Subsequent to sterilization it is possible to check the calibration of the pH probe by comparing the pH of a sample of medium withdrawn from the fermentor using a calibrated external pH meter. The function of the pH control loop will be discussed in more detail later.
- Dissolved oxygen control is achieved in a number of different ways. It is possible to rely totally on increasing agitator speed to shear air bubbles rising through the broth to increase surface area and thus mass transfer of oxygen from the gaseous to liquid phase. In addition, however, it is also possible to cascade the control output from the dissolved oxygen controller to agitator speed, airflow, and pressure or any combination of these. In a particularly high oxygen-demanding fermentation, all three outputs may be specified, care has to be exercised in the use of independent control loops such as agitator speed that this loop will be the 'slave' to the DOT 'master' control loop function. The function of the DOT control loop will be discussed in more detail later.
- Foam is controlled by the addition of antifoam agents via the feed addition system in response to a contact probe detecting rising foam.

13.3   Controllers

Control instrumentation has developed rapidly in recent years and the range of options available for fermentation control is extensive (see Chapter 10 for more details). The development of integrated circuits in the last 20 years or so has meant that complex control functions can be devolved to cheaper instruments, or conversely more sophisticated control options have become increasingly available to the fermentation scientist.

13.3.1   *Types of control*

Two fundamental types of control may be incorporated into a fermentation system: sequence control and loop control. Sequence control is that part of control that permits automation of the fermentor operation such as sterilization and other valve automation sequences, i.e., for the most part providing an ordered array of digital (on/off) signal control. Loop control is generally associated with that part of control dealing with combinations of digital and analogue control signals and although used in sequence control (particularly automated sterilization sequences) is most often associated with control of incubation. The industrial fermentation scientist will frequently wish to adapt the control strategy for optimal process performance and in order to ensure versatility, bespoke control programs may be specified although the costs and potential delays associated with this approach are likely to be significant.

Control of fermentation systems can be achieved by the use of discrete single-loop controllers, by programmable logic controllers controlling both sequence and loop functions, and by the use of specific software packages to control all aspects of the fermentation. This type of control is sometimes called Distributed Digital Control (DDC) and defines boundaries of operation whereby whole plant control can be achieved using computer-based systems.

13.3.2 *Control algorithms*

Whatever type of controller is selected for fermentation control, effective action will depend on the response of the controller. This response is determined by the nature of the control algorithms programmed into the system. A control algorithm is a mathematical representation of the steps required to achieve effective control, most often programmed as equations in which the controller output is a function of signal deviation away from a set-point. As indicated in Figure 13.1, most controllers will function as feedback control systems, i.e., the deviation of measured variable compared with the desired set-point will determine how large the control effect should be on the fermentor system via the appropriate control algorithm.

13.3.3 *PID*

The types of control algorithm most frequently encountered are three-term or PID controllers. The PID controller is made up of three elements P: proportional, I: integral, D: derivative/differential; the purpose of these functions is to provide a fast-acting response to process deviation and scale the response to the output to achieve smooth control action. The characteristics of PID control are

- Proportional control provides an output, the magnitude of which is proportional to the deviation between measured variable and set-point.
- Integral control tends to reduce the effect of proportional control alone, helping to bring the measured variable back to set-point faster by minimizing the integral of control error.

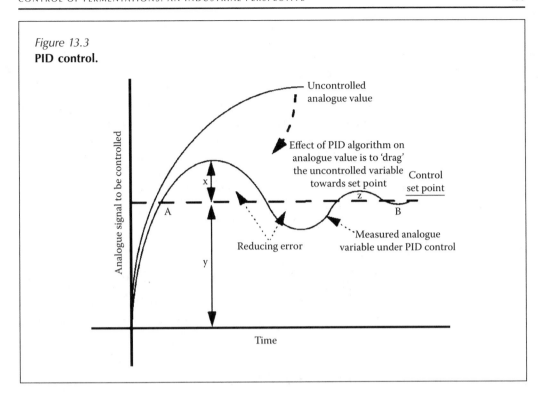

Figure 13.3
PID control.

- Derivative action also tends to reduce the effect of proportional control alone, this time by estimating the slope of measured variable with time and maximizing the slope of the measured variable compared with the set-point.

The effect of PID control on an uncontrolled variable is shown in Figure 13.3. As indicated, the algorithm tends to "drag" the analogue value towards the set-point, the speed and effectiveness of this control response are a function of the parameters specified by the PID algorithm, the process of setting these parameters is termed, tuning the loop. The method is generally carried out by trial and error with the loop on-line. There are three parameters associated with PID loop tuning; these are the proportional band (usually set as a percentage of full-scale deflection of the analogue signal), the integral time, and the derivative time. The method chosen for loop tuning will be recommended in control instrument vendors' instructions, the sequence generally follows:

1. Set the proportional band first with no integral (maximum value) or derivative function (minimum value).
2. When oscillations occur with small process perturbations, decrease the integral time (from its maximum) until oscillations reoccur, then reset the integral time to minimize oscillations.

3. Establish whether control is adequate with PI control alone, if not satisfactory following process perturbations, increase the derivative time constant (from its minimum) to minimize oscillations.

PID loop performance can be assessed by measuring the initial overshoot as a ratio of x:y. A measure of PID loop effectiveness can be assessed by examining the control error decay ratio of z:x, in addition, the "gain" of the controller, i.e., how fast the measured variable attains the set-point when full control is applied is described by point A. Point B describes the 'settling time' for the loop, i.e. when the measured variable is within a fixed percentage of the set-point value, e.g., $\pm 5\%$.

13.4   Design of a fermentation control system

The first stage in designing a fermentation control system is to clearly establish what the process objectives are for the automation project.

13.4.1   Control system objectives

Objectives may be typically divided into various "categories" of control, for example:

1. Control of basic incubation functions only, typically airflow, agitator speed, and temperature.
2. Automation of the control of incubation only relying on manual operation of valves associated with sterilization.
3. Full automation of the fermentor system including all sterilization and auxiliary vessel control.
4. Advanced control options including event-based control.
5. Advanced computing methods for inferential control.

Whatever type of controller is considered necessary, the process scientist will have to specify the type of control and then make this specification available to the manufacturer of various control systems. The more complex the fermentor system to be controlled, the more detailed the specification will need to be.

1. Basic control will typically be achieved using single-loop controllers and is frequently associated with autoclavable fermentors. These controllers tend to combine both amplification of process signals and control function in one "box," this functionality may also include a local readout function in engineering units, i.e., values recognizable to the operator. There is a wide range of commercially available single-loop controllers, some of which have been

developed with control of fermentation processes in mind. Configuration of the controllers will follow manufacturers' guidelines, but as with all control options, factors such as signal type (current or voltage) and output device will have to be considered before final control specifications can be issued to potential vendors.

2. Incubation control may include other features including operation of feed delivery systems and control of more complex loops such as dissolved oxygen by multiple outputs which may be specified by the operator. The specification in this case may be more complex and be subject to negotiation with the instrument vendor. Sterilization of this type of vessel may be by autoclaving or by in situ steam sterilization but this will be done manually, probably to minimize cost.

3. Full incubation and sterilization control of the fermentor will require a control function for valve sequencing (digital control) as well as control of analogue parameters for incubation (process variables and control of proportional valves). The requirement for valve sequencing as well as more complex patterns of analogue control is met by the use of Programmable Logic Controllers (PLC) or often fermentor manufacturer's own control software running on a personal computer. This type of control is generally required for larger fermentors (> 20 liters) or where a number of identical vessels are to be purchased and manual sterilization will be too manpower-intensive.

   The need for a detailed specification is determined by whether the vendor's own software meets the needs of the fermentation scientist. If it does, then the basic requirements of sterilization and incubation control may be easily identified from the vendor's own specification. If, "however." pre-existing software does not meet the needs of the scientist, then a much more complex and detailed specification may be required because it is most likely that the software will have to be written specifically for the purpose. It is in the best interests of both customer and vendor to agree to the objectives of the bespoke software prior to code being written, costs and timescales can very quickly escalate out of control. The benefit of using vendor's own software is speed of implementation and probably lower costs; the disadvantages are that you get what the vendor thinks you want, not what you might actually need. The opposite is generally the case with PLCs; costs and timescales will be higher but if you get the specification right you get exactly what you want and need.

4. Advanced incubation control regimes may be required where complex fermentation patterns are required and changes to set-points may be needed on-line, particularly in response to specified "events." This will be discussed in

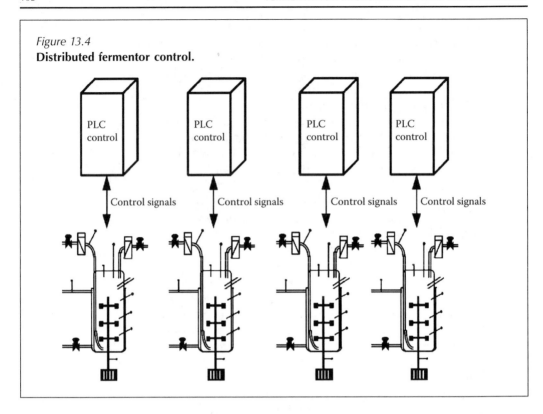

Figure 13.4
Distributed fermentor control.

more detail later. The need for a detailed specification is very high under these circumstances, and clear milestones and checkpoints must be agreed during contract negotiations. It is quite likely that stage payments will be part of a contract. Large fermentor control tasks may be related to both complexity of control and numbers of fermentors within a system. If the task involves large numbers of fermentors to be controlled, then a system of distributed control will probably be required. The architecture for distributed control is shown in Figure 13.4.

Figure 13.4 illustrates one possible control option for the control of four fermentors. In this configuration each fermentor is independent of the next and control is affected by each fermentor being equipped with its own controller (in this case a PLC). This type of distributed control has the advantage that failure of a control system will only affect one fermentor, the experimental design may be ruined by such an occurrence but without a distributed system, failure of a single controller will result in failure of the entire run. The disadvantage of distributed control is that the cost may be prohibitive to implement one controller per vessel. In the author's laboratory a compromise has been reached with vessels < 100 liters working volume where a single PLC controls a pair of identical fermentors.

5. Advanced computing methods may be required where analytical systems are inadequate to optimize the fermentation process and cost of goods or value of product warrant investment in "next-generation" computing methods. Inference methods are available which algorithmic sensors can function to estimate for analytes and processes for which no other sensor or probe exists. The sort of computing methods being developed include expert systems, artificial neural networks, and model-based systems. These self-learning or decision-making systems rely on process pattern recognition to identify regions of the fermentation process more or less susceptible to perturbations. Once identified remedial action beyond the scope of experienced operators or less sophisticated controllers can be initiated to reduce cost, avoid catastrophes, or simply improve the process. These systems can often only be justified in production environments where the value of the product warrants large investment in time and money to maximize productivity. As these methods become more widely available, the costs of implementation of advanced computing methods will decrease and wider applications will be sought and introduced.

13.4.2 *Fermentation computer control system architecture*

In defining the overall strategy for computer control of fermentations, system architecture will emerge to indicate how the system will be operated, how data will flow around the system, and how that data can be captured, stored, and interrogated when required. The control system when configured is an information exchange system.

Information about the progress of the fermentation is detected by sensors, transmitted to amplification/signal conditioning units that forward this information to the controller. The controller receives the information, compares it to pre-existing information about how the fermentation should proceed (setpoints), and then after generating new information in the form of an algorithmic output, transmits information back to the fermentor, to a control unit capable of receiving information and translating that signal into a control action. Over and above this control function the operator can observe the flow of information (data), intervene if necessary (new information), and recall past information from a data storage system. Figure 13.5 shows an example of a fermentor control system architecture where the information exchange is indicated.

In Figure 13.5 there is flow of analogue and digital sensor data from the fermentor vessel to the control cabinet. The input

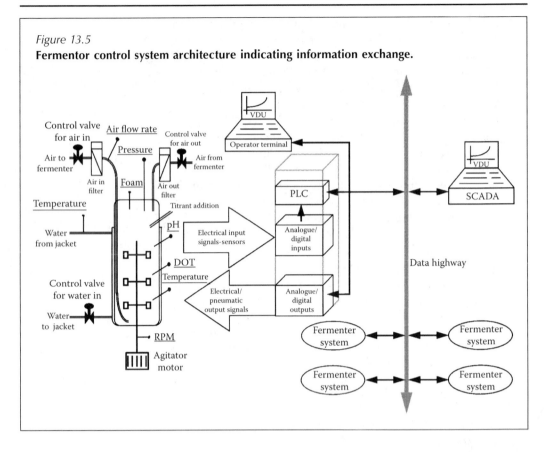

Figure 13.5

Fermentor control system architecture indicating information exchange.

data are marshaled in I/O (input/output) cards whose main function is to convert a mixture of signals to a common electrical signal type for the computer to interpret. These signals are passed to the programmable logic controller (in this case) where control strategies (set-points, etc.) are programmed. In the PLC the input data are fed into PID algorithms and other control functions, an output signal is fed back to the fermentor via I/O cards. The control signal returning to the fermentor may have to be transduced to a more usable format for the process plant; this may take the form of pneumatic signals scaled from the frequently used 4–20 mA to the equivalent 0–1 bar pressure. The PLC in Figure 13.5 is connected to the operator terminal, which may be a personal computer with a VDU or a standalone plant terminal. If located on the process plant floor, then the whole cabinet has to be splash- and dust-proof, often referred to as an IP55 rating.

The PLC is also connected to a Data Highway onto which other fermentor systems are connected. In Figure 13.5 all the fermentors are in communication with a Supervisory and Data Acquisition System (SCADA) whose function is to collect all the plant data and store it for later retrieval and possible interrogation. It should be understood that the SCADA is not a database, it generally cannot give answers to database type queries but using advanced graphics and trending graphs can present the operator with comprehensive information on plant status.

13.4.3 *Fermentation plant safety*

One of the overriding functions of a fermentation control system is monitoring and action relevant to safety. A fermentor is a potentially dangerous piece of equipment using as it does live steam (to 135°C), over pressure (to 2 bar gauge +), acids and bases, as well as the micro-organism and its products. The automated control system can ensure by program interlocks, etc., that the plant is operated with the maximum safety margins on every function, and specifying this is perhaps the most important part of the acquisition of a new control system.

13.5   Fermentor control specification

When planning for a new fermentor control system the specification for the system is crucial to the success of the project. The specification will be used to judge vendor's response and to subsequently quote and oversee the project to completion. Time and effort spent generating a comprehensive system specification are rewarded many times over during implementation and commissioning.

13.5.1 *Specifying sequence control*

One of the most important tasks when planning to automate a fermentor system, particularly for sterilization and other automated sequences, is to accurately specify what the sequences are to be automated and how this will be achieved. There are fixed unit operations associated with fermentor control and the specification will define these and indicate how this control will be achieved.

13.5.2 *Fermentation unit operations*

Taking batch fermentation as a typical example, the unit operations to be selected might include

- Blank sterilization: generally employed as a prebatching cleansing sequence when all steam and drain valves are opened.
- Medium batching: the fermentor is usually in a safe state for opening and preparation. In the author's laboratory the safe state is defined by "Standby." In standby all valves

are shut with the exception of valves venting the vessel interior to atmosphere. During batching, probe calibration and insertion into the vessel must be complete before water is added to the vessel!

- Medium sterilization: the sterilization sequence will be partly dictated by the fermentor vessel configuration and geometry. Steam sterilization of the medium can be achieved either by direct steam injection or by indirect heating via a heat exchanger or vessel jacket. Whatever the mode of sterilization chosen, it is crucial that the correct valves are operated in the correct sequence, not just for process integrity but also for safety.

- Medium hold: it is useful to maintain the fermentation medium in a safe state post-sterilization. It is at this time that final adjustments to the fermentor set-up can be made, e.g., pH adjustment via a tartan feed system or temperature alteration prior to inoculation.

- Fermentor inoculation: to introduce inoculum into the sterilized fermentor requires opening the vessel in a controlled manner and introducing inoculum using strict aseptic techniques throughout. If introducing the inoculum via peristaltic pumps, then a reduction in vessel pressure will be required.

- Incubation: although not strictly a sequence, there are clearly certain valves that must be opened to permit incubation of the culture to proceed under specified conditions. The principal function of the incubation sequence is to start the clock counting the hours elapsed since inoculation. During incubation the full functionality of the control system may be used to program the correct fermentation "trajectory" for the run. This will involve specifying setpoints for all the main controllers including agitator speed, airflow, temperature and feeds if available as well as controls responding to more metabolic influences such as pH and dissolved oxygen.

- Harvest: It may be necessary to specify a sequence dealing with harvest operations. This could entail prechilling or suitable pH adjustment of the broth to permit easier product recovery operations. Harvest could also include "killing" the vessel, i.e., sterilizing the vessel interior (with or without cells) prior to safe opening and cleaning.

- Cleaning: A number of options may be available here including full sterilization or heating in the presence of caustic detergents to fully automated Clean In Place (CIP) systems with complex valve operations of their own.

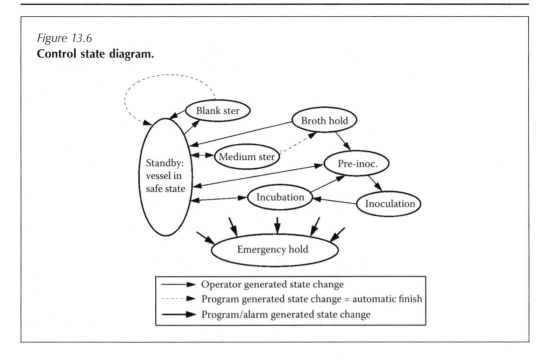

Figure 13.6
Control state diagram.

13.5.3 *Vessel states*

Fermentation unit operations may be termed vessel states and the transition between them must be strictly regulated in an automated plant; for example, initiating automatic sterilization of the batch during the incubation should be avoided at all costs. State changes can be summarized in a state diagram (Figure 13.6).

Each one of the states in the diagram defines a part of the control program specifying sequences to be executed for effective control of the fermentation system. Navigating around the state diagram defines the safe operation of the vessel under automatic control. Certain transitions are under operator control such as from Standby to Medium Sterilization. Other states are attained automatically as a result of a particular sequence finishing such as the transition from sterilization of the Medium Sterilization to Broth Hold. There is within an automated control system the opportunity to initiate emergency action which could come about as a result of alarm settings or operator intervention. This is indicated in Figure 13.6, where a state called Emergency Hold may be attained from anywhere in the program. Such a state can be programmed to do whatever is deemed appropriate for safe containment of the process; in the author's laboratory Emergency Hold constitutes a safe universal shutdown of the fermentation process sealing the vessel but retaining set-points for possible restart of Incubation.

13.5.4 *Sequence logic*

Control of multiple valves on an automated fermentor requires the program to control the opening and closing of valves such that the state operation is effectively and safely completed. Each of the vessel states indicated previously and possibly many others have to be programmed taking best human operator practice and engineering constraints into account. Defining the sequence follows a pattern of working that ensures ambiguities are minimized and objectives are clearly stated. Taking one of the states in Figure 13.6, sterilization of the fermentor plus medium, the system developer must start with a complete description of the fermentor and the valves associated with sterilization (which is likely to be most of them). A comprehensive description will come from accurate drawings of the plant; these drawings are often referred to as piping and instrumentation drawings (P&ID, not to be confused with PID for control). The drawing will identify the units of control to be defined in the program. Modern operating systems will tend to function with structured code and it is possible to consider individual blocks of code controlling individual units of control.

This can be illustrated with reference to Figure 13.7, a typical fermentor configuration (as a notional P&ID) is shown for sterilization, only those valves are shown that are relevant to this highly simplified representation.

The P&ID identifies the key elements for the following groups of valves:

- Air in group V1.X
- Air out group V2.X
- Jacket group V3.X

SEQUENCE LOGIC for Sterilization operations
(key words in BOLD)

1. START: all valves closed
2. Drain jacket
3. Heat-up phase
4. Direct steam injection
5. Air filter sterilization
6. Sterilization temperature
7. Air filter pressurization
8. Crash cool
9. Ballast air
10. Broth hold

13.5.5 *Flow charting*

P&ID drawings, valve descriptions (Table 13.1), and status charts (Table 13.2) are essential in defining the operation to be

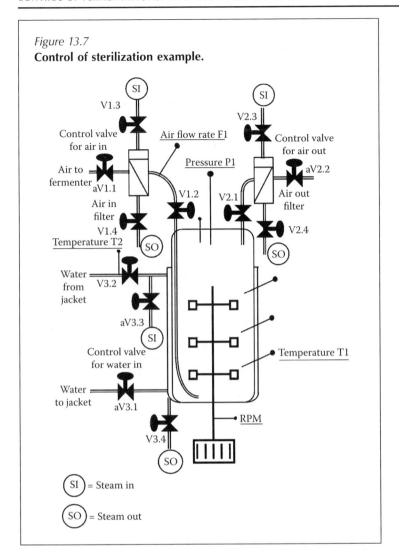

Figure 13.7
Control of sterilization example.

automated. However, before code can be written, the operation is translated into a flow chart of valve operation and decision gates that must be followed. The flow of code to accomplish the task will follow the chart (Figure 13.8).

The "English" translation of the flow chart is approximately as follows:

- START: All valves are closed and remain closed unless instructed otherwise.
- DRAIN jacket: The jacket drain valves are opened for a specified time (dependent on jacket geometry).
- HEAT: Cycle commences by opening air out valves to allow for the escape of air and by opening steam valves to the jacket.

Table 13.1 Valve descriptions for example fermentor (Figure 13.7)

Air in group	Air out group	Jacket group
V2.1	Digital air out valve	
aV2.2	Analogue control valve for pressure	
V2.3	Digital Steam to air out filter	
V2.4	Digital Steam out from air out filter	
aV3.1	Analogue water to jacket valve	
V3.2	Digital water from jacket valve	
aV3.3	Analogue control valve steam to jacket	
V3.4	Digital jacket drain/ condensate out valve	
aV1.1	Analogue control valve: air to vessel	
V1.2	Digital Air ex: filter block valve	
V1.3	Digital Steam to air in filter	
V1.4	Digital Steam out from air in filter	
	BOTH: analogue and time	
	i.e., pH > 6.5 AND < 24 h true	
	EITHER or BOTH:	
	i.e., pH > 6.5 true OR time < 24 h true OR both true	
	EITHER true (but not both):	
	either pH > 6.5 true OR time < 24 h true	
	EITHER or NEITHER (but not both):	
	i.e., neither pH > 6.5 OR time < 24 h, OR pH, OR time, but NOT pH AND time true	
	NEITHER:	
	neither pH > 6.5 OR time < 24 h true	

- FILTER: The sterilization of the air inlet filters must be complete before the medium so that sterile air can be admitted to break the vacuum caused by collapsing steam after sterilization is complete. The commencement of filter sterilization may be dependent on the temperature of the medium in the fermentor.
- DIRECT: To assist in heating the medium to sterilization temperature, steam can be admitted via the air in filter and down the air sparge pipe. The point at which this occurs can be determined by the medium temperature.
- PRESSURIZE: At some point during the sterilization of the medium (0.5 total sterilization time?) the air in filters are pressurized to prevent ingress of contaminants prior to ballast air.

Table 13.2 Valve status chart for sterilization sequence logic

Valves	Valve status									
	START	Drain	Heat	Filter	Direct	STER	Press	Cool	Ballast	Hold
Air in										
aV1.1	0	0	0	0	0	0	a = 1	a = 1	a	a
V1.2	0	0	0	0	1	1	0	0	1	1
V1.3	0	0	0	1	1	1	0	0	0	0
V1.4	0	0	0	1	0	0	$1 \Rightarrow 0$	0	0	0
Air out										
V2.1	0	0	1	1	1	1	1	1	1	1
aV2.2	0	0	a = 1	a	a	a	a	a	a	a
V2.3	0	0	0	0	0	1	0	0	0	0
V2.4	0	0	0	0	0	1	0	0	0	0
Jacket										
aV3.1	0	0	0	0	0	0	0	a = 1	a	a
V3.2	0	0	0	0	0	0	0	1	1	1
aV3.3	0	0	a = 1	a	a	a	a	0	0	0
V3.4	0	1	1	1	1	1	1	0	0	0

0: closed valve
1: open valve
a: analogue control 0 to 100% open
a = 1: analogue valve fully open
$1 \Rightarrow 0$: open to closed valve.

- CRASH COOL: When the sterilization time is complete the fermentor and its contents are "crash-cooled" (to minimize time at elevated temperatures) by shutting off the supply of steam and admitting cold water to the heat exchanger system.
- BALLAST AIR: Whilst the temperature of the medium is still above boiling, the vessel is pressurized with sterile air to prevent the vessel pulling a vacuum and thus permitting possible ingress of contamination micro-organisms.
- BROTH HOLD: When the temperature of the medium is at some predetermined level (35°C), the vessel is held in a quasi-incubation mode, ready for inoculation and/or other treatment.

In this simple example many different factors have been taken into account but the flow chart is far from complete, there is for example no mechanism in this for operators to enter set-points. Again with this example there are only 12 valves, on a 4500-liter pilot scale fermentor used for research and production there may be 70+ valves organized in many functional groups, hence programming such a system will require many days/weeks of

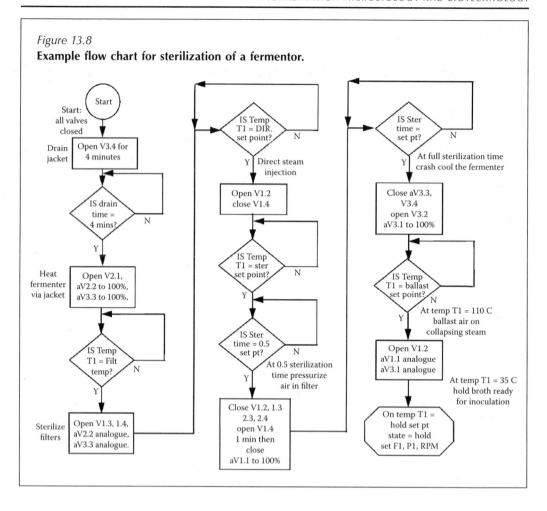

Figure 13.8
Example flow chart for sterilization of a fermentor.

specifying, programming, and testing before the fermentor control system may be commissioned.

13.6 Control of incubation

Specifying the sequence logic of a fermentor control system is only part of the control task, arguably the most important function of the fermentor control system is to control the incubation of the micro-organism of interest. Returning to the case made for the control of fermentations at the start of the chapter, the purpose of the control system is to provide an optimal environment for growth and expression of an attribute associated with that growth. This is very difficult; one can pose the rhetorical question, what is the natural habitat for *Escherichia coli*? The answer given by many might be "the intestinal tract of most vertebrate animals," but enteric bacteria can be isolated from many mesophillic environments and evolved probably billions of years before the advent of colons!

Therefore, when faced with the prospect of defining the environmental conditions to maximize microbial productivity, how does the fermentation scientist decide what conditions to apply? In answering this question a distinction should be made between fermentation development strategies for research and development organizations and manufacturing. Each function imposes different constraints. Within the manufacturing sector fermentation development will be investigating the factors associated with a well-established fermentation process where the objectives of a development fermentor will be to obtain relatively modest increases in productivity (leading to substantial financial savings on large-scale production) or to achieve reduction in "cost of goods." In a research and development environment the constraints usually come from working with a wide range of culture types, limited/no knowledge of those cultures, limited/no knowledge of the fermentation systems, and with very short development times required.

13.6.1 *Specification for incubation control*

Specifying the control options for a fermentor system must then take into account not only the type of organisms to be grown but also what sort of control options will actually be required. The control options for a typical R&D fermentor system are described below.

13.6.1.1 *Temperature*

Control of incubation temperature will be achieved by the use of system of addition of heat and cooling. This control loop is fundamental to fermentor systems. The degree of accuracy of control during incubation will probably be in the order of a setpoint in the range 20°C to 50°C ± 0.1°C. Response times may also be important and controlled ramp rates of better than 1°C per minute may be necessary.

13.6.1.2 *Aeration*

For most industrially significant processes the micro-organisms will be aerobic. Supply of air to the fermentor will need to be controlled typically in the range 0 to 2 VVM. The accuracy of the control system will need to be in the order of ±1% of full scale deflection (FSD). This is generally a fast-acting loop and obtaining adequate ramp rates is not a problem.

13.6.1.3 Agitation

The stirrer for a traditional fermentor is used to minimize gradients within the bulk broth, the larger the vessel the greater the potential for gradient (mass transfer) problems. Sufficient motor power has to be available to obtain uniform mixing times to be in the order of 1 to 2 minutes or less (determined from the addition of a marker into the bulk liquid and the time taken to homogeneity recorded). These fast mixing times may not be achievable in pilot scale vessels (3000 liters + working volumes) and these system constraints must be identified and resolved for example, in scaled down experimental designs (modeling large-scale parameters in small-scale vessels). Agitation ramp rates may be high and the control system should be capable of ramp rates equivalent to 10% of FSD per minute.

13.6.1.4 Pressure

This control loop only applies to pressure-rated vessels. The control of pressure is required for accurate control of *in situ* steam sterilization as well as for some incubation regimes. The control requirement during incubation may come from maintaining adequate pressure for high oxygen transfer or to simulate hydrostatic head pressure in large-scale vessels in scale down experiments in smaller scale vessels. Control of pressure is critically linked to safety, and safeguards within the control system must ensure that the vessel cannot be overpressurized. Pressure can be a very fast-acting loop, and obtaining adequate ramp rates is not a problem. The function of this loop may be in conflict with the aeration loop and the fermentation scientist has to bear in mind that control of these two loops may be incompatible. For example, a set-point for pressure near the safe operating limit of the vessel may be impossible to achieve with extremely low air flow rates (0.1 VVM).

13.6.1.5 Hydrogen ion concentrations, the pH

Most industrial fermentations need to be run within a certain range of pH values for maximal productivity. Although media are formulated in such a way that ensures a certain level of buffering capacity "built in," there is often a requirement to control the pH away from this "natural" value. This is achieved most often by the addition of acid or base titrants. Addition of acid or base titrants requires a feed system of some description, under these circumstances the control of the feed system is subordinated to the pH controller, in other circumstances the feed system may act under the direction of another controller or independently; this will be discussed in more detail below. When tuning a pH control loop it is very difficult to define the

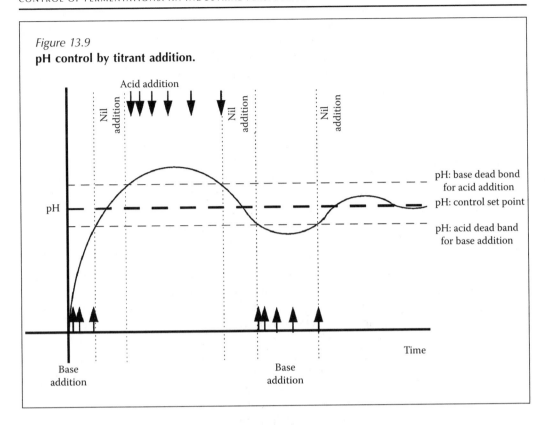

Figure 13.9
pH control by titrant addition.

"correct" PID settings because the strength of the titrants will vary. Therefore, the "gain" of the controller (i.e., how quickly a response will be achieved) will also vary greatly. This problem of adequate pH control can be addressed by user definable "dead bands" around the set-point where no control action is initiated. The purpose of the dead band is to allow for the natural buffering capacity of the medium to have sufficient time to act so that titrants are not added close to the set-point on a detuned loop, which will be the case with variable titrant concentrations.

In Figure 13.9 the response of a pH control loop following acid or base addition is shown. It will be noted that there is a proportional action on the additions of titrant made, i.e., the larger the "error" the more frequent the control action. The closer the measured pH gets to the set-point the longer the delay between each addition of titrant; however, when the dead band threshold is crossed, all titrant addition stops and the natural buffering capacity of the medium results in an approach to the set-point within the dead band limits. The setting of the dead band limits may be asymmetrical about the set-point depending on the criticality or tendency of the fermentation to be acid or base tolerant. Mineral acid titrants may be substituted for organic acids or the principal carbohydrate energy source such as glucose. In this case the buffering capacity or delay equates to the metabolic rate for the consumption of the sugar and protons generated by catabolism will require a delay of some minutes before a noticeable effect on the pH will be detected. This type of control of pH by metabolic action is really a means of permitting the micro-organism to autoregulate the supply of carbon.

13.6.1.6 *Dissolved oxygen*

The control of dissolved oxygen in aerobic fermentations may be critical to the successful outcome of that fermentation. The requirement for oxygen may be very high during the rapid growth phase of a batch culture and oxygen limitation may result in inadequate growth and incomplete oxidation of the primary energy source. The control of dissolved oxygen is generally achieved by increasing the "driving force" for the mass transfer of oxygen from the gaseous phase to the liquid phase. The methods or outputs available to the control system to achieve this include increasing the agitator speed, which generally increases the shear on the air bubbles, making them smaller and increasing their surface area available for gaseous exchange.

The second output that may be programmed is to simply increase the air flow rate, ensuring a greater volumetric airflow through the liquid. The last output that may be incorporated into a dissolved oxygen controller is vessel overpressure, which will have the effect of increasing the hydrostatic head pressure and preventing rapid flushing of "precious" air from the fermentor.

All these outputs can be specified for a dissolved oxygen controller either individually or in any combination of all three. The output of the dissolved oxygen controller (master) must be cascaded onto the closed-loop controllers for agitator, airflow, and pressure (slaves) as required. The consequence of this is that the function of the closed-loop controller will be subordinated to that of the dissolved oxygen controller to achieve the desired dissolved oxygen tension in the fermentation broth.

This type of control may be considered a replenishment system where the micro-organisms' metabolism is the oxygen sink and the outputs of the controller replenish the resulting deficiency. In setting up a dissolved oxygen controller it may be necessary to set default set-points for each of the closed-loop controllers such that at the start of the fermentation a baseline level of agitator speed, airflow rate, and vessel pressure will permit adequate mass transfer of oxygen for microbial metabolic action to ensue. As the microbe respires, the depletion of dissolved oxygen will occur and the output of the dissolved oxygen controller will increase to compensate. Depending how the DOT control loop is specified, the outputs can be scaled to one or more of the specified closed-loop controllers.

In Figure 13.10 the dissolved oxygen controller (master) is linked to all three closed-loop controllers (slaves). As the DOT falls the oxygen controller output will rise and this will cause an

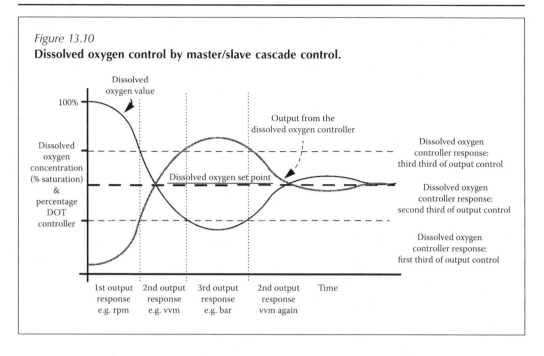

Figure 13.10
Dissolved oxygen control by master/slave cascade control.

increase in agitator speed from a default closed-loop set-point to a maximum rpm specified for the DOT output. In the example the DOT continues to fall as the output of the controller passes a threshold equivalent to one-third the total output range of the DOT controller. At this point the second specified closed-loop output is subordinated to the DOT controller and takes over from agitator speed leaving the set-point for the first output at its specified maximum. The second output, in this case airflow, increases to try and compensate for the fall in dissolved oxygen and when this output limit is reached (second third of the DOT controller output), then the third closed-loop (in the example, vessel pressure) takes over, leaving airflow at a maximum value. Eventually the rate of replenishment equals the rate of oxygen depletion and DOT is controlled in the example with rpm at its maximum set-point, vessel pressure at its minimum, and airflow responding as the principal output of the DOT controller as a PID loop.

As well as control by oxygen replenishment, it is also possible to initiate dissolved oxygen control by "depletion." With the slave outputs set to relatively high output values, medium batched with little or no energy source present can have the dissolved oxygen tension controlled by the regulated feed of, say, carbohydrate. Under these circumstances the feed system becomes the cascaded slave output and only delivers glucose at a rate equivalent to the metabolic rate of the organism under fixed oxygen mass transfer limitations set by the closed-loop outputs (speed, air, and pressure).

13.6.1.7   Feed systems and antifoam

Most fermentor systems will be equipped with a mechanism for introducing various feed solutions under aseptic conditions. The method for achieving this will depend largely on the size and configuration of the fermentor itself. On smaller laboratory scale vessels, feeds will be controlled by the use of peristaltic pumps working on flexible walled tubing. On larger vessels such as those found in the pilot plant (> 10 to 20 liters), it is possible that purpose designed and constructed addition vessels independently sterilizable will deliver feeds using shot-wise additions of feed solutions.

Whatever the method of addition chosen it is very likely that feeds will be pulsed in some way and therefore the variables available for control are pulse width (i.e., size of liquid 'shot') and interval between pulses. In the author's laboratory it has been found that fixed pulse width or liquid volume shot size and variable intervals provide a satisfactory control option for the addition of feeds. The set-point for feed controller can be a number of shots in a specified interval of control or the feed controller can be cascaded onto a master controller such as pH or dissolved oxygen. The output of the feed controller can be linked to the interval such that the larger the output of the controller (further away from the desired number of shots) the shorter the interval and hence the more rapid the feed rate. Calibration of a feed system is difficult and will be affected by tubing age and vessel back-pressure. It is possible to introduce another sensor to take over the control of feed additions, i.e., a balance for the feed reservoir, this type of control loop for feeds can provide very accurate and nonpulsed control of feed flows, but costs may be high and it is only really applicable to smaller-scale vessels (< 20 liters).

Antifoam addition is variation of the feed control loop where the sensor is contact probe detecting rising foam. The control variables here may be time related. To prevent over addition of antifoam agent (which may destroy dissolved oxygen control by bursting air bubbles) a splash time interval can be specified whereby the contact probe would need to be covered for more than a few seconds to initiate antifoam addition. Similarly, if the probe remains covered following antifoam pumps being switched on, then secondary control action can be specified whereby the airflow rate and/or agitation speed may be reduced to minimize the generation of foam. Clearly, if this happens then the dissolved oxygen controller will be affected, but the output of the antifoam controller may be the ultimate master controller of closed loops to prevent loss of broth by foam-out.

13.7 Advanced incubation control

The control options specified above represent the basic control elements for most fermentation development purposes. However, many fermentation development programs will require more than just a system of fixed set-point control, and that is where we can return to some of the observations made at the start of the chapter. To close this chapter there follows a description of an advanced programmable logic controller in the author's laboratory that is capable of more advanced fermentation control options.

The wild-type micro-organism in its natural habitat (whatever that is) as previously discussed, is subject to transient environments. When a fermentation scientist attempts to grow the microbe and encourages it to express the desired attribute, a fermentation regime will be imposed that will only represent a tiny fraction of all the influences that the organism will encounter. Given that most fermentations will be run under fixed set-point regimes, the environment that is thus imposed will effectively be a 'snapshot' of the full range of conditions that have brought about the expression, through evolutionary pressure, of the desired phenotype. Another option for control therefore is to provide a mechanism whereby events beyond the fixed set-point control regime can be specified.

13.7.1 Fermentation profiles

A typical fermentation control profile is shown below (Figure 13.11). The main features of the profile are captured in Table 13.3. When analyzing the progress of fermentation, it is sometimes useful to identify the key features and in which phase of the growth of the organism they occur. For this purpose it is useful to use standard definitions for the growth phases although their strict interpretation is obviously open to debate. In the example given, the following features become apparent:

- Harvest time is a long time after product accretion has ceased (this is frequently the case with old processes that have been inherited).
- Carbohydrate uptake is largely linear over the growth phase of the organism and is depleted by 60 hours.
- Biomass accretion occurs between 0 and 60 hours and then following substrate depletion goes into the decline phase.
- Dissolved oxygen profile is the mirror image of the growth profile.
- Product accretion has its onset at 24 hours and is complete by 72 hours.
- Harvest time using this protocol can be at 72 to 84 hours.

Figure 13.11
Typical fermentation profile for a filamentous micro-organism producing a secondary metabolite.

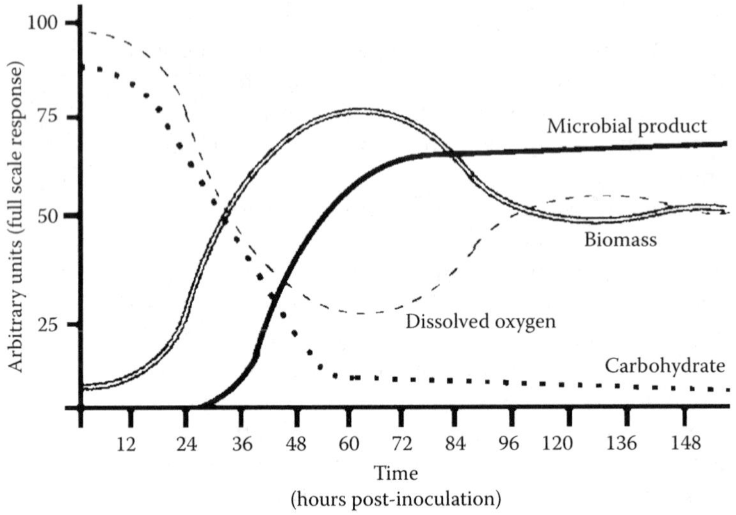

- Dissolved oxygen is not limiting; therefore, under this fixed set-point regime would higher biomass yield more product (increasing fermentor vessel volumetric productivity)?
- Carbohydrate is depleted by 60 hours, a further carbohydrate feed at either this point or when the DOT was less than 40% of saturation may promote a further product accretion phase.
- pH is not shown here, but the level is likely to fall between 12 and 60 hours and this may require pH control by titrant addition.

This type of profile is typical of the kind that would be obtained with fixed set-point control for the principal closed-loop controllers (not including DOT control).

13.7.2 *Event-tracking control*

As indicated above, with the carbohydrate feed option it is possible to identify an event in the fermentation which serves to trigger another control action. Hence an event is a specific change in state or time or any combination of changes that can initiate a new event. In the case cited, a carbohydrate feed could be initiated at 60 hours post-inoculation (time-based event) or if the dissolved oxygen measurement fell below 40% of saturation (analogue value event). Computer control systems, particularly those specified by the customer, can easily accommodate

Table 13.3 Fermentation profile data for key analytes at various growth phases ("snapshots")

Analyte	Time (h)	Value (arbitrary)	Growth phase
Biomass	12	10	lag
	24	25	acceleration
	36	50	exponential
	48	70	deceleration
	60	75	stationary
	84	60	decline
	148	50	harvest
Carbohydrate	0	85	inoculation
	24	70	acceleration
	36	35	exponential
	48	20	deceleration
	60	10	stationary
	148	5	harvest
Dissolved oxygen	0	100%	inoculation
	24	80%	acceleration
	36	40%	exponential
	48	30%	deceleration
	60	25%	stationary
Microbial product	24	0	acceleration
	36	10	exponential
	48	35	deceleration
	60	50	stationary
	72	60	decline
	148	65	harvest

program decision gates that will test the status of the fermentation and apply another control action on the fermentation if the decision gate criteria are met. To define what the decision gates should be, and what values of analyte or combinations of analyte values should be used, requires the fermentation scientist to establish set-points for control loops that he or she can control and then observe the effect on analytes that do not have on-line sensors (biomass, carbohydrate, and product in this example).

As soon as the fermentation scientist wishes to use event-based control, then further options may be sought to extend this capability. In the author's laboratory the PLC fermentor control system has been programmed with four types of user-definable events:

- Time-based events: become true at a specified number of hours post-inoculation.
- Analogue value events: become true when a process value or combination of two process values exceed a threshold limit.

- Elapsed time events: become true at a defined time after another event has occurred.
- Boolean events: logical combinations of any two other events using standard Boolean operators

The trigger events for the most part can be almost any process signal or event "flag" indicating the status of that event, in addition, system events or alarm levels can be specified (this apart from warning alarms is a very powerful use of alarms signals). The events can be organized in any combination giving virtually limitless fermentation control strategies. The events themselves can initiate new set-points, ramp rates (rates of change between set-points), or new events. It is possible to "latch" initiating events such that dependent events will remain true even if the initiating event is no longer true.

Figure 13.12 shows three graphs for event-based control. Graph 3 shows the status (true/false) of three fermentation process events. Graph 2 shows set-points for two output controllers (these could be any type of controller as discussed previously) to be driven by the specified events (1, 2, and 3). Graph 1 is the fermentation profile against which the new set-point control regime is to be imposed. A summary of the main changes is given below:

- Event 2 becomes true first (e.g., time value = 12 hours) and the set-point for controller #1 is changed from the default set-point (arbitrary value 25) to the event 2 set-point (50) control action initiated at end of lag phase.
- The rate of change from one set-point to the other can also be specified and in the graph it is comparatively slow — ramp rate control action over acceleration phase.
- At approximately 40 hours event 1 becomes true. The set-point of controller #1 changes from event set-point 2 to event set-point 1 (arbitrary value 75) with a faster ramp rate — control action at end of the exponential phase, start of deceleration phase.
- Events 1 and 2 are both true at this time. This initiates event 3 also becoming true, the set-point for controller #2 changes from arbitrary value 10 to 100 (could be a feed rate for example) — example feed rate response could be in response to deceleration phase.
- At approximately 50 hours event 1 then no longer is true and set-point controller #1 returns to event set-point 2 level (50). Note event 3 set-point is latched and remains true for the remainder of the batch — start of stationary phase.
- At approximately 70 hours both events 1 and 2 are no longer true and the default set-point for controller #1 returns to the default level of 25 (approximately 84 hours) — 70 hours end of product accretion phase.

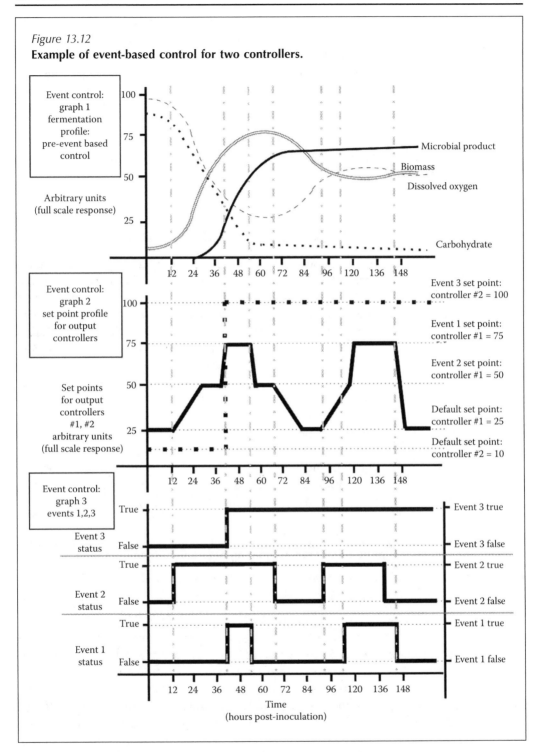

Figure 13.12
Example of event-based control for two controllers.

• At approximately 90 hours event 2 becomes true again, and the event 2 set-point (50) for controller #1 is used. Before the event 2 set-point is reached, event 1 becomes true and the event 1 set-point (75) is used again (faster ramp rate). Event 1 set-point is in use for controller #1 by 120 hours — recovery from decline/death phase by 90 hours (scavenged substrates?).

- At approximately 140 hours event 2 is no longer true, and a short time later event 1 is no longer true; at this point the controller #1 set-point is reduced from 100 to 25 (default set-point) — harvest time.

The results of this event-based control will be judged by comparing the fermentation profile (as shown in Figure 13.12: Graph 1) with the one generated for the event control regime.

13.7.3 Boolean control and rule generation

It may be seen from the example in Figure 13.12 that complex patterns of control can be imposed on the fermentation and more or less significant changes to the growth environment and expression of phenotype will follow. It is, however, difficult for the fermentation scientist to know which fermentation control regime to impose. This lack of knowledge comes from not knowing how the organism will respond to changes in individual control loops but also what the interactions with other control loops will be. To address this, Boolean logic can be employed to introduce an element of control by "choice." With Boolean logic it is possible to present options or choices from which the control system can select a preprogrammed path for control. This type of experiment is then generating a new kind of response variable, one based on the control path selected by the metabolism or response of the total fermentations system, where:

Fermentation system = stainless steel vessel + valves + sensors + services (air, electricity, etc.) + medium + micro-organism + microbial metabolism + expressed phenotype

The five Boolean operators that can be used are AND, OR, XOR, NAND, and NOR. On their own these terms are rather impenetrable. Table 13.4 summarizes what they mean; examples given all represent truth statements affecting just two fermentation events — a pH value and a time value.

If Boolean options or choices are presented to the system, then the path selected by the total system can represent a "Rule" by which the response of the micro-organism to an imposed environment can be described. To illustrate the principle of control by rules, Figure 13.13 shows a simple XOR rule statement.

The rule defined in Figure 13.13 is remarkably simple and helps to define glucose feed regime by pH and dissolved oxygen control. Having established this rule a fermentation experiment

Table 13.4 Boolean truth table for fermentation control

Boolean operator	Event 'a' (pH > 6.5) status	Event 'b' (time < 24 h) status	Comments (example event combinations)
AND	Yes	Yes	
OR	Yes	No	
	No	Yes	
	Yes	Yes	
XOR	Yes	No	
	No	Yes	
NAND	No	No	
	No	Yes	
	Yes	No	
NOR	No	No	

can then test which route is chosen under defined conditions of medium or mass transfer for oxygen, and so on. Another powerful use for control by rules comes with culture or mutant evaluation where the chosen route may indicate a propensity towards one pattern of metabolism or another, which may have been induced within a putative mutant culture. Clearly this is just one of many rules that could be introduced, and libraries of rules or elements that make up rules can be constructed to provide an array of control options to explore microbial metabolism under controlled conditions.

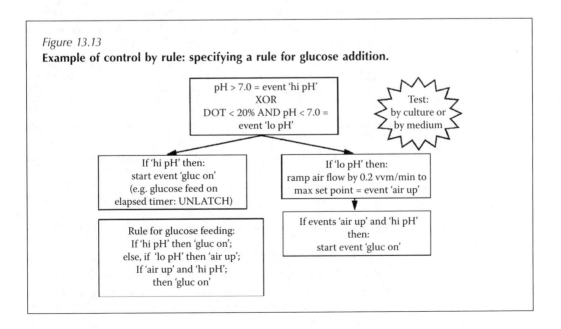

Figure 13.13

Example of control by rule: specifying a rule for glucose addition.

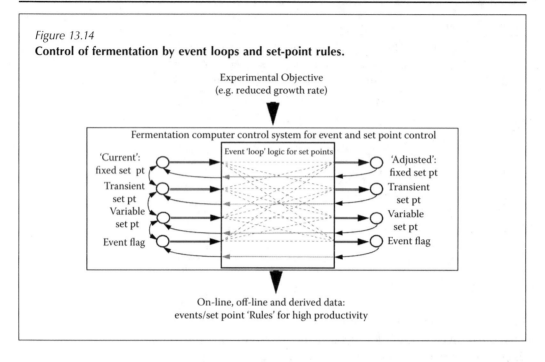

Figure 13.14

Control of fermentation by event loops and set-point rules.

13.7.4 *Summary of event and nonstable set-point control*

A summation of the possible interactions for event-based control is shown in Figure 13.14; here control options are defined as possible inputs and outputs to an experimental system where events become control loops themselves evaluating whether conditions defined by rules are satisfied. This type of control system is an extension of the standard media and control recipes normally prepared and fermentation response data will produce a data set of on-line, off-line, and derived data that generate rules for high productivity.

Returning to some of the opening discussion points in this chapter, the fermentation scientist is being asked to establish the correct conditions for a micro-organism to express a desirable attribute or phenotype in a totally artificial environment; event and rule-based control together with transient and variable set-point control (nonstable) may help establish what unique set of factors and their interactions support the expression of a rare and valuable phenotype.

13.8 Other advanced fermentation control options

This chapter has focused on a tiny part of control technology available for fermentation systems. The reader should be aware that other control philosophies are in use and this technology is certain to continue to change and develop. Among some of the control philosophies currently in use or in advanced development are described below.

13.8.1 Knowledge-based systems (KBSs)

These systems can be considered an extension of the sort of control capability described previously where previously held "expert knowledge" is captured in a control system often referred to as an expert system. With this type of control already established, facts and data associated with a process can be held in a database of knowledge against which future decisions can be made. The principal difference with expert systems and the event-based control described here is that knowledge-based systems will only work on existing knowledge and won't necessarily generate new knowledge. The power of expert systems lies in their ability to execute potentially hundreds of rules per second controlling many functions associated with plant operation simultaneously. Early warning patterns can be recognized from this previous knowledge to alert either human or computer control operations of an excursion from normal operating parameters.

13.8.2 Artificial neural networks (ANNs)

Much discussion has been made in recent years of "intelligent" computer systems. By intelligent systems we mean systems capable of learning. The mechanism by which this occurs is beyond the scope of this chapter but for the most part is very similar to the control loop principle whereby an input to a system is modified by attempting to minimize the error between desired and actual outcome. The use of neural networks in fermentation will come from the ability of these systems to recognize patterns and take actions on unknown data based on those learned patterns. It is possible that the event-based system described here will be augmented by neural net ideas to help with obtaining data or process inferences not available by direct measurement. ANNs do not in themselves assist in providing data on the exact nature of process input interactions because the training and optimization of an ANN is a pure black box model of the process.

13.8.3 Genetic algorithms (GAs)

In this case control philosophy has been influenced by modern biological genetics. Again it is not possible to describe how these systems work here, but the principle that should be understood is that Gas are capable of searching vast areas of experimental space and by a process of natural selection of rules or algorithms that come closest to minimizing some process cost function tend to optimize on a process goal. This type of system may assist in rule induction by eliminating rules not fit for the process environment.

13.8.4 *Modeling*

For many years there has been a disparity between what modeling could apparently offer (see Chapter 7 for more details) and the acceptance and use of modeling by the process community for everyday control and optimization problems. This may now be changing with the advent of advanced user-friendly software to assist in forming and testing model predictions. There are many kinds of models that may be constructed, in attempting to describe a complex system the key to approximation is simple models, i.e., relating carbon dioxide evolution to biomass or growth estimations. These types of relationships can be defined relatively easily by curve fitting raw data and then predicting on the basis of alteration of parameters defined in function generated from curve fits. When several functions describing microbial growth and productivity have been created, then they can be combined in a mathematical relationship potentially useful in predicting fermentation outcomes. If models are successful, then thousands of fermentations can be run in software before committing to expensive stainless steel vessels.

Editorial update

13.9 Recent trends in fermentation Control

Advances in several fields have allowed new possibilities in the areas of measurement and control:

- New sensor technologies such as the use of fluorescence. This has allowed also shrinking of sensor systems for use a wide range of subfermentor applications, e.g., in microtitre plates.
- Use of Direct Digital Control (DDC) by a PC for instrumentation/ links to external software by, e.g., Object Linking and Embedding (OLE).
- Increasing use of common communication standards such as Modbus and OPC to quickly and easily expand measurement and control options.

13.9.1 *New sensor technology*

The ability to measure yield of product directly on line has long been a fondly held objective of fermentation technologists and researchers alike. Use of new sensors (see Chapter 11 for more details) with simple optical fiber connections that fit in the fermentor vessels and detect changing amounts of fluorescence in a dye make this possible. Light of a specific wavelength is used to excite the fluorescence, which can then be measured fluorimetrically.

Linking the expression of the desired protein with a fluorescent detection system can provide direct, on-line measurements related to yield. A simple analogue output signal from the detector system allows integration of the new data with process control software. This, in turn, can modify other measured parameters to maximize production.

This system can also be used as a new way to detect standard parameters, e.g., pH, dissolved oxygen and dissolved carbon dioxide concentration. The fluorescent detection systems can be incorporated in the sensor probes or separately as tiny "patches." These can be placed in simple disposables such as microtitre plate wells or added to incubation flasks. This enables rapid screening based on small volumes with some indication of the prevailing environmental conditions in the culture. This data can influence judgments about which isolates to use for a particular process and generate data of potential use in scale-up.

Another factor aiding real time measurement of novel process parameters is the development of simple micro-filtration systems to allow separation of cells from the supernatant culture medium which can then be taken to an external analyzer. In this case, the separation system is key, rather than the analytical processes. The common feature, however, is the acquisition of data in real time to inform control profiles designed to optimize yield and/or extend productivity in terms of metabolites or biomass.

13.9.2 Expansion of the capability of DDC instrumentation

Virtually all commercial fermentors now use control instrumentation (see Chapter 14 for more details) based on Direct Digital Control (DDC). A few manufacturers have developed systems with all control functions directly handled by the computer. This has some advantages or additional of new capabilities and library routines if a standard process control software package is used. Also, modern computers have more than adequate processing power to cope with real-time process control. The downside of this approach is, of course, the potential for a computer failure leaving the fermentor with no control whatsoever.

A "halfway" position is to allow external software to interact directly with supervisory process control software. Calculation of some control characteristics is made externally and the result passed to the fermentor software as, e.g., a revised set-point for one or more parameters. Windows*-based operating systems allow for just this process through Object Linking and Embedding (OLE). A short script in Visual Basic*, for example, could link supervisory software with an Excel* spreadsheet making calculations and generating set-points in real time based on fermentation data.

This introduces another layer in the control process but has the advantage of one or more independent microprocessors providing control at the fermentor.

An increasing trend regarding usability is the use of color touch screens, even for bench-scale fermentors. They typically use a graphical user interface and incorporate some of the features previously only found in PC-based software, such as trend graphing. Alternative approaches involve enhancements to navigation between menus and the use of tabs for rapid access to options. Often, the more advanced functions of the controller, such as configuration of PID loops, are now placed in restricted sections where the casual user cannot accidentally make fundamental changes.

13.9.3　*Use of common communication protocols*

A general trend in fermentation control is to add more peripheral items such as balances, pumps, and additional gases, almost irrespective of the scale of the process. A number of manufacturers of components such as mass flow control valves and peristaltic pumps are offering them for use with the Modbus serial protocol. This allows for great flexibility in adding these items to a fermentor as the physical connection and firmware changes are very easy to implement. As the Modbus protocol is a genuine common standard allowing the potential range of equipment, which can be added to a standard fermentor to be greatly increased.

The trend for interactivity is also shown at a higher level of process control. Fermentor instrumentation and supervisory software can be linked to other devices or programs using the OLE for Process Control (OPC) protocol. This requires each device has a driver written in its firmware or software and this allows two-way communication with any other enabled systems. Both client and server systems are catered for within the specifications of the OPC standards.

The potential use for this system in industrial process control of fermentation is to realise another long-held desire – linking of upstream and downstream processes together with the actual incubation phase in the fermentor for optimization which is truly process-wide and fully integrated. Of course, Microsoft Office* applications are OPC enabled so, for example, a controller could transfer data directly to an Excel* spreadsheet via OPC.

Summary

- This chapter has attempted to elucidate the complexity of fermentation systems and the difficult task facing the fermentation scientist who has to elucidate the key features of a fermentation to significantly enhance microbial productivity with a minimum expenditure of resource.
- Fermentation development has traditionally used a heuristic approach to refine empirical knowledge.
- The requirement for empiricism is still very much present, but modern fermentation control systems and techniques should allow for a more systematic approach to fermentation optimization (see Chapters 6 and 14 for more details).
- It is likely that modern computing techniques such as event-tracking and rule-based controls will augment the application of microbial physiology to solving problems of applied microbiology.
- Microbial physiology knowledge itself in turn may benefit from advanced control ideas by increasing understanding how a micro-organism can successfully respond to its environment.

Acknowledgments

The update of this chapter was kindly carried out by Dr. Tony Allman, an associate editor.

Suggested reading

Bailey, J. E., D. F. Ollis (1986). *Instrumentation and Control, in Biochemical Engineering Fundamentals*, 2nd edn, Maidenhead: McGraw-Hill, pp. 658–722.

Beluhan, D., D. Gosak, N. Pavlovic, M. Vampola (1995). Biomass estimation and optimal control of the fermentation process, *Comp. Chem. Eng.*, 19(Suppl.), 387–392.

Blackmore, R. S., J. S. Blome, J. O. Neway (1996). A complete computer monitoring and control system using commercially available, configurable software for laboratory and pilot plant *Escherichia coli* fermentations, *J. Ind. Microbiol.*, 16, 383–389.

Chen, W., C. Graham, R. B. Ciccarelli (1997). Automated fed-batch fermentation with feed-back controls based on dissolved oxygen (DO) and pH for production of DNA vaccines, *J. Ind. Microbiol. Biotechnol.*, 18, 43–48.

Diaz, C., P. Dieu, C. Feuillerat, P. Lelong, M. Salome (1995). Adaptive control of dissolved oxygen concentration in a laboratory-scale bioreactor, *J. Biotechnol.*, 43, 21–32.

Gregory, M. E., P. J. Keay, P. Dean, M. Bulmer, N. F. Thornhill. (1994). A visual programming environment for bioprocess control, *J. Biotechnol.*, 33, 233–241.

Kong, D. Y., R. Gentz, J. L. Zhang (1998), Development of a versatile computer integrated control system for bioprocess controls, *Cytotechnology*, 26, 227–236.

Kurtanjek, Z. (1994). Modelling and control by artificial neural networks in biotechnology, *Comp. Chem. Eng.*, 18(Suppl.) 627–631.

Omstead, D. R. (ed.) (1990). Computer *Control of Fermentation Processes*, Boca Raton, USA, CRC Press.

Onken, U., P. Weiland (1985). Control and optimisation, in Rehm, H.-J. and Reed, G. (eds.) *Biotechnology*, Vol. 2, Weinheim: VCR Verlag, pp. 787–806.

Romeu, F. J. (1995). Development of biotechnology control systems, *ISA Trans.*, 34, 3–19.

Sys, J., A. Prell, I. Havlik (1993). Application of the distributed control system in fermentation experiments, *Folia Microbiol.*, 38, 235–241.

Wang, H. Y. (1986) Bioinstrumentation and computer control of fermentation processes, in Demain, A.L. and Solomon, N.A. (eds.) *Manual of Industrial Microbiology and Biotechnology*, Washington DC: ASM, pp. 308–320.

14

Modeling, Software Sensors, Control, and Supervision of Fermentation Processes

Boutaieb Dahhou,
Gilles Roux and Y. Nakkabi

14.1 Introduction

Fermentation processes are dynamic, i.e., nonlinear and non-stationary, in nature, thus leading to many difficulties in modeling. In addition, the lack of specific sensors makes certain variables inaccessible by on-line estimation. Measuring the state variables is difficult, especially on-line, and its reliability is uncertain. The difficulties encountered by the control engineer, when studying this type of process, result from the fact that the process model is based on nonlinear algebraic relations and differential equations. These difficulties make the biotechnological process an excellent field for application of advanced automatic control tools (Dahhou et al., 1991a). All available information on the process, e.g., environmental variables (temperature, pH, stirring, etc.) and control loops signals, must be exploited to ensure that the process is operating reliably under the given conditions. In this way and for fermentation processes, supervision becomes an issue of primary importance to increase the reliability, availability, and safety of these systems. Unfortunately, the lack of process knowledge, the absence of reliable sensors, and unexpected behavior of micro-organisms make this task very difficult, if not impossible, to be fulfilled by the human operator. Several schemes of supervision applied to various domains have been proposed (Aguilar-Martin, 1996; Antsaklis and Passino, 1993; Dojat et al., 1998). However, from a general point of view, the architectures presented in the literature are similar and globally can be described by the scheme shown in the Figure 14.1. According to (Kotch, 1993), two main tasks

Figure 14.1
Supervision scheme.

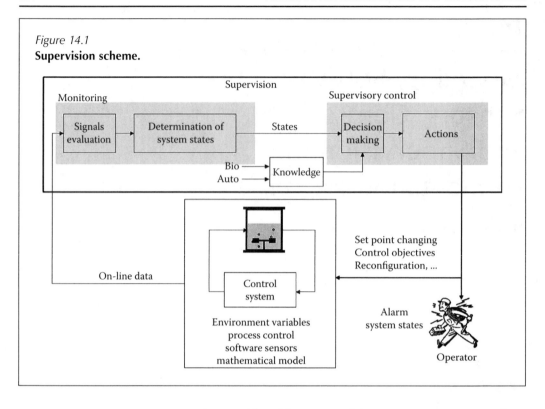

are to be considered in a supervision system: process monitoring and supervised control. The process monitoring part utilizes data collection and signal processing in order to decide firstly, if the process is in an abnormal state, and secondly, if a corrective action must be taken. Such intervention, along with diagnosis, is regarded under the supervised control task.

Within this framework, the development of an integrated methodology for the control of the fermentation processes results in approaching the following aspects:

- Modeling
- Adaptive techniques
- Supervision for process control

In this section, the integrated methodology of the process control is formulated by adding a block of supervision in which we exploit all information resulting from the modeling, the estimation, the observation, the control loop, the environment and the expert knowledge of the processes.

14.2   The model system

A commercial strain of *Saccharomyces cerevisiae* UG5 was used, growing on mineral salts glucose medium. The composition of

Table 14.1   Composition of growth medium

KH_2PO_4	$3 \ g.dm^{-3}$
$(NH_4)_2SO_4$	$3 \ g.dm^{-3}$
Sodiumglutamate	$1 \ g.dm^{-3}$
$Na_2HPO_4.12H_2O$	$3 \ g.dm^{-3}$
$CaCl_2.12H_2O$	$0.25 \ g.dm^{-3}$
$MgSO_4.7H_2O$	$0.25 \ g.dm^{-3}$
$ZnSO_4.7H_2O$	$5 \ mg.dm^{-3}$
$(NH_4)_2FeSO_4.6H_2O$	$1.5 \ mg.dm^{-3}$

Table 14.2   Operating conditions of the variable reactor types used

	Batch	Fed-batch	Continuous
Temperature (°C)	30	30	30
pH	3.8	3.8	3.8
Working Volume (dm³)	1.5	6–16	1.15
Stirrer Speed (rev.min⁻¹)	200	300	200
Aeration (dm³.h⁻¹)	3	32	3
Flow rate (dm³.h⁻¹)	—	0–2	0.01–0.36
Inpute Substrate (g.dm⁻³)	—	160	20–300

the growth medium is listed in Table 14.1. The carbon source was glucose monohydrate (cerelose). The bioreactor was filled with tap water and the medium steam sterilized at 120°C for 20 min in the bioreactor. After cooling, the vitamin solution was added aseptically by filtration. The organism was cultivated in $2dm^3$, $20dm^3$, $2dm^3$, and KH_2PO_4 bioreactors for, respectively, batch, fed-batch, and continuous fermentations. The bioreactors were equipped with instrumentation to control pH, temperature, and stirred speed. Aqueous ammonia was used for pH adjustment.

For fed-batch and continuous modes, fresh medium was supplied to the reactor when and as required. The input flow rate of nutrient was controlled by means of a peristaltic pump and calculated from the glucose concentration measurement. For continuous mode, a level sensor was used to maintain the bioreactor at constant volume. The environment variables, such as temperature, pH, and stirred speed, were regulated at the specified set-points with hardware controllers. Typical operating conditions and parameters of the experimental process are summarized in Table 14.2.

14.2.1   Off-line measurements

Samples were taken during the fermentation for the determination of substrate, ethanol, and biomass. Glucose concentration was measured with an industrial enzymatic analyzer YSI 27A (Yellow Springs Instrument Co, Ohio), which consisted of an immobilized glucose oxidase membrane with an oxygen sensor. Ethanol concentration was determined by gas chromatography using isopropanol as an internal standard. Biomass concentration was estimated from dry weight and turbidity measurements at 620 min.

14.2.2 *On-line measurements*

On-line glucose concentration was determined using the YSI enzymatic analyzer (Queinnec et al., 1992); a set of pneumatic jacks around a calibrated syringe ensured sampling, injection into the glucose analyzer, and rinsing of the syringe before the next measurement. By first selecting the syringe travel, the working scale for glucose measurement could be set; such a system is capable of detecting up to 20–300 glucose could be measured with an accuracy of approximately 95%. A programmable logic computer (PLC) was used to manage the procedure for rinsing of the enzyme membrane, rinsing of the syringe, and injection of the sample and control of the glucose measurement.

14.3   Modeling

The unstructured model is based on the observation of the macroscopic kinetics within the reactor. We consider that the biomass activity is sufficiently specified only by one variable. Generally the micro-organisms concentration in the medium and the biomass composition changes are completely ignored. These models are generally based on Monod equation (Monod, 1942) or equations of the same type including the various enzymatic reactions. The constants appearing in these mathematical equations are empirical and often determined by optimization based on experimental data. Other equations speed describing consumption of the substrate and the production of the products are necessary to write the model.

In our case, the process behavior is described by a model developed by using a fuzzy classification method (Piera et al., 1989). This type of model does not require any mathematical knowledge of the process. It builds a representation of the process in physiological states form. The physiological states of a biological process can be defined as being a lasting interval which the principal physiological variables remain constant. This model does not try to describe the evolution of the process variables but to determine the current situation based on these variables. The goal of this methodology is to design a process model in which supervision and diagnosis can be exercised effectively.

14.3.1 *Unstructured models*

14.3.1.1 *Methodology*

We used our own software, which was developed under MAT-LAB (Bâati et al., 2004). To solve the system of nonlinear algebraic equations representing the culture, the Gauss-Newton method with a mixed quadratic and cubic line search procedure was applied. For numerical integration, Runge-Kutta algorithms with different order according to the difficulty of the modeling equations were used (which check for integrability, and thus prevents the frequent numerical problems). The optimization runs were carried out using a multitask Pentium computer.

As an example of learning process (by learning process we mean a heuristic search strategy that allows us to select the most appropriate model from those models stored in a data base) a fermentation process, which is nonlinear, can be modeled by the following dynamic equations.

$$X(t) = \Omega(X(t), u(t), \eta(t))$$
$$Y(t) = HX(t)$$

(14.1)

where $X(t)$ is the state vector generally including biomass, substrate, and product concentration; $Y(t)$ is the observation vector, which can be measured; $u(t)$ is the input vector, which can be used to take into account the effect of environmental variables; $\eta(t)$ is the kinetic vector, which contains the main biological parameters of the fermentation reaction. It is known that $\eta(t)$ is composed of complex functions of the state variables and of several biological constants; its expression is varied for different fermentation processes. Therefore, the primary task of modeling is to identify which model of $\eta(t)$ is suited to the specificities of a process and then to determine the corresponding biological constants. A set Ξ of models of $\eta(t)$ is stored in the database: $\Xi = \{f_1(\theta_1, X(t)), f_1(\theta_1, X(t)), \cdots\}$ where, θ_i is the unknown parameter vector (biological constants). The minimization of the criterion between the output of the model $Y^m(t)$ and the output of the process $Y(t)$ allows the best match parameter vector θ_i^* for i selection of the model of $\eta(t)$ to be obtained.

$$J_i = \min_{\theta_i \to \theta_i^*} \int_0^{t_f} (Y^m(t) - Y(t))^T Q(Y^m(t) - Y(t))$$

(14.2)

The parameter adjustment rule facilitates this task, and an optimisation algorithm can be used. The minimization of the criterion $J = \min_{k \to k^*}\{J_k\}$ allows the suitable model k^* and the corresponding parameter vector θ^* for the real fermentation process to be obtained, i.e., $\eta(t) = f_{k^*}(\theta^*, X(t))$. A heuristic search strategy from the database containing the different models of

$\eta(t)$, allows the "model match" task to be realized, that is, from any k to k^*.

Fermentation processes are characterized by biological degradation of substrate (glucose) S by a population of micro-organisms (biomass) C into metabolites, such as alcohol (ethanol) P. The physical model of the process is usually described by a set of nonlinear differential equations derived from the material mass-balances and involving modeling of the growth rate. These equations are

$$\begin{cases} \dfrac{dC(t)}{dt} = \mu(t)C(t) - D(t)C(t) \\[2mm] \dfrac{dS(t)}{dt} = -\dfrac{1}{Y_{c/s}}\mu(t)C(t) + D(t)S_{in}(t) - D(t)S(t) \\[2mm] \dfrac{dP(t)}{dt} = -\dfrac{Y_{p/s}}{Y_{c/s}}\mu(t)C(t) - D(t)P(t) \end{cases} \quad (14.3)$$

In a batch fermentation, where all nutrients are initially introduced into the bioreactor the dilution rate $D(t) = 0$.

In fed-batch fermentation, fresh nutrient in the culture medium is added as and when required. Here, the dilution rate $D(t)$ is given by:

$$\begin{cases} D(t) = \dfrac{F(t)}{V(t)} \\[2mm] \dfrac{dV(t)}{dt} = F(t) \end{cases} \quad (14.4)$$

The specific growth rate $\mu(t)$ is a function of the process state and several biological parameters $\theta(\theta \in R^n)$. These parameters are time-varying and depend on the environmental conditions, which are then held fixed.

The usual approach in bioprocess modeling adopts particular analytical structure for specific growth rate and calibrates the kinetic coefficients from experimental data. However, this modeling is often hazardous because the reproducibility of experiments is often uncertain as the same environmental conditions may be difficult to obtain to prevent changes in the internal state of the organism.

Many analytical laws have been suggested in the literature for specific growth rate modeling which take into consideration the limitation and/or inhibition of the growth by certain variables of the process.

14.3.1.2 Model validation

We propose to seek a model of the specific speed of growth able to take into account specificities of alcoholic fermentation,

Table 14.3 The process model parameters

μ_m	$0.38h^{-1}$
K_s	$5 \ g.dm^{-3}.h^{-1}$
$Y_{c/s}$	0.07
$Y_{p/s}$	0.16

remaining mathematically simple. This model will be used in the design of estimation, control, fault-detection and isolation algorithms for the automatic control of the fermentation process. The selected model is:

$$\mu(t) = \mu(\theta, S) = \mu_m \frac{S(t)}{K_s + S(t)} \tag{14.5}$$

The specific growth rate is of the Monod type. The influence of the substrate concentration on the growth is translated by a time limitation of a substrate lack. Specific rates of degradation and production are coupled with the growth by yield coefficients. In this choice, a simple model in which the parameters are easy to determine is privileged. Experimental data were used to calculate the numerical values of the parameters μ_m, K_s, and the yield coefficients $Y_{c/s}$ and $Y_{p/s}$. The results obtained are given in Table 14.3.

The determination of the model is carried out by the research of the best minimization of the criterion (14.2) by using experimental data available of fermentation in discontinuous mode. The discontinuous mode is much richer in kinetic information than the continuous one, which provides only stationary states. The obtained results by application of the selected model on an experiment are presented in Figure 14.2. This figure represents the comparison of the measured values and those given by the model. These results show that the selected model simulates correctly the process through the experimental data. The validation of the model parameters was carried out by using other experiments. This model (cf. Eqs. 14.3 and 14.5) will be used in others sections for the development of estimation, control and fault detection and isolation algorithms.

14.3.2 *Behavioral models*

14.3.2.1 *Methodology*

The complexity of this task imposes the combination of the classification techniques and the expert knowledge. The process states, their causal relations, and the transition conditions are identified starting from classification. However, expert knowledge is necessary to validate the results in agreement with the nature of the process. It is possible, starting from expert knowledge, to give a valid semantics to the model of supervision.

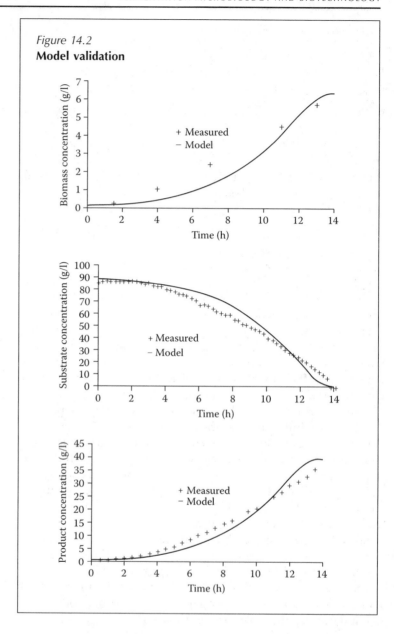

Figure 14.2
Model validation

Suggested methodology uses classification under the supervision of the expert to obtain a process model. It is based on iterative application of fuzzy techniques. The objective of the method is to identify a set of significant states for the expert. It can be summarized by the following stages as shown in Figure 14.3 (Waissman-Vilanova et al., 1999; Waissman-Vilanova, 2000).

In the beginning, given a set of measured data, no knowledge on the process states is available (i.e,. all data correspond to the same general state). An unsupervised learning of data is applied and a set of classes is obtained. Then, the expert must map the set of classes to a set of physiological states. Three situations are

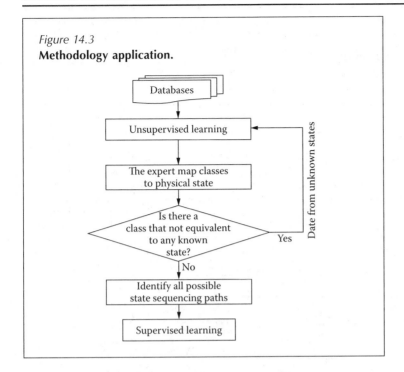

Figure 14.3
Methodology application.

possible: one class is equivalent to a physiological state, a set of classes is equivalent to a physiological state or any class is equivalent to any known state. If there exists a class in the last situation, we apply a new unsupervised learning considering just the data assigned to that class. This procedure is called data refinement. When all data are classified in known states, all possible sequencing paths of the state are identified. This is accomplished by looking on every possible temporal state orderings observed on all data available.

14.3.2.2 *Application*

The database considered here is constructed by real data extracted from batch functioning mode of biotechnological process. Several signals are available by on-line measurement. Among them, the expert chooses a subset of four signals which contain the most relevant information to determinate the physiological state in the process.

- Percentage of dissolved oxygen pO_2
- pH
- Percentage of rO_2 in output gas
- Percentage of rCO_2 in output gas

For the methodology application, four data sets records are considered. Three data sets are considered for learning and

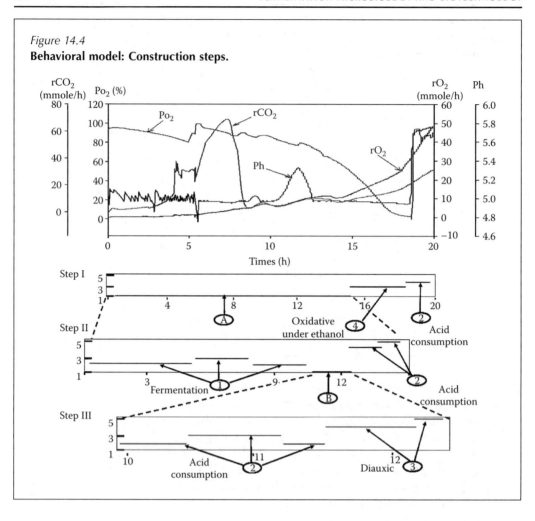

Figure 14.4
Behavioral model: Construction steps.

one data set for testing the results obtained by the methodology, all the physiological states are determined by the expert with the help of off-line measurements analysis (i.e., intra- and extracellular analysis). Figure 14.4 presents the different steps for extracting a process behavioral model. So, in step I, with an unsupervised learning of all the data available we obtain a set of three classes, and two physiological states are identified: consumption of acid (2) and oxidative under ethanol (4), the class (A) does not present a physiological state but a set of several states. In order to refine the data in class (A), an unsupervised learning is used (Step II). Among the six extracted classes the expert identifies two states. The state one (1) (fermentation) is identified by union of three classes and state (2) (acid consumption) by union of two classes. In step III, the set of data in (B) is used. With means of an unsupervised learning, the states (3) and (2) are recognized. The direct observation of all state sequencing leads us to build an automaton structure representing the

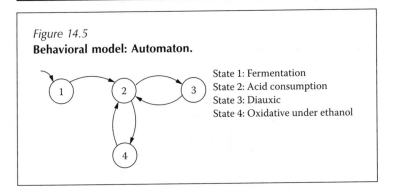

Figure 14.5
Behavioral model: Automaton.

State 1: Fermentation
State 2: Acid consumption
State 3: Diauxic
State 4: Oxidative under ethanol

Table 14.4 Behavioral model: Transitions

Transitions	Conditions
Transition 1 to 2	$-1.037*pH+0.118*PO2-0.008*rO2>0$
Transition 2 to 3	$1.082*pH-0.104*PO2+0.04*rO2+0.001*rCO2>0$
Transition 3 to 2	$-0.88*pH+0.018*PO2+0.001*rO2-0.05*rCO2>0$
Transition 2 to 4	$1.33*pH-0.21*PO2+0.06*rO2+0.067*rCO2>0$
Transition 4 to 2	$60.13*pH+17.59*PO2-0.027*rO2-0.027*rCO2>0$

process model (Nakkabi et al., 2002) as shown in Figure 14.5. In the Table 14.4, the transitions conditions are presented.

14.4 Adaptive techniques

The achievement of a good control and an effective monitoring of fermentation processes require the availability in real time of information on the physiological states of the process. However, the internal operating and the bioprocesses dynamics are still badly understood and many methodological problems of modeling are to be solved. In the laboratory, certain variables can be evaluated, using techniques of off-line analysis, which, for the majority, require too long a time of treatment for controlling in real time the biological reaction. Moreover, the market of the instrumentation of bioprocesses suffers from a crucial lack of reliable and inexpensive direct sensors.

The nonstationary character of these processes encourages elaboration of methods of adaptive control order, linear or nonlinear, making it possible to represent nonlinearity of the model, but also to take into account the variations of their parameters. It's the nature of the studied processes that led us to develop algorithms that exploit the inherent nonlinear structure of their models.

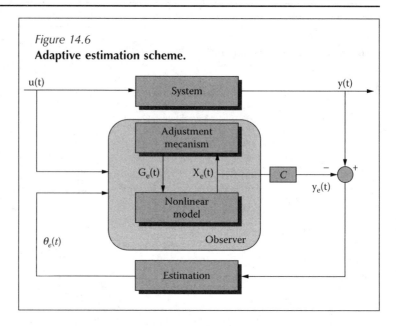

Figure 14.6
Adaptive estimation scheme.

14.4.1 *Estimation and software sensors*

In this problematic context, the techniques of adaptive filtering and estimate, commonly called "software sensors," seem to be an alternative impossible to circumvent (Ben Youssef, 1996; Ben Youssef et al., 1996; Nejjari et al., 1999a, 1999b). Their objective is to rebuild the unavailable variables by using the on-line measurements of off-line experimental data and the physico-chemical model of the process.

The nonlinear and nonstationary character of these processes, bringing into play the living micro-organisms, constitute a considerable limitation of the performances obtained by diagrams of estimate based on a linear approach. Being given the strongly nonlinear character of the process, we choose the use of nonlinear adaptive techniques based on the structure of the model of this type of processes. The used model for developing estimation and software sensors algorithms is that obtained from experimental data as explained in the modeling section. As mentioned above, the two principal problems that appear in the control of the fermentation processes are the inaccessibility of certain biological variables and the important temporal variability of the kinetic parameters. Our motivation is to know how to use measurements and the equations of modeling to minimize the errors introduced by the measurements noises. For that, we use the techniques of filtering, which take into account the measurements available (substrate concentration) and the information, resulting from unstructured models. Thus, we make a joint estimate of states and parameters. The diagram of the developed algorithm is represented by Figure 14.6.

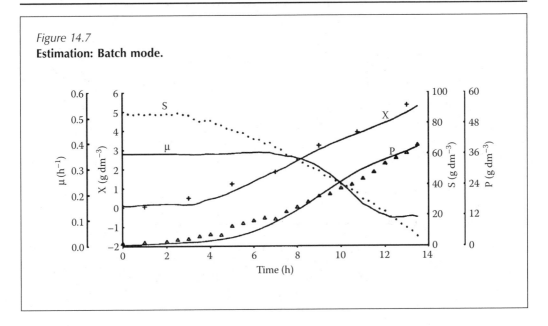

Figure 14.7
Estimation: Batch mode.

The software sensor proposed uses the available data to rebuild the evolution of the state variables and/or the estimated kinetic parameters.

The estimator described in (Zeng et al., 1993b) is applied to three fermentations under various type of operations (batch, fed-batch, and continuous modes). Figure 14.7 shows the estimated biomass concentration, the estimated product concentration, and the estimated specific growth rate obtained by the adaptive estimation algorithm. The estimation results are satisfactory compared with the off-line analysis (symbols) of biomass and product concentrations.

Controlled fed-batch fermentation is shown in Figure 14.8. The substrate feed flow rate controlled to regulate the substrate concentration in the reactor at $44g.dm^{-3}$. The input substrate concentration $S_{in}(t)$ was set at $160g.dm^{-3}$. The estimation results of biomass concentration, product concentration, and specific growth rate are given in Figure 14.8. Good agreement was found between the on-line estimations and measurements obtained by off-line analysis (symbols).

The results obtained from a controlled continuous fermentation process are given in Figure 14.9. In this application, the substrate concentration $S(t)$ and the biomass concentration $C(t)$ were controlled by manipulating the dilution rate $D(t)$ and the input substrate concentration $S_{in}(t)$. The evolution of estimated biomass concentration, product concentration, and specific growth rate are illustrated in Figure 14.9. These estimated results are again satisfactory, as shown by the off-line biomass analysis (symbols).

Figure 14.8
Estimation: Fed-batch mode.

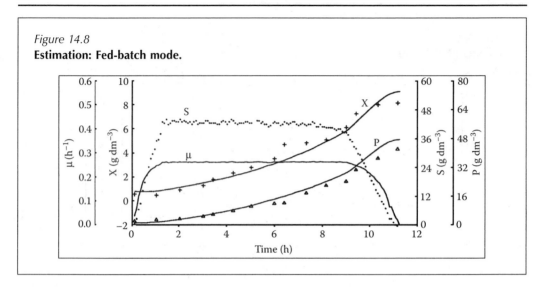

Figure 14.9
Estimation: Continuous mode.

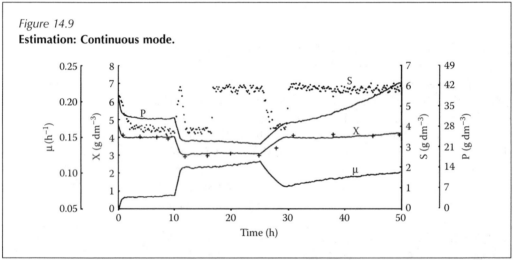

14.4.2 *Control*

In this section, we present results obtained from the linear
approach where we used the nonlinear model of the fermentation
process operating in continuous mode (see modeling section) to
determine the synthesis parameters of the linear controller.

We used the approach of the indirect adaptive control scheme
(Dahhou et al., 1991a,, 1991b, 1991c); the system parameters
are estimated directly, as shown in Figure 14.10. These estimates
are used for the readjustment of the regulator parameters. In
this type of algorithm, the adaptation loop comprises an identification block of the inaccessible parameters. These parameters
are considered by the control law as if they were true parameters. This manner of proceeding is founded in the certainty
equivalence principle.

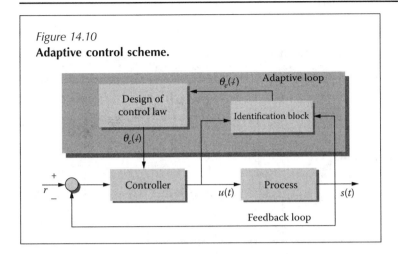

Figure 14.10
Adaptive control scheme.

We have chosen to present the experimental results obtained from the control of alcoholic fermentation in continuous mode. The principal objective is the regulation and the tracking of the substrate concentration inside the bioreactor as the dilution rate changes (Dahhou et al., 1993). The control objective is the minimisation of a quadratic criterion in a receding horizon sense with a "partial state model reference control" strategy (M'Saad et al., 1990). The control law is coupled with a robust estimator. Input and output data are first filtered and normalized, which reduces unmodeled dynamics and noise measurements effects. The parameter adaptation algorithm is of the standard recursive least-squares form with forgetting factor, data normalization factor, and adaptation freezing associated with the definition of an available information signal (M'Saad et al., 1989). The results obtained are represented by the graphs of Figure 14.11. The substrate concentration profile follows the requested model reference output. The influent substrate concentration change generates a perturbation on the output, which has been rejected after two hours. The second graph of Figure 14.11 shows the corresponding manipulated input profile, which does not saturate. The estimated model is represented by parameters whose evolutions are given by the graphs of Figure 14.12. We can note that the freezing parameter is sometimes activated (zones 1, 2, and 3) and this depends on the available information signal. At this time, we can believe that the activation of the freezing parameter is justified according the measurement of the information signal, i.e., the condition of persistent excitation is not satisfied, but we can suppose also that the system dynamic is changing. For us, the fact that this condition is not checked can cause a drift of the parameters and thus it can have a risk of instability and the freezing of the parameters was a means to mitigate this problem. However, this drift of the parameters can be justified by a change of system dynamics. This change can be due to the nature of the process, i.e., for example the aging

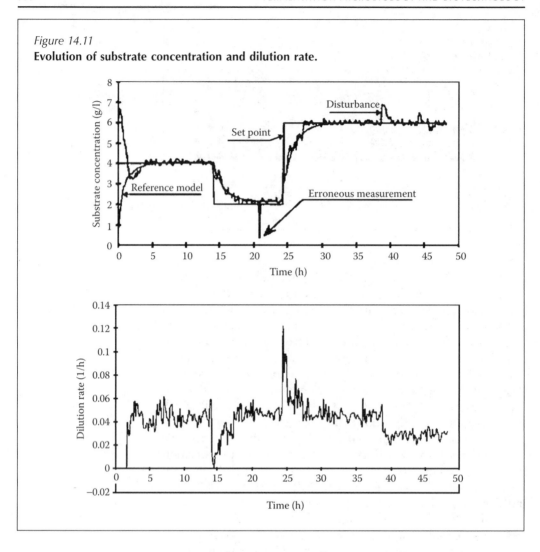

Figure 14.11
Evolution of substrate concentration and dilution rate.

of the micro-organisms, contamination, and then the fact of freezing the parameters can be a serious error of appreciation and comprehension. These reasons take us along to think that it should be considered that the process can be subjected to faults of operation that lead us to abnormal situations.

14.5 Supervision for process control

Modeling, estimation, filtering and control aspects mentioned previously constitute a layer of lower level that does not hold account of all information resulting from the process as well as those provided by the specialists in this one. We adopted a strategy that consists of adding a layer of higher level or supervision block in which one will exploit all this information.

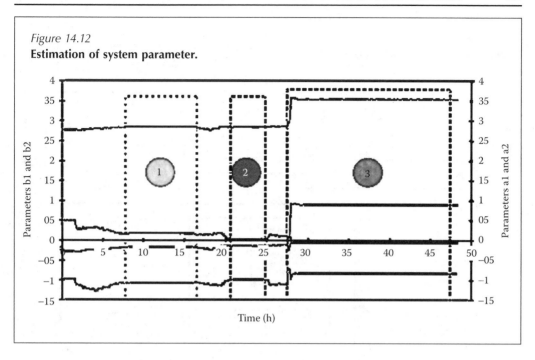

Figure 14.12
Estimation of system parameter.

If a hierarchical order is established, the feedback loop and the adaptation loop will take the level low. We considered the layer immediately above the adaptive loop, which is the supervision layer. In this level, the evolution of the signals coming from the adaptation and feedback loops are used in order to recognize specific situations and to act on the parameters of the different algorithms of control and estimation. The general idea of the supervision is the evaluation of the significant signals of the system in order to test its performances based on certain predefined criteria, which can be inherent to the controller or with the process. The violation of these criteria starts a second task of the supervisor: it must act on the system in a way not envisaged by the controller to improve his performances or to help the operator to make a decision at the time of the appearance of dysfunctions. These anomalies can be related to the biological reaction or ascribable to the operation of the material (actuators, sensors, etc.). Figure 14.13 illustrates an example of supervision block.

14.5.1 *Classification*

We use the obtained model in modeling section for supervision purpose (Nakkabi et al., 2002). This model is validated by some experiments to confirm the expected optimal production. Thereafter, we use this model like a reference in the supervision system for the monitoring in real time of an alcoholic

Figure 14.13
Physiological states and fault detection.

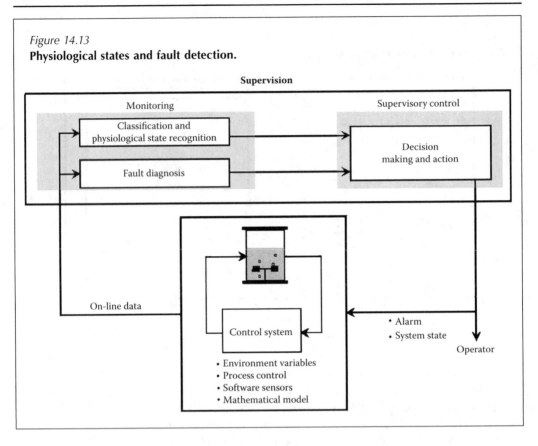

fermentation process. The results of the first experiment of the supervised process are presented in Figure 14.14.

Table 14.5 presents the different membership's functions concerning the acknowledged four physiological states. Note that the difference between *t+1* and *t* corresponds to one acquisition. By using the automaton obtained in Figure 14.15 and Table 14.5, it is possible to know the state of the system and this tendency. Actually, the membership's function of the state 1 (fermentation), during the instant *t*, is the highest followed by the one of state 2 (consumption of acid). This order indicates that the processes are in functional state 1 with tendency to switch

Table 14.5   Membership's function for good learning

		Time			
		t	t+1	t+2	t+3
States	Fermentation	0.29	0.35	0.32	0.24
	Acid consumption	0.26	0.32	0.31	0.29
	Diauxic	0.22	0.11	0.15	0.28
	Oxidative under ethanol	0.23	0.22	0.22	0.19
		normal	normal	normal	normal
		Situation			

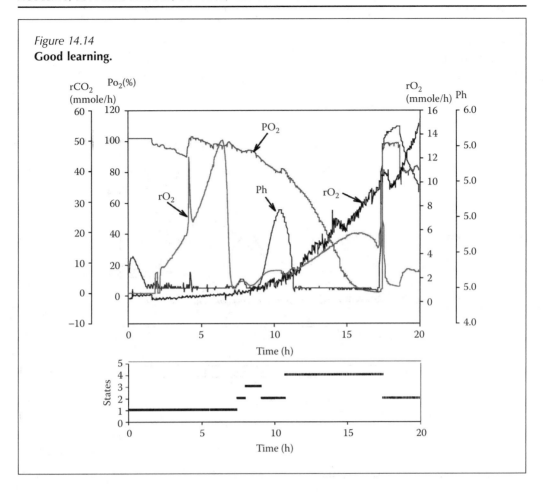

Figure 14.14
Good learning.

to state 2; this transition is authorized by the global model of process (Figure 14.15). So, the situation is considered as normal. The same analysis keeps its validity for the instants $t+1$ and $t+2$. At $t+3$ the membership's function is the greatest; the system is, then, in state 2 with an inclination to shift to state 3. The situation is always considered normal because this transition is authorized by the process model (Figure 14.15). Furthermore, the supervision system did not detect any unusual physiological behavior to the reference model, suggesting that the system is operationally satisfactory.

Table 14.6 shows the different membership's function. At the instant t, the process is at state 3 (diauxic state), which represents the highest value of the membership function. The membership's function of state 2 (consumption of acid), comes after, indicating the process tendency. The situation is considered as normal because the transition is allowed by the process model. At the instant $t+1$, the process remains at the state 3 but changes its tendency to state 4 (oxidative under ethanol state). According to the process model (Figure 14.15), this transition is not authorized. In this case, an alarm is set off. This analysis remains

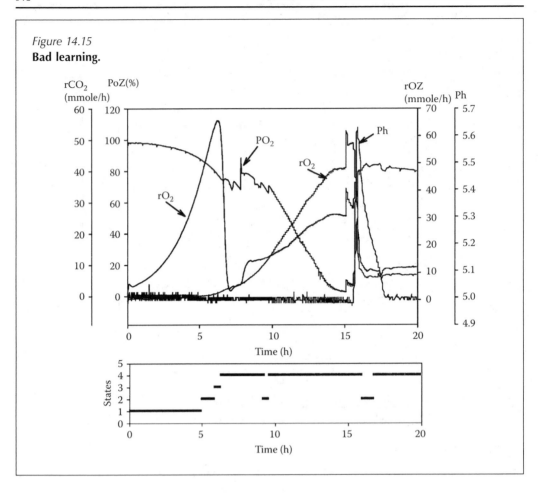

Figure 14.15
Bad learning.

Table 14.6   Membership's function for bad learning

		Time			
		t	**t+1**	**t+2**	**t+3**
States	Fermentation	0.01	0.03	0.01	0.01
	Acid consumption	0.32	0.12	0.12	0.21
	Diauxic	0.41	0.46	0.46	0.37
	Oxidative under ethanol	0.26	0.41	0.41	0.41
		normal	alarm	alarm	abnormal
			Situation		

true for the instant *t+2*. At the instant *t+3* the tendency of the
process is confirmed and the system switches from the state 3 to
state 4. The situation is abnormal pointing out an abnormality
in the process. Then, the supervision system makes a good detection of a different physiological behavior from those usual ones;
this is explained by the fact that the phase of diauxic is shorter.
Indeed, one notices in the figure that there is no intermediate

consumption of acid (there is no increase in pH) between the ethanol consumption and the glucose consumption. Maybe we have a better capacity of oxidation.

14.5.2   Fault detection and isolation (FDI)

Fault detection and isolation is one of the most important tasks assigned to intelligent supervisory control systems. A fault is understood as any kind of dysfunction in the actual dynamic system that leads to an unacceptable anomaly in the overall system performance. Such malfunctions may occur either in the sensors, in the actuators, or in the components of the process.

In this section, we are interested with faults in process dynamics. In model-based fault detection methods, a fault is considered as a variation of one or several parameters compared to a reference value. The problem is, then, to detect these parameters' variations, to distinguish between those resulting from faults and those resulting from normal behaviors, and to decide if these variations are indeed significant compared to uncertainties on the model and the noise on the measured data. As is well-known, the FDI procedure is explicitly divided into two stages: residual generation and residual evaluation. The principle is to use the measurements of output and input signals and the process mathematical model in order to generate residuals or indicators. After evaluation, the fault is detected and isolated. In the model-based methods the residuals can be generated using observers, parameter estimation, and parity relations.

We develop here a method based on adaptive observers for fault detection and isolation (Zhang, 2000) in an alcoholic fermentation process (Kabbaj et al., 2001). The faults are modeled as changes in the system parameters $\theta = [\theta_1 \quad \theta_2 \quad ... \quad \theta_n]^T$. The fault-free operating mode is characterized by the nominal vector θ^0, which is supposed to be known. The residuals are generated using a state observer, based on nominal parameters θ^0, and a set of adaptive observers as shown by Figure 14.16. Each adaptive observer estimates only one parameter of the supervised system, in addition to the state variables. The residuals $\gamma_0, \gamma_1, ..., \gamma_n$ are defined as being the prediction error of each observer.

We would like to point out the considerations behind our motivation of fault diagnosis in biotechnological processes (Kabbaj, 2004). More precisely, faults in process system (no fault in actuators or sensors) will be considered. As is well-known, in biotechnologies the growth rate is one of most important parameters to describe the evolution of the microbial population. The variation of this parameter is sensitive to the operating conditions (pH, temperature, agitation, oxygen ...). The modeling of these various influences within a single mathematical expression is very complex. In order to supervise the variables of the

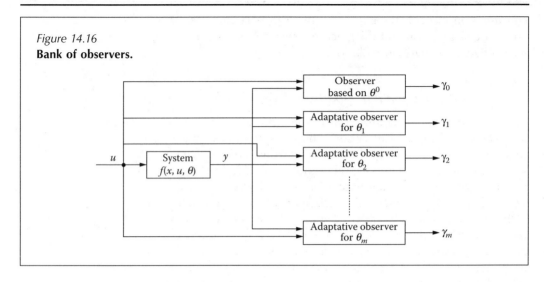

Figure 14.16
Bank of observers.

environment we can rather supervise the structural parameters of the growth rate (Eq. 14.5). Faults are so modeled as changes in the system parameters $\theta = [\mu_m \; K_s]$ (the maximum growth rate and the saturation constant). The faultfree operating mode is characterized by the nominal vector $\theta^0 = [\mu_m^0 \; K_s^0]$, which is supposed to be known.

14.5.2.1 *Adaptive observers for FDI*

As explained above, the scheme consists in developing two adaptive observers denoted *observer^j* $(j = 1, 2)$ and a state observer, based on nominal parameters θ^0, called *observer^0*. Each adaptive observer estimates only one parameter of the supervised system, in addition to the state variables. We have opted for an estimator using the nonlinearity of the process and the approach of the model reference to reconstitute the state variables and the parameters of the model (Zeng et al., 1993a). Instead of estimating the growth rate like a time varying parameter, we estimate rather the structural parameters. This makes the estimation algorithm more adapted for the used fault detection approach.

Let $\hat{S}^0(t)$, $\hat{S}^1(t)$, and $\hat{S}^2(t)$ be, respectively, the estimated outputs given by the *observer^0*, *observer^1*, and *observer^2*. The residuals may be defined as being the corresponding estimation errors:

$$\gamma_0 = \hat{S}^0(t) - S(t)$$
$$\gamma_1 = \hat{S}^1(t) - S(t) \qquad\qquad (14.6)$$
$$\gamma_2 = \hat{S}^2(t) - S(t)$$

The residual γ_1 is associated with the maximum growth rate $\theta_1 \equiv \mu_m$ and the residual γ_2 with the saturation constant $\theta_2 \equiv K_s$.

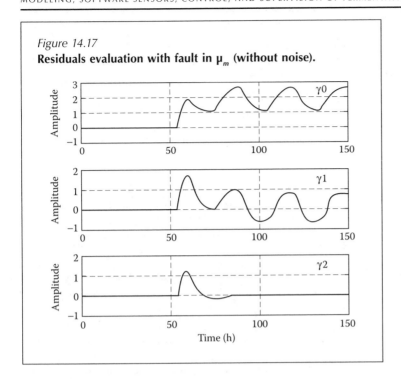

Figure 14.17
Residuals evaluation with fault in μ_m (without noise).

14.5.2.2 Residual behaviors

In the faultfree operating mode, all the residuals γ_0, γ_1, and γ_2 are practically zero and as such any departures from zero will be detected as a fault. If the fault corresponds to a change in a single parameter, all the residuals except one will persistently differ from zero. If, after a transient, γ_j ($j = 1, 2$) converges back to zero, the fault corresponds to a change in θ_j. Figures 14.17 and 14.18 show, respectively, the residuals behavior when the parameter μ_m changes from $0.38h^{-1}$ to $0.40h^{-1}$ and K_s from $5g.l^{-1}$ to $4.7g.l^{-1}$.

Microbiologically speaking, any changes in growth conditions in the bioreactor, e.g., temperature, pH, or aeration, will have an adverse effect on substrate consumption rate (K_s) which, in turn, diminishes growth rate (μ).

This method of detection and isolation of faults based on the adaptive observers gives satisfactory results without presence of noise. However, following (Kabbaj et al., 2002), we noted that it very is difficult to detect and isolate faults in the presence of the noise. To mitigate this problem, we propose a new methodology which combines analytical and knowledge for fault detection and isolation.

This combines the knowledge about process and the information hidden in data to extract a behavioural model under form of decision tree. This model is then used for an automatic residual evaluation. This approach proves to be remarkably robust.

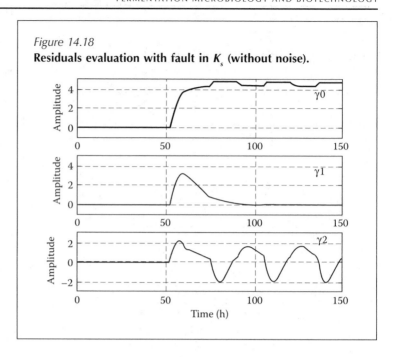

Figure 14.18
Residuals evaluation with fault in K_s (without noise).

Thus, considered faults are detected and isolated even in presence of measurement noise (Nakkabi et al., 2003).

The generated residuals are used in classification to recognize the process state (state 1: no faults, state 2: faults in μ_m, or state 3: faults in K_s). Besides, these residuals (γ_1, γ_2) we propose to use some composite residuals defined by:

$$s_1 = \gamma_0 - \gamma_1$$
$$s_2 = \gamma_0 - \gamma_2$$

(14.7)

These kinds of residuals are very significant and helpful in the classification procedure since they are robust to measurement noise. The goal is to diagnose the kind of fault by evaluating the given residuals. Therefore, two behavior models ($M1$, $M2$) are implemented. Each of them is designed to be sensitive to one fault, as shown in Figure 14.19. These models are generated as explained in Figure 14.3 (Section 14.3.2.1). The models output (f_1, f_2) denote the presence or not of fault. For each fault the training data are collected in presence of measurement Gaussian noise with zero mean and variance equal to 0.5. The effect of this noise on residuals is very important compared to faults. Thus, residual evaluation is delicate. The validation has been done on multiple data with different noise of variance 0.2, 0.3, 0.5, and 1. The results obtained with variance equal to 1 are presented.

As can be seen from Figure 14.20, residuals are generated as growth rate (μ_m) changes from $0.38h^{-1}$ to $0.40h^{-1}$. As dictated by the design, a fault is detected should any residuals, e.g.,

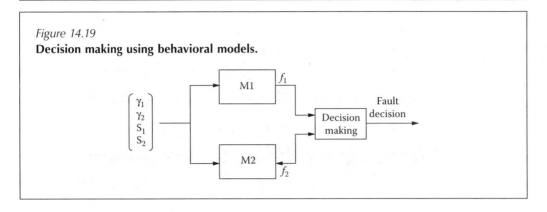

Figure 14.19
Decision making using behavioral models.

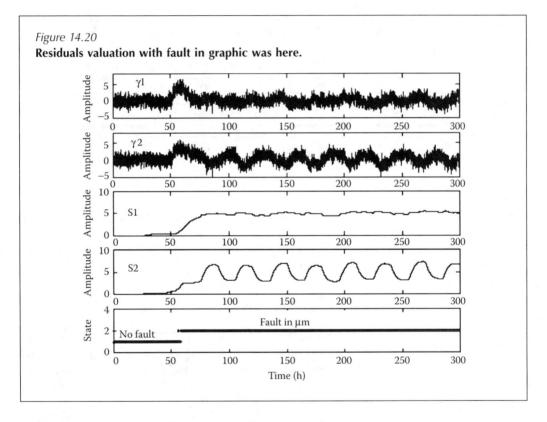

Figure 14.20
Residuals valuation with fault in graphic was here.

γ_1, depart from zero value. Unfortunately, this behavior is not clearly observed in the residuals γ_1 and γ_2 plotted in Figure 14.20 due to high noise in the measurements. In this case the composite residuals can be used, but just for fault detection. The residuals evaluation using behavioral models allows us to detect and isolate fault as shown in the bottom of Figure 14.20 by the transient from state 1 (No fault) to state 2 (fault in μ_m).

The fault is now simulated by changing the parameter K_s from $5gl^{-1}$ to $4.7gl^{-1}$ at $t = 130h$ and from $4.7gl^{-1}$ to $5gl^{-1}$ at $t = 250h$. The corresponding residuals are illustrated in Figure 14.21.

Figure 14.21
Residuals valuation with fault in graphic was here.

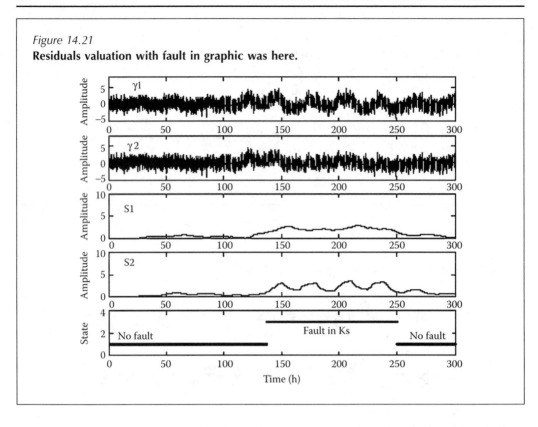

Similarly, the behavioral models based on residual evaluation shows clearly the transient from state 1 (No fault) to state 3 (Fault in K_s) at $t = 130h$ and from state 3 to state 1 at $t = 250h$.

14.6   Conclusions

In this work, we tried to highlight the stages to be followed and the difficulties engineers can meet to ensure good control of the biotechnological processes and in particular the alcoholic fermentation processes. We developed two approaches of modeling according to the objective considered: one is quantitative and the other is qualitative. The models obtained from mass balance considerations are called nonstructured models and those obtained from the methods of classification are called behavioral models. The nonstructured models were used for the development of the algorithms of the software sensors, control and faults detection and isolation. The behavioral models were used for the determination of physiological states and the fault detection and isolation problem. These behavioral models use the environmental variables such as the temperature, the pH, the stirring velocity... and the expert's knowledge of the processes to determine the physiological states.

The validation of the obtained nonstructured model was carried out by the research for the best minimization of a mathematical criterion by using experimental data available of fermentation in discontinuous mode. We preferred this mode because it is richer in kinetic information than the continuous mode, which provides only stationary states.

The classification method was applied to alcoholic fermentation process. It was supposed that the behavior during batch processing influences the phenomena observed during the continuous phase. Then a good knowledge of the physiological states during the batch phase is of capital importance for the biologists. In agreement with the experts of the process, a total of four signals was selected, which according to its knowledge, contain the most relevant information to determine the physiological states.

Considering the lack of reliable and inexpensive sensors in the field of biotechnological processes, we were motivated by the development of the software sensors. We used the nonstructured model for the development of this type of algorithms. The experimental results showed that from the measurement of the substrate concentration, one can estimate the biomass and product concentrations with the help of an adaptive observer which jointly estimates the state and the parameters. This algorithm was validated on three types of experiment (batch, fed-batch, and continuous modes) and gave satisfactory results.

Experimental results obtained from the control of alcoholic fermentation in continuous mode are presented. The objective was the regulation and the tracking of the substrate concentration inside the bioreactor while acting the dilution rate. The obtained results are also satisfactory. We obtained these results by the application of an indirect adaptive control scheme using an estimator with freezing parameter when the condition of persistent excitation is not satisfied. This situation can be caused by a change of the system dynamics, and thus we concluded that it was necessary to develop a supervision system to supervise the control algorithms and to detect and isolate at the same time the faults coming from process dynamics.

In the supervision system, the two approaches of modeling are used for the determination of physiological states and for faults detection and isolation. For the determination of physiological states, the classification method based on fuzzy techniques was used. This method enabled us to construct the behavioral model, which was used in supervision system as a reference model for on-line monitoring task. The obtained nonstructured model was used for development of faults detection and isolation algorithms. The used method is based on adaptive observers. This method gives satisfactory results without presence of noise, but it proves limited in the presence of noise. To solve this problem, a combined analytical and knowledge-based method was proposed for fault detection and isolation.

Summary

- The development of an integrated process control methodology requires the designing of a supervision block containing all available information and procedures.
- This supervision block recognizes specific indicators/parameters and act, if necessary, on the process or by informing the operator.
- Two approaches of modeling: the behavioral model and the nonstructured model were implemented.
  - While the former is based on the use of a fuzzy classification technique, the latter is based on mass balance considerations.
  - The behavioral model is obtained by using the on-line measurement of environmental variables (temperature, pH, stirring velocity, etc.) and describes the physiological states of the process.
  - The nonstructural model, however, is obtained from mass balance considerations and is used for the development of software sensors, control scheme, and fault detection and isolation algorithms.
- The whole of the results obtained and the developed algorithms using these modeling approaches are used in a supervision block to ensure an effective monitoring of the process.
- The application of this methodology to fermentation processes gave satisfactory results.

References

Aguilar-Martin, J. (1996). Knowledge-based real time supervision. *Tempus-Modify* workshop. Budapest, Hungary.
In: *Tempus-Modify workshop*. Budapest, Hungary.

Antsaklis, P. J., K. M. Passino, eds. (1993). *An Introduction to Intelligent and Autonomous Control*. Kluwer Academic Publishers, Norwell, MA.

Bâati, L., G. Roux, B. Dahhou, J. L. Uribelarrea (2004). Unstructured modeling growth of Lactobacilius acidophilus as a function of the temperature. *Mathematics and Computer in Simulation*, vol. 65, pp. 137–145.

Ben-Youssef, C. (1996). Filtrage, estimation et commande adaptative d'un procédé de traitement des eaux usées. PhD thesis. Institut National Polytechnique de Toulouse.

Ben-Youssef, C., B. Dahhou, F. Y. Zeng, J. L. Rols (1996). Estimation and filtering of non-linear systems: application to a wastewater treatment process. *International Journal of Systems Science*, vol. 27, n° 5, pp. 497–505.

Dahhou B., G. Chamilothoris, G. Roux (1991a). Adaptive predictive control of a continuous fermentation process. *International Journal of Adaptive Control and Signal Processing*, vol. 5, n° 6, pp. 351–362.

Dahhou, B., G. Roux, A. Cheruy (1993). Linear and nonlinear adaptive control of alcoholic fermentation process. *International Journal of Adaptive Control and Signal Processing*, vol. 7, n° 3, pp. 213–233.

Dahhou, B., G. Roux, I. Queinnec (1991b). Adaptive control of a continuous fermentation process. Symposium on Modeling and Control of Technological Systems, Lille, France, pp. 738–743.

Dahhou, B., G. Roux, I. Queinnec, J. B. Pourciel (1991c). Adaptive pole placement control of a continuous fermentation process. *International Journal of Systems Sciences*, vol 22, n° 12, pp. 2625–2638.

Dojat, M., N. Ramaux, D. Fontaine (1998). Scenario recognition for temporal reasoning in medical domains. *Artificial Intelligence in Medicine*, vol. 14, pp. 139–155.

Kabbaj, N. (2004). Développement d'algorithmes de détection et d'isolation de défauts pour la supervision des bioprocédés. PhD thesis. Université de Perpignan.

Kabbaj, N., A. Doncescu, B. Dahhou, G. Roux (2002). Wavelet based residual evaluation for fault detection and isolation. 17th IEEE International Symposium on Intelligent Control. Vancouver, British Columbia, Canada.

Kabbaj, N., M. Polit, B. Dahhou, G. Roux (2001). Adaptive observers based fault detection and isolation for an alcoholic fermentation process. 8th IEEE International Conference on Emerging Technologies and Factory Automation. Antibes – Juan les Pins, France.

Kotch, G-G. (1993). Modular reasoning. A new approach towards intelligent control. Ph.D thesis. Swiss Federal Institute of Technology. Zurich, Suisse.

Ph.D thesis. Swiss Federal Institute of Technology. Zurich, Suisse.

Monod, J. (1942). *Recherche sur la Croissance des Cultures Bactériennes*. Edition Hermes. Paris.

M'Saad, M., I-D. Landau, M. Duque (1989). Example applications of the partial state reference model adaptive control design technique. *International Journal of Adaptive Control and Signal Processing*, vol. 3, pp. 155–165.

M'Saad, M., I-D. Landau, M. Samaan (1990). Further evaluation of the partial state reference model adaptive control design. *International Journal of Adaptive Control and Signal Processing*, vol. 4, pp. 133–146.

Nakkabi, Y., N. Kabbaj, B. Dahhou, G. Roux, J. Aguilar (2003). A combined analytical and knowledge based method for fault detection and isolation. 9th IEEE International Conference on Emerging Technologies and Factory Automation. Vol. 2. Lisbon, Portugal. pp. 161–166.

Nakkabi,Y., A. Doncescu, G. Roux, M. Polit, V. Guillou (2002). Application of data mining in biotechnological process. The second IEEE International Conference on Systems, Man and Cybernetics Hammamet, Tunisia.

Nejjari, F., B. Dahhou, A. Benhammou, G. Roux (1999a). Nonlinear multivariable adaptive control of an activated sludge wastewater treatment process. *International Journal of Adaptive Control and Signal Processing*, vol. 13, pp. 347–365.

Nejjari, F., G. Roux, B. Dahhou, A. Benhammou (1999b). Estimation and optimal control design of a biological and wastewater treatment process. *International Journal of Mathematics and Computers in Simulation*, vol. 48, n° 3, pp. 269–280.

Piera, N., P. Desroches, J. Aguilar-Martin (1989). Lamda: An incremental conceptual clustering system. Report 89420. LAAS/CNRS.

Queinnec, I., C. Destruhaut, J-B. Pourciel, G. Goma (1992). An effective automated glucose sensor for fermentation monitoring and control. *World Journal of Microbiol. Biotechnol.*, vol. 8, pp. 7–13.

Waissman-Vilanova, J. (2000). Construction d'un modèle compartemental pour la supervision de procédés: Application à une station de traitement des eaux. PhD thesis. Institut National Polytechnique de Toulouse.

Waissman-Vilanova, J., R. Sarrate-Estruch, B. Dahhou, J. Aguilar-Martin (1999). Building an automaton for condition monitoring in a biotechnological process. *5th* European Control Conference. Karlsruhe, Alemagne.

Zeng, F. Y., B. Dahhou, M. T. Nihtila (1993a). Adaptive control of nonlinear fermentation process via MRAC technique. *Journal of Applied Mathematic Modeling*, vol. 17, pp. 58–69.

Zeng, F. Y., B. Dahhou, M. T. Nihtila, G. Goma (1993b). Microbial specific growth rate control via MRAC method. *International Journal of Systems Science*, vol. 24, 1973–1985.

Zhang, Q. (2000). A new residual generation and evaluation method for detection and isolation of faults in nonlinear systems. *International Journal of Adaptive Control and Signal Processing*, vol. 14, pp. 759–773.

Appendix: Suppliers List

Not exhaustive and no recommendation implied by inclusion. Compiled by A.R. Allman.

Instrumentation, Sensors and Software

http://www.aber-instruments.co.uk/	biomass monitor
http://www.adc-service.co.uk/page5.html	gas analysers
http://www.broadleyjames.com/	pH and pO_2 electrodes
http://www.buglab.com/	external optical density sensor
http://www.foxylogic.com/	control software
http://www.cerexinc.com/	optical density monitor with sample chamber
http://www.wireworkswest.com/fermworks/	fermentation software for process plants
http://www.fluorometrix.com/products/cellphase/	external pO_2 analyzer
http://www.flownamics.com/	sample probes and analysers
http://www.hamiltoncomp.com	pO_2 sensors
http://www.ksr-kuebler.com/website/index.php	level sensors for larger fermentors
http://www.red-y.com/en/	mass flow measurement and control
http://uk.mt.com/home	Mettler Toldeo – sensors, balances etc.
http://www.magellaninstruments.com/	perfusion/filtration system
http://www.millipore.com	filters, tangential flow filtration
http://www.processmeasurement.uk.com/	Monitek optical density systems
http://www.ptiinstruments.com/products	methanol sensor
http://www.oceanoptics.com/homepage.asp	optical sensors for pH, pO_2 etc.
http://www.pall.com/ Filters,	micro-filtration, etc.
http://www.hitec-zang.com/html/biotech.htm	RQ monitoring
http://ravenbiotech.com/	methanol sensing and control
http://www.russellph.com/foam2.htm	antifoam sensor
http://www.schleicher-schuell.com	tangential flow filtration
http://www.watson-marlow.com/	peristaltic pumps
http://www.ysi.com/lifesciences.htm	glucose, CO_2 sensors, etc.

Fermentation Equipment

http://www.abec.com/design.htm

http://www.applikon.com/

http://www.bioxplore.net/home-en.html

http://www.bbraunbiotech.com/
    (Sartorious http://www.sartorius.com)

http://www.belach.se/kent_series.html

http://www.biolafitte.com/Fermentors.htm (Pierre Guerin)

http://www.broadleyjames.com/

http://www.bioengineering.ch/

http://www.dasgip.de/

http://www.electrolab.co.uk/

http://www.infors-ht.com/

http://www.labkorea.com/products/fermenter/ (BioG range)

http://www.bemarubishi.co.jp/

http://www.newmbr.ch/

http://www.nbsc.com/Main.asp (New Brunswick Scientific)

http://www.novaferm.se/

http://www.wavebiotech.com/products/

Index

A

ABE, *see* acetone, butanol, ethanol (ABE)
Accu-Check Advantage, 397, 398f
acetaldehyde, 4
acetate
   Clostridium, 126
   Escherichia coli
      central metabolic pathway, 195f, 196f, 200t
      skeletal model, 197t, 198t
      W3110, 264
acetic acid
   production, 122
Acetobacter
   vinegar fermentation, 122
Acetobacter aceti
   vinegar fermentation, 122
acetone
   Clostridium, 126
acetone, butanol, ethanol (ABE), 128
acetone-butanol fermentation, 5–6
acetyl-CoA, 236f
Actinobacillus succinogenes
   succinic acid, 125
adaptive control scheme, 507f
adaptive estimation scheme, 504f
adaptive techniques, 503
adherent monolayers, 290
adhesines, 313
adsorption, 373
aeration
   incubation control, 473
Aerobacter aerogenes
   chemostat, 83f
   diauxic growth, 29
   glucose
      Ks values, 80t
   growth limitations, 31
aerobic fermentation
   oxygen transfer, 325
affinity-based assays, 371f
agar, 295
agarose, 295
age distribution model, 94
aggregation phenomena, 313
agitation
   incubation control, 474
agitator speed control
   action, 456
airflow control
   action, 456
airflow rate
   aerobic fermentation systems, 455
airlift fermentors, 412f, 444
   Gluconacetobacter oxydans, 382
air lift reactors, 326t, 328f

AK, *see* aspartate kinases (AK)
alarmone, 104
alcohol-free beer
   ICT, 338t
   immobilized cells, 337
   Saccharomyces cerevisiae, 337
alginate-poly-L-lysine (PLL)-polyethyleneimine (PEI)
      microcapsule, 308, 374
alginates, 292–293
alkaline endopeptidases
   Bacillus, 169
alpha ketoglutarate, 14
amidases, 101
amino acids
   animal cell culture
      metabolic analysis, 177f
   aromatic, 115–116
      Corynebacterium glutamicum, 115
   auxotrophic mutants, 107
   metabolism
      proteases, 171f
   production
      primary metabolite microbial synthesis,
         106–116
      stereospecificity, 107
ammonia
   Escherichia coli, 101
   repression, 101
amperometric glucose
   blood glucose monitoring
      case study, 388–398
amperometric transducers, 366
amperometry, 396
ampicillin
   beta-lactamase, 153
amylases, 219
   starch hydrolysis, 221
analogue instruments, 414
analogue value events
   fermentor control, 481
anaplerotic pathways
   bacterial protoplasm/biomass chemical synthesis,
      14–15
anchorage-dependent cells, 290
ancient Egypt
   beer fermentation, 1
   bread fermentation, 1
animal cell(s)
   bioreactor recombinant protein production,
      176–179
animal cell bioprocesses
   metabolic analysis and optimization,
      159–170
animal cell culture
   amino acids
      metabolic analysis, 177f

LIVERPOOL
JOHN MOORES UNIVERSITY
AVRIL ROBARTS LRC
TITHEBARN STREET
LIVERPOOL L2 2ER
TEL. 0151 231 4022